# Combinatorics and Random Matrix Theory

GRADUATE STUDIES IN MATHEMATICS 172

# Combinatorics and Random Matrix Theory

Jinho Baik
Percy Deift
Toufic Suidan

American Mathematical Society
Providence, Rhode Island

2010 *Mathematics Subject Classification.* Primary 05A15, 15B52, 33E17, 35Q15, 41A60, 47B35, 52C20, 60B20, 60K35, 82C23.

For additional information and updates on this book, visit
**www.ams.org/bookpages/gsm-172**

**Library of Congress Cataloging-in-Publication Data**

Names: Baik, Jinho, 1973- | Deift, Percy, 1945- | Suidan, Toufic Mubadda, 1975-
Title: Combinatorics and random matrix theory / Jinho Baik, Percy Deift, Toufic Suidan.
Description: Providence, Rhode Island : American Mathematical Society, 2016. | Series: Graduate studies in mathematics ; volume 172 | Includes bibliographical references and index.
Identifiers: LCCN 2015051274 | ISBN 9780821848418 (alk. paper)
Subjects: LCSH: Random matrices. | Combinatorial analysis. | AMS: Combinatorics – Enumerative combinatorics – Exact enumeration problems, generating functions. msc | Linear and multilinear algebra: matrix theory – Special matrices – Random matrices. msc | Special functions (33-XX deals with the properties of functions as functions) – Other special functions – Painlevé-type functions. msc | Partial differential equations – Equations of mathematical physics and other areas of application – Riemann-Hilbert problems. msc | Approximations and expansions – Approximations and expansions – Asymptotic approximations, asymptotic expansions (steepest descent, etc.). msc | Operator theory – Special classes of linear operators – Toeplitz operators, Hankel operators, Wiener-Hopf operators. msc | Convex and discrete geometry – Discrete geometry – Tilings in 2 dimensions. msc | Probability theory and stochastic processes – Probability theory on algebraic and topological structures – Random matrices (probabilistic aspects; for algebraic aspects see 15B52). msc | Probability theory and stochastic processes - Special processes – Interacting random processes; statistical mechanics type models; percolation theory. msc | Statistical mechanics, structure of matter – Time-dependent statistical mechanics (dynamic and nonequilibrium) – Exactly solvable dynamic models. msc
Classification: LCC QA188.B3345 2016 | DDC 511/.6–DC23 LC record available at http://lccn.loc.gov/2015051274





♾ The paper used in this book is acid-free and falls within the guidelines
established to ensure permanence and durability.
Visit the AMS home page at `http://www.ams.org/`

10 9 8 7 6 5 4 3 2 1 21 20 19 18 17 16

To my wife Hyunsuk and my daughter Haesue
To my wife Rebecca and my daughter Abby
To my parents Mubadda and Aida Suidan

# Contents

# Preface

As a consequence of certain independent developments in mathematics in recent years, a wide variety of problems in combinatorics, some of long standing, can now be solved in terms of random matrix theory (RMT). The goal of this book is to describe in detail these developments and some of their applications to problems in combinatorics. The book is based on courses on two key examples from combinatorial theory, viz., Ulam's increasing subsequence problem, and the Aztec diamond. These courses were given at the Courant Institute and the University of Michigan by two of the authors (P.D. and J.B., respectively) some ten years ago.

The authors are pleased to acknowledge the suggestions, help, and information they obtained from many colleagues: Eitan Bachmat, Gerard Ben Arous, Alexei Borodin, Thomas Kriecherbauer, Eric Nordenstam, Andrew Odlyzko, Eric Rains, Raghu Varadhan, and Ofer Zeitouni. In particular, Eitan Bachmat and Thomas Kriecherbauer took on the task of reading the manuscript in full, catching typos, and suggesting many very helpful changes to the text. The authors would also like to acknowledge the support of NSF over the years when this book was written in the form of Grants DMS-0457335, DMS-0757709, DMS-1068646, and DMS-1361782 for J.B., DMS-0500923, DMS-1001886, and DMS-1300965 for P.D., and DMS-0553403 and DMS-0202530 for T.S. The first author (J.B.) and the third author (T.S.) would, in addition, like to acknowledge the support of an AMS Centennial Fellowship (2004–2005) and a Sloan Research Fellowship (2008-2010), respectively.

November 2015

*Chapter 1*

# Introduction

Over the last fifteen years, various problems in combinatorics have been solved in terms of random matrix theory (RMT). More precisely, the situation is as follows: The problems at hand are probabilistic in nature, and in an appropriate scaling limit, it turns out that certain key quantities associated with these problems behave statistically like the eigenvalues of a (large) random matrix. Said differently, RMT provides a "stochastic special function theory" for a (broad and growing) class of problems in combinatorics. The goal of this book is to analyze in detail two key examples of this phenomenon, viz., Ulam's problem for increasing subsequences of random permutations, and domino tilings of the Aztec diamond. We will also describe other examples along the way, but in less detail.

Techniques from many different areas in mathematics are needed to analyze these problems. These areas include combinatorics, probability theory, functional analysis, complex analysis, and the theory of integrable systems. Specialized techniques from these areas, such as the Robinson-Schensted-Knuth (RSK) correspondence in combinatorics (see Chapters 3 and 10) and the classical steepest descent method for the asymptotic analysis of integrals in the plane, play key roles. The Riemann-Hilbert problem and the nonlinear steepest descent method, also play a role. We have tried to make this book self-contained, and along the way we develop enough of the theory that we need from each area so that a general reader with, say, two or three years experience in graduate school, can learn the subject directly from the text.

## 1.1. Ulam's Problem: Random Permutations

**1.1.1. Longest increasing subsequences.** Ulam's problem concerns the symmetric group, $S_N$, of permutations of the numbers $1, 2, \ldots, N$. Let $\pi \in S_N$. For

$$1 \leq i_1 < i_2 < \cdots < i_k \leq N, \tag{1.1}$$

we say that $\pi(i_1), \ldots, \pi(i_k)$ is an *increasing* (resp., *decreasing*) *subsequence* of $\pi$ of *length* $k$ if

$$\pi(i_1) < \pi(i_2) < \cdots < \pi(i_k), \tag{1.2}$$

(resp., $\pi(i_1) > \pi(i_2) > \cdots > \pi(i_k)$). Let $\ell_N(\pi)$ denote the maximal length of all increasing subsequences of $\pi$. Any increasing subsequence of $\pi$ of length $\ell_N(\pi)$ is called a *longest increasing subsequence* of $\pi$. For example, for $N = 5$, let $\pi$ be the permutation $\pi(1) = 5$, $\pi(2) = 1$, $\pi(3) = 3$, $\pi(4) = 2$, $\pi(5) = 4$. Alternatively, in so-called one-line notation, $\pi$ has the form

$$\pi = 51324. \tag{1.3}$$

Then 13 and 12 are increasing subsequences of length 2, and by inspection $\ell_5(\pi) = 3$. Moreover, 134 and 124 are longest increasing subsequences for $\pi$. As another example, the permutation

$$\pi = 7\ 3\ 1\ 17\ 10\ 18\ 9\ 20\ 6\ 12\ 16\ 13\ 2\ 8\ 19\ 4\ 15\ 11\ 14\ 5 \tag{1.4}$$

has $\ell_{20}(\pi) = 5$. A longest increasing subsequence is

$$1\ 6\ 8\ 11\ 14. \tag{1.5}$$

An increasing subsequence has a natural graphical representation, reflecting the fact that a permutation is a bijection on a finite set. For example, the bijection corresponding to the permutation (1.4) is in Figure 1.1. The increasing subsequence (1.5) corresponds to an "up/right path" in this picture.

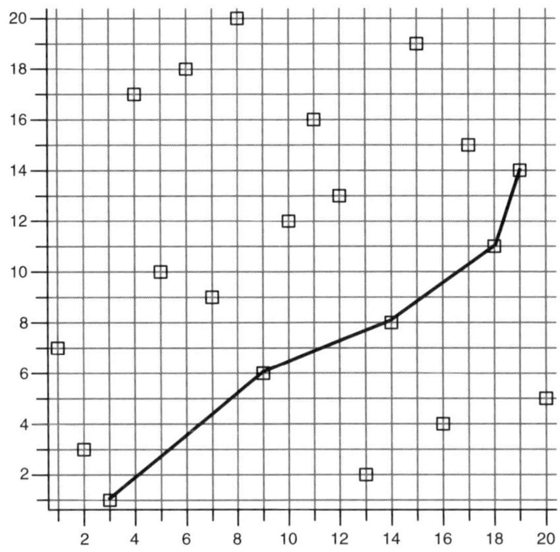


**Figure 1.1.** Graphical representation of a permutation in $S_{20}$

Equip $S_N$ with uniform measure:

$$\mathbb{P}\{\pi\} = \frac{1}{N!}, \qquad \text{for all } \pi \in S_N. \tag{1.6}$$

A basic question addressed in this book is: how does $\ell_N$ behave statistically as $N \to \infty$?

The simplest statistic to consider is $\mathbb{E}[\ell_N]$, the expectation of $\ell_N$. In 1961, using Monte Carlo simulations for $N$ up to 101, Ulam [**Ula61**] found that

$$\mathbb{E}[\ell_N] \sim 1.7\sqrt{N}, \qquad N \uparrow 100. \tag{1.7}$$

Ulam then conjectured that

$$\lim_{N\to\infty} \frac{\mathbb{E}[\ell_N]}{\sqrt{N}} = c \tag{1.8}$$

exists for some constant $c$. What came to be known as *Ulam's problem* was to prove Ulam's conjecture and to identify the constant $c$. Elementary arguments, using an old result of Erdős and Szekeres in 1935 [**ES35**], show[1] that if the limit (1.8) exists, then $\frac{1}{2} \leq c \leq e$. Subsequent numerical work of Baer and Brock in 1968, using Monte Carlo simulation up to $N = 10000$ and exact evaluations up to $N = 36$, suggested that $c = 2$ [**BB68**].

In 1972, Hammersley [**Ham72**] introduced a "Poissonized" version of $\ell_N$ (see Chapter 2) and used results from subadditive ergodic theory to prove

[1]See Chapter 4, especially Corollary 4.3 and Proposition 4.9, for more details.

that the limit in (1.8) indeed exists (see Proposition 4.8 below). However, he was not able to identify $c$.

In 1977, Logan and Shepp [**LS77**] showed that $c \geq 2$, and independently Vershik and Kerov [**VK77, VK85**] showed that $c = 2$, thus solving Ulam's problem. Many alternative proofs of this result were given later in the 1990s [**AD95, Sep96, Joh98**]. These various proofs used a variety of techniques foreshadowing the multifaceted development of the subject that was to come. In particular, Logan and Shepp, and Vershik and Kerov used RSK and variational techniques. Aldous and Diaconis [**AD95**] employed Hammersely's ergodic viewpoint together with the theory of partial differential equations. Seppäläinen [**Sep96**] used techniques from stochastic partial differential equations. Finally, Johansson [**Joh98**] used Gessel's formula[2] [**Ges90**] from algebraic combinatorics together with the asymptotic theory of Toeplitz determinants. For a history of these developments, see [**AD99**] and [**OR00**].

**1.1.2. Limit theorem and Tracy-Widom distribution.** The next statistic to be considered was the variance, $\mathrm{Var}(\ell_N)$, which was conjectured to grow as

$$\mathrm{Var}(\ell_N) \sim N^{\alpha} \qquad \text{as} \quad N \to \infty \tag{1.9}$$

for some $\alpha > 0$. In particular, using arguments from first passage percolation theory, Kesten conjectured that $\alpha = \frac{1}{3}$. In the early 1990s, Odlyzko became interested in the problem and, together with Rains, performed Monte Carlo simulations using values of $N \leq 10^{10}$ and exact computations up to $N = 120$. Table 1.1 below is the number of $\pi$'s in $S_{30}$ for which $\ell_{30}(\pi)$ equals $k$, $k = 1, 2, \ldots, 30$. The reader is referred to the webpage `http://www.dtc.umn.edu/~odlyzko/tables/index.html` of Odlyzko for the outcomes of various simulations. These simulations indicated that

$$\lim_{N\to\infty} \frac{1}{N^{1/3}} \mathrm{Var}(\ell_N) = c_0 \sim 0.819\cdots, \tag{1.10}$$

which is consistent with Kesten's $\alpha = \frac{1}{3}$. Odlyzko and Rains also simulated $\mathbb{E}[\ell_N]$ to higher order and found that

$$\lim_{N\to\infty} \frac{\mathbb{E}[\ell_N] - 2\sqrt{N}}{N^{1/6}} = c_1 \sim -1.758\cdots. \tag{1.11}$$

Apart from some concentration results and certain large deviation results [**Fri91, BB92, Kim96, Sep98, DZ99**], this was where the matter stood in the late 1990s.

[2]See Chapter 7 below.

**Table 1.1.** Longest increasing subsequences in $S_{30}$. This table is due to Odlyzko and Rains.

| $k$ | number of $\pi$'s such that $\ell_{30}(\pi) = k$ |
|---|---|
| 1 | 1 |
| 2 | 3814986502092303 |
| 3 | 122238896672891001069665 |
| 4 | 1790036582998939530743648877 |
| 5 | 449044243619862872721423598179 |
| 6 | 10236819433951393776243660748875 |
| 7 | 50241067877038219983230124657600 |
| 8 | 86511371455863277882723853476200 |
| 9 | 70971582765623356071324810857700 |
| 10 | 33700117351593715495661064101700 |
| 11 | 10447178628714722178634866396630 |
| 12 | 2277900847905046253535807880680 |
| 13 | 366440157064983378222220318530 |
| 14 | 44912755712412555783652789980 |
| 15 | 4289203871330156652985437480 |
| 16 | 324301002215082697285357800 |
| 17 | 19633107355949074371195000 |
| 18 | 959064229546178387532600 |
| 19 | 37982369568044622191625 |
| 20 | 1222055891584247185425 |
| 21 | 31925927141978856309 |
| 22 | 675007128155925069 |
| 23 | 11475430101232224 |
| 24 | 155228816648544 |
| 25 | 1644397829384 |
| 26 | 13319151176 |
| 27 | 79490741 |
| 28 | 328861 |
| 29 | 841 |
| 30 | 1 |

In [**BDJ99**], the authors considered the full distribution $\mathbb{P}\{\ell_N \leq n\}$. Let $\mathrm{Ai}(x)$ denote the Airy function which is the unique solution of the Airy equation with prescribed asymptotics

$$\mathrm{Ai}''(x) = x\,\mathrm{Ai}(x), \qquad \mathrm{Ai}(x) \sim \frac{e^{-\frac{2}{3}x^{3/2}}}{2\sqrt{\pi}x^{1/4}} \quad \text{as } x \to +\infty. \tag{1.12}$$

Let $u(x)$ be the Hastings-McLeod solution of the *Painlevé II* equation,

$$u''(x) = 2(u(x))^3 + xu(x), \tag{1.13}$$

with asymptotics

$$u(x) \sim \mathrm{Ai}(x) \qquad \text{as } x \to +\infty. \tag{1.14}$$

The proof of the global existence and uniqueness of this solution is due to Hastings and McLeod [**HM80**], who also showed that

$$u(x) \sim -\sqrt{\frac{-x}{2}}(1 + O(x^{-2})) \qquad \text{as } x \to -\infty. \tag{1.15}$$

Relations (1.14) and (1.15) show that

$$F(x) \equiv e^{-\int_x^\infty (y-x)(u(y))^2 dy} \tag{1.16}$$

is a distribution function, i.e., $F(+\infty) = 1$, $F(-\infty) = 0$, and $F(x)$ is monotone increasing. The distribution function $F(x)$ is very different from the Gaussian distribution as we can see from the following properties (see [**TW94, DIK08, BBD08**]):

- The expected value and the variance are $-1.7711\cdots$ and $0.8132\cdots$, respectively.
- As $x \to +\infty$,

$$F(x) = 1 - \frac{e^{-\frac{4}{3}x^{3/2}}}{16\pi x^{3/2}}\left(1 - \frac{35}{24x^{3/2}} + O(x^{-3})\right). \tag{1.17}$$

- As $x \to -\infty$,

$$F(x) = \tau \frac{e^{-\frac{1}{12}|x|^3}}{|x|^{1/8}}\left(1 + \frac{3}{2^6|x|^3} + \frac{2025}{2^{13}|x|^6} + \frac{2470825}{2^{19}|x|^9} + O(x^{-12})\right), \tag{1.18}$$

  where

$$\tau = 2^{1/24}e^{\zeta'(-1)} = 0.8723714149\cdots. \tag{1.19}$$

  Here $\zeta(z)$ is the Riemann-zeta function. It is interesting to note that in the exponential in (1.18), $-\frac{1}{12} = \zeta(-1)$.

See Figure 1.2 for the graph of $F'(x)$. This distribution is called the *Tracy-Widom* distribution (see Theorem 1.4) and plays a central role in all the results that follow in this book.

**Remark 1.1.** The Painlevé equations—there are six of them, PI–PVI—are fundamental nonlinear ODE's with many remarkable properties [**Inc44, FIKN06**]. The Painlevé equations, whose solutions $P_I(x), P_{II}(x), \ldots$ are called Painlevé functions, are regarded as forming the core of "modern special function theory", as opposed to the classical special function theory of the 19th century, involving the Airy function $\mathrm{Ai}(x)$, the Bessel functions $J_\nu(x)$, the Legendre functions $P_\lambda(x)$, etc. Classical special functions have

**1.1.4. Plancherel measure on the set of partitions.** Let $\lambda=(\lambda_1,\ldots,\lambda_r)$ be a partition of $N$, i.e., $\lambda_1,\ldots,\lambda_r$ are positive integers satisfying

- $\lambda_1 \geq \lambda_2 \geq \cdots \geq \lambda_r > 0$,
- $\mid \lambda \mid \equiv \sum_{i=1}^{r} \lambda_i = N$.

We call $r = r(\lambda)$ the *length* of the partition and use $\lambda \vdash N$ to indicate that $\lambda$ is a partition of $N$. It is convenient to display a partition $\lambda = (\lambda_1,\ldots,\lambda_\ell)$ as a *Young diagram* (or Ferrer's diagram). For example

$$\begin{array}{|c|c|c|c|c|} \hline \ & \ & \ & \ & \ \\ \hline \ & \ \\ \cline{1-2} \ & \ \\ \cline{1-2} \end{array} \qquad \text{represents } \lambda = (\lambda_1,\lambda_2,\lambda_3) = (5,2,2) \vdash 9. \tag{1.27}$$

We will often abuse notation by referring to a Young diagram as a partition and vice versa. The set of all Young diagrams (or partitions) of $N$ is denoted by $\mathcal{Y}_N$. We also set $\mathcal{Y}_0 \equiv \{\emptyset\}$.

A *Young tableau* is a Young diagram $\lambda = \{\lambda_1,\ldots,\lambda_r\} \vdash N$ together with a bijective assignment of the numbers $1,\ldots,N$ to the boxes of the diagram, one number per box. For example,

$$\begin{array}{|c|c|c|c|c|} \hline 7 & 8 & 4 & 5 & 6 \\ \hline 1 & 2 \\ \cline{1-2} 3 & 9 \\ \cline{1-2} \end{array} \qquad \begin{array}{|c|c|c|c|c|} \hline 5 & 4 & 3 & 1 & 9 \\ \hline 8 & 6 \\ \cline{1-2} 7 & 2 \\ \cline{1-2} \end{array} \tag{1.28}$$

are two of the $N! = 9!$ Young tableaux associated with the diagram $\lambda = (5,2,2) \vdash 9$. A *standard Young Tableau* (SYT) is a Young tableau with the property that the entries increase along the rows and down the columns. For example,

$$\begin{array}{|c|c|c|c|c|} \hline 1 & 2 & 3 & 4 & 5 \\ \hline 6 & 7 \\ \cline{1-2} 8 & 9 \\ \cline{1-2} \end{array} \qquad \begin{array}{|c|c|c|c|c|} \hline 1 & 3 & 5 & 7 & 9 \\ \hline 2 & 4 \\ \cline{1-2} 6 & 8 \\ \cline{1-2} \end{array} \tag{1.29}$$

are two SYT's associated with $\lambda = (5,2,2)$. Note that neither of the tableaux in (1.28) is standard. The set of all SYT's of size $N$ is denoted by $\mathrm{SYT}_N$.

If $T$ is a tableau associated with partition $\lambda$, we write $\lambda = \mathrm{sh}(T)$ (=shape of $T$). It is a fundamental result in the representation theory of symmetric groups that $S_N$ is in bijection with the set of all pairs $(P,Q)$ of SYT's where

- $\mathrm{sh}(P) = \mathrm{sh}(Q)$,
- $\mathrm{sh}(P) \vdash N$,

and therefore

$$\sum_{\lambda \vdash N} F_\lambda^2 = N!, \tag{1.30}$$

where

$$F_\lambda = \#\{P \in \mathrm{SYT}_N : \mathrm{sh}(P) = \lambda\}. \tag{1.31}$$

The identity (1.30) is a result of Frobenius and Young; see, for example [**Sag01, Sta99**]. For $N = 3$, for example, there are three partitions

$$\lambda^{(1)} = (3), \qquad \lambda^{(2)} = (2,1), \qquad \lambda^{(3)} = (1,1,1) \tag{1.32}$$

giving rise to four SYT's

$$\underset{T_1}{\begin{array}{|c|c|c|}\hline 1 & 2 & 3 \\ \hline \end{array}} \quad \underset{T_2}{\begin{array}{|c|c|}\hline 1 & 2 \\ \hline 3 \\ \cline{1-1} \end{array}} \quad \underset{T_3}{\begin{array}{|c|c|}\hline 1 & 3 \\ \hline 2 \\ \cline{1-1} \end{array}} \quad \underset{T_4}{\begin{array}{|c|}\hline 1 \\ \hline 2 \\ \hline 3 \\ \hline \end{array}} \;. \tag{1.33}$$

Thus there are $1 + 2 \times 2 + 1 = 6$ possible pairs of SYT's $(P,Q)$ of the same shape, and this agrees with the order of the group $S_3$, which is $3!$.

A particular bijection from $S_N$ to the pairs $(P,Q)$ is obtained by a remarkable algorithm due to Robinson and Schensted which constructs $P$ and $Q$ from $\pi \in S_N$ explicitly. Let us denote this bijection by

$$\mathrm{RS} : S_N \to \{(P,Q) \in \mathrm{SYT}_N \times \mathrm{SYT}_N : \mathrm{sh}(P) = \mathrm{sh}(Q)\}. \tag{1.34}$$

This algorithm shows the following important connection between the longest increasing subsequence and the first row of the shape of $P$. We denote $\mathrm{sh}(P)$ by $\lambda(\pi) = (\lambda_1(\pi), \lambda_2(\pi), \cdots)$ when $\mathrm{RS}(\pi) = (P,Q)$. A surprising result due to Schensted [**Sch61**] is that

$$\ell_N(\pi) = \lambda_1(\pi). \tag{1.35}$$

Thus the length of the longest increasing subsequence of a permutation, $\pi$, is equal to the number of boxes in the first row of either of the associated SYT's, $P(\pi)$ or $Q(\pi)$.

For example, as follows by applying the Robinson-Schensted algorithm (see Subsection 3.1.1), we find

$$\mathrm{RS}(4236517) = \left( \begin{array}{|c|c|c|c|}\hline 1 & 3 & 5 & 7 \\ \hline 2 & 6 \\ \cline{1-2} 4 \\ \cline{1-1} \end{array} \;,\; \begin{array}{|c|c|c|c|}\hline 1 & 3 & 4 & 7 \\ \hline 2 & 5 \\ \cline{1-2} 6 \\ \cline{1-1} \end{array} \right). \tag{1.36}$$

We can check directly that the increasing subsequences of the permutation 4236517 are 2367 and 2357, and their length is 4. This is same as the length of the first row of $P$ and $Q$.

Under RS, the uniform distribution on $S_N$ pushes forward onto $\mathrm{RS}(S_N) = \{(P,Q) \in \mathrm{SYT}_N^2 : \mathrm{sh}(P) = \mathrm{sh}(Q)\}$, giving equal weight to each of the pairs $(P,Q)$, and hence onto the set $\mathcal{Y}_N$ of all partitions of $N$. Moreover, the random variable $\lambda_1$ on $\mathcal{Y}_N$ has the same distribution as $\ell_N$ on $S_N$. The induced probability measure on $\mathcal{Y}_N$ is given by

$$\mathbb{P}\{\lambda\} = \frac{F_\lambda \times F_\lambda}{N!} = \frac{F_\lambda^2}{N!} \tag{1.37}$$

for a fixed $\lambda \vdash N$, where $F_\lambda$ is the number of SYT's of shape $\lambda$ defined in (1.31).

For example, for $N = 3$, we find from (1.33) that

$$\begin{aligned} &\mathbb{P}\{(3)\} = \mathbb{P}\{(P,Q) \in \mathrm{RS}(S_3) : \mathrm{sh}(P) = (3)\} = \frac{1}{6}, \\ (1.38) \qquad &\mathbb{P}\{(2,1)\} = \mathbb{P}\{(P,Q) \in \mathrm{RS}(S_3) : \mathrm{sh}(P) = (2,1)\} = \frac{2 \times 2}{6} = \frac{2}{3}, \\ &\mathbb{P}\{(1,1,1)\} = \mathbb{P}\{(P,Q) \in \mathrm{RS}(S_3) : \mathrm{sh}(P) = (1,1,1)\} = \frac{1}{6}, \end{aligned}$$

and $\mathbb{P}\{\ell_3 = 2\} = \mathbb{P}\{(2,1)\} = \frac{2}{3}$.

**Definition 1.5.** *Plancherel measure* on the set $\mathcal{Y}_N$ of partitions of $N$ is defined by

$$(1.39) \qquad \mathbb{P}\{\lambda\} = \frac{F_\lambda^2}{N!}, \qquad \lambda \in \mathcal{Y}_N,$$

where $F_\lambda$ denotes the number of SYT's of shape $\lambda$ given in equation (1.31). Here we set $F_\emptyset \equiv 1$.

The above discussion implies

$$(1.40) \qquad \mathbb{P}\{\ell_N \le n\} = \frac{1}{N!} \sum_{\lambda \in \mathcal{Y}_N; \lambda_1 \le n} F_\lambda^2.$$

The above considerations can be assembled as follows: With respect to Plancherel measure, the number $\lambda_1$ of boxes in the first row of a random partition $\lambda \vdash N$ under Plancherel measure behaves statistically like the largest eigenvalue of a GUE matrix, as $N \to \infty$. In answer to the **question** raised above, the authors in [**BDJ99**] conjectured that the number of boxes in the second, third, ... rows of $\lambda$ should behave like the second, third, ... largest eigenvalues of a GUE matrix. This result for the second row, $\lambda_2$, was subsequently proved by [**BDJ00**] in 2000. The full conjecture that the number of boxes $(\lambda_1, \dots, \lambda_k)$ in the first $k$ rows of $\lambda = (\lambda_1, \lambda_2, \dots)$ behaves statistically as $N \to \infty$, like the first $k$ eigenvalues of a random GUE matrix, was proved shortly thereafter, independently by Okounkov, Borodin-Olshanski-Okounkov, and Johansson [**Oko00, BOO00, Joh01a**].

One of the primary goals of this book is to show how to prove the results in [**BDJ99**]. Along the way, we discuss many related problems and results which are of interest, but not strictly necessary to prove the results in [**BDJ99**].

**Remark 1.6.** An elegant description of the results in [**BDJ99**], and more, can also be found in Romik's recent text [**Rom15**].

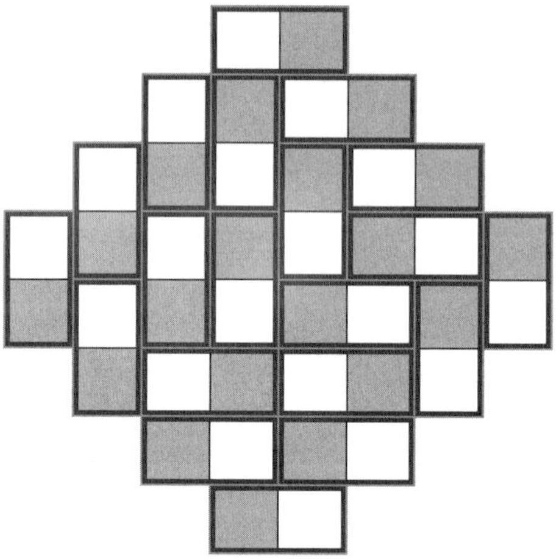

**Figure 1.3.** Sample tiling for $n = 4$

**Remark 1.7.** Actually, in [**Ula61**] Ulam considered the length $\hat{\ell}_N$ of the longest monotone subsequence, increasing or decreasing. The leading asymptotics for $\mathbb{E}[\hat{\ell}_N]$, however, can be shown to be the same as the leading asymptotics of $\mathbb{E}[\ell_N]$. In addition to the value (1.7), Ulam's best, but inaccurate, guess for the constant $c$ in (1.8), Ulam also reported, prematurely, in [**Ula61**] that the limiting distribution looks like a Gaussian distribution.

## 1.2. Random Tilings of the Aztec Diamond

The second basic combinatorial problem that provides a model for the phenomena described in this book is the problem of random tilings of the so-called Aztec diamond. Consider the planar region

$$\Omega = \{(x, y) \in \mathbb{R}^2 : |x| + |y| \leq n + 1\}.$$

Tile $\Omega$ with $2 \times 1$ dominoes, placed vertically and horizontally, such that

- the corners lie in $\mathbb{Z} \times \mathbb{Z}$, and
- the dominoes lie entirely within $\Omega$.

Figure 1.3 provides an example for $n = 4$ and Figure 1.4 provides an example for $n = 40$. Because of the obvious resemblance of the upper part of the tilings to a Mexican pyramid, such arrangements of the dominoes are called tilings of the Aztec diamond [**EKLP92a, EKLP92b, Joh02**]. For any $n$, there are[4] precisely $2^{\frac{n(n+1)}{2}}$ such tilings [**EKLP92a, EKLP92b**]. Under the assumption that all tilings are equally likely, the authors in [**JPS**] prove the following remarkable theorem. We quote from [**JPS**]:

[4]See Remark 12.26 below.

*Every tiling determines a partition of the Aztec diamond into five subregions: In the four outer regions, every tile lines up with nearby tiles, while in the fifth, central subregion, differently oriented tiles coexist side by side. We show that when n is sufficiently large, the boundary of the central subregion becomes arbitrarily close to a perfect circle of radius $\frac{n}{2}$ for all but a negligible portion of the tilings. The circle is referred to as the "Arctic circle": The four outer regions in which the tiling is "frozen" are referred to as the "polar zones" and the central subregion is called the "temperate zone".*

**Figure 1.4.** A random tiling of the Aztec diamond of size 40. Horizontal tiles are shaded. This figure is due to Eric Nordenstam (used with permission).

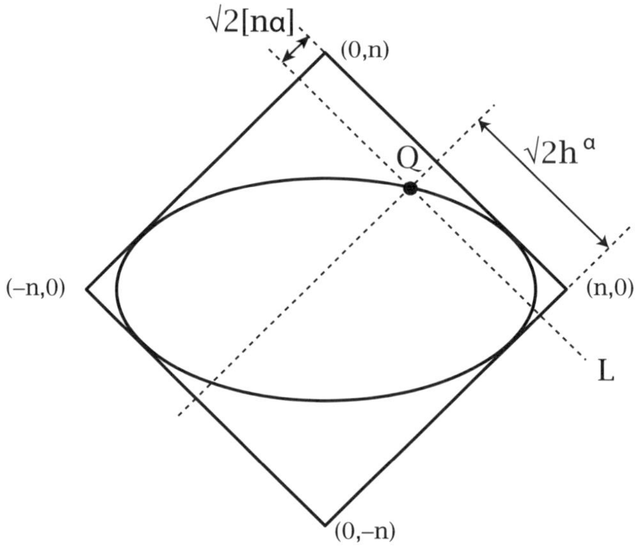


**Figure 1.5.** Arctic ellipse

In this book we will consider a class of distributions on the tilings that generalize the uniform distribution. For such distributions, the arctic circle becomes an ellipse as in Figure 1.5.

But this is not the end of the story. In a *tour de force*, and here we make contact with the theme of this book, Johansson [**Joh02**] showed that the fluctuations of the Arctic circle as $n \to \infty$ are described by RMT. More precisely, fix $0 < \alpha < \frac{1}{2}$, and consider the upper point $Q$ where the line $x + y = n - \lfloor n\alpha \rfloor$ intersects the boundary of the (fluctuating) central subregion. Let $h^\alpha$ denote $\frac{1}{\sqrt{2}}$ times the Euclidean distance of $Q$ to the line $L = \{y = x - n\}$. Then the following result describes the fluctuations of $Q$ about the arctic ellipse: there exist explicit positive constants $\beta, \rho$ such that for any $t \in \mathbb{R}$

$$\lim_{n\to\infty} \mathbb{P}\left\{h^\alpha \le \beta n + t\rho n^{\frac{1}{3}}\right\} = F(t), \tag{1.41}$$

where $F(t)$ is the Tracy-Widom distribution function defined in (1.16) above. This result (Theorem 12.28), which is the second main goal of the book, is proved in Chapter 12 below.

## 1.3. General Remarks

In the two combinatorial problems discussed above, we see that certain key statistics are described in some asymptotic limit by the eigenvalues of large random matrices. This illustrates the central theme of this book. The eigenvalues of large random matrices typically accumulate in an interval

$[a,b]$. A point $E \in [a,b]$ is called an energy level and is said to be in the bulk of the spectrum if $E$ is in the open interval $(a,b)$. Quoting from [**Dei07**],

> *Loosely speaking, we say that a system is modeled by random matrix theory (RMT) if it behaves statistically like the eigenvalues of a "large" (random) matrix. In analyzing such systems, there is something known as the standard procedure: Suppose we wish to compare some statistical quantities $\{a_k\}$ in the neighborhood of some point $A$ with the eigenvalues $\{\lambda_k\}$ of some matrix in the neighborhood of some energy $E$, say, in the bulk of the spectrum. Then we always center and scale the $a_k$'s and the $\lambda_k$'s,*

$$a_k \mapsto \overline{a}_k = \gamma_a(a_k - A), \qquad \lambda_k \mapsto \overline{\lambda}_k = \gamma_\lambda(\lambda_k - E) \tag{1.42}$$

> *so that*

$$\mathbb{E}[\#\{\overline{a}_k' s \text{ per unit interval}\}] = \mathbb{E}[\#\{\overline{\lambda}_k' s \text{ per unit interval}\}] = 1. \tag{1.43}$$

> *For energies $E$ at the edge of the spectrum, the above procedure must be modified slightly.*
>
> *This procedure can be viewed as follows: A scientist wishes to investigate some statistical phenomenon. What s'he has at hand is a microscope and a handbook of matrix ensembles. The data $\{a_k\}$ are embedded on a slide which can be inserted into the microscope. The only freedom that the scientist has is to center the slide, $a_k \mapsto a_k - A$, and then adjust the focus $a_k - A \mapsto \overline{a}_k = \gamma_a(a_k - A)$ so that on average one data point $\overline{a}_k$ appears per unit length on the slide. At that point the scientist takes out his'r handbook, and then tries to match the statistics of the $\overline{a}_k$'s with those of the eigenvalues of some ensemble. If the fit is good, the scientist then says that the system is well modeled by RMT.*

In this text we only consider (a selection of) combinatorial problems that are modeled by RMT. However, quite remarkably, RMT provides a model for a variety of mathematical and physical problems across the scientific spectrum, from nuclear scattering on the one hand, to the distribution of the zeros of the Riemann-zeta function on the critical line on the other. We refer the reader to [**Odl, Bee97, Meh04, KS99, DG07, For10, ABDF11**] for more information.

The statistical quantities $\{a_k\}$ described above typically arise in applications as realizations of a point process. A point process (see Chapter 11) is a probability measure $\mathbb{P}$ on the space of point configurations $\{\mathcal{X}\}$ on a complete separable metric space $\Lambda$. Typically, $\Lambda = \mathbb{R}^n$ or $\mathbb{Z}^n$. We are particularly interested in so-called *determinantal point processes.* These are processes for which the correlation functions (see Section 6.3 and Chapter 11), $\rho_k(x_1, \ldots, x_k)$, $x_i \in \Lambda$, are expressible in determinantal form with a *correlation kernel* $K(x, y)$:

$$\rho_k(x_1, \ldots, x_k) = \det(K(x_i, x_j))_{1 \le i,j \le k}. \tag{1.44}$$

An important subclass of finite point processes are those processes with joint probabilities of the form

$$p(x_1, \ldots, x_k) = \frac{1}{\mathcal{Z}_k} \prod_{1 \le i < j \le k} (x_j - x_i)^2 \prod_{\ell=1}^{k} w(x_\ell) \tag{1.45}$$

in the cases $\Lambda = \mathbb{Z}$ or $\Lambda = \mathbb{R}$. Here $w(x) \ge 0$ and $\mathcal{Z}_k$ is a normalizing constant. It is a theorem that such processes are determinantal (see Chapter 6). Many of the determinantal processes that arise in this text are scaling limits of such finite determinantal point processes (1.45).

There is an intimate connection between RMT and determinantal point processes of the type (1.45), and it is this connection which ultimately makes it possible to describe a variety of combinatorial problems in terms of RMT. In particular, the probability distribution on $N \times N$ Hermitian matrices, $M = M^*$, in GUE has the form (see (1.25))

$$P_N(M)dM = \frac{1}{\mathcal{Z}_N} e^{-\operatorname{Tr} M^2} \prod_{i=1}^{N} dM_{ii} \prod_{1 \le i < j \le N} dM_{ij}^R dM_{ij}^I, \tag{1.46}$$

which induces the distribution

$$\hat{P}_N(\xi) d\xi_1 \cdots d\xi_N = \frac{1}{\hat{\mathcal{Z}}_N} \prod_{1 \le j < k \le N} (\xi_i - \xi_j)^2 \prod_{i=1}^{N} e^{-\xi_i^2} d\xi_1 \cdots d\xi_N \tag{1.47}$$

on the eigenvalues $\xi_1, \ldots, \xi_N$ of $M$ (see Chapter 6). Clearly (1.47) is a special case of (1.45). In particular, the eigenvalues of a GUE matrix give rise to a determinantal point process.

It turns out that the notion of ensembles of nonintersecting paths plays an enabling role in connecting combinatorial problems to random matrices. The connection arises in the following way. Starting with the work of Karlin and McGregor [**KM59**] in probabilistic contexts and also with the work of Lindström [**Lin73**] and Gessel and Viennot [**GV85**] in combinatorial contexts, it was recognized that collections of nonintersecting paths give rise to determinantal point processes of the type (1.45). Then, by the above discussion, such distributions can be interpreted in terms of distributions of eigenvalues of random matrices. The fact on the ground is that there is a class of combinatorial problems of great interest that can be rephrased as problems concerning ensembles on nonintersecting paths, and so by the above discussion, such combinatorial problems are then connected to random matrix theory. This is the case, in particular, for Ulam's problem and for tilings of the Aztec diamond (see [**Joh02**]), which are the main subjects of this book.

In Chapter 2, we discuss the Poissonization of Ulam's problem due to Hammersley. In Chapter 3, we present basic information from combinatorics and also describe some combinatorial models which are equivalent to Ulam's problem. In Chapter 4, we begin the asymptotic analysis of $\ell_N$ by considering the expectation $\mathbb{E}[\ell_N]$ as $N \to \infty$. We also show how Young diagrams can be viewed as states in a Markov process following [**VK85**]. The chapter ends with a description of the statistics of conjugacy classes of random permutations. In Chapter 5, we introduce basic information about orthogonal polynomials and Toeplitz matrices, including the strong Szegő limit theorem. In Chapter 6 we present the basics of random matrix theory including the Tracy-Widom distribution. The main analysis of the distribution of $\ell_N$ is in Chapters 7, 8, and 9. In Chapter 7 we derive Gessel's formula which relates Ulam's problem to the theory of Toeplitz determinants. Chapter 8 contains a Fredholm determinant formula for the distribution function of the Poissonized version of $\ell_N$. Two proofs of the Fredholm determinant formula are provided: one by the Borodin-Okounkov-Case-Geronimo identity which converts Toeplitz determinants to Fredholm determinants and the other by a direct derivation. In Chapter 9, we analyze the formula obtained in Chapter 8 asymptotically and obtain various limiting results. In particular, we prove Theorems 1.2 and 1.3 above. We also give upper and lower tail estimates and show how to evaluate transverse fluctuations. In Chapter 10, we introduce Schur measure on partitions and describe Plancherel measure as a particular specialization of Schur measure. Much of the theory for Ulam's problem, including the asymptotics, goes through for general Schur measures. This chapter also contains a description of other equivalent models such as the last passage percolation model and the totally asymmetric exclusion process. In Chapter 11, we describe the basic theory of determinantal point processes. In Chapter 12, we analyze the Aztec diamond, first for finite $n$, and then as $n \to \infty$. For finite $n$, nonintersecting lattice paths play an essential role. Finally, in Chapter 13, we discuss certain dynamical versions of random matrix theory, viz., the Dyson process and Dyson Brownian motion. While we will not use these processes directly in the book, they have many similarities with various processes that are considered in our text. In addition, in limiting form, they describe the so-called Airy process which arises in many combinatorial problems, such as, in particular, the tiling problem for the Aztec diamond. In recent years, the Dyson process has also emerged as an extremely powerful tool in proving universality for various noninvariant random matrix ensembles (see [**Joh01b, EPR⁺10**]). There are also three appendices which provide additional technical information.

The theory described in this text for the longest increasing subsequence problem and the Aztec diamond has been extensively developed and generalized in recent years through the efforts of many researchers, and many new combinatorial problems have been shown to display RMT statistics. We refer the reader to [**Joh06, KK10, FS11, BC14, Cor12, BG**] and the many references therein.

Chapter 2

# Poissonization and De-Poissonization

Hammersley introduced a "Poissonized" version $L(t)$ of Ulam's random variable $\ell_N$ which is readily amenable to ergodic theoretic analysis and which, in addition, produces formulae that are particularly suitable for asymptotic analysis. Many of the applications of the longest increasing subsequence problem have more natural interpretations in terms of $L(t)$ (see Section 3.3). Once the relevant asymptotic behavior of $L(t)$ as $t \to \infty$ is determined, one needs to "de-Poissonize" the process to recover the asymptotic results for $\ell_N$ as $N \to \infty$. In this (rather technical) chapter we describe in detail the algebraic and analytical relationships between $L(t)$ and $\ell_N$.

## 2.1. Hammersley's Poissonization of Ulam's Problem

Before introducing Poissonization, we make some remarks concerning $\ell_N$. Consider the graph of $\pi \in S_N$, regarded as a map from $\{1, 2, \dots, N\}$ to $\{1, 2, \dots, N\}$. Given $\pi \in S_N$, an *up/right path* $p$ from $(0, 0)$ to $(N, N)$ is a piecewise linear up/right function from $(0, 0)$ to $(N, N)$ with nodes at $\{(i_j, \pi(i_j)) : 1 \le j \le k\}$ for some $1 \le k \le N$, where $i_j < i_{j+1}$ and $\pi(i_j) < \pi(i_{j+1})$, $1 \le j \le k - 1$ (cf. Figure 1.1). For any $\pi \in S_N$, we observe that

$$\ell_N(\pi) = \max_p \{\# \text{ of nodes on } p\}, \tag{2.1}$$

where the maximum is taken over all up/right paths $p$ with nodes in $\{(i, \pi(i))\}_{i=1}^N$. Note that the point $(0, 0)$ is not a node and the point $(N, N)$ is counted as a node only when $\pi(N) = N$.

Note that the analogue of $\ell_N$ can be defined for an arbitrary $N$-tuple of (not necessary integral) points $z = \{(x_i, y_i)\}_{i=1}^N$ in a square $[0, t] \times [0, s]$ provided that $x_i \neq x_j$, $i \neq j$, and $y_i \neq y_j$, $i \neq j$. By considering the relative orders of the $x$-coordinates and $y$-coordinates of points, a collection of such $N$ points defines a unique permutation $\pi = \pi(z) \in S_N$. The length $L_N(z)$ of the longest up/right path with nodes in $z$ from $(0, 0)$ to $(t, s)$ defined analogously to (2.1) is precisely equal to $\ell_N(\pi) = \ell_N(\pi(z))$. Suppose that we pick $N$ points in $[0, t] \times [0, s]$ uniformly at random. With probability 1, all the $x$- and $y$-coordinates of points are distinct. It is an exercise to show that such random $N$ point configurations generate random permutations uniformly in $S_N$. Hence, $L_N$ and $\ell_N$ have the same distribution.

To define Hammersley's Poissonization, we introduce a rate 1 Poisson process in the quarter plane $\mathbb{R}_+^2$ (see, e.g., Durrett [**Dur96**]): For any bounded region $A \subset \mathbb{R}_+^2$, there is a configuration composed of a random number $\mathcal{N}_A$ of points which are uniformly and independently distributed in $A$ where $\mathcal{N}_A$ is a Poisson random variable with parameter $|A|$, the area of $A$, i.e.,

$$\mathbb{P}\left(\mathcal{N}_A = k\right) = e^{-|A|} \frac{|A|^k}{k!}, \qquad k = 0, 1, 2, \dots. \tag{2.2}$$

For any two disjoint sets $A$ and $B$ the configurations of points in $A$ and $B$ are independent. *Define $L(t, s)$ as the length of the longest up/right path with nodes in the Poisson points from $(0, 0)$ to $(t, s)$*, and set $L(t) = L(t, t)$. From (2.2), it is straightforward to see that $L(t, s)$ and $L(t', s')$ have the same distribution if $ts = t's'$.

**Remark 2.1.** Since the Poissonized random variables $L(t) = L(t, \omega)$ are defined on a fixed probability space $\{\omega\}$, one can entertain the question of almost sure convergence of $\frac{L(t)}{t}$ as $t \to \infty$. In contrast, the variables $\ell_N$ are defined on different probability spaces, one for each $N$. It is easy to see (see, e.g., [**Dur96**] Section 6.7) that $\ell_N$ is equal in distribution to $\tilde{\ell}_N \equiv L(\tau_N)$, where $\tau_N = \inf\{t > 0 : [0, t] \times [0, t]$ contains exactly $N$ Poisson points$\}$, so one could define almost sure convergence of $\ell_N$ to mean almost sure convergence of $\tilde{\ell}_N$ as $N \to \infty$. However, as this is simply a rephrasing of the properties of $L(t)$, we refrain from any statements about the almost sure convergence of $\ell_N$.

If we denote by $\mathcal{N}$ a Poisson random variable with parameter $t^2$, then

$$\begin{aligned} \mathbb{P}\{L(t) \leq n\} &= \sum_{N=0}^{\infty} \mathbb{P}\{L(t) \leq n | \mathcal{N} = N\} \, \mathbb{P}\{\mathcal{N} = N\} \\ &= \sum_{N=0}^{\infty} e^{-t^2} \frac{(t^2)^N}{N!} \, \mathbb{P}\{\ell_N \leq n\}, \end{aligned} \tag{2.3}$$

where

$$\ell_0 \equiv 0. \tag{2.4}$$

In combinatorial terminology, the probability distribution $\mathbb{P}\{L(t) \leq n\}$ is the exponential generating function of the sequences $\{\mathbb{P}\{\ell_N \leq n\}\}_{N=0}^{\infty}$ times $e^{-t^2}$. On the other hand, in statistical physics terminology, the Poissonization corresponds to a grand canonical ensemble in which the number of particles itself is a random variable. In this terminology, $\ell_N$ corresponds to the microcanonical ensemble.

As noted in Section 1.1.4, $\ell_N$ has the same distribution as the length of the first part of a random partition of $\mathcal{Y}_N$ equipped with Plancherel measure (see Definition 1.5). Plancherel measure also has a Poissonized version. Let $\mathcal{Y} \equiv \bigcup_{N=0}^{\infty} \mathcal{Y}_N$ be the set of all partitions of arbitrary size. Recall that we set $\mathcal{Y}_0 \equiv \{\emptyset\}$.

**Definition 2.2.** Fix $t > 0$. The *Poissonized Plancherel measure* (with parameter $t$) on $\mathcal{Y}$ is defined by

$$\mathbb{P}^{P.P.}\{\lambda\} = e^{-t^2} \frac{t^{2|\lambda|} F_\lambda^2}{(|\lambda|!)^2}, \qquad \lambda = (\lambda_1, \lambda_2, \dots) \in \mathcal{Y}. \tag{2.5}$$

Here when $\lambda = \emptyset$, $F_\emptyset \equiv 1$ and $\mathbb{P}^{P.P.}\{\emptyset\} = e^{-t^2}$.

It follows immediately from (2.3) and Definition 1.5 that

$$\mathbb{P}^{P.P.}\{\lambda_1 \leq n\} = \mathbb{P}\{L(t) \leq n\}. \tag{2.6}$$

## 2.2. De-Poissonization Lemmas

As mentioned earlier, we will see in the next sections that the Poissonized random variable $L(t)$ has algebraic structure which is particularly amenable to asymptotic analysis. Once the asymptotic behavior of $L(t)$ is obtained, we extract the asymptotic properties of $\ell_N$ itself which is accomplished by means of the following de-Poissonization lemma.

Since a Poisson random variable of parameter $a$ has expected value $a$ and standard deviation $\sqrt{a}$, the number of Poisson points in the square $[0, t] \times [0, t]$ is typically of order $t^2$, and as the parameter $t$ becomes large, this number concentrates around $t^2$. Hence, one may suspect that $\ell_N$ is "close to $L(\sqrt{N})$". Corollary 2.5 and Lemma 2.6 coupled with Lemma 2.7 provide explicit comparisons of these two quantities. We note from the following calculations that these estimates hold for a broader class of generating functions, not just for $L(t)$.

Given a sequence $\{a_N\}_{N=0}^{\infty}$, let

$$\phi(t) = \phi(t; \{a_N\}) \equiv \sum_{N=0}^{\infty} e^{-t^2} \frac{t^{2N}}{N!} a_N. \tag{2.7}$$

Then $e^t \phi(\sqrt{t})$ is the usual exponential generating function of the sequence.

**Lemma 2.3.** *Suppose that $\{a_N\}_{N=0}^{\infty}$ satisfies*

$$0 \le a_N \le 1 \quad \text{and} \quad a_{N+1} \le a_N \tag{2.8}$$

*for all $N = 0, 1, 2, \ldots$. Then, for all $0 < \epsilon < 1$ and $N \ge 1$,*

$$a_N \ge \phi(\sqrt{(1+\epsilon)N})) - e^{-\frac{1}{4}\epsilon^2 N} \tag{2.9}$$

*and*

$$a_N \le \phi(\sqrt{(1-\epsilon)N}) + \frac{1-\epsilon}{\epsilon\sqrt{2\pi N}} e^{-\frac{1}{2}\epsilon^2 N}. \tag{2.10}$$

**Proof.** From the monotonicity, we have

$$\begin{aligned} \phi(\sqrt{(1+\epsilon)N}) - a_N &= \sum_{k=0}^{\infty} \frac{e^{-(1+\epsilon)N}((1+\epsilon)N)^k}{k!} (a_k - a_N) \\ &\le \sum_{k=0}^{N-1} \frac{e^{-(1+\epsilon)N}((1+\epsilon)N)^k}{k!} (a_k - a_N). \end{aligned} \tag{2.11}$$

Since $a_k - a_N \le 1$, the right-hand side of (2.11) is less than or equal to

$$\begin{aligned} \sum_{k=0}^{N-1} \frac{e^{-(1+\epsilon)N}((1+\epsilon)N)^k}{k!} &\le e^{-(1+\epsilon)N}(1+\epsilon)^N \sum_{k=0}^{N-1} \frac{N^k}{k!} \\ &\le e^{-\epsilon N}(1+\epsilon)^N. \end{aligned} \tag{2.12}$$

This implies (2.9) since $\log(1+\epsilon) \le \epsilon - \frac{1}{4}\epsilon^2$ for $0 \le \epsilon \le 1$.

For the estimate (2.10), by a similar argument,

$$\begin{aligned} a_N - \phi(\sqrt{(1-\epsilon)N}) &= \sum_{k=0}^{\infty} \frac{e^{-(1-\epsilon)N}((1-\epsilon)N)^k}{k!} (a_N - a_k) \\ &\le \sum_{k=N+1}^{\infty} \frac{e^{-(1-\epsilon)N}((1-\epsilon)N)^k}{k!} (a_N - a_k) \\ &\le e^{-(1-\epsilon)N} \sum_{k=N+1}^{\infty} \frac{((1-\epsilon)N)^k}{k!} \\ &= e^{-(1-\epsilon)N}((1-\epsilon)N)^N \sum_{m=1}^{\infty} \frac{((1-\epsilon)N)^m}{(m+N)!}. \end{aligned} \tag{2.13}$$

Since $(m+N)! \geq N!N^m$ for $m \geq 0$, this is less than or equal to

$$e^{-(1-\epsilon)N}\frac{((1-\epsilon)N)^N}{N!}\sum_{m=1}^{\infty}(1-\epsilon)^m = e^{-(1-\epsilon)N}\frac{((1-\epsilon)N)^N}{N!}\frac{1-\epsilon}{\epsilon}. \tag{2.14}$$

Thus, by using Stirling's formula,

$$N! = \sqrt{2\pi}N^{N+\frac{1}{2}}e^{-N}e^{\frac{\theta}{12N}}, \qquad N \geq 1, \tag{2.15}$$

for some $0 < \theta = \theta(N) < 1$, we obtain

$$a_N - \phi(\sqrt{(1-\epsilon)N}) \leq e^{\epsilon N}(1-\epsilon)^N\frac{1-\epsilon}{\epsilon\sqrt{2\pi N}}. \tag{2.16}$$

Since $\log(1-\epsilon) \leq -\epsilon - \frac{1}{2}\epsilon^2$, $0 \leq \epsilon < 1$, (2.10) follows. □

**Remark 2.4.** Note that the above proof can be adapted to other cases, such as $a_N \leq a_{N+1}$ or $0 \leq a_N \leq N$, leading to appropriate estimates.

**Corollary 2.5.** *Suppose that* $\{a_N\}_{N=0}^{\infty}$ *satisfies*

$$0 \leq a_N \leq 1 \quad \text{and} \quad a_{N+1} \leq a_N, \tag{2.17}$$

*for all* $N = 0, 1, 2, \ldots$. *Set*

$$\mu_N \equiv N + 2\sqrt{sN\log N}, \qquad \nu_N \equiv N - 2\sqrt{sN\log N} \tag{2.18}$$

*for* $s \geq 1$ *and* $N \geq 2$. *Then, for* $4s\log N < N$,

$$\phi(\sqrt{\mu_N}) - \frac{1}{N^s} \leq a_N \leq \phi(\sqrt{\nu_N}) + \frac{1}{N^s}. \tag{2.19}$$

**Proof.** This follows from Lemma 2.3 by taking $\epsilon = 2\sqrt{\frac{s\log N}{N}} < 1$ and noting that

$$\frac{1-\epsilon}{\epsilon\sqrt{2\pi N}}e^{-\frac{1}{2}\epsilon^2 N} \leq \frac{1}{\epsilon\sqrt{N}}e^{-\frac{1}{2}\epsilon^2 N} = \frac{1}{2\sqrt{s\log N}N^{2s}} \leq \frac{1}{N^{2s}}$$

for $s \geq 1$ and $N \geq 2$. □

Even though we can prove all the necessary de-Poissonization arguments needed in this book using Corollary 2.5 above, it is sometimes more convenient to use the following estimates.

**Lemma 2.6.** *There is a positive integer* $N_0$ *such that for* $\{a_N\}_{N=0}^{\infty}$ *satisfying* (2.17),

$$a_N \leq 5\phi((N-\sqrt{N})^{1/2}) \tag{2.20}$$

*and*

$$1 - a_N \leq 5(1 - \phi((N+\sqrt{N})^{1/2})) \tag{2.21}$$

*for all* $N \geq N_0$.

**Proof.** Using (2.17),

$$\begin{aligned}\phi((N-\sqrt{N})^{1/2}) &= \sum_{m=0}^{\infty} e^{-(N-\sqrt{N})}\frac{(N-\sqrt{N})^m}{m!}a_m\\ &\geq \sum_{N-\sqrt{N}\leq m\leq N} e^{-(N-\sqrt{N})}\frac{(N-\sqrt{N})^m}{m!}a_m \qquad (2.22)\\ &\geq a_N \sum_{N-\sqrt{N}\leq m\leq N} e^{-(N-\sqrt{N})}\frac{(N-\sqrt{N})^m}{m!}.\end{aligned}$$

The summand is a decreasing function in $m \geq N-\sqrt{N}$, and hence the summation is bounded below by $\sqrt{N}$ times the summand computed at $m = N$. Using Stirling's formula,

$$\phi((N-\sqrt{N})^{1/2}) \geq \frac{a_N e^{O(\frac{1}{N})}}{\sqrt{2\pi}} e^{\sqrt{N}+N\log(1-\frac{1}{\sqrt{N}})} \geq \frac{1}{5}a_N \tag{2.23}$$

if $N$ is large enough. The other inequality is similar. □

In order to be able to apply the above de-Poissonization lemmas to the sequence $a_N \equiv \mathbb{P}(\ell_N \leq n)$, we need to check the conditions (2.17). It is clear that $0 \leq \mathbb{P}\{\ell_N \leq n\} \leq 1$. The following lemma shows the monotonicity.

**Lemma 2.7.** *For all $n \geq 0$ and $N \geq 0$,*

$$\mathbb{P}\{\ell_{N+1} \leq n\} \leq \mathbb{P}\{\ell_N \leq n\}. \tag{2.24}$$

**Proof.** For $n \geq N$, (2.24) is clear since $\mathbb{P}\{\ell_N \leq n\} = 1$. Assume that $0 \leq n \leq N$. Define the characteristic function $g_N : S_N \to \{0,1\}$ by

$$g_N(\sigma) = \begin{cases} 1, & \ell_N(\sigma) \leq n,\\ 0, & \ell_N(\sigma) > n.\end{cases} \tag{2.25}$$

Then

$$\mathbb{P}\{\ell_N \leq n\} = \frac{1}{N!}\sum_{\sigma\in S_N} g_N(\sigma). \tag{2.26}$$

For $k = 1, 2, \ldots, N+1$, set $S_{N+1}^{(k)} \equiv \{\pi \in S_{N+1} : \pi(1) = k\}$, the set of permutations of size $N+1$ which map 1 to $k$. Then $S_{N+1}$ is the disjoint union of the sets $S_{N+1}^{(k)}$, $k = 1, \ldots, N+1$. Define the bijection $F_k : S_{N+1}^{(k)} \to S_N$ by

$$F_k(\pi)(m) = \begin{cases} \pi(m+1), & \pi(m+1) \leq k-1,\\ \pi(m+1)-1, & \pi(m+1) \geq k+1,\end{cases} \tag{2.27}$$

for $\pi \in S_{N+1}^{(k)}$. For example, if $N = 3$, $k = 2$, and $\pi = 2413 \in S_4$, then $F_2(\pi) = 312$ (see Figure 2.1). Clearly, $F_k$ is a bijection by construction.

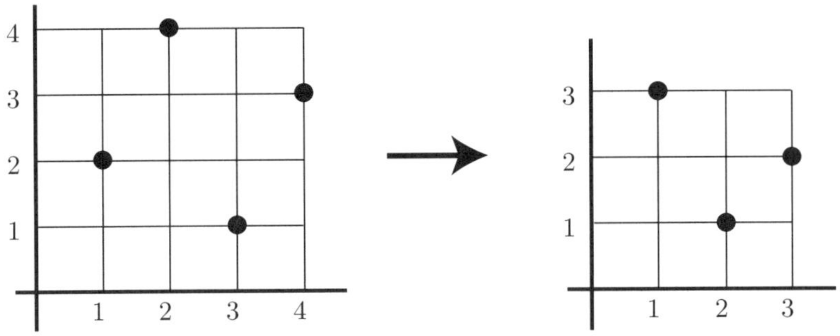


**Figure 2.1.** Map $F_2 : S_4^{(2)} \to S_3$ when $\pi = 2143$

Note that for $\pi \in S_{N+1}^{(k)}$, if $\ell_{N+1}(\pi) \le n$, then $\ell_N(F_k(\pi)) \le n$. Hence, $g_{N+1}(\pi) \le g_N(F_k(\pi))$ and

$$\begin{aligned}\mathbb{P}\{\ell_{N+1} \le n\} &= \frac{1}{(N+1)!} \sum_{\pi \in S_{N+1}} g_{N+1}(\pi) \\ &= \frac{1}{(N+1)!} \sum_{k=1}^{N+1} \sum_{\pi \in S_{N+1}^{(k)}} g_{N+1}(\pi) \\ &\le \frac{1}{(N+1)!} \sum_{k=1}^{N+1} \sum_{\pi \in S_{N+1}^{(k)}} g_N(F_k(\pi)) \\ &= \frac{1}{(N+1)!} \sum_{k=1}^{N+1} N!\, \mathbb{P}\{\ell_N \le n\} = \mathbb{P}\{\ell_N \le n\}. \end{aligned} \tag{2.28}$$ □

**Corollary 2.8.** *Recall* (2.3)*. Corollary* 2.5 *and Lemma* 2.6 *apply to* $a_N \equiv \mathbb{P}\{\ell_N \le n\}$ *and* $\phi(t) = \mathbb{P}\{L(t) \le n\}$ *for all* $n$.

A consequence of Lemma 2.7 is the monotonicity of $\mathbb{E}[\ell_N]$. We leave the proof as an exercise.

**Corollary 2.9.** *For all* $N \ge 0$,

$$\mathbb{E}[\ell_{N+1}] \ge \mathbb{E}[\ell_N]. \tag{2.29}$$

Chapter 3

# Permutations and Young Tableaux

In this chapter, we will present some basic properties of the symmetric group $S_N$. Our presentation is far from complete, and we focus almost exclusively on the parts of the theory that are needed in what follows. More detailed information and additional topics can be found, in particular, in the books of Sagan [**Sag01**], Fulton [**Ful97**], and Stanley [**Sta99**]. The material we present is taken mostly from [**Sag01**], and we use his examples and many of his arguments without further comment.

As each permutation $\pi \in S_N$ is a bijection from the set $\{1, \ldots, N\}$ onto itself, $S_N$ inherits a natural group structure by composition of mappings i.e., if $\pi, \sigma \in S_N$, then $\pi\sigma$ is the permutation corresponding to the mapping which takes $i \mapsto \pi(\sigma(i))$ for $i = 1, \ldots, N$. The trivial permutation $\pi(i) = i$, $i = 1, \ldots, N$ is the identity for the group. The study of $S_N$ as a group has been and continues to be a fundamental and extremely fruitful endeavor in the general theory of groups. Nevertheless, as we will see, the group structure of $S_N$ *per se* plays no role in the calculations and phenomena in this text. Rather, what will be important are various presentations of $S_N$ as geometric shapes, such as pairs of standard Young tableaux as described in Chapter 1. A tantalizing hint that the group structure of $S_N$ could play a larger role is that the irreducible representations of $S_N$ are in one-to-one correspondence with the Young diagrams of size $N$ or, equivalently, partitions $\lambda \vdash N$; see for example [**Sag01**]. However, at this stage, the role of the group structure of $S_N$ is unclear.

We write a permutation in either *one-line* notation or *two-line* notation. For example, one-line notation and two-line notation for the permutation

$$\pi(1) = 2, \quad \pi(2) = 3, \quad \pi(3) = 1, \quad \pi(4) = 4, \quad \pi(5) = 5$$

are

$$\pi = 23145,$$

and

$$\pi = \begin{matrix} 12345 \\ 23145 \end{matrix},$$

respectively.

## 3.1. The Robinson-Schensted Correspondence

As stated in Chapter 1 (see (1.34)), there is a particular bijection, due to Robinson [**Rob38**] and Schensted [**Sch61**], between permutations and pairs of standard Young tableaux (SYT's) of the same shape. This Robinson-Schensted correspondence is a central result in combinatorics and will play a fundamental role in the analysis of Ulam's longest increasing subsequence problem. We define the Robinson-Schensted algorithm that defines this correspondence and discuss various properties of this algorithm. A key result is the theorem due to Schensted (Theorem 3.7) that relates the length of the longest increasing subsequence and the first row of the corresponding SYT.

**3.1.1. The Robinson-Schensted Algorithm.** Let $\mathrm{SYT}_N$ denote the set of SYT's of size $N$. We now construct a map due to Robinson-Schensted

$$\mathrm{RS} : S_N \to \{(P, Q) \in \mathrm{SYT}_N \times \mathrm{SYT}_N : \mathrm{sh}(P) = \mathrm{sh}(Q)\}. \tag{3.1}$$

Write $\pi \in S_N$ in two-line notation:

$$\pi = \begin{matrix} 1 & 2 & \cdots & N \\ x_1 & x_2 & \cdots & x_N. \end{matrix}$$

A *partial tableau* (PT) is a Young diagram with distinct integer entries which increase along the rows and down the columns. For example,

$$\begin{array}{|c|c|c|}\hline 1 & 3 & 8 \\ \hline 9 & 12 \\ \cline{1-2} 14 \\ \cline{1-1}\end{array} \quad \text{and} \quad \begin{array}{|c|c|c|c|}\hline 1 & 2 & 5 & 7 \\ \hline 8 & 9 \\ \cline{1-2} 10 \\ \cline{1-1}\end{array}$$

are partial tableau. Clearly a PT is an SYT if its elements are precisely $\{1, 2, \ldots, N\}$. For $\pi \in S_N$, $\mathrm{RS}(\pi)$ is obtained by constructing a sequence of pairs of PT's:

$$(P_0, Q_0) = (\emptyset, \emptyset), \ (P_1, Q_1), \ \ldots, \ (P_N, Q_N) = (P, Q) \equiv \mathrm{RS}(\pi) \tag{3.2}$$

where $x_1, x_2, \ldots, x_N$ are *inserted* into the $P_k$'s and $1, 2, \ldots, N$ are *placed* in the $Q_k$'s so that $\mathrm{sh}\, P_k = \mathrm{sh}\, Q_k \vdash k$ for each $k = 1, \ldots, N$.

In the presentation that follows, we frequently drop the box structure for PT's: For example, in place of the PT

| | | |
|---|---|---|
| 1 | 3 | 8 |
| 9 | 12 | |
| 14 | | |

above, we will simply write

$$\begin{matrix} 1 & 3 & 8 \\ 9 & 12 & \\ 14 & & \end{matrix} \quad .$$

Also, by the $(i, j)$-th element of a PT, we mean the element in the $i$th row counted downwards and the $j$th column counted from the left. Thus, 9 is the $(2, 1)$ element in the preceding PT.

We now define *row insertion*: Suppose $P$ is a PT of size $k$, say, and suppose $x$ is not in $P$. We *row insert* $x$ into $P$ by executing the following do-loop:

RS1: Set $R :=$ first row of $P$.

RS2: While $x$ is less than some element of row $R$, do

RSa: Let $y$ be the smallest element of $R$ greater than $x$, and replace $y$ by $x$ in $R$ (denoted by $R \leftarrow x$).

RSb: Set $x := y$ and $R :=$ next row down.

RS3: Now $x$ is greater than every element of $R$, so place $x$ at the end of row $R$ and stop.

For example, suppose $x = 3$ and

$$P = \begin{matrix} 1 & 2 & 5 & 8 \\ 4 & 7 & & \\ 6 & & & \\ 9 & & & \end{matrix} \quad . \tag{3.3}$$

The *path of insertion* of $x$ into $P$ is as follows:

$$\begin{matrix} 1 & 2 & 5 & 8 & \leftarrow 3 \\ 4 & 7 & & & \\ 6 & & & & \\ 9 & & & & \end{matrix} \qquad \begin{matrix} 1 & 2 & 3 & 8 & \\ 4 & 7 & & & \leftarrow 5 \\ 6 & & & & \\ 9 & & & & \end{matrix} \qquad \begin{matrix} 1 & 2 & 3 & 8 & \\ 4 & 5 & & & \\ 6 & & & & \leftarrow 7 \\ 9 & & & & \end{matrix}$$

$$\begin{matrix} 1 & 2 & 3 & 8 \\ 4 & 5 & & \\ 6 & 7 & & \\ 9 & & & \end{matrix} = P'.$$

We see that 5 and 7 are displaced (or bumped) during the insertion.

If the result of row insertion $x \notin P$ into $P$ is $P'$, we write

$$r_x(P) = P'. \tag{3.4}$$

Observe that

$$\text{a) if } P \text{ is of size } k, \text{ then } P' \text{ is of size } k+1, \text{ and} \tag{3.5}$$

$$\text{b) if } P \text{ is a PT, then } P' \text{ is a PT.} \tag{3.6}$$

The first fact (3.5) is of course obvious, but the second fact (3.6) needs some proof: Write

$$P = \begin{array}{llllll} a_1 & a_2 & a_3 & \cdots & a_\ell \\ b_1 & b_2 & \cdots & b_m & \\ c_1 & c_2 & \cdots & c_n & \\ \vdots & & & & \end{array} \tag{3.7}$$

for some $\ell \geq m \geq n \geq \cdots$. As $P$ is a PT, we must have $a_1 < a_2 < \cdots < a_\ell$, $b_1 < b_2 < \cdots < b_m$, $\cdots$ and $a_1 < b_1 < c_1 < \cdots$, $a_2 < b_2 < c_2 < \cdots$, etc. Now consider the insertion $x \notin P$ into $P$. By the above procedure, $x$ is inserted into the first row

$$R = a_1 a_2 \cdots a_\ell \tag{3.8}$$

of $P$. If $x > a_\ell$, we must simply place $x$ at the end of $R$, and the insertion procedure ends with

$$P' = \begin{array}{llllll} a_1 & a_2 & \cdots & a_\ell & x \\ b_1 & \cdots & b_m & & \\ \vdots & & & & \end{array} \tag{3.9}$$

which is clearly a PT. (Note: An essential part of this assertion is that $\mathrm{sh}(P')$ is still a partition, i.e., $\lambda_1' \geq \lambda_2' \geq \lambda_3' \geq \cdots$ for the lengths of the rows.) If $x < a_i$ for some $1 \leq i \leq \ell$ but $x > a_{i-1}$ (this condition is of course vacuous if $i = 1$), then $a_i$ is bumped by $x$ and $P$ takes the form $P^{(1)}$.

$$P^{(1)} \equiv \begin{array}{lllllllll} a_1 & a_2 & \cdots & a_{i-1} & x & a_{i+1} & \cdots & a_\ell \\ b_1 & b_2 & \cdots & b_{i-1} & b_i & b_{i+1} & \cdots & b_m \\ c_1 & c_2 & \cdots & & & & & \\ \cdots & & & & & & & \end{array} \tag{3.10}$$

But, as $a_{i-1} < x < a_i < a_{i+1}$ and as $a_i < b_i$, it follows that $P^{(1)}$ is also a PT. Of course, if $m < i$, the condition $x < b_i$ is vacuous. Note also that $\mathrm{sh}\, P^{(1)} = \mathrm{sh}\, P$, a partition. Now we must insert $a_i$ into the second row

$$b_1 b_2 \cdots b_{i-1} b_i b_{i+1} \cdots b_m \tag{3.11}$$

of $P^{(1)}$, obtaining $P^{(2)}$. In view of the preceding argument, all we have to do to verify that $P^{(2)}$ is a PT is to show that after insertion of $a_i$, the element in the first row that lies above $a_i$ is less than $a_i$, and to verify that $\mathrm{sh}\, P^{(2)}$

remains a Young diagram. But if $m \geq i$, we have $a_i < b_i < b_{i+1} < \cdots$, so $a_i$ must be inserted into the second row under one of $a_1, a_2, \ldots, a_{i-1}, x$. The same if clearly true if $m < i$. This implies in particular that the length of the second row of $P^{(2)}$ is always less than or equal to the length of the first row, so $\operatorname{sh} P^{(2)}$ is still a partition. Also, as $a_1 < a_2 < \cdots < a_{i-1} < x < a_i$, it follows that the element in the first row of $P^{(2)}$ above $a_i$ is less than $a_i$. Hence, $P^{(2)}$ is a PT.

**Remark 3.1.** The insertion path for $x$ moves downwards and weakly to the left; i.e., if $x$ is inserted in the first row into position $i$, then $a_i$ is inserted into position $i'$ in the second row where $i' \leq i$, and $b_{i'}$ is inserted into position $i''$ in the third row where $i'' \leq i' \leq i$, etc.

If $\lambda \vdash N$ of length $\ell$,

$$\lambda = (\lambda_1, \lambda_2, \ldots, \lambda_{i_1}, \lambda_{i_1+1}, \ldots, \lambda_{i_1+i_2}, \ldots, \lambda_{i_1+\cdots+i_k}, \ldots, \lambda_\ell), \tag{3.12}$$

where $\lambda_1 = \lambda_2 = \cdots = \lambda_{i_1} > \lambda_{i_1+1} = \cdots = \lambda_{i_1+i_2} > \cdots > \cdots$, then the *outer corners* of $\lambda$ are $(1, \lambda_1+1)$, $(i_1+1, \lambda_{i_1+1}+1)$,..., $(\ell+1, 1)$, and the *inner corners* of $\lambda$ are $(i_1, \lambda_{i_1})$,$(i_1+i_2, \lambda_{i_1+i_2}), \ldots, (\ell, \lambda_\ell)$. For example, if $\lambda = (4,4,3,2,2,1)$, then $x = (1,5), (3,4), (4,3), (6,2), (7,1)$ are the outer corners of $\lambda$ and $x = (2,4), (3,3), (5,2), (6,1)$ are the inner corners of $\lambda$. In the picture

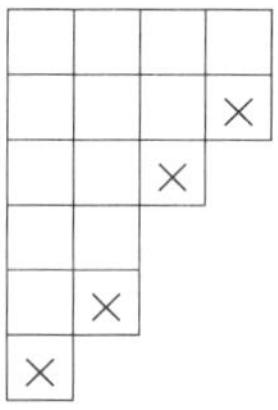

only the inner corners are marked with the $\times$ symbol. The geometric motivation for this notation is clear.

**Remark 3.2.** Observe that an insertion procedure $x \to P$ always terminates at an outer corner of $P$. To see this, note that for each step in the insertion procedure, except for the final step, the shape of the PT's does not change. The final step involves the addition of an element $x$ to a row. Suppose this step takes place in a row, say $i$, which does not correspond to an outer corner, then $\lambda_i = \lambda_{i-1}$.

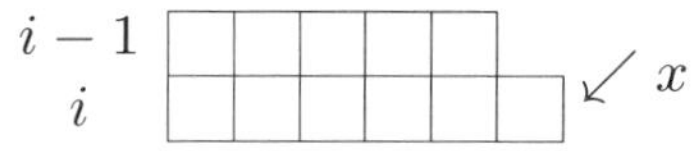


$$\lambda_i = \lambda_{i-1}$$

In this case, after $x$ is inserted, row $i$ is longer than row $i-1$, so the diagram is no longer a Young diagram, contradicting (3.6). Thus $x \to P$ always terminates at an outer corner of $P$.

*Placement* of an element $x$ in a tableau $Q$ is even easier than insertion and is only defined if $x$ is greater than every element of $Q$. Suppose $Q$ is a PT of shape $\lambda$ and that $(m, n)$ is an outer corner of $\lambda$. *To place $x$ in $Q$ at box $(m, n)$*, we set $Q'_{m,n} = x$. Otherwise, set $Q'_{i',j'} \equiv Q_{i',j'}$ for all boxes $(i', j')$ in $Q$. For example if

$$Q = \begin{matrix} 1 & 2 & 5 \\ 4 & 7 & \\ 6 & & \\ 8 & & \end{matrix},$$

then placing $x = 9$ in outer box $(i, j) = (2, 3)$ yields

$$Q' = \begin{matrix} 1 & 2 & 5 \\ 4 & 7 & 9 \\ 6 & & \\ 8 & & \end{matrix},$$

If $Q$ is a PT of size $k$, it is clear that $Q'$ will be a PT of size $k + 1$.

**Definition 3.3** (Robinson-Schensted algorithm). For permutation

$$\pi = \begin{matrix} 1 & 2 & \cdots & N \\ x_1 & x_2 & \cdots & x_N \end{matrix},$$

we define the sequence (3.2) as follows. Set $(P_0, Q_0) = (\emptyset, \emptyset)$. Assume $(P_{k-1}, Q_{k-1})$ has been constructed as a pair of PT's of size $k - 1$ with $\operatorname{sh} P_{k-1} = \operatorname{sh} Q_{k-1}$. Define

$$P_k = r_{x_k}(P_{k-1}), \tag{3.13}$$

and define $Q_k$ by placing $k$ into $Q_{k-1}$ at the box $(i, j)$ where the insertion $x_k \to P_{k-1}$ terminates.

Note that by Remark 3.2, this procedure is always well defined. Moreover, we have $\operatorname{sh} P_k = \operatorname{sh} Q_k$. Set $P = P_N$ and $Q = Q_N$. Clearly, $P(\pi)$ and $Q(\pi)$ are SYT's. The $P$-tableau is called the *insertion tableau*, and the $Q$-tableau is called the *recording tableau*.

**Remark 3.4.** $Q_k$ is in fact an SYT of size $k$ for all $1 \leq k \leq N$.

The following example illustrates the algorithm. For $N = 7$ and

$$\pi = \begin{matrix} 1 & 2 & 3 & 4 & 5 & 6 & 7 \\ 4 & 2 & 3 & 6 & 5 & 1 & 7 \end{matrix} \tag{3.14}$$

as in (3.20), we have

$$P'_k s: \quad \emptyset \quad \begin{array}{c}4\end{array} \quad \begin{array}{c}2\\4\end{array} \quad \begin{array}{cc}2&3\\4&\end{array} \quad \begin{array}{ccc}2&3&6\\4&&\end{array} \quad \begin{array}{ccc}2&3&5\\4&6&\end{array} \quad \begin{array}{ccc}1&3&5\\2&6&\\4&&\end{array} \quad \begin{array}{cccc}1&3&5&7\\2&6&&\\4&&&\end{array},$$

$$Q'_k s: \quad \emptyset \quad \begin{array}{c}1\end{array} \quad \begin{array}{c}1\\2\end{array} \quad \begin{array}{cc}1&3\\2&\end{array} \quad \begin{array}{ccc}1&3&4\\2&&\end{array} \quad \begin{array}{ccc}1&3&4\\2&5&\end{array} \quad \begin{array}{ccc}1&3&4\\2&5&\\6&&\end{array} \quad \begin{array}{cccc}1&3&4&7\\2&5&&\\6&&&\end{array}.$$

**3.1.2. Properties of the Robinson-Schensted Correspondence.** We prove in Section 3.1.3 that the map RS constructed in the previous section is a bijection.

**Theorem 3.5** (Robinson-Schensted correspondence). *For any $N$, the Robinson-Schensted mapping defined in Section* 3.1.1,

$$\mathrm{RS}: S_N \to \{(P,Q) \in \mathrm{SYT}_N \times \mathrm{SYT}_N : \mathrm{sh}(P) = \mathrm{sh}(Q)\}, \tag{3.15}$$

*is a bijection.*

We will use the notation $\mathrm{RS}(\pi) = (P(\pi), Q(\pi))$. The SYT's $P(\pi)$ and $Q(\pi)$ are called the *$P$-tableau* and the *$Q$-tableau* of $\pi$, respectively.

Let

$$F_\lambda \equiv \#\{P \in \mathrm{SYT}_N : \mathrm{sh}(P) = \lambda\} \tag{3.16}$$

denote the number of SYT's with shape $\lambda$, as in (1.31). An immediate consequence of the bijectivity of Theorem 3.5 is (another) proof of the following classical result of Frobenius and Young (see, e.g., [**Sag01, Sta99**]):

$$\sum_{\lambda \vdash N} F_\lambda^2 = N!. \tag{3.17}$$

**Remark 3.6.** This result also has group theoretic significance as follows. The irreducible representations of $S_N$ are in one-to-one correspondence with the Young diagrams of size $N$. For any Young diagram $\lambda$, the dimension of the corresponding irreducible representation is precisely $F_\lambda$. Thus, equation (3.17) is just the familiar fact [**Ser77**] in the representation theory of finite groups that the sum of the squares of the dimensions of the irreducible representations equals the order of the group.

The following property due to Schensted [**Sch61**] was noted in Chapter 1 and plays an *absolutely central* role in this book. Recall that $\ell_N(\pi)$ is the length of the longest increasing subsequence for $\pi$. For $\pi \in S_N$, let $\lambda(\pi) = (\lambda_1(\pi), \lambda_2(\pi), \dots)$ denote $\mathrm{sh}(P(\pi))$.

**Theorem 3.7** (Schensted). *For $\pi \in S_N$,*

$$\ell_N(\pi) = \lambda_1(\pi). \tag{3.18}$$

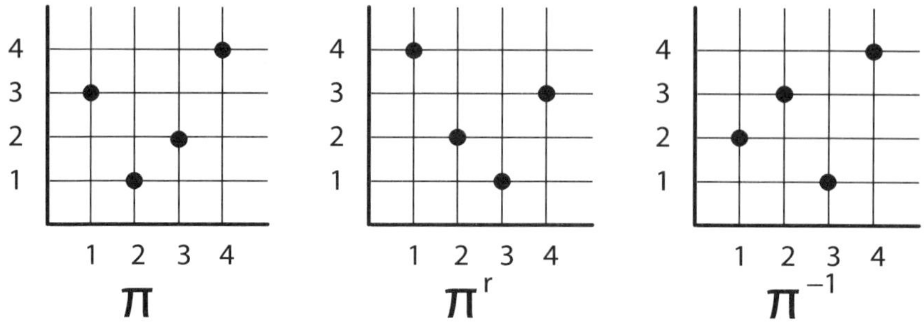


**Figure 3.1.** $\pi = 3124$, $\pi^r = 4213$, and $\pi^{-1} = 2314$

For example, for the permutation $\pi = 4236517$ in (3.14), $\ell_7(4236517) = 4$ (2357 and 2367 are longest increasing subsequences) and $\lambda_1(4236517) = 4$.

As discussed in Chapter 1, this result enables us to reformulate the distribution of $\ell_N$ in terms of the distribution of the number $\lambda_1$ of boxes in the first row of a random Young diagram $\lambda$ under the Plancherel measure (1.39). Theorem 3.7 can be further generalized to give an interpretation of $\lambda_2(\pi)$, $\lambda_3(\pi)$, ... in terms of certain generalized increasing subsequences for $\pi \in S_N$ (see Theorem 3.19 below).

The proof of Schensted's theorem is given in Subsection 3.1.4. Greene's generalization of Schensted's theorem is also discussed in the same subsection, but without proof.

The map RS has many other remarkable properties. We mention two of them. For a permutation $\pi = a_1 a_2 \cdots a_N \in S_N$, the *reversal* of $\pi$, denoted $\pi^r$, is the permutation $\pi^r = a_N \cdots a_2 a_1$. For example, if $\pi = 3124 \in S_4$, then $\pi^r = 4213$. Graphically, this is the mid-vertical reflection of the map $\pi$ as in Figure 3.1. Given an SYT $P$, $P^t$ denotes the *transposition* obtained by the reflection of $P$ about the line $y = -x$. For example,

$$P = \begin{array}{|c|c|c|c|} \hline 1 & 2 & 6 & 7 \\ \hline 3 & 5 & 8 \\ \cline{1-3} 4 \\ \cline{1-1} \end{array}, \qquad P^t = \begin{array}{|c|c|c|} \hline 1 & 3 & 4 \\ \hline 2 & 5 \\ \cline{1-2} 6 & 8 \\ \cline{1-2} 7 \\ \cline{1-1} \end{array}.$$

If $P \in \mathrm{SYT}_N$, then clearly $P^t \in \mathrm{SYT}_N$. If $\lambda = \mathrm{sh}(P)$, we define $\lambda^t = \mathrm{sh}(P^t)$. Under the map RS, the image of $\pi^r$ has the following property.

**Theorem 3.8.** *For* $\pi \in S_N$,

$$P(\pi^r) = (P(\pi))^t. \tag{3.19}$$

The proof of this elegant result, which we omit, is rather technical and can be found, for example, in [**Sag01**]. We will not use the result in any essential way in what follows.

**Remark 3.9.** $Q(\pi^r)$, however, is not simply related to $Q(\pi)$. See [**Sag01**].

As an example, for the permutation $\pi = 4236517$, we have $\pi^r = 7156324$. We see from the Robinson-Schensted algorithm that

$$\mathrm{RS}(\pi) = \mathrm{RS}(4236517) = \left( \begin{array}{|c|c|c|c|} \hline 1 & 3 & 5 & 7 \\ \hline 2 & 6 \\ \cline{1-2} 4 \\ \cline{1-1} \end{array} \, , \quad \begin{array}{|c|c|c|c|} \hline 1 & 3 & 4 & 7 \\ \hline 2 & 5 \\ \cline{1-2} 6 \\ \cline{1-1} \end{array} \right) \tag{3.20}$$

and

$$P(\pi^r) = P(7156324) = \begin{array}{|c|c|c|} \hline 1 & 2 & 4 \\ \hline 3 & 6 \\ \cline{1-2} 5 \\ \cline{1-1} 7 \\ \cline{1-1} \end{array} = P(\pi)^t.$$

However,

$$Q(\pi^t) = Q(7156324) = \begin{array}{|c|c|c|} \hline 1 & 3 & 4 \\ \hline 2 & 7 \\ \cline{1-2} 5 \\ \cline{1-1} 6 \\ \cline{1-1} \end{array} \, ,$$

which is not simply related to $Q(\pi)$ (or $P(\pi)$).

The above theorem and Schensted's theorem immediately imply the following corollary.

**Corollary 3.10** (Schensted's theorem for longest decreasing subsequences). *For $\pi \in S_N$, the length of the longest decreasing subsequence of $\pi$ equals $(\lambda(\pi)^t)_1$, the number of boxes in the first row of $\lambda(\pi)^t$.*

For the example (3.14), $(\lambda(4236517)^t)_1 = 3$, the length of a longest decreasing subsequence of $\pi$ (421, 431, or 651 are longest decreasing subsequences).

The next property about $\pi^{-1}$ is due to Schützenberger [**Sch63**].

**Theorem 3.11** (Schützenberger). *For $\pi \in S_N$,*

$$P(\pi^{-1}) = Q(\pi) \quad \textit{and} \quad Q(\pi^{-1}) = P(\pi). \tag{3.21}$$

For the example (3.20), $\pi^{-1} = 6231547$ and the Robinson-Schensted algorithm implies that

$$\mathrm{RS}(\pi^{-1}) = \mathrm{RS}(6231547) = \left( \begin{array}{|c|c|c|c|} \hline 1 & 3 & 4 & 7 \\ \hline 2 & 5 \\ \cline{1-2} 6 \\ \cline{1-1} \end{array} \, , \quad \begin{array}{|c|c|c|c|} \hline 1 & 3 & 5 & 7 \\ \hline 2 & 6 \\ \cline{1-2} 4 \\ \cline{1-1} \end{array} \right) = (Q(\pi), P(\pi)).$$

We prove Schützenberger's Theorem in Subsection 3.1.5 using Viennot's shadow lines, which are of independent interest.

The following corollary is an immediate consequence of Theorem 3.11.

**Corollary 3.12.** *For* $\pi \in S_N$,

(3.22) $$P(\pi) = Q(\pi) \quad \Leftrightarrow \quad \pi \text{ is an involution, i.e., } \pi^2 = \mathrm{id}\,.$$

Thus, we have

(3.23) $$\sum_{\lambda \vdash N} F_\lambda = \#\{\pi \in S_N : \pi^2 = \mathrm{id}\}.$$

For example, when $N = 3$,

$$\begin{array}{|c|c|c|}\hline 1 & 2 & 3 \\ \hline \end{array}\,, \quad \begin{array}{|c|c|}\hline 1 & 2 \\ \hline 3 \\ \cline{1-1} \end{array}\,, \quad \begin{array}{|c|c|}\hline 1 & 3 \\ \hline 2 \\ \cline{1-1} \end{array}\,, \quad \begin{array}{|c|}\hline 1 \\ \hline 2 \\ \hline 3 \\ \hline \end{array}$$

are all the SYT's of size 3, hence $\sum_{\lambda \vdash 3} F_\lambda = 4$. On the other hand,

$$\pi = 123, \quad 213, \quad 321, \quad 132$$

are all the involutions of $S_3$, verifying (3.23) for the case $N = 3$.

### 3.1.3. Bijectivity of the Robinson-Schensted Correspondence.

We prove Theorem 3.5 in this subsection by constructing the inverse map for RS,

(3.24) $$\mathrm{SR} : \{(P, Q) \in \mathrm{SYT}_N^2 : \mathrm{sh}(P) = \mathrm{sh}(Q)\} \to S_N.$$

This is done by reversing the Robinson-Schensted algorithm.

Given a pair of PT's $(P, Q)$ with $\mathrm{sh}\, P = \mathrm{sh}\, Q \vdash k$, we define a *deletion procedure*

(3.25) $$(P, Q) \to (P', Q'),$$

where

(3.26)
$$P', Q' \text{ are PT's of size } k-1,\ \mathrm{sh}\, P' = \mathrm{sh}\, Q',\ \text{and } P' \subset P,\ Q' \subset Q \text{ as sets.}$$

The (unique) element $\hat{x} \in P \backslash P'$ is said to be *deleted* from $P$.

Let the largest element in $Q$ be located in the $(i, j)$ box of $Q$ which is necessarily an inner corner. For convenience, we assume the existence of an empty zeroth row above the first row of $P$. Now execute the following do-loop which *deletes box* $(i, j)$ *from* $P$ *and* $Q$:

SR1: Set $x := P_{i,j}$, and erase $P_{i,j}$ from $P$ (denoted by $P \to x$). Set $R := (i-1)$st row of $P$.

SR2: While $R$ is not the zeroth row of $P$, do

SRa: Let $y$ be the largest element of $R$ smaller than $x$, and replace $y$ by $x$ in $R$.

SRb: Set $x := y$ and $R :=$ the next row up.

SR3: Now an element in the first row has been removed. Stop.

Let $Q' = Q\backslash\{Q_{i,j}\}$, and let $P'$ be the remaining tableau after SR is executed.

For example, if

$$P = \begin{array}{|c|c|c|}\hline 1 & 2 & 4 \\ \hline 3 & 6 \\ \cline{1-2} 5 \\ \cline{1-1} 7 \\ \cline{1-1}\end{array} \qquad Q = \begin{array}{|c|c|c|}\hline 1 & 3 & 4 \\ \hline 2 & 7 \\ \cline{1-2} 5 \\ \cline{1-1} 6 \\ \cline{1-1}\end{array},$$

then $(i,j) = (2,2)$. The construction of $P'$ proceeds as follows:

$$\text{SR1:} \quad \begin{array}{c} x = 6 \\ R = \text{first row} \end{array} \qquad \begin{array}{|c|c|c|}\hline 1 & 2 & 4 \\ \hline 3 \\ \cline{1-1} 5 \\ \cline{1-1} 7 \\ \cline{1-1}\end{array},$$

$$\text{SR2:} \quad \text{SRa: } y = 4 \qquad \begin{array}{|c|c|c|}\hline 1 & 2 & 6 \\ \hline 3 \\ \cline{1-1} 5 \\ \cline{1-1} 7 \\ \cline{1-1}\end{array},$$

$$\text{SRb: } x = 4 \quad \text{R} = \text{zeroth row},$$

$$\text{SR3:} \quad \text{4 has been removed} \quad P' = \begin{array}{|c|c|c|}\hline 1 & 2 & 6 \\ \hline 3 \\ \cline{1-1} 5 \\ \cline{1-1} 7 \\ \cline{1-1}\end{array} \qquad Q' = \begin{array}{|c|c|c|}\hline 1 & 3 & 4 \\ \hline 2 \\ \cline{1-1} 5 \\ \cline{1-1} 6 \\ \cline{1-1}\end{array}.$$

We must verify (3.26). The fact that $(i,j)$ is an inner corner implies that $\operatorname{sh} Q'$ is a Young diagram. The fact that $\operatorname{sh} P' = \operatorname{sh} Q'$ and that size $P' =$ size $Q' = k-1$ is immediate from the definitions. Also, $P' \subset P$, $Q' \subset Q$ as sets and $Q'$ is a PT, in fact an SYT if $Q$ is an SYT. We only need to check that $P'$ is a PT. But, if we have

$$P = \begin{array}{c} \vdots \\ i-1 \\ i \\ i+1 \\ \vdots \end{array} \left[\begin{array}{lllll} \cdots & & & & \\ b_1 & \cdots & b_m & \cdots & b_\ell \\ c_1 & \cdots & \boxed{c_m} & & \\ d_1 & \cdots d_n & & & \\ \cdots & & & & \end{array}\right] \to c_m, \qquad \ell \geq m > n,$$

for $i > 1$, then the outcome of the first iteration of the step SR2 is

$$P_1' = \begin{array}{c} \vdots \\ i-1 \\ i \\ i+1 \\ \vdots \end{array} \left[\begin{array}{llllllll} \cdots & & & & & & & \\ b_1 & \cdots & b_{m-1} & \cdots b_{r-1} & \boxed{c_m} & b_{r+1} & \cdots b_\ell \\ c_1 & \cdots & c_{m-1} & & & & \\ d_1 & \cdots d_n & & & & & \\ \cdots & & & & & & \end{array}\right] \to b_r,$$

for some $r$ such that $m \leq r \leq \ell$. As $b_1 < b_2 < \cdots < b_r < c_m < b_{r+1}$, we see that the order is preserved for the rows $i-1$, $i$, $i+1, \ldots$. We still have to show that $a_r$, the $r$th element in row $i-2$, is less than $c_m$ (if $i \leq 2$, this step is of course vacuous). As $a_r < b_r < c_m$, $P_1'$ is a PT. Now in the second iteration of SR2, $b_r$ has to be inserted in row $i-2$. Observe that the underlined elements in row $i-2$

$$\begin{array}{ll} i-2 & \underline{a_1 \cdots a_{r-1} a_r} a_{r+1} \cdots a_p \\ i-1 & b_1 \cdots b_{r-1} \underline{c_m b_{r+1} \cdots b_\ell} \end{array}$$

are less than $b_r$ while the underlined elements in row $i-1$ are greater than $b_r$. It follows that $b_r$ must replace one of $a_r$, $a_{r+1}, \ldots$, $a_p$, say $a_q$. Using $a_q < b_r$ and the property of the underlined elements, we see that the diagram remains a PT. Continuing up the rows by induction, we conclude that $P'$ is a PT and we have proved (3.26).

**Remark 3.13.** Note that the deletion path moves weakly to the right.

The inverse map $\mathrm{SR} : \{(P, Q) \in \mathrm{SYT}_N^2 : \mathrm{sh}(P) = \mathrm{sh}(Q)\} \to S_N$ is constructed as follows. Suppose $(P, Q) \in \{(P, Q) \in \mathrm{SYT}_N^2 : \mathrm{sh}(P) = \mathrm{sh}(Q)\}$. Begin by setting

$$(P_N, Q_N) \equiv (P, Q). \tag{3.27}$$

Assuming that $(P_k, Q_k)$ has been constructed, let $(P_{k-1}, Q_{k-1})$ be the result of applying the deletion procedure to $(P_k, Q_k)$ and let $x_k$ be the element deleted from $P_k$. In this way we obtain a sequence of PT's, $\{P_k\}$, and SYT's, $\{Q_k\}$, such that $\mathrm{sh}\, P_k = \mathrm{sh}\, Q_k$, size $P_k$ = size $Q_k = k$,

$$(P, Q) = (P_N, Q_N) \to (P_{N-1}, Q_{N-1}) \to \cdots \to (P_1, Q_1) \to (\emptyset, \emptyset) \tag{3.28}$$

and an ordered sequence of numbers

$$x_N \quad x_{N-2} \quad \cdots \quad x_1. \tag{3.29}$$

Define $\mathrm{SR}(P, Q)$ to be the permutation

$$\pi \equiv x_1 \cdots x_N \in S_N. \tag{3.30}$$

We claim that $\mathrm{SR} = \mathrm{RS}^{-1}$.

For a partial tableau $P$ and an SYT $Q$ such that $\mathrm{sh}\, P = \mathrm{sh}\, Q$ of size $k$, let $\mathcal{G}_{k,x}(P, Q)$ denote the result of inserting $x \notin P$ into $P$ and placing

$k + 1$ into $Q$ as described above. With this notation we clearly have for $\pi = x_1 x_2 \cdots x_N$,

$$\text{RS}(\pi) = \mathcal{G}_{N-1,x_N} \circ \cdots \circ \mathcal{G}_{1,x_2} \circ \mathcal{G}_{0,x_1}(\emptyset, \emptyset). \tag{3.31}$$

Conversely, for a partial tableau $P$ and an SYT $Q$ such that $\operatorname{sh} P = \operatorname{sh} Q$, let $x(P, Q)$ denote the element deleted from $P$ as described above. When $P$ and $Q$ are of size $k$, let $\mathcal{D}_k(P, Q)$ denote the result of applying the deletion procedure to $(P, Q)$. Given SYT's $P$, $Q$ such that $\operatorname{sh} P = \operatorname{sh} Q$ of size $N$, set

$$\begin{aligned} x_N &= x(P, Q), \\ x_k &= x(\mathcal{D}_{k+1} \circ \cdots \circ \mathcal{D}_N(P, Q)), \qquad 1 \le k \le N - 1. \end{aligned} \tag{3.32}$$

Clearly, we have

$$\pi = \text{SR}(P, Q) = x_1 x_2 \cdots x_N. \tag{3.33}$$

To prove that $\text{SR} = \text{RS}^{-1}$, we show that $\text{SR} \circ \text{RS} = \text{id}$ and $\text{RS} \circ \text{SR} = \text{id}$. This means that we must show that for any $\pi = x_1 \cdots x_N \in S_N$,

$$\begin{aligned} & x(\text{RS}(\pi)) = x_N, \\ & x(\mathcal{D}_{k+1} \circ \cdots \circ \mathcal{D}_N(\text{RS}(\pi))) = x_k, \text{ for all } 1 \le k \le N - 1, \end{aligned} \tag{3.34}$$

and for any pair of SYT's $(P, Q)$ of size $N$,

$$\mathcal{G}_{N-1,x_N} \circ \cdots \circ \mathcal{G}_{0,x_1}(\emptyset, \emptyset) = (P, Q), \tag{3.35}$$

where $x_N = x(P, Q)$ and $x_k = x(\mathcal{D}_{k+1} \circ \cdots \circ \mathcal{D}_N(P, Q))$ for $1 \le k \le N - 1$. But as $\mathcal{D}_{k+1} \circ \cdots \circ \mathcal{D}_N(\text{RS}(\pi))$ equals

$$\mathcal{D}_{k+1} \circ \cdots \circ \mathcal{D}_N \circ (\mathcal{G}_{N-1,x_N} \circ \cdots \circ \mathcal{G}_{k,x_{k+1}} \circ (\mathcal{G}_{k-1,x_k} \circ \cdots \circ \mathcal{G}_{0,x_1}(\emptyset, \emptyset))),$$

for any $k$, $1 \le k \le N - 1$, it is clearly enough to establish the following two relationships in order to prove (3.34): if $\tilde{P}$ is a PT and $\tilde{Q}$ an SYT, $\operatorname{sh} \tilde{P} = \operatorname{sh} \tilde{Q}$, of size $k$ and $\hat{x} \notin \tilde{P}$, then

$$\mathcal{D}_k \circ \mathcal{G}_{k-1,\hat{x}}(\tilde{P}, \tilde{Q}) = (\tilde{P}, \tilde{Q}) \quad \text{and} \quad x(\mathcal{G}_{k-1,\hat{x}}(\tilde{P}, \tilde{Q})) = \hat{x}. \tag{3.36}$$

Conversely, as $(\emptyset, \emptyset) = \mathcal{D}_1 \circ \cdots \circ \mathcal{D}_N(P, Q)$, (3.35) follows from the following two relationships: if $\tilde{P}, \tilde{Q}$ are as above

$$\mathcal{G}_{k-1,\hat{x}} \circ \mathcal{D}_k(\tilde{P}, \tilde{Q}) = (\tilde{P}, \tilde{Q}) \quad \text{where } \hat{x} = x(\tilde{P}, \tilde{Q}). \tag{3.37}$$

It is an easy exercise, which we leave to the reader, to verify that (3.36) and (3.37) are true. This proves the bijectivity of RS.

**Remark 3.14.** If we knew *a priori* the Frobenius-Young formula (3.17), then (3.35) would of course follow from (3.34) and vice versa (see the discussion of the Frobenius-Young formula (3.17) following Theorem 3.5).

**3.1.4. Schensted's Theorem and Greene's Theorem.** Schensted's Theorem 3.7 is a direct consequence of the following stronger result. As in the previous subsection, $P_k$ denotes the PT formed after $k-1$ insertions of the RS algorithm.

**Lemma 3.15.** *If $\pi = x_1x_2\cdots x_n$ and $x_k$ enters $P_{k-1}$ in column $j$, then the longest increasing subsequence of $\pi$ ending in $x_k$ has length $j$.*

For example, for

$$\pi = \ 4 \ \ 2 \ \ 3 \ \ 6 \ \ 5 \ \ 1 \ \ 7 \tag{3.38}$$

we have the insertion sequence

$$\emptyset \quad \begin{array}{|c|}\hline 4\\\hline\end{array} \quad \begin{array}{|c|}\hline 2\\\hline 4\\\hline\end{array} \quad \begin{array}{|c|c|}\hline 2&3\\\hline 4\\\cline{1-1}\end{array} \quad \begin{array}{|c|c|c|}\hline 2&3&6\\\hline 4\\\cline{1-1}\end{array} \quad \begin{array}{|c|c|c|}\hline 2&3&5\\\hline 4&6\\\cline{1-2}\end{array} \quad \begin{array}{|c|c|c|}\hline 1&3&5\\\hline 2&6\\\cline{1-2} 4\\\cline{1-1}\end{array} \quad \begin{array}{|c|c|c|c|}\hline 1&3&5&7\\\hline 2&6\\\cline{1-2} 4\\\cline{1-1}\end{array} \ . \tag{3.39}$$

The above lemma then implies that the longest increasing subsequences of $\pi$ ending in

$$4 \ \ 2 \ \ 3 \ \ 6 \ \ 5 \ \ 1 \ \ 7 \tag{3.40}$$

have length

$$1 \ \ 1 \ \ 2 \ \ 3 \ \ 3 \ \ 1 \ \ 4, \tag{3.41}$$

respectively.

**Proof of Lemma 3.15.** We use induction on $k$. The case $k=1$ is trivial. Suppose the result holds for all values up to $k-1$. Let $x_k$ be inserted in column $j$:

$$\begin{array}{cccc} & & j & \\ & & \downarrow & \\ 1 \rightarrow & \cdots & x_k & \cdots . \end{array}$$

We first show that there exists an increasing subsequence of length $j$ ending in $x_k$. Let $x_r$ for some $1 \le r \le k-1$ be the element in $P_{k-1}$ in box $(1, j-1)$. Necessarily, $x_r < x_k$, since $x_k$ enters in column $j$. But, by induction, there exists an increasing subsequence $\sigma$ of length $j-1$ ending in $x_r$. Thus $\sigma x_k$ is the desired subsequence.

We must show that there does not exist a longer increasing subsequence ending in $x_k$. If there were such a subsequence, let $x_i$, $i<k$, be the element preceding $x_k$ in that subsequence $\underbrace{x_\ell \cdots x_i x_k}_{\text{length}>j}$. Now when $x_i$ is inserted at step $i$, since there is an increasing subsequence of length $\ge j$ ending at $x_i$, it follows by induction that it must be inserted in some column weakly to the right of column $j$. Consider the element $x_q$ in the box $(1, j)$ of $P_i$. It satisfies

$$x_q \le x_i < x_k.$$

Now let $z$ be the element in box $(1, j)$ of $P_{k-1}$. Since the entries in a given position of a tableau never increase with subsequent insertions, we have $z \le x_q < x_k$. This contradicts the fact that $x_k$ displaces $z$ at step $k$. This completes the proof of the lemma. □

**Remark 3.16.** Note that the first row of $P(\pi)$ need not be an increasing subsequence, even though its length equals the length of the longest increasing subsequence. For example, for the permutation given in (3.38), the first row of $P(\pi)$ is 1357 which is clearly not an increasing subsequence of $\pi = 4236517$.

**Exercise 3.17.** Show how an increasing subsequence of $\pi$ of maximal length can be constructed from the RS algorithm.

The lengths of the other rows and columns of $P(\pi)$ can be interpreted as follows.

**Definition 3.18.** Let $\pi$ be a permutation. A subsequence $\sigma$ of $\pi$ is $k$-*increasing* if, as a set ("$\uplus$" denotes disjoint union),

$$\sigma = \sigma_1 \uplus \sigma_2 \uplus \cdots \uplus \sigma_k, \tag{3.42}$$

where the $\sigma_i$ are all increasing subsequences of $\pi$. If the $\sigma_i$ are all decreasing, then we say that $\sigma$ is $k$-*decreasing*. The *length* of a $k$-increasing (or $k$-decreasing) subsequence $\sigma = \sigma_1 \uplus \cdots \uplus \sigma_k$ of $\pi$ is defined to be the sum of the lengths of its components $\sigma_i$, $1 \le i \le k$. Let

$$i_k(\pi) \equiv \text{ length of the longest } k\text{-increasing subsequence of } \pi$$

and

$$d_k(\pi) \equiv \text{ length of the longest } k\text{-decreasing subsequence of } \pi.$$

Note that $k = 1$ corresponds to the usual increasing/decreasing subsequences. Using our familiar example

$$\pi = \; 4 \;\; 2 \;\; 3 \;\; 6 \;\; 5 \;\; 1 \;\; 7 \;,$$

we see that the longest 1-, 2-, and 3-increasing subsequences are given by

$$2 \;\; 3 \;\; 5 \;\; 7$$

and

$$4 \;\; 2 \;\; 3 \;\; 6 \;\; 5 \;\; 7 \;=\; 2 \;\; 3 \;\; 5 \;\; 7 \;\uplus\; 4 \;\; 6$$

and

$$4 \;\; 2 \;\; 3 \;\; 6 \;\; 5 \;\; 1 \;\; 7 \;=\; 2 \;\; 3 \;\; 5 \;\; 7 \;\uplus\; 4 \;\; 6 \;\uplus\; 1 \;,$$

respectively. Thus

$$i_1(\pi) = 4, \qquad i_2(\pi) = 6, \qquad i_3(\pi) = 7.$$

On the other hand, from RS, we have

$$P(\pi) = \begin{array}{|c|c|c|c|}\hline 1 & 3 & 5 & 7 \\ \hline 2 & 6 \\ \cline{1-2} 4 \\ \cline{1-1} \end{array},$$

so that $\lambda_1(\pi) = 4$, $\lambda_1(\pi) + \lambda_2(\pi) = 6$, $\lambda_1(\pi) + \lambda_2(\pi) + \lambda_3(\pi) = 7$, and we see a pattern.

The basic result on $k$-increasing or $k$-decreasing sequences, which generalizes Theorem 3.7, is due to Greene [**Gre74**].

**Theorem 3.19** (Greene). *Given* $\pi \in S_N$, *let* $\operatorname{sh} P(\pi) = \lambda = (\lambda_1, \lambda_2, \ldots, \lambda_\ell)$. *Denote* $\lambda^t = (\lambda_1^t, \lambda_2^t, \ldots, \lambda_m^t)$. *Then, for any* $k$,

$$i_k(\pi) = \lambda_1 + \cdots + \lambda_k \tag{3.43}$$

*and*

$$d_k(\pi) = \lambda_1^t + \cdots + \lambda_k^t. \tag{3.44}$$

This result, however, will not be used in this text. The proof is omitted, but can be found, for example, in [**Sag01, Sta99**].

**Remark 3.20.** We note that most of the definitions and theorems above involving standard tableaux and permutations make equally good sense when applied to PT's and *partial permutations*, which are bijections $\pi\colon K \to L$ between two sets of positive integers. If $K = \{k_1, k_2, \ldots, k_m\}$, $L = \{\ell_1, \ldots, \ell_m\}$, where $k_1 < k_2 < \cdots < k_m$ and $\ell_1 < \cdots < \ell_m$, then we can write $\pi$ in two-line notation

$$\pi = \begin{matrix} k_1 & k_2 & \cdots & k_m \\ \ell_{\sigma(1)} & \ell_{\sigma(2)} & \cdots & \ell_{\sigma(m)} \end{matrix},$$

where $\sigma \in S_m$. Insertion of $\ell_i$ and placement of the $k_i$ set up a bijection $\pi \to (P, Q)$ of partial permutations with PT's. For example, for $K = \{2, 4, 5, 9\}$, $L = \{3, 6, 7, 8\}$ and

$$\pi = \begin{matrix} 2 & 4 & 5 & 9 \\ 6 & 8 & 7 & 3 \end{matrix},$$

we have

$$P's: \quad \emptyset \quad \begin{array}{|c|}\hline 6 \\ \hline \end{array} \quad \begin{array}{|c|c|}\hline 6 & 8 \\ \hline \end{array} \quad \begin{array}{|c|c|}\hline 6 & 7 \\ \hline 8 \\ \cline{1-1} \end{array} \quad \begin{array}{|c|c|}\hline 3 & 7 \\ \hline 6 \\ \cline{1-1} 8 \\ \cline{1-1} \end{array} \quad P(\pi),$$

$$Q's: \quad \emptyset \quad \begin{array}{|c|}\hline 2 \\ \hline \end{array} \quad \begin{array}{|c|c|}\hline 2 & 4 \\ \hline \end{array} \quad \begin{array}{|c|c|}\hline 2 & 4 \\ \hline 5 \\ \cline{1-1} \end{array} \quad \begin{array}{|c|c|}\hline 2 & 4 \\ \hline 5 \\ \cline{1-1} 9 \\ \cline{1-1} \end{array} \quad Q(\pi).$$

Under reversal $\pi^r = \begin{matrix} 2 & 4 & 5 & 9 \\ 3 & 7 & 8 & 6 \end{matrix}$, we have

$$P's: \quad \emptyset \quad \boxed{3} \quad \boxed{3\,|\,7} \quad \boxed{3\,|\,7\,|\,8} \quad \begin{matrix} 3 & 6 & 8 \\ 7 & & \end{matrix} = P(\pi^r) = P(\pi)^t,$$

$$Q's: \quad \emptyset \quad \boxed{2} \quad \boxed{2\,|\,4} \quad \boxed{2\,|\,4\,|\,5} \quad \begin{matrix} 2 & 4 & 5 \\ 9 & & \end{matrix} = Q(\pi^r).$$

The preceding theory goes through for bijections from $X$ to $Y$ where $X$ and $Y$ are arbitrary finite ordered sets.

**3.1.5. The Inverses of Permutations and Shadow Diagrams.** We now prove Schützenberger's Theorem 3.11. To this end we introduce a beautiful geometric construction of Viennot [**Vie77**] that is of independent interest.

Consider the first quadrant of the Cartesian plane. Represent a given permutation $\pi = x_1 x_2 \cdots x_N$ as a graph by a collection of points with coordinates $(i, x_i)$ (cf. with the permutation matrix of $\pi$). Thus, for our running example

$$\pi = 4 \quad 2 \quad 3 \quad 6 \quad 5 \quad 1 \quad 7, \tag{3.45}$$

we obtain Figure 3.2. Imagine shining lights from the origin so that the region south-west of each point casts a shadow parallel to the coordinate axes. For example, the shadow cast by $(4, 6)$ looks like Figure 3.3. Consider those points for $\pi$ that are in the shadow of no other point. In our case, these points are $(1, 4)$, $(2, 2)$, and $(6,1)$. The *first shadow line* $L_1$ is the boundary of the combined shadows of these points. Graphically, we have Figure 3.4. Note that this is a broken line consisting of horizontal and vertical line segments and exactly one horizontal and one vertical ray. To

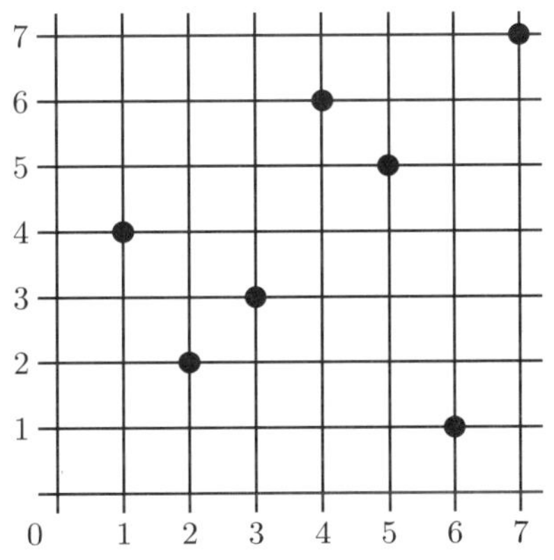


**Figure 3.2.** Graph of permutation $\pi = 4236517$

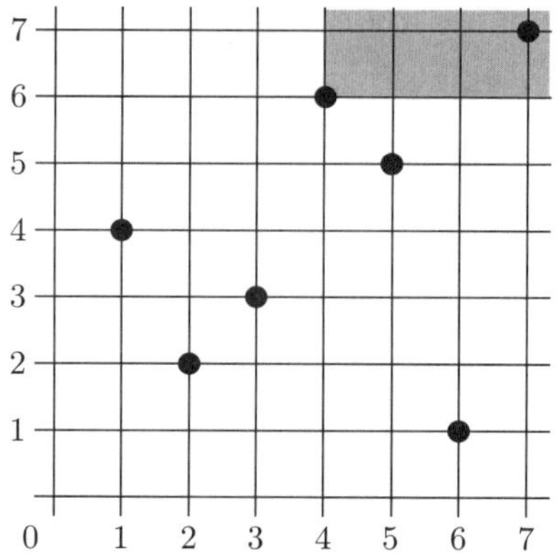


**Figure 3.3.** Shadow cast by $(4, 6)$

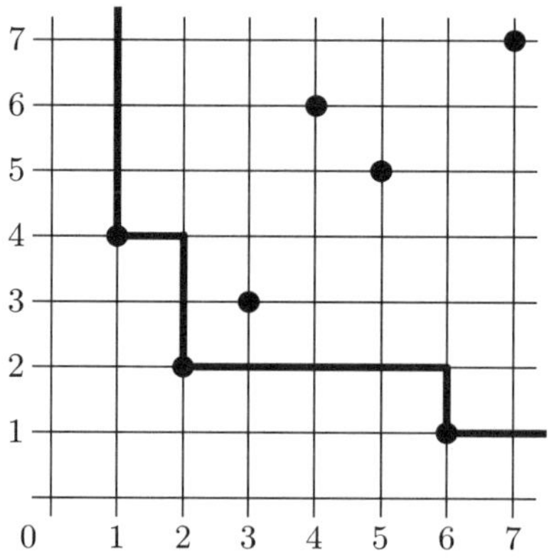


**Figure 3.4.** First shadow line

form the *second shadow line* $L_2$, one removes the points on the first shadow line and repeats the procedure. In this way, we construct the *shadow lines* $L_1, L_2, \ldots$ of the permutation $\pi$. The *x-coordinate of* $L_i$ is

$$x_{L_i} \equiv \text{the } x\text{-coordinate of } L_i\text{'s vertical ray} \tag{3.46}$$

and the *y-coordinate of* $L_i$ is

$$y_{L_i} \equiv \text{the } y\text{-coordinate of } L_i\text{'s horizontal ray.} \tag{3.47}$$

In our example (3.45) we have four shadow lines; see Figure 3.5.

**Remark 3.21.** Note that

(i) shadow lines are "decreasing" functions in a natural sense;

(ii) $L_{i+1}$ lies "above" $L_i$ for each $i$;

(iii) each point $(\ell, x_\ell)$ lies on some $L_j$ and forms a "lower corner" for $L_j$;

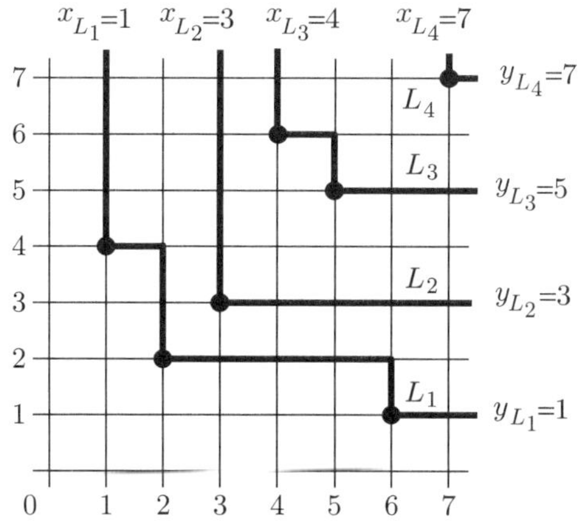


**Figure 3.5.** Shadow lines of $\pi = 4236517$

(iv) for each $k$, the vertical line $x = k$ intersects precisely one shadow line in a ray or segment; all other intersections with a shadow line are in single points.

For the permutation $\pi = 4236517$, the SYT's under RS are

$$P = P(\pi) = \begin{array}{|c|c|c|c|}\hline 1 & 3 & 5 & 7 \\ \hline 2 & 6 \\ \cline{1-2} 4 \\ \cline{1-1}\end{array}\,, \qquad Q = Q(\pi) = \begin{array}{|c|c|c|c|}\hline 1 & 3 & 4 & 7 \\ \hline 2 & 5 \\ \cline{1-2} 6 \\ \cline{1-1}\end{array}\,. \tag{3.48}$$

For this example we see that

$$P_{1,j} = y_{L_j} \qquad \text{and} \qquad Q_{1,j} = x_{L_j}. \tag{3.49}$$

As we will now show, this and more is true in general.

**Lemma 3.22.** *Let the shadow diagram of $\pi = x_1 x_2 \cdots x_N$ be constructed as above. Suppose the vertical line $x = k$ intersects the first $i$ shadow lines, $L_1, \ldots, L_i$. For each $1 \le j \le i$, let $y_j$ be the $y$-coordinate of the lowest point of the intersection with $L_j$. Then, the first row of $P_k = P(x_1, \ldots, x_k)$, the $P$-diagram after $k$ RS insertions, is*

$$R_1 = y_1 y_2 \cdots y_i. \tag{3.50}$$

**Remark 3.23.** It follows that points on line $L_j$ are precisely those elements passing through the point $(1, j)$ during the construction of $P$.

**Proof of Lemma 3.22.** We use induction on $k$. The lemma is obvious for $k = 1$. Now assume the lemma is true for $x = k$, and consider $x = k + 1$. Assume that the vertical $x = k$ intersects $i$ shadow lines. There are two cases: $x_{k+1} > y_i$ and $x_{k+1} < y_i$. Note that the induction hypothesis implies that $\{y_1, \ldots, y_i\} \subset \{x_1, \ldots, x_k\}$, therefore $x_{k+1} \neq y_i$.

If

$$x_{k+1} > y_i, \tag{3.51}$$

then the point $(k+1, x_{k+1})$ starts a new shadow line since $(k+1, x_{k+1})$ is strictly inside each of the shadows cast by $(1, y_1), \ldots, (i, y_i)$. Moreover, since the line $x = k$ intersects precisely $i$ shadow lines, the shadow line containing $(k+1, x_{k+1})$ does not contain a point $(k, z)$ for some $z > y_i$. This implies that the shadow line containing $(k+1, x_{k+1})$ is $L_{i+1}$ and contains a vertical ray emerging from the point $(k+1, x_{k+1})$. Hence,

$$y_{i+1} = x_{k+1}. \tag{3.52}$$

But by (3.50) and (3.51), the $(k+1)$-st insertion merely results in $x_{k+1}$ at the end of the first row, i.e., $R_1 \to y_1 \cdots y_i x_{k+1} = y_1 \cdots y_i y_{i+1}$, and the lemma is true for $k+1$.

If, on the other hand, $x_{k+1} < y_i$, assume that

$$y_1 < \cdots < y_{j-1} < x_{k+1} < y_j < \cdots < y_i. \tag{3.53}$$

Then $(k+1, x_{k+1})$ is added to line $L_j$. Indeed, $(k+1, x_{k+1})$ lies above $L_1, \ldots, L_{j-1}$ as $L_{j-1}$ is decreasing. On the other hand, it cannot lie in the shadow of any other point remaining after removing $L_1, \ldots, L_{j-1}$: For if there were such a point, say $(p, z)$, then necessarily $p < k+1$ and $z < x_{k+1}$. But then there would be a shadow line passing through line $k$ at a height between $y_{j-1}$ and $x_{k+1}$, which cannot be since $x_{k+1} < y_j$. Thus, $(k+1, x_{k+1})$ must be added to line $L_j$. The lowest coordinate on $L_j$ intersecting $x = k+1$ is given by

$$y_j' = x_{k+1}, \tag{3.54}$$

and all the other $y$-values stay the same. This is because (see Remark 3.21 (iv)), for any step $k \to k+1$, at most 1 of the $L_m$'s can "step down"; otherwise, there would be more than one point at $k+1$. Furthermore, by (3.50) and (3.53), insertion of $x_{k+1}$ results in the first row of $P_{k+1}$ having the form

$$y_1 \cdots y_{j-1} x_{k+1} y_{j+1} \cdots y_i = y_1 \cdots y_j, y_j' y_{j+1} \cdots y_i, \tag{3.55}$$

which completes the proof of the lemma. □

It follows from the proof of the lemma that the shadow diagram of $\pi$ can be read from left to right like a time line recording the construction of $P(\pi)$. Recalling Remark 3.21 (iv), we see that in terms of the first row of $P_k$, a ray corresponds to placing an element at the end, a line segment corresponds to displacing an element, and the single points correspond to elements that are unchanged.

We can now prove (3.49).

**Corollary 3.24** (Viennot). *If the permutation $\pi$ has RS tableaux $(P, Q)$ and shadow lines $L_j$, then for all $j$,*

$$P_{1j} = y_{L_j} \qquad \textit{and} \qquad Q_{1j} = x_{L_j}. \tag{3.56}$$

**Proof.** The statement for $P$ is just the case $k = n$ of the lemma. For $Q$, the entry $k$ is added to $Q$ in point $(1, j)$ precisely when $x_k$ is greater than every element of the first row of $P_{k-1}$. But the proof of the previous lemma shows that this happens precisely when the line $x = k$ intersects the shadow line $L_j$ in a vertical ray. In other words, $x_{L_j} = k = Q_{1,j}$. □

For the example $\pi = 4236517$ in (3.14),

(3.57)

| $k$ | first row of $P_k$ | first row of $Q_k$ |
|---|---|---|
| 1 | 4 | 1 |
| 2 | 2 | 1 |
| 3 | 23 | 13 |
| 4 | 236 | 134 |
| 5 | 235 | 134 |
| 6 | 135 | 134 |
| 7 | 1357 | 1347 |

which may be checked against Figure 3.5 and Corollary 3.24.

We now see how to recover the rest of the $P$ and $Q$ tableaux from the shadow diagram of $\pi$. Consider the north-east corners of the shadow lines, which are marked with a white square in Figure 3.6. If such a corner has coordinates $(k, x')$, then, by the proof of the above lemma, $x'$ must be displaced from the first row of $P_{k-1}$ by the insertion of $x_k$. So, the white square corresponds to the elements inserted into the second row during the construction of $P$. Thus, we can get the rest of the two tableaux by iterating

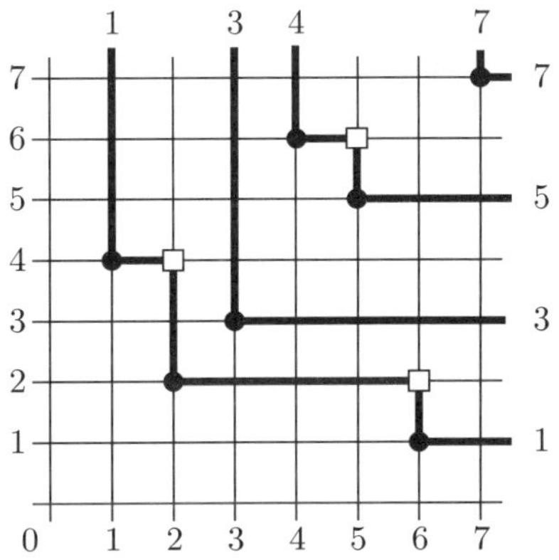


**Figure 3.6.** North-east corners

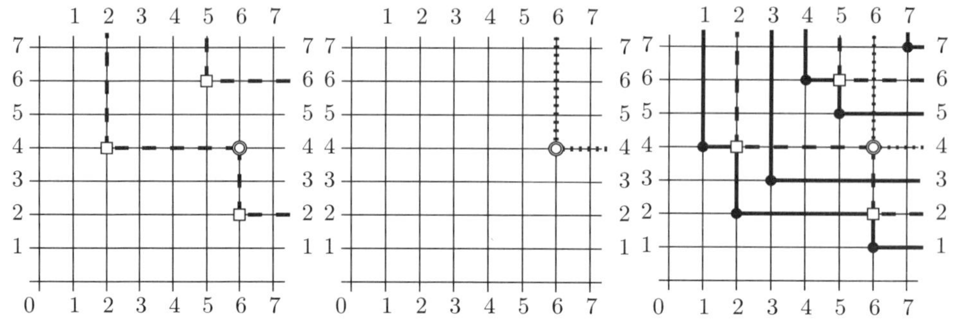


**Figure 3.7.** Shadow lines

the shadow diagram construction. In our example, the second and third rows come from the dashed and dotted lines, respectively, of Figure 3.7.

Formally, we have the following definition.

**Definition 3.25.** The *ith skeleton* of $\pi \in S_N$, $\pi^{(i)}$, is defined inductively by $\pi^{(1)} = \pi$ and

$$\pi^{(i)} = \begin{matrix} k_1 & k_2 & \cdots & k_m \\ \ell_1 & \ell_2 & \cdots & \ell_m \end{matrix}, \tag{3.58}$$

where $(k_1, \ell_1), \ldots, (k_m, \ell_m)$ are the north-east corners of the shadow lines of $\pi^{(i-1)}$ listed such that $k_1 < \cdots < k_m$. The shadow lines for $\pi^{(i)}$ are denoted by $L_j^{(i)}$.

The proof of the following result is clear from the preceding discussion and results.

**Proposition 3.26.** *Suppose* $\mathrm{RS}(\pi) = (P, Q)$. *Then* $\pi^{(i)}$ *is a partial permutation such that*

$$\mathrm{RS}(\pi^{(i)}) = (P^{(i)}, Q^{(i)}), \tag{3.59}$$

*where* $P^{(i)}$ *(respectively,* $Q^{(i)}$*) consists of the rows below the ith row of* $P$ *(respectively,* $Q$*), including the ith row. Furthermore, for all* $i, j$,

$$P_{i,j} = y_{L_j^{(i)}} \quad \textit{and} \quad Q_{i,j} = x_{L_j^{(i)}}. \tag{3.60}$$

The proof of Schützenberger's Theorem 3.11 is now immediate.

**Proof of Theorem 3.11.** Taking the inverse of a permutation corresponds to reflecting the shadow diagram in the line $y = x$. The result now follows from Proposition 3.26. □

**Remark 3.27.** If we represent the elements of $S_N$ as permutation matrices as in Figure 3.2 above, then the dihedral group of the square $D_4$ (all reflections and rotations bringing a square back onto itself) acts on these matrices naturally. The group $D_4$ has eight elements and is generated, in particular, by reflection in the diagonal and by reflection in the mid-vertical axis (exercise). We have shown that reflection across the diagonal corresponds to $P(\pi^{-1}) = Q(\pi)$, $Q(\pi^{-1}) = P(\pi)$, and reflection across the mid-vertical corresponds to $P(\pi^r) = (P(\pi))^t$. Thus, the action of the whole of $D_4$ can be computed as simple actions on the SYT's under RS. For the effect on $Q(\pi)$ by the action of reflection across the vertical, see Sagan [**Sag01**].

## 3.2. The Number of Standard Young Tableaux

The notation

$$F_\lambda \equiv \#\{\text{SYT's with shape } \lambda\} \tag{3.61}$$

was defined in (1.31). We now present three formulae of $F_\lambda$ (Theorem 3.28, Corollary 3.31, and Theorem 3.42 below). These formulae can be derived from each other by elementary means, so it is enough to prove any one of them.

The first is the so-called *Determinantal Formula*, which was known already to Frobenius and Young [**Fro00, You02**] at the beginning of the 20th century. In the following, $\frac{1}{r!} \equiv 0$ if $r < 0$ as usual.

**Theorem 3.28** (Determinantal Formula). *For* $\lambda = (\lambda_1, \dots, \lambda_r) \vdash N$,

$$F_\lambda = N! \det \left( \frac{1}{(\lambda_i - i + j)!} \right)_{1 \le i,j \le r}. \tag{3.62}$$

**Example 3.29.** When $\lambda = (2, 2, 1) \vdash 5$,

$$F_{(2,2,1)} = 5! \times \det \begin{pmatrix} \frac{1}{2!} & \frac{1}{3!} & \frac{1}{4!} \\ \frac{1}{1!} & \frac{1}{2!} & \frac{1}{3!} \\ 0 & \frac{1}{0!} & \frac{1}{1!} \end{pmatrix} = 5, \tag{3.63}$$

and

$$\begin{array}{|c|c|}\hline 1 & 2 \\ \hline 3 & 4 \\ \hline 5 \\ \cline{1-1} \end{array} \quad \begin{array}{|c|c|}\hline 1 & 2 \\ \hline 3 & 5 \\ \hline 4 \\ \cline{1-1} \end{array} \quad \begin{array}{|c|c|}\hline 1 & 3 \\ \hline 2 & 4 \\ \hline 5 \\ \cline{1-1} \end{array} \quad \begin{array}{|c|c|}\hline 1 & 3 \\ \hline 2 & 5 \\ \hline 4 \\ \cline{1-1} \end{array} \quad \begin{array}{|c|c|}\hline 1 & 4 \\ \hline 2 & 5 \\ \hline 3 \\ \cline{1-1} \end{array} \tag{3.64}$$

are the five SYT's with shape $(2, 2, 1)$.

**Remark 3.30.** If the partition in Theorem 3.28 is represented as $\lambda = (\lambda_1, \dots, \lambda_n)$ with $n > r$ and $\lambda_{r+1} = \cdots = \lambda_n = 0$, then the determinantal formula still holds with $r$ replaced by $n$. Indeed, for rows $i > r$, since $\lambda_i - i + j = -i + j < 0$ for $j < i$ and $\lambda_i - i + j = 0$ for $j = i$, the lower-left $(n - r) \times r$ block of the matrix is zero and the lower-right $(n - r) \times (n - r)$ block of the matrix is upper-triangular with 1's on the diagonal. Hence, the determinant of the $n \times n$ matrix equals the determinant of the $r \times r$ matrix. For example, if we represent $\lambda = (2,2,1,0)$, then

$$F_{(2,2,1,0)} = 5! \times \det \begin{pmatrix} \frac{1}{2!} & \frac{1}{3!} & \frac{1}{4!} & \frac{1}{5!} \\ \frac{1}{1!} & \frac{1}{2!} & \frac{1}{3!} & \frac{1}{4!} \\ 0 & \frac{1}{0!} & \frac{1}{1!} & \frac{1}{2!} \\ 0 & 0 & 0 & \frac{1}{0!} \end{pmatrix} = 5. \tag{3.65}$$

The above determinantal formula can also be written as follows.

**Corollary 3.31.** *Given* $\lambda = (\lambda_1, \lambda_2, \ldots, \lambda_r) \vdash N$, *set*

$$\ell_i = \lambda_i + r - i, \qquad 1 \le i \le r. \tag{3.66}$$

*Then*

$$F_\lambda = N! \frac{\prod_{1\le i<j\le r}(\ell_i - \ell_j)}{\ell_1! \cdots \ell_r!}. \tag{3.67}$$

**Proof.** As $\lambda_i - i + j = \ell_i - r + j$,

$$\begin{aligned} &\left[\prod_{s=1}^{r} \ell_s!\right] \cdot \det\left(\frac{1}{(\lambda_i - i + j)!}\right) \\ &= \left[\prod_{s=1}^{r} \ell_s!\right] \cdot \begin{vmatrix} \frac{1}{(\ell_1 - r+1)!} & \frac{1}{(\ell_1 - r+2)!} & \cdots & \frac{1}{(\ell_1 - r+r)!} \\ \vdots & \vdots & \ddots & \vdots \\ \frac{1}{(\ell_r - r+1)!} & \frac{1}{(\ell_r - r+2)!} & \cdots & \frac{1}{(\ell_r - r+r)!} \end{vmatrix} \\ &= \begin{vmatrix} \{\ell_1(\ell_1 - 1)\cdots(\ell_1 - r+2)\} & \cdots & \ell_1(\ell_1 - 1) & \ell_1 & 1 \\ \vdots & \ddots & \vdots & \vdots & \vdots \\ \{\ell_r(\ell_r - 1)\cdots(\ell_r - r+2)\} & \cdots & \ell_r(\ell_r - 1) & \ell_r & 1 \end{vmatrix} \\ &= \begin{vmatrix} \ell_1^{r-1} & \cdots & \ell_1^2 & \ell_1 & 1 \\ \vdots & \ddots & \vdots & \vdots & \vdots \\ \ell_r^{r-1} & \cdots & \ell_r^2 & \ell_r & 1 \end{vmatrix} = \prod_{1\le i<j\le r} (\ell_i - \ell_j). \end{aligned}$$

Thus (3.67) follows from (3.62). □

**Remark 3.32.** As in Remark 3.30, formula (3.67) still holds if the partition is written such that some of $\lambda_j$'s are 0.

In the next two subsections we give two proofs of Theorem 3.28.

**3.2.1. First Proof: Ballot Sequences.** This proof is due to Zeilberger [**Zei83**]. First we need some notation. A *ballot sequence* with $N$ voters and $r$ candidates, and with *initial standing* $u = (a_1, \ldots, a_r) \in \mathbb{Z}_+^r$ and *outcome*

$v = (b_1, \dots, b_r) \in \mathbb{Z}_+^r$ is a sequence of vectors $u_0, u_1, \dots, u_N$, such that $u_0 = u$, $u_N = v$, and $u_i = (u_{i,1}, \dots, u_{i,r}) \in \mathbb{Z}_+^r$ satisfy

$$\begin{aligned} u_{i+1,j} &= u_{i,j} + 1 && \text{for some (unique) } j = j(i) \in \{1, \dots, r\}, \\ u_{i+1,\ell} &= u_{i,\ell} && \text{for all } \ell \in \{1, \dots, r\} \backslash \{j\}. \end{aligned}$$

Necessarily, $\sum_{i=1}^r b_i - \sum_{i=1}^r a_i = N$. For example,

$$\begin{aligned} (0,0,0) &= u_0 = u \\ (1,0,0) &= u_1 \\ (1,1,0) &= u_2 \\ (1,2,0) &= u_3 \\ (1,2,1) &= u_4 = v \end{aligned}$$

is a ballot sequence with four voters and three candidates where all the candidates start with no voter and, at the end of the balloting process, the first candidate has one vote, the second candidate has two votes, and the third candidate has one vote.

We say a vector $u = (c_1, \dots, c_r) \in \mathbb{Z}_+^r$ is *(monotone) decreasing* if $c_1 \geq c_2 \geq \cdots \geq c_r$. If $u$ and $v$ are decreasing, we say that the ballot sequence $u_0$, $\dots$, $u_N$ with initial standing $u_0 = u$ and outcome $v$ is *order preserving* if each $u_i$, $1 \leq i \leq N-1$, is also decreasing. Thus

$$\begin{aligned} (0,0,0) &= u_0 = u \\ (1,0,0) &= u_1 \\ (1,1,0) &= u_2 \\ (2,1,0) &= u_3 = v \end{aligned}$$

is an *order preserving ballot sequence* (*OPBS*). To emphasize the initial and final points, we sometimes write $(\text{OPBS})_{u,v}$.

**Lemma 3.33.** *Let $\lambda = (\lambda_1, \dots, \lambda_r) \vdash N$ be given. Then the set of SYT's with shape $\lambda$ is in one-to-one correspondence with the OPBS's from $u = (0, \dots, 0)$ to $v = \lambda = (\lambda_1, \dots, \lambda_r)$.*

**Proof.** Given an OPBS, $u_0 = u, \dots, u_N = v$, we show how to construct an SYT $P$ with shape $\lambda$. We have $u_0 = (0, 0, \dots, 0)$ and we must have $u_1 = (1, 0, \dots, 0)$. Set

$$P_0 = \emptyset, \qquad P_1 = \boxed{1}.$$

Now, either $u_2 = (1, 1, 0, \dots, 0)$ or $u_2 = (2, 0, 0, \dots, 0)$. Set

$$\begin{aligned} P_2 &= \begin{array}{|c|}\hline 1 \\ \hline 2 \\ \hline\end{array} && \text{if } u_2 = (1,1,0,\dots,0), \\ P_2 &= \begin{array}{|c|c|}\hline 1 & 2 \\ \hline\end{array} && \text{if } u_2 = (2,0,0,\dots,0). \end{aligned}$$

Suppose that at step $k$ we have constructed an SYT where $\mathrm{sh}(P_k) \vdash k$, and suppose for some $j^* \in \{1, \dots, r\}$,

$$\begin{cases} u_{k+1,j^*} = u_{k,j^*} + 1, & \\ u_{k+1,\ell} = u_{k,\ell}, & \text{for } \ell \neq j^*. \end{cases} \tag{3.68}$$

Then $P_{k+1}$ is constructed by adding in a box to row $j^*$ and inserting $k+1$ into that box. We must show that $P_{k+1}$ is an SYT with $\mathrm{sh}(P_{k+1}) \vdash k+1$. Clearly, $u_{k,j}$, $1 \le j \le r$, is equal to the number of votes that candidate $j$ has received after $k$ voters have cast their ballot and is also the number of boxes in the $j$th row of $P_k$. As $u_{k+1}$ is decreasing it is necessary that $u_{k(j^*-1)} > u_{kj^*}$. Thus row $j^*-1$ of $P_k$ is longer than row $j^*$, hence $\mathrm{sh}\, P_{k+1}$ is a Young diagram. Also, as we are inserting $k+1$ into an outer corner of $P_k$, all of whose entries are less than or equal to $k$, it follows that $P_{k+1}$ is a SYT with $\mathrm{sh}\, P_{k+1} \vdash k+1$. Set $P = P_N$. Clearly, $\mathrm{sh}(P) = \lambda$. Thus, we have constructed a map

$$h\colon (\mathrm{OPBS})_{(0,\lambda)} \to \{P : P \text{ is an SYT, } \mathrm{sh}(P) = \lambda\}, \tag{3.69}$$

$h(u_0, u_1, \dots, u_N) = P$ where $u_0 = (0, \dots, 0)$, $u_N = \lambda$. Clearly, if $\lambda = (\lambda_1, \dots, \lambda_r) \vdash N$, then $P$ has $r$ rows and $N$ boxes.

Now, $h$ is one-to-one. Indeed, if $u_0' = (0, \dots, 0)$, $u_1', \dots, u_N' = \lambda$ and $u_0 = (0, \dots, 0)$, $u_1, \dots, u_N = \lambda$ are two distinct ballot sequences in $(\mathrm{OPBS})_{(0,\lambda)}$, then there exists an $\ell, 1 \le \ell \le N-1$, such that $u_i' = u_i$ for $i < q$, but $u_q' \neq u_q$. Then $P_{q-1}' = P_{q-1}$, but $q$ is inserted into *different* rows of $P_q'$ and $P_q$ and so $P_N' \neq P_N$.

Finally, $h$ is surjective. By induction on $N$ (trivial for $N = 1$), assume this is true if $\lambda \vdash N$. Let $P$ be any SYT with $\lambda = (\lambda_1, \dots, \lambda_r)$ with $\mathrm{sh}\, P \vdash N+1$. If we remove the box containing $N+1$ from $P$, we obtain an SYT, $P'$ say, with $\lambda' = \mathrm{sh}\, P' \vdash N$. $P'$ is indeed an SYT as $N+1$ must lie in an inner corner of $P$. By the induction hypothesis, there exists a OPBS $u_0 = (0, \dots, 0)$, $u_1, \dots, u_N = (\lambda_1', \dots, \lambda_r')$ (here $\lambda_r'$ may be 0, which, although a slight abuse of notation, does not cause any confusion). Clearly, $u_0, u_1, \dots, u_N, u_{N+1} = (\lambda_1, \dots, \lambda_r)$ is an OPBS from 0 to $\lambda = (\lambda_1, \dots, \lambda_r)$ which maps under $h$ onto $P$. This establishes the desired correspondence. □

We say that a vector $u = (c_1, \dots, c_r) \in \mathbb{Z}_+^r$ is *strictly (monotone) decreasing* if $c_1 > c_2 > \cdots > c_r$. We say that the ballot sequence $u_0, u_1, \dots, u_N$ is *strictly order preserving* if each $u_i$, $0 \le i \le N$, is strictly decreasing. We denote the *strictly order preserving ballot sequences* from $u$ to $v$ by $(\mathrm{SOPBS})_{u,v}$.

Now, the shift

$$u = (c_1, c_2, \dots, c_r) \to u' = (c_1 + r - 1, c_2 + r - 2, \dots, c_r) \tag{3.70}$$

maps the decreasing sequences on $\mathbb{Z}_+^r$ onto the strictly decreasing sequences on $\mathbb{Z}_+^r$. Under this map the OPBS's $u_0 = 0, u_1, \ldots, u_N = v$ from $u = 0$ to $v = (v_1, \ldots, v_r)$ are mapped bijectively onto SOPBS's $u_0', u_1', \ldots, u_n' = v'$ from $u_0' = (r-1, r-2, \ldots, 0)$ to $v' = (v_1 + r - 1, v_2 + r - 2, \ldots, v_r)$.

For a given $\lambda = (\lambda_1, \ldots, \lambda_r) \vdash N$, set

$$\ell_j \equiv \lambda_j + r - j, \qquad 1 \le j \le r. \tag{3.71}$$

Then

$$v = (\ell_1, \ldots, \ell_r) \tag{3.72}$$

is strictly decreasing, and by Lemma 3.33, $F_\lambda$, the number of SYT's with shape $\lambda$, equals the number of SOPBS's $u_0$, $u_1$, ..., $u_N$ from $u_0 = (r-1, r-2, \ldots, 0)$ to $v$.

For $\sigma \in S_r$, we define

$$v_\sigma = (\ell_{\sigma(1)}, \ldots, \ell_{\sigma(r)}).$$

Let

$$\begin{aligned} \mathcal{P}_v = &\{P\colon P \text{ is a ballot sequence } u_0,\ u_1, \ldots, u_N \text{ from} \\ &u_0 = (r-1, r-2, \ldots, 0) \text{ to } u_N = v_\sigma \text{ for some } \sigma \in S_r\}. \end{aligned}$$

For $P \in \mathcal{P}_v$, set

$$\operatorname{sgn} P \equiv \operatorname{sgn} \sigma, \qquad \text{if } u_N = v_\sigma.$$

Here of course, as the $\ell_i's$ are distinct, the relation $u_N = v_\sigma$ determines $\sigma \in S_r$ uniquely.

**Remark 3.34.** It could of course happen that for a particular $\sigma \in S_r$ there is no ballot sequence with $u_N = v_\sigma$. For example, for $r = 3$, $u_0 = (2,1,0)$, $v = (3,2,1)$, and $\sigma = 321 \in S_3$, clearly there cannot be a ballot sequence from $u_0$ to $v_\sigma = (1,2,3)$.

Set

$$A = \sum_{P \in \mathcal{P}_v} \operatorname{sgn} P. \tag{3.73}$$

We will now evaluate $A$ in two ways. First, let

$$\begin{aligned} \mathcal{P}_v' = &\{P \in \mathcal{P}_v \colon P \text{ is not a SOPBS}\}, \\ \mathcal{P}_v'' = &\{P \in \mathcal{P}_v \colon P \text{ is a SOPBS}\}. \end{aligned}$$

We define an involution of $\phi : \mathcal{P}_v' \to \mathcal{P}_v'$ as follows. Observe that if $P \in \mathcal{P}_v'$, $P = u_0, u_1, \ldots, u_N$, then for some $i$, $u_{i,j} = u_{i,j'}$ for some $j \neq j'$. Let $i_p$ be the last such $i$: as $u_N$ has distinct elements, $i_p$ is necessarily $< N$. Now $u_{i_p}$ has precisely two entries, say $u_{i_p j_p} = u_{i_p j_p'}$, $j_p \neq j_p'$, which are equal, otherwise $u_{i_p+1}$ would have at least two equal elements from the construction (3.68), which contradicts the definition of $i_p$. Now, set $\phi(P) = \tilde{P} = \tilde{u}_0\tilde{u}_1\cdots\tilde{u}_N$, where $\tilde{u}_q = u_q$ for $q \le i_p$, and for $i_p < q \le N$,

$$\begin{cases} \tilde{u}_{q,j} = u_{q,j}, & \text{for } j \ne j_p, j'_p, \\ \tilde{u}_{q,j_p} = u_{q,j'_p}, \\ \tilde{u}_{q,j'_p} = u_{q,j_p}. \end{cases}$$

Thus we just flip the $j_p$ and $j'_p$ columns for all ballots after $i_p$. An example with $N = 6$ and $u_6 = v_\sigma = (3, 4, 2)$ is

$$\begin{array}{ll} u_0 = (2,1,0) & \tilde{u}_0 = (2,1,0) \\ u_1 = (2,2,0) & \tilde{u}_1 = (2,2,0) \\ u_2 = (2,2,1) & \tilde{u}_2 = (2,2,1) \\ u_3 = (3,2,1) & \tilde{u}_3 = (3,2,1) \\ u_4 = (3,2,2) & \tilde{u}_4 = (3,2,2) \\ u_5 = (3,3,2) & \tilde{u}_5 = (3,3,2) \\ u_6 = (3,4,2) & \tilde{u}_6 = (4,3,2) \end{array}$$

where $i_p = 5$, $j_p = 1$, $j'_p = 2$. Clearly, $\phi$ maps $\mathcal{P}'_v$ into itself and also satisfies $\phi \circ \phi = $id by construction. So $\phi$ is bijective from $\mathcal{P}'_v \to \mathcal{P}'_v$ and is an involution on $\mathcal{P}'_v$. As

$$\operatorname{sgn} P = -\operatorname{sgn} \tilde{P}, \tag{3.74}$$

we find that

$$\sum_{P \in \mathcal{P}'_v} \operatorname{sgn} P = 0. \tag{3.75}$$

This implies that $A = \sum_{P \in \mathcal{P}''_v} \operatorname{sgn} P$. But if $P \in \mathcal{P}''_v$, necessarily $u_n = v$, i.e., $\sigma = $ id, and so $\operatorname{sgn} P = 1$. Thus $A = \#\mathcal{P}''_v$, hence

$$A = \#\{P : P \text{ is a SOPBS from } u_0 = (r-1, \ldots, 0) \text{ to } v\} = F_\lambda, \tag{3.76}$$

where $\lambda$ and $v$ are related by (3.71) and (3.72).

We now evaluate $A$ in a different way: The number of ballot sequences $P$ from $u_0$ to $v_\sigma$ is clearly the same as the number of ways of lining up $\ell_{\sigma(1)} - r + 1$ balls of color 1, $\ell_{\sigma(2)} - r + 2$ balls of color 2, etc. Thus,

$$\begin{aligned} &\#\text{ of ballot sequences P from } u_0 \to v_\sigma \\ &= \frac{N!}{(\ell_{\sigma(1)} - r + 1)!(\ell_{\sigma(2)} - r + 2)! \cdots \ell_{\sigma(r)}!}, \end{aligned} \tag{3.77}$$

where $N = \sum_{k=1}^{r} (\ell_{\sigma(k)} - r + k)$ (note that, by convention, $\frac{1}{m!} = 0$ for $m < 0$). Inserting (3.77) into (3.73), we find that

$$\begin{aligned} A &= \sum_{\sigma \in S_r} \sum_{P \text{ is from } u_0 \text{ to } v_\sigma} \operatorname{sgn} P \\ &= \sum_{\sigma \in S_r} \operatorname{sgn} \sigma \frac{N!}{(\ell_{\sigma(1)} - r + 1)!(\ell_{\sigma(2)} - r + 2)! \cdots \ell_{\sigma(r)}!}. \end{aligned} \tag{3.78}$$

But by the definition of a determinant, this equals

$$N! \det \left( \frac{1}{(\ell_i - r + j)!} \right)_{i,j=1}^{r}. \tag{3.79}$$

Therefore, combined with (3.76) and (3.71), we obtain

$$F_\lambda = N! \det \left( \frac{1}{(\ell_i - i + j)!} \right)_{i,j=1}^{r}. \tag{3.80}$$

This completes the proof of Theorem 3.28.

**Remark 3.35.** Zeilberger's proof given above is a generalization of André's celebrated proof [**And87**] of *Bertrand's ballot problem* which can be stated as follows. Suppose that A and B are in an election and A receives $p$ votes while B receives $q$ votes, $p > q$. How many ways can the $p + q$ votes be arranged so that A maintains a lead throughout the counting of the ballots? In the language of the above proof, Bertrand's problem is equivalent to counting the number of SOPBS's from $u_0 = (1, 0)$ to $v = (p, q)$. But by equation (3.70), this is just the number of OPBS's from $(0, 0)$ to $(\lambda_1, \lambda_2)$, where $p = \lambda_1 + 2 - 1$ and $q = \lambda_2 + 2 - 2$, i.e., $\lambda_1 = p - 1, \lambda_2 = q$. From (3.67), we see that the answer to Bertrand's problems is given by

$$((p-1)+q)! \frac{((p-1)+2-1)-(q+2-2))}{(p-1+2-1)!(q+2-2)!} = \binom{p+q}{p} \frac{p-q}{p+q}, \tag{3.81}$$

which is Bertrand's result.

**Remark 3.36.** Formula (3.67) also solves the following "Subway Problem" which is a restatement of a problem posed to one of the authors by C. Peskin. Consider a subway line going, say, from north to south. Along the line there are stations. At each station, one may take a train going either north or south. One boards the train at the end of the line corresponding to the station that is furthest south. The train stops at each station and one has the choice of either remaining on the train or crossing the platform and taking the train in the opposite direction. One returns home for the first time at the end of the day after taking $N$ stops. (Necessarily, $N$ is even.) *Question*: How many different journeys can one make in a day with $N$ stops? As the first and last legs of the journey are necessarily northward and southward, respectively, the number of such journeys is clearly the same as

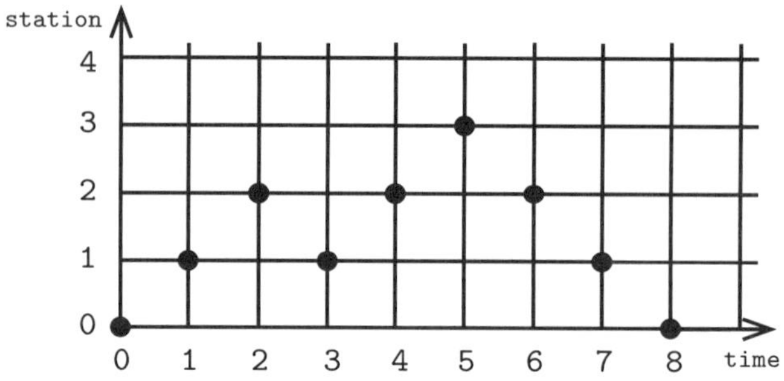


**Figure 3.8.** Simple random walks and OPBS

the number of OPBS's from $(0,0)$ to $(n,n)$ where $n = \frac{N-2}{2}$. Then by (3.67), we see that the number of different journeys is given by (with $(n,n) \vdash 2n$, $\ell_1 = n+1$, and $\ell_2 = n$)

$$(2n)!\frac{((n+1)-n)}{(n+1)!n!} = \frac{1}{n+1}\binom{2n}{n}. \tag{3.82}$$

For example, if $N = 8$, the number of journeys is $\frac{6!}{4!3!} = 5$. This result is easily verified by inspection. Indeed, labeling the stations $0, 1, 2, \ldots$ from south to north, one sees that the possible journeys are given by

$$0 \to 1 \to 2 \to 3 \to 4 \to 3 \to 2 \to 1 \to 0$$
$$0 \to 1 \to 2 \to 3 \to 2 \to 3 \to 2 \to 1 \to 0$$
$$0 \to 1 \to 2 \to 1 \to 2 \to 3 \to 2 \to 1 \to 0$$
$$0 \to 1 \to 2 \to 1 \to 2 \to 1 \to 2 \to 1 \to 0$$
$$0 \to 1 \to 2 \to 3 \to 2 \to 1 \to 2 \to 1 \to 0.$$

Alternatively, these journeys are parametrized by the graphs of simple random walks starting at zero and ending at zero at time $N = 8$ conditioned to stay strictly above the $x$-axis for the times $1, 2, \ldots, 7 = N-1$ in between. For example, the third journey corresponds to Figure 3.8. The numbers (3.82) are called the Catalan numbers. In addition to the OPBS's with two candidates and the walks staying above the $x$-axis, there are many, many more combinatorial objects which are enumerated via the Catalan numbers. The website of Richard Stanley (at the time of publication) records 205 such examples.

**3.2.2. Second Proof: Nonintersecting Poisson Processes.** This subsection employs a nonintersecting continuous time random walk argument [**Kar88**] to give a probabilistic proof of the determinantal formula (3.62) for $F_\lambda$.

We first state a theorem of Karlin and McGregor. Let $W^{(n)} = \{z = (z_1, \ldots, z_n) \in \mathbb{R}^n : z_1 > \cdots > z_n\}$ denote the Weyl chamber.

**Theorem 3.37** (Karlin-McGregor formula). *Let $N_t = (N_t^1, \ldots, N_t^n)$ be a process where all coordinates are independent strong Markov processes (see Appendix C sharing the same transition probability density (and in the discrete case, the same transition probability) $p(t_1, t_2, x, y)$. In addition, assume that for any $0 \leq s < t \leq T$, $N_s \in W^{(n)}$ and $N_t \notin W^{(n)}$ implies that there exists an $r \in (s, t]$ such that $N_r \in \partial W^{(n)}$, the boundary of $W^{(n)}$. Let $x, y \in W^{(n)}$. Given $N_0 = x$, the probability (or probability density) that $N_T = y$ and $N_t \in W^{(n)}$ for all $t \in [0, T]$, i.e., $N_t^1, \ldots, N_t^n$ do not intersect is given by*

$$\begin{aligned}\mathbb{P}^x\{N_T = y; N_t \in W^{(n)}\ \forall t \in [0, T]\} &= \mathbb{P}^x\{N_T = y;\ \textit{no intersection}\} \\ &= \det\left(p(0, T, x_i, y_j)\right)_{i,j=1}^n . \end{aligned} \tag{3.83}$$

**Proof.** For $i \in \{1, \ldots, n-1\}$, define the stopping times

$$\tau_i = T \wedge \inf\{s \in [0, T] : N_s^i = N_s^{i+1}\}.$$

Define $\tau = \tau_1 \wedge \cdots \wedge \tau_{n-1}$, the minimum of $\tau_1, \ldots, \tau_{n-1}$. Then by using the independence of the $N_t^i$'s,

(3.84)
$$\begin{aligned}\det(p_T(x_i, y_j))_{i,j=1}^n &= \sum_{\sigma \in S_n} \operatorname{sgn}(\sigma) p(0, T, x_1, y_{\sigma(1)}) \cdots p(0, T, x_n, y_{\sigma(n)}) \\ &= \sum_{\sigma \in S_n} \operatorname{sgn}(\sigma) \prod_{i=1}^n \mathbb{E}^{x_i}[\chi_{\{N_T^i = y_{\sigma(i)}\}}] \\ &= \sum_{\sigma \in S_n} \operatorname{sgn}(\sigma) \mathbb{E}^x\left[\prod_{i=1}^n \chi_{\{N_T^i = y_{\sigma(i)}\}}\right].\end{aligned}$$

From the definition of $\tau$, the right-hand side of the last line above equals

$$\begin{aligned}&\sum_{\sigma \in S_n} \operatorname{sgn}(\sigma) \mathbb{E}^x\left[\prod_{i=1}^n \chi_{\{N_T^i = y_{\sigma(i)}\}} \left(\chi_{\{\tau = T\}} + \sum_{j=1}^{n-1} \chi_{\{\tau = \tau_j < T\}}\right)\right] \\ &= \sum_{\sigma \in S_n} \operatorname{sgn}(\sigma) \mathbb{E}^x\left[\chi_{\{\tau = T\}} \prod_{i=1}^n \chi_{\{N_T^i = y_{\sigma(i)}\}}\right] \\ &\quad + \sum_{j=1}^{n-1} \sum_{\sigma \in S_n} \operatorname{sgn}(\sigma) \mathbb{E}^x\left[\chi_{\{\tau = \tau_j < T\}} \prod_{i=1}^n \chi_{\{N_T^i = y_{\sigma(i)}\}}\right].\end{aligned} \tag{3.85}$$

The first sum in the last line of the above equality is reduced to one term corresponding to the identity permutation $\sigma = \text{id}$ since (by the second assumption on $N_t$) $\tau = T$ implies that $N_t \in W^{(n)}$ for all $t \in [0, T]$. Hence, the first sum equals $\mathbb{E}^x\left[\chi_{\{N_T=y;\tau=T\}}\right] = \mathbb{P}^x\{N_T = y;\text{ no intersection}\}$. It remains to show that the second term in the last line of the above equality vanishes.

If $\tau = \tau_j < T$, then the processes $N^j$ and $N^{j+1}$ intersect at a time before $T$. Due to the strong Markov property of $N_t$, the probability of the event that $N^j$ and $N^{j+1}$ intersect before terminating at $y_{\sigma(j)}$ and $y_{\sigma(j+1)}$ at time $T$, respectively, equals the probability that $N^j$ and $N^{j+1}$ intersect before terminating at $y_{\sigma(j+1)}$ and $y_{\sigma(j)}$ at time $T$, respectively, by switching the processes after the intersection time. For $j \in \{1, \ldots, n-1\}$, let $\rho_j \in S_n$ be the transposition $\rho_j(j) = j+1$, $\rho_j(j+1) = j$ and $\rho_j(i) = i$ for $i \neq j, j+1$. We have for any $\sigma \in S_n$,

(3.86)

$$\mathbb{E}^x\left[\chi_{\{\tau=\tau_j<T\}}\prod_{i=1}^{n}\chi_{\{N_T^i=y_{\sigma(i)}\}}\right] = \mathbb{E}^x\left[\chi_{\{\tau=\tau_j<T\}}\prod_{i=1}^{n}\chi_{\{N_T^i=y_{\sigma\circ\rho_j(i)}\}}\right].$$

As the map $\sigma \mapsto \sigma\circ\rho_j$ is an involution on $S_n$, for each $j \in \{1, 2, \ldots, n-1\}$,

$$\sum_{\sigma\in S_n}\operatorname{sgn}(\sigma)\mathbb{E}^x\left[\chi_{\{\tau=\tau_j<T\}}\left(\prod_{i=1}^{n}\chi_{\{N_T^i=y_{\sigma(i)}\}}\right)\right] = 0. \tag{3.87}$$

Hence, the second sum in (3.85) vanishes and the proof is complete. □

**Remark 3.38.** Diffusions with continuous sample paths, all of whose coordinates are independent with identical transition laws as well as Poisson processes with initial conditions on the integer lattice, clearly satisfy the conditions of the theorem.

Now we give the second proof of the determinantal formula (3.62) of $F_\lambda$. Fix a partition $\lambda = (\lambda_1, \ldots, \lambda_r) \vdash n$. Let $N_t^1, \ldots, N_t^r$ be independent rate 1 Poisson processes with initial conditions $N_0^i = 1-i$ for $i = 1, 2, \ldots, r$. Define $A_\lambda$ to be the event that at time $t = 1$, $N_1^i = \lambda_i + (1-i)$ for all $i = 1, 2, \ldots, r$. A *row increasing Young tableau* is defined to be a Young tableau whose rows increase from left to right, but whose columns are unconstrained by ordering. For almost every element of $A_\lambda$ (the elements of $A_\lambda$ where no two jumps of these processes occur at the same time) there is a natural map to a row increasing Young tableau defined as follows: If $N^i$ jumps first, then place a 1 in the leftmost box in row $i$; if $N^j$ jumps second, then place a 2 in the first box of row $j$ if $j \neq i$ and a 2 in the second box of row $i$ if $j = i$. Continue in this fashion to produce a row increasing Young tableau of shape $\lambda$. For example, the realization in Figure 3.9 is mapped to the row

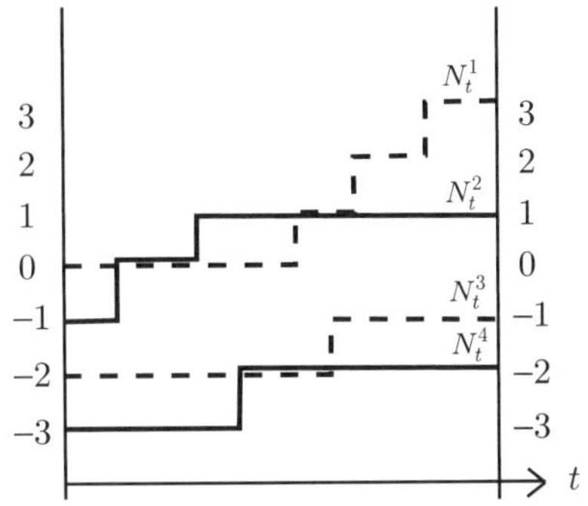


**Figure 3.9.** A realization of $A_\lambda$, $\lambda = (3, 2, 1, 1)$

increasing Young tableau

$$\begin{array}{|c|c|c|}\hline 4 & 6 & 7 \\ \hline 1 & 2 \\ \cline{1-2} 5 \\ \cline{1-1} 3 \\ \cline{1-1}\end{array} \tag{3.88}$$

since the jumps occur in the following order of $i$'s:

$$\begin{array}{cc} \text{order of jump} & 1234567 \\ i & 2241311 \end{array}.$$

It is not hard to show that this map induces the uniform probability measure (when properly normalized by $\mathbb{P}\{A_\lambda\}$) on the row increasing Young tableaux of shape $\lambda$. Indeed, we may imagine that each walker carries an independent clock which rings at an exponential random time and the walker moves when the clock rings. Here all exponential random times are independent and have the same rate. It is a simple exercise to check that, due to the memoryless property of the exponential random variables, the probability that the $r$ clocks ring in any particular order during a time interval is the same irrespective of the ordering.

Now the subset $B_\lambda \subset A_\lambda$ which is mapped to the standard Young tableaux of shape $\lambda$ corresponds to the realizations whose paths do not intersect each other for all $t \in [0, 1]$. Since the mapping described induces uniform measure on the row increasing Young tableaux of shape $\lambda$ and the standard Young tableaux corresponding to nonintersecting path realizations $B_\lambda$, the number of standard Young tableaux of shape $\lambda$ can be computed by evaluating

$$F_\lambda = \#\{\text{row increasing Young tableaux of shape } \lambda\} \times \frac{\mathbb{P}\{B_\lambda\}}{\mathbb{P}\{A_\lambda\}}. \tag{3.89}$$

Recall that by definition of a Poisson process $p_t(x,y) = p(0,t,x,y) = \frac{e^{-t}t^{y-x}}{(y-x)!}$ for $x,y \in \mathbb{Z}$. Hence, using the independence of the $N^i$,

$$\mathbb{P}\{A_\lambda\} = \prod_{i=1}^{r} e^{-1}\frac{1^{\lambda_i}}{\lambda_i!} = e^{-r}\prod_{i=1}^{r}\frac{1}{\lambda_i!}. \tag{3.90}$$

On the other hand, by elementary combinatorics, the number of row increasing Young tableaux of shape $\lambda$ is

$$\binom{n}{\lambda_1,\ldots,\lambda_r} = \frac{n!}{\lambda_1!\cdots\lambda_r!}. \tag{3.91}$$

Finally, the Karlin-McGregor formula (3.83) implies that

$$\mathbb{P}\{B_\lambda\} = \det\left(\frac{e^{-1}}{(\lambda_i - i + j)!}\right)_{i,j=1}^{r}. \tag{3.92}$$

Hence,

$$F_\lambda = n!\det\left(\frac{1}{(\lambda_i - i + j)!}\right)_{i,j=1}^{r}. \tag{3.93}$$

This completes the second proof of the determinantal formula (3.62) of $F_\lambda$.

**Exercise 3.39.** This exercise is an application of the Karlin-McGregor theorem to symmetric simple random walks. Let $X(k)$, $k = 0,1,\ldots$, be discrete-time symmetric simple random walks, i.e., $X(k+1) - X(k) = \pm 1$ each with probability $1/2$. Let $X^i(k)$, $i = 1,\ldots,n$, be $n$ independent symmetric simple random walks, and assume that the initial condition for the walkers is $X^i(0) = 2i - 2$. First show that the Karlin-McGregor theorem applies in this situation. Then, for nonnegative integers $x_1 < x_2 < \cdots < x_n$, evaluate the probability that $X^1(k) < X^2(k) < \cdots < X^n(k)$ for all $k = 0,1,\ldots,2m$, and $X^i(2m) = -2m + 2x_i$. (Note that if the walker $X^1(k)$ always moves downward, then it ends at $-2m$ at time $2m$. Hence, $2x_i$ are the position of the walkers measured from the bottom-most possible position at time $2m$.) Show that this probability equals

$$C_{m,n}\prod_{1\le i<j\le n}(x_j - x_i)\prod_{i=1}^{n}\frac{1}{x_i!(2m+n-1-x_i)!}, \tag{3.94}$$

where the constant

$$C_{m,n} = \left(\frac{(2m)!}{2^{2m}}\right)^{n}\prod_{i=1}^{n-1}(2m+i)^{n-i}. \tag{3.95}$$

From this, conclude that conditional on the event that the random walks $\{X^i(k)\}_{i=1}^{n}$ do not intersect during time $k = 0,\ldots,2m$ (and start at $2i-2$), the probability that they arrive at $-2m + 2x_i$, $i = 1,\ldots,n$, after $2m$ time steps equals (3.94) with a new constant $C_{n,m}$.

**Exercise 3.40.** This exercise is an application of the Karlin-McGregor theorem to Brownian motion. Let $a_1 < \cdots < a_n$ be real numbers. Let $\{B^i(t; a_i)\}_{i=1}^n$ be $n$ independent Brownian motions starting at $a_i$, i.e., $B^i(0; a_i) = a_i$. Let $x_1 < \cdots < x_n$ be real numbers. Set $x = (x_1, \ldots, x_n)$ and $a = (a_1, \ldots, a_n)$. Evaluate the probability density, $p_T(x; a)$, that $B^1(t; a_1) < B^2(t; a_2) < \cdots < B^n(t; a_n)$ for all $t \in [0, T]$ and $B^i(T; a_i) = x_i$. Check that

$$\lim_{a \to 0} \frac{\det(e^{x_i a_j})}{\prod_{1 \leq i < j \leq n}(a_j - a_i)} = \prod_{1 \leq i < j \leq n} (x_j - x_i) \prod_{i=1}^{n-1} \frac{1}{i!}. \tag{3.96}$$

Using this result, show that the density satisfies

$$\lim_{a \to 0} \frac{p_T(x; a)}{\prod_{1 \leq i < j \leq n}(a_j - a_i)} = \frac{1}{(2\pi)^{n/2} T^{n^2/2} \prod_{i=1}^{n-1} i!} \prod_{1 \leq i < j \leq n} (x_j - x_i) \prod_{i=1}^{n} e^{-\frac{x_i^2}{2T}}. \tag{3.97}$$

Now, consider the $n$ nonintersecting Brownian bridges starting at 0 and ending at 0 after time $2T$ (i.e., the "watermelon process"). The density of this process at time $T$ may be interpreted as

$$\lim_{a \to 0} \frac{p_T(x; a) p_T(a; x)}{p_{2T}(a; a)}. \tag{3.98}$$

From (3.97), this equals

$$\frac{1}{(2\pi)^{n/2} (T/2)^{n^2/2} \prod_{i=1}^{n-1} i!} \prod_{1 \leq i < j \leq n} (x_j - x_i)^2 \prod_{i=1}^{n} e^{-\frac{x_i^2}{T}}. \tag{3.99}$$

See Section 13 for further applications of the Karlin-McGregor theorem.

### 3.2.3. Hook Formulae for $F_\lambda$.

We present an alternate formula for $F_\lambda$. This is called the *hook formula* and is due to Frame, Robinson, and Thrall [**FRT54**].

**Definition 3.41.** If $v = (i, j)$ is a box in the diagram $\lambda = (\lambda_1, \ldots, \lambda_r)$ (denoted by $(i, j) \in \lambda$), then it has a *hook*

$$H_v = H_{i,j} = \{(i, j') \colon j' \geq j\} \cup \{(i', j) \colon i' \geq i\} \tag{3.100}$$

with corresponding *hook length* $h_v = h_{i,j} = \# H_{i,j}$.

To illustrate, if $\lambda = (4, 4, 3, 3, 3, 1)$, then $H_{2,2}$ is given by the crossed boxes in

(3.101)

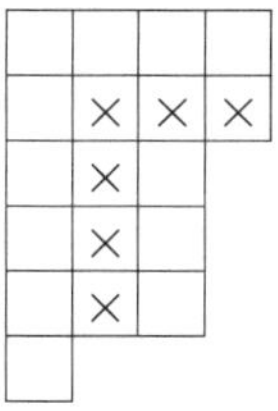

and $h_{2,2} = 6$.

**Theorem 3.42** (Hook formula). *If* $\lambda = (\lambda_1, \dots, \lambda_r) \vdash N$, *then*

$$F_\lambda = \frac{N!}{\prod_{(i,j)\in\lambda} h_{i,j}}. \tag{3.102}$$

For example, if $\lambda = (2,2,1) \vdash 5$, then

$$h_{1,1} = 4, \quad h_{1,2} = 2, \quad h_{2,1} = 3, \quad h_{2,2} = 1, \quad h_{3,2} = 1,$$

and by (3.102), $F_\lambda = \frac{5!}{4\cdot 2\cdot 3\cdot 1\cdot 1} = 5$ as in (3.63).

**Proof of Theorem 3.42.** Set $\ell_i = \lambda_i + r - i$. From Corollary 3.31,

$$F_\lambda = N! \frac{\prod_{1\le i<j\le r}(\ell_i - \ell_j)}{\ell_1! \cdots \ell_r!}. \tag{3.103}$$

Let $\tilde\lambda = (\lambda_1 - 1, \lambda_2 - 1, \dots, \lambda_r - 1) \vdash N - r$. Then from Corollary 3.31 and Remark 3.32,

$$F_{\tilde\lambda} = (N-r)! \frac{\prod_{1\le i<j\le r}(\tilde\ell_i - \tilde\ell_j)}{\tilde\ell_1! \cdots \tilde\ell_r!}, \tag{3.104}$$

where $\tilde\ell_i = \tilde\lambda_i + r - i = \ell_i - 1$. Hence

$$\frac{F_\lambda}{F_{\tilde\lambda}} = \frac{N(N-1)\cdots(N-r+1)}{\prod_{i=1}^r \ell_i}. \tag{3.105}$$

Observe that $h_{i,1} = \lambda_i + r - i = \ell_i$. Thus

$$F_\lambda = \frac{N(N-1)\cdots(N-r+1)}{\prod_{i=1}^r h_{i,1}} F_{\tilde\lambda}. \tag{3.106}$$

By applying the same argument for $F_{\tilde\lambda}$, we find

$$F_{\tilde\lambda} = \frac{(N-r)(N-r-1)\cdots(N-r-\tilde r+1)}{\prod_{i=1}^{\tilde r} h_{i,2}} F_{\tilde{\tilde\lambda}}, \tag{3.107}$$

where $\tilde r$ is the # of rows in the second column of $\lambda$, etc.

Repeating the argument $\lambda_1$ times, we obtain the hook formula (3.102). □

**Remark 3.43.** The original argument of Frame-Robinson-Thrall proceeds, as above, from a slightly modified version of the determinantal formula to the hook formula.

**Remark 3.44.** A *bijective* combinatorial proof of the hook formula is given in [**Sag01**, §3.10], which also contains some history of the proof. The proof proceeds by writing (3.102) as

$$N! = F_\lambda \prod_{(i,j)\in\lambda} h_{i,j} \tag{3.108}$$

and then constructing a bijection

$$\Psi\colon T \mapsto (P, J), \tag{3.109}$$

where $\mathrm{sh}(T) = \mathrm{sh}(P) = \mathrm{sh}(J)$, $T$ is an arbitrary Young tableau, $P$ is an SYT, and the entries of the array $J$ are restricted in a way we now describe. Define the *arm* and *leg* of $H_{i,j}$ to be

$$A_{i,j} = \{(i, j')\colon j' > j\} \qquad \text{and} \qquad L_{i,j}, = \{(i', j)\colon i' > i\}, \tag{3.110}$$

respectively, with corresponding *arm length* and *leg length*

$$a\ell_{i,j} = |A_{i,j}| \qquad \text{and} \qquad \ell\ell_{i,j} = |L_{i,j}|. \tag{3.111}$$

For example, for $\lambda = (4, 4, 3, 3, 3, 1)$, as above in (3.101),

$$\begin{aligned} A_{2,2} &= \{(2,3), (2,4)\} & a\ell_{2,2} &= 2, \\ L_{2,2} &= \{(3,2), (4,2), (5,2)\} & \ell\ell_{2,2} &= 3. \end{aligned} \tag{3.112}$$

For each box we require

$$-\ell\ell_{i,j} \le J_{i,j} \le a\ell_{i,j}. \tag{3.113}$$

As $h_{i,j} = a\ell_{i,j} + \ell\ell_{i,j} + 1$, we see that the number of possible choices of $J_{i,j}$ is $h_{i,j}$. Such a $J$ is called a *hook tableau*. The proof that $\Psi$ is indeed a bijection, is, however, far from trivial (see [**Sag01**]).

**Remark 3.45.** There is also a probabilistic proof of the hook formula due to Greene, Nijenhuis, and Wilf [**GNW79**]; see in [**Sag01**, exercise 17, pp. 137–138].

## 3.3. Applications and Equivalent Models

The length $\ell_N(\pi)$ of the longest increasing subsequence of $\pi \in S_N$ arises in several statistical models. In this section, we consider a few of them.

**3.3.1. Patience Sorting.** The name of this game is due to Mallows; see [**AD99**] for some history of the problem. Suppose we have $N$ cards, labeled $1, 2, \ldots, N$ for convenience. The game is played as follows. We quote [**AD99**]:

> The deck is shuffled, cards are turned up one at a time and dealt into piles on the table, according to the following rules.
>
> - A low card may be placed on a higher card (e.g., 2 may be placed on 7), or may be put into a new pile to the right of the existing piles.
>
> At each stage we see the top card on each pile. If the turned up card is higher than the cards showing, then it *must* be placed into a new pile to the right of the others.

We follow the *greedy strategy*: place a card as far to the left as possible.

| | | | | | | |
|---|---|---|---|---|---|---|
| | | | | | 1 | 1 |
| | 2 | 2 | 2 | 2 5 | 2 5 | 2 5 |
| 4 | 4 | 4 3 | 4 3 6 | 4 3 6 | 4 3 6 | 4 3 6 7 |

**Figure 3.10.** Patience sorting of the permutation $\pi = 4236517$

For example, suppose we have $N = 7$ cards and suppose we shuffle the deck and obtain the "hand" $\pi = 4236517$. The game proceeds as in Figure 3.10. Note that, for the case $\pi = 4236517$, patience sorting sorts the deck into four decreasing piles and $\ell_7(\pi) = 4$. We have verified the following general fact for the particular case $\pi = 4236517$.

**Theorem 3.46.** *Let $ps_N(\pi)$ be the number of piles obtained in playing patience sorting with $N$ cards starting from a shuffle, $\pi \in S_N$. Then*

$$ps_N(\pi) = \ell_N(\pi). \tag{3.114}$$

**Proof.** Denote the permutation by $\pi = a_1 a_2 \cdots a_N$. Note that if $a_{i_1} < a_{i_2} < \cdots < a_{i_k}$ is an increasing subsequence of $\pi$ for some $1 \leq i_1 < i_2 < \cdots < i_k \leq N$, then those numbers lie in distinct piles after patience sorting: the rules ensure that if $a_{i_k}$ and $a_{i_\ell}$, $\ell > k$, are in the same pile, then $a_{i_\ell}$ must lie above $a_{i_k}$ and $a_{i_k} > a_{i_\ell}$, which is a contradiction. Thus $ps_N(\pi) \geq \ell_N(\pi)$.

Conversely, choose any element, $a_i$ say, in the last pile. When $a_i$ was inserted into the last pile, there must have been an element, $a_{i'}$ say, in the second to last pile with $a_{i'} < a_i$. But then there must have been an element, $a_{i''}$ say in the third to last pile when $a_{i'}$ was inserted, where $a_{i''} < a_{i'}$, $i'' < i'$, etc. In this way we obtain an increasing subsequence $\cdots < a_{i''} < a_{i'} < a_i$ for $\pi$ with length $= ps_N(\pi)$. Then $ps_N(\pi) \leq \ell_N(\pi)$, and the proof is complete. □

It is clear from the definition of patience sorting and the RS algorithm that the top cards on the final pile obtained from $\pi$ are exactly same as the entries of the first row of the $P$-tableau obtained by applying RS to $\pi$. For example, this can be seen in Figure 3.10 and (3.20). The reader will observe that the proof of the above theorem is very similar to the proof of Schensted's formula (3.18).

**Remark 3.47.** The proof of Theorem 3.46 shows how to obtain a longest increasing subsequence. For example, if $\pi = 4236517$, Figure 3.11 gives a longest increasing subsequence 2367. But Figure 3.12 also gives a longest increasing subsequence, viz. 2357.

Patience sorting is so-named because it gives a convenient way to sort a deck of cards. Indeed, it is clear that after playing out the game, the card

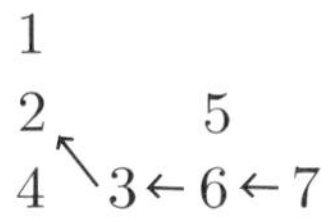


**Figure 3.11.** Patience sorting of the permutation $\pi = 4236517$

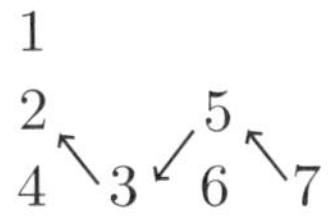


**Figure 3.12.** Patience sorting of the permutation $\pi = 4236517$

1 must be on the top of the left-most pile. (Why?) If we remove 1, then 2 must be on top of some other pile, etc.

From our previous comments in Chapter 1, we conclude that as $N \to \infty$ the number of piles obtained in patience sorting $N$ cards behaves statistically like the largest eigenvalues of a random matrix. In particular, from (1.11),

$$\mathbb{E}[ps_N] = 2\sqrt{N} - (1.758 + o(1))N^{1/6} \tag{3.115}$$

as $N \to \infty$. Said differently, you can make the following bet with your friends: How big a table do you need (say each card is 1 unit wide) to play patience sorting with a deck of $N$ cards? For a regular deck with $N = 52$, we obtain

$$\begin{aligned} \mathbb{E}[\ell_N] &\approx 2\sqrt{N} - 1.758N^{1/6} + o(N^{1/6}) \\ &\approx 14.40 - 3.40 \\ &= 11. \end{aligned} \tag{3.116}$$

Moreover, from the steep decay of the Tracy-Widom distribution, the typical shuffle will not be too far off from 11 units. Give it a try!

### 3.3.2. Airplane Boarding.

The following observation is due to Bachmat et al. [**BBS$^+$06**].

Consider the following simplified model of airplane boarding:

- There is one seat per row.
- The bulk of the boarding time is taken up by passengers blocking the passage way while they stow their bags in the bin above their seat and sit down. Suppose storage takes 1 unit of time.
- The passengers are very thin and they move quickly compared to the storage time.
- The plane is fully booked. There are $N$ seats and $N$ passengers.

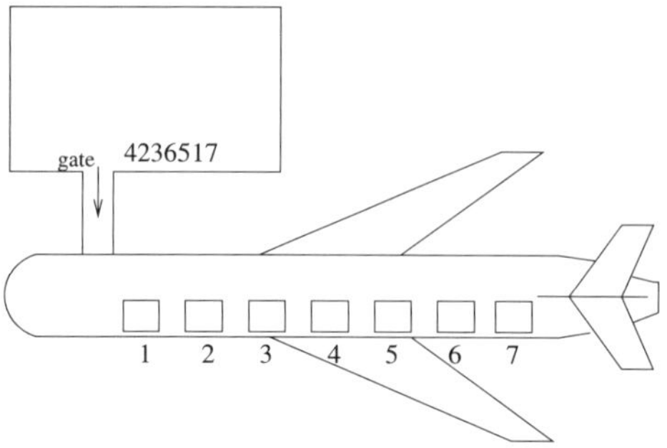


**Figure 3.13.** Airplane boarding

Boarding proceeds as follows:

(1) The attendant announces that the plane is about to board.

(2) The passengers line up at the gate in some random order. For example for $N = 7$ and $\pi = 4236517$ (see Figure 3.13). Figure 3.14 illustrates the boarding procedure.

(3) The passengers enter the plane and proceed to their seats. If their way is blocked, they must wait until the passenger in front has stowed his or her bags. Clearly, in the first unit of time, passengers 4, 2, and 1 stow their bags, while 365 and 7 are blocked at 2 and 1, respectively, and wait to proceed further into the passageway.

(4) At the end of the first unit of time, passengers 4, 2, and 1 sit down and 3657 continue to their seats.

(5) During the second unit of time, passenger 3 stows his or her bag, while 657 wait, etc.

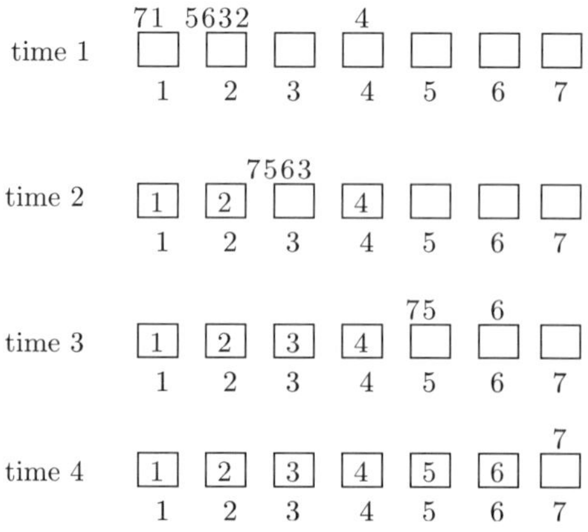


**Figure 3.14.** Airplane boarding

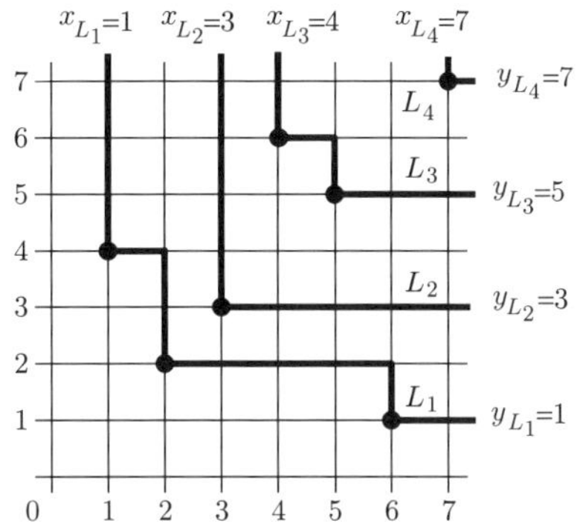


**Figure 3.15.** Shadow lines of $\pi = 4236517$

We see that seating takes four units of time and the groups (421)(3)(65)(7) are seated sequentially, which is exactly what happens in patience sorting! Thus, the boarding time is equal to $\ell_7(\pi)$.

This situation is true in general. Note that for $\pi = 421657$, the seating process can be read off the Viennot diagram (see Subsection 2.1.5) for $\pi$; see Figure 3.15. Indeed, (421) are the coordinates of the vertices of the first shadow line, etc. We have the following result.

**Theorem 3.48.** *Let $BT_N(\pi)$ denote the boarding time for passengers who are $\pi$-arranged at the gate. Then*

$$BT_N(\pi) = \#\ \textit{shadow lines} = \ell_N(\pi). \tag{3.117}$$

*Moreover for $1 \leq i \leq \ell_N(\pi)$,*

$$\begin{aligned}&\{\textit{passengers that are seated in the ith period}\}\\ &\quad = \{y\textit{-coordinates of the boxes on line } L_i\}.\end{aligned} \tag{3.118}$$

**Proof.** Given $\pi = x_1 x_2 \cdots x_N$, let $x_{i_1}$, $i_1 > 1$, be the first element in $\pi$ which is less than $x_1$, i.e., $x_k > x_1$ for $k = 2, \ldots, i_1 - 1$, but $x_{i_1} < x_1$. Now let $i_2 > i_1$ be the first element in $\pi$ after $x_{i_1}$ with $x_{i_2} < x_{i_1}$, etc. Thus, for $\pi = 4236517$, we have $i_1 = 2$, $i_2 = 6$, and there is no $i_3$, etc.

Now in the airplane boarding problem it is clear that while $x_1$ is being seated $x_2, x_3, \ldots, x_{i_1-1}$, must all wait to get to their seats, but $x_{i_1}$ can proceed to be seated, while $x_{i_1+1}, \ldots, x_{i_2-1}$, must wait, etc. Thus

$$x_1, x_{i_1}, x_{i_2}, \ldots \tag{3.119}$$

are precisely the first set of passengers to be seated. On the other hand, $(2, x_2), \ldots, (i_1 - 1, x_{i_1-1})$ are all in the shadow of $(1, x_1)$. Similarly $(i_1 + 1, x_{i_1+1}), \ldots, (i_2 - 1, x_{i_2-1})$ are in the shadow of $(i_1, x_{i_1})$, etc. None of

$(1, x_1)$, $(i_1, x_{i_1})$, $(i_2, x_{i_2})$, ... is in the shadow of any other. Also, by transitivity, they cannot be in the shadow of any other $(j, x_j)$, $j \notin \{1, i_1, i_2, \dots\}$. Thus $(1, x_1)$, $(i_1, x_{i_1})$, $(i_2, x_{i_2}), \dots$ gives the first shadow line $L_1$ for $\pi$. From (3.119) we see that the $y$-coordinates of the boxes on $L_1$ are precisely the first set of passengers to be seated. Repeating this argument for the partial permutation $x_2 \cdots x_{i_1-1} x_{i_1+1} \cdots x_{i_2-1} \cdots$ , etc., it is clear that we have proved (3.117) and (3.118). □

From our previous discussion, it follows that as $N \to \infty$, the boarding time for $N$ passengers behaves statistically like the largest eigenvalue of a random matrix.

**Remark 3.49.** The airplane boarding model described above is the simplest model discussed in [**BBS+06, BBSS07**]. The authors also describe more sophisticated models which take into account the finite size of the passengers, etc. These models have a connection to $1 + 1$ Lorentzian geometry.

The reader is referred to the book [**Bac14**] of Bachmat for a very comprehensive treatment of the airplane boarding problem.

**3.3.3. The Random Turn Vicious Walker Model.** This is one of the so-called *walker models* introduced by M. Fisher [**Fis84**]. Suppose that at time $t = 0$ countably many walkers are located at positions $1, 2, \dots$ in $\mathbb{Z}$. The rules for a walk are the following:

- At each (discrete) time $t = k$, one of the walkers takes one step to the left.
- No two walkers can occupy the same position (in other words, the walkers are *vicious*). In particular, only the first walker can move at time $t = 1$, etc.

Figure 3.16 is an example of a possible walk $W$ from time $t = 0$ to time $t = 7$. Note that at any given time there are only finitely many walkers which can move without violating the noncollision condition.

Let $d_N$ denote the number of walks $W$ that can occur up to time $N$. Equip the set of walks $\{W\}$ of duration $N$ with uniform probability. Let $W_i$ be the distance travelled by the $i$th walker at time $N$. Thus, $W_1 = 3$ in Figure 3.16. We are interested in the quantity

$$V_{n,N} = \mathbb{P}(W_1 \leq n) = \frac{\#\{W : W_1 \leq n\}}{d_N}. \tag{3.120}$$

**Remark 3.50.** The above uniform probability on the walks is different from choosing a walker uniformly randomly among possible walkers at each time, and then moving the walker one step to the left. For example, when $N = 4$, if we let $W$ be the walk obtained by moving the first walker at each time,

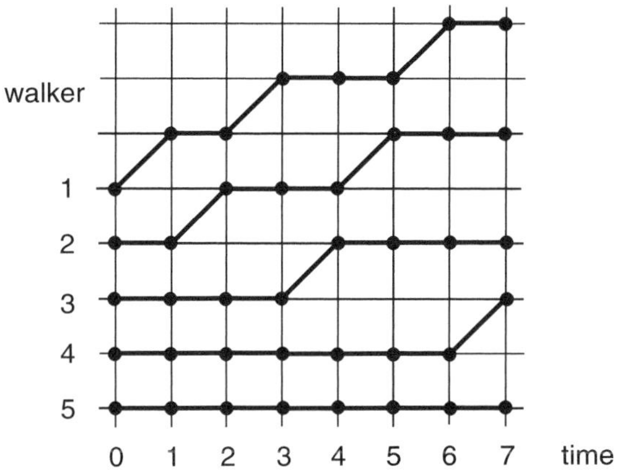


**Figure 3.16.** Vicious random walk

the probability of this walk is $\frac{1}{8}$ if, at each time, a random possible walker is chosen uniformly. On the other hand, there are ten different walks when $N = 4$, so that the probability of $W$ is $\frac{1}{10}$ if we take uniform probability on the set of all walks.

As the walkers are vicious, we must have $W_1 \geq W_2 \geq W_3 \geq \cdots$ and for any walk, $W$, $\sum_i W_i = N$. In particular, as noted above, only a finite number of $W_i$'s are greater than or equal to 1. Denote this number by $r = r(W)$. In other words, we see that every walk of duration $N$ gives rise to a partition of $N$

$$W \mapsto (W_1, W_2, \ldots, W_r) \vdash N. \tag{3.121}$$

Thus, we may write

$$d_N = \sum_{\mu \vdash N} \#\{\text{W: } W_i = \mu_i,\ 1 \leq i \leq r, \text{ where } \mu = (\mu_1, \ldots, \mu_r)\}. \tag{3.122}$$

But it is clear that walks $W$ with $W_1 = \mu_1, \ldots, W_r = \mu_r$ are just another way of representing order preserving ballot sequences (OPBS's) with $r$ candidates and $N = \sum_{i=1}^r \mu_i$ voters, from $(0, \ldots, 0)$ to $(\mu_1, \mu_2, \ldots, \mu_r)$ (see Subsection 3.2.1).

For example, the walk in Figure 3.16 above corresponds to $r = 4$, $N = 7$, and is equivalent to the following OPBS:

$$\begin{array}{ll} t=0 & (0,0,0,0) \\ t=1 & (1,0,0,0) \\ t=2 & (1,1,0,0) \\ t=3 & (2,1,0,0) \\ t=4 & (2,1,1,0) \\ t=5 & (2,2,1,0) \\ t=6 & (3,2,1,0) \\ t=7 & (3,2,1,1). \end{array} \tag{3.123}$$

Here $(\mu_1, \mu_2, \mu_3, \mu_4) = (3, 2, 1, 1) \vdash 7$. This ballot sequence corresponds to the SYT

$$\begin{array}{|c|c|c|} \hline 1 & 3 & 6 \\ \hline 2 & 5 \\ \cline{1-2} 4 \\ \cline{1-1} 7 \\ \cline{1-1} \end{array}\;. \tag{3.124}$$

The map from $W$ to the SYT is simple to describe: If the $i$th walker moves at time $k$, then a box with entry $k$ is added to the $i$th column. Thus $d_N = \sum_{\mu \vdash N} F_\mu$ where, again, $F_\mu$ is the number of SYT's of shape $\mu$. Hence,

$$V_{n,N} = \frac{\sum_{\mu \vdash N, \mu_1 \le n} F_\mu}{\sum_{\mu \vdash N} F_\mu}. \tag{3.125}$$

Note this is different from $\mathbb{P}\{\ell_N \le n\}$ which involves Plancherel measure: here $F_\mu$ appears in the sums, not $F_\mu^2$. From Schützenberger's Theorem 3.11 (see also Corollary 3.12 and (3.23)), we have:

**Theorem 3.51.** *$V_{n,N}$ equals the probability that the length of the longest increasing subsequence of a random* involution *$\pi$ of size $N$ is less than or equal to $n$.*

We can modify the walker model so that it is related to Plancherel measure and, in turn, random permutations. Suppose that during the first $N$ time steps, the walkers move as described above. During the next $N$ time steps, the walkers move instead to the *right*, assuming all the rules as before, and finally come back to the initial positions $0, 1, 2, \ldots$. The walks, during the first $N$ time steps, correspond to an OPBS as before, and the walks during the second $N$ time steps correspond to another OPBS. It follows that the (complete) walks are in one-to-one correspondence with pairs of SYT's of the same shape. For example, the walk in Figure 3.17 corresponds

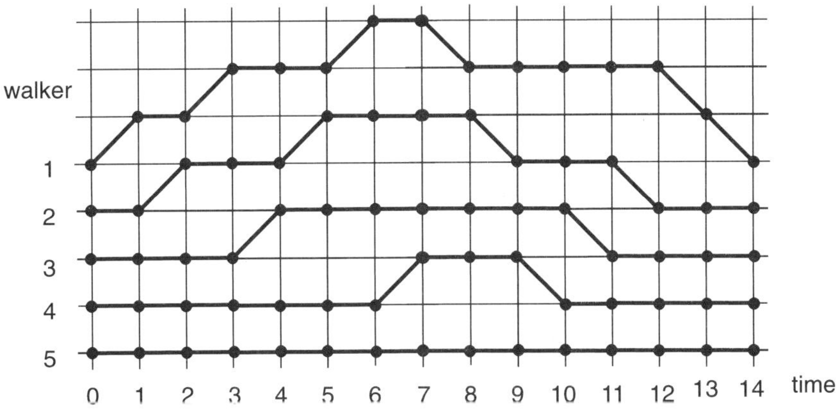


**Figure 3.17.** Watermelon-like random turn vicious walk

to the pair

$$\left( \begin{array}{|c|c|c|} \hline 1 & 3 & 6 \\ \hline 2 & 5 \\ \cline{1-2} 4 \\ \cline{1-1} 7 \\ \cline{1-1} \end{array} \; , \; \begin{array}{|c|c|c|} \hline 1 & 2 & 7 \\ \hline 3 & 6 \\ \cline{1-2} 4 \\ \cline{1-1} 5 \\ \cline{1-1} \end{array} \right). \tag{3.126}$$

Hence, the displacement of the first walker at time $N$ has the same distribution as $\ell_N$.

The asymptotics of $W_1$ for the above (watermelon-like) random turn vicious walker model was studied in [**For01**] using the above connection to random permutations. The asymptotics of the first walker model of Theorem 3.51 was considered in [**BR01b**] using the connection to random involutions [**BR01a**].

**3.3.4. The Polynuclear Growth (PNG) Model.** The polynuclear growth model is an example of random growth models. Wetting, burning, and melting are all examples of randomly growing interfaces, and there is a large body of literature on the study of random growth models in both probability and statistical physics (see, for example, [**KS92, BS95, Mea98, Spo06, FS11, Cor12**]).

The connection between random permutations and a particular random growth model, the PNG model, was first observed by Prähofer and Spohn [**PS00**]. The PNG model is a model for layer by layer growth of crystals on a one-dimensional substrate. Particles nucleate on existing plateaus of the crystal and form new layers, or islands. These islands are assumed to spread laterally with constant speed through steady condensation of material at the edge of the islands. When two adjacent islands at the same level meet, they coalesce.

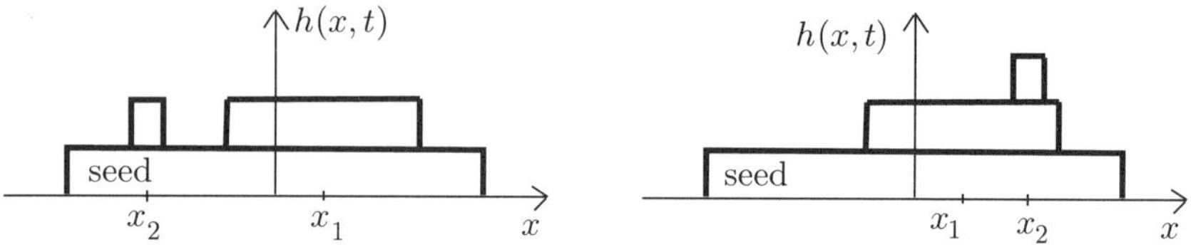


**Figure 3.18.** PNG with droplet initial condition

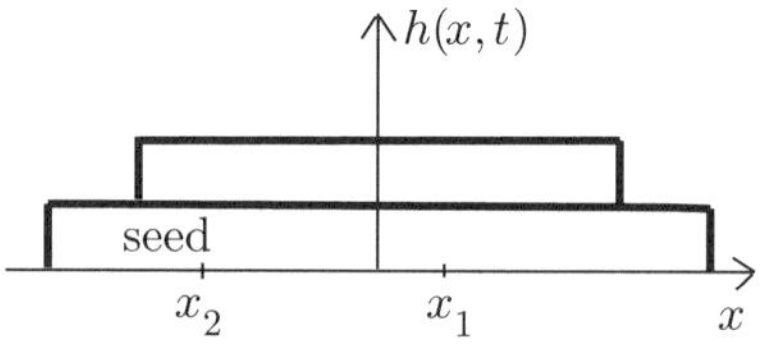


**Figure 3.19.** Two islands coalesce in PNG model

Consider the so-called *droplet initial condition* in which the crystal grows from a single seed at $x = 0$, $t = 0$. The seed at $x = 0$, $t = 0$ spreads laterally at speed 1, say, until at some random time $t_1$ another droplet condenses somewhere, say at $x_1$, at some random point on top of it. The droplets spread laterally until a random time $t_2 > t_1$ when another droplet condenses somewhere, say at $x_2$. See Figure 3.18 for an example. The left figure is the case in which the $x_2 \notin (x_1 - t_2 + t_1, x_1 + t_2 - t_1)$ and the right figure is the case in which $x_2 \in (x_1 - t_2 + t_1, x_1 + t_2 - t_1)$. For the left figure, assuming that no further droplets are formed, at some time (precisely at time $(x_1 - x_2 + t_2 + t_1)/2$), the two islands on top of the seed island meet and coalesce as in Figure 3.19.

We now describe the PNG model more precisely. Let $h(x,t)$ be the height of the droplet above $x$ at time $t$ (we regard the seed as height 0). We will focus on $h(T) = h(0,T)$ at $x = 0$ for convenience. Assume that the condensation of the nucleations take place at points $(x,t)$ in space-time governed by a rate 1 Poisson process. As condensations take place, by assumption, only over the seed-droplet, and as this droplet expands with unit speed, it is clear that condensations can only take place at points $(x,t)$ with $|x| < t$, i.e., inside the forward light-cone at $(0,0)$. On the other hand, the only condensations that can affect $h(T)$ are those points in the backward light-cone at $(0,T)$, i.e., $(x,t)$ with $|x| \leq T - t$. Thus, we see that $h(T)$ is completely determined by the condensations that take place in the square given by $\{(x,t) \in \mathbb{R} \times \mathbb{R}_+ : |x| \leq \min\{t, T-t\}\}$ of side $\frac{T}{\sqrt{2}}$ and area $\frac{T^2}{2}$ (see

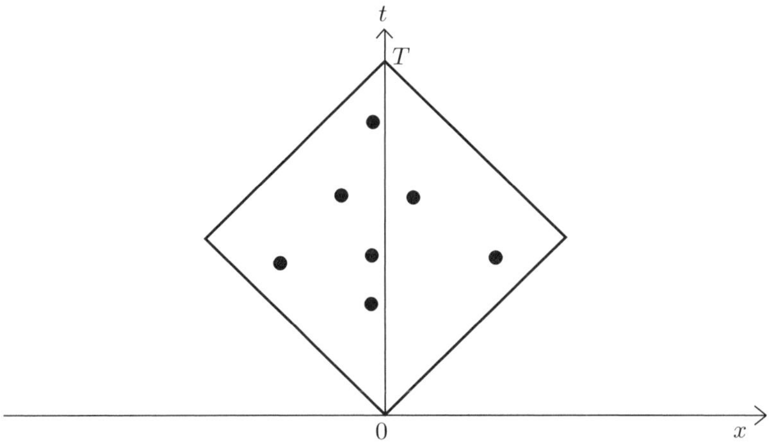


**Figure 3.20.** Condensations in the square $|x| \leq \min\{t, T - t\}$

Figure 3.20). In particular,

$$\mathbb{P}\{\#\text{ condensations} = N\} = e^{-T^2/2}\frac{(T^2/2)^N}{N!}. \tag{3.127}$$

The height, $h(T)$, can be obtained from a given condensation portrait by drawing forward light-cones at each condensation point as follows. First note that each point $(x_i, t_i)$ affects events only in its forward light-cone. Given a condensation point $(x_1, t_1)$, consider the forward light-cone at $(x_1, t_1)$ as in Figure 3.21. For $t > t_1$ with $t - t_1$ small enough (so that no other forward light-cones are present), the coordinates $x$ of the points $(x, t)$ on the forward light-cone at $(x_1, t_1)$ denote the edges of the islands at time $t$ which started at $x_1$ at time $t_1$. Hence, the height of the droplet increases by 1 upon crossing the forward light-cone "upwards". Sometimes two forward light-cones meet

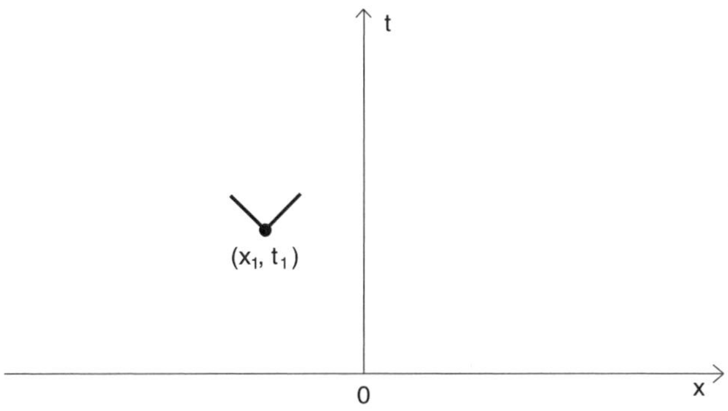


**Figure 3.21.** Forward light cone

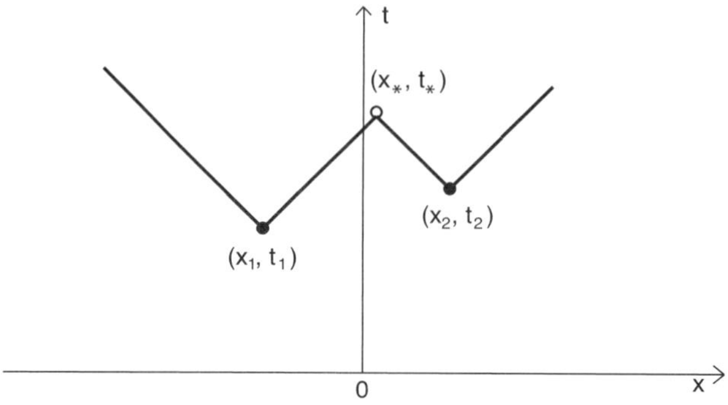


**Figure 3.22.** Two forward light cones meet

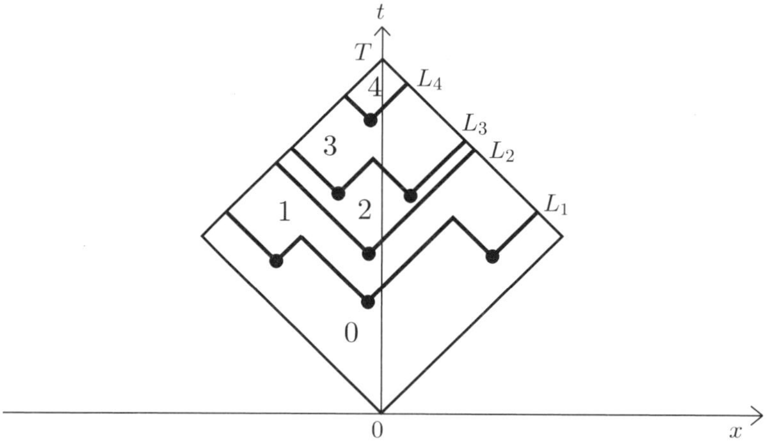


**Figure 3.23.** Lines $L_1, \ldots, L_4$ and height $h(x, t)$

at some $(x_*, t_*)$ as in Figure 3.22. (We consider here "first" meetings, i.e., for $t < t_*$, the two forward light-cones do not intersect any other light-cones.) This happens precisely when two islands of the same height meet at location $x_*$ at time $t_*$. Once the islands coalesce, we remove the forward light-cones emanating from $(x_*, t_*)$ from the figure. Connecting the forward light-cones in this way, we are left with a set of nonintersecting lines $L_1, L_2, \ldots$. For the example in Figure 3.20, the resulting lines are $L_1, \ldots, L_4$ in Figure 3.23. The square in the space-time plane is divided into regions by the lines $L_1, \ldots, L_4$ and the height $h(x, t)$ is the same for points $(x, t)$ in the same region. The height increases by 1 as we go "upwards" in the space-time plane: The heights $h(x, t)$ in each region in Figure 3.23 are displayed in the picture. Thus,

$$h(0, T) = \# \text{ of lines } L_i. \tag{3.128}$$

Clearly, the lines $L_i$ are just the Viennot shadow lines corresponding to the permutation $\pi$ induced by the points $\{(x_1, t_1), (x_2, t_2), \dots\}$ after rotating the figure by 45°. But we know from Subsection 3.1.5 that the number of shadow lines equals $\ell_N(\pi)$. Since the number of points $N$ in the square is a Poisson random variable given in (3.127), we obtain the following result.

**Theorem 3.52.** *Let $h(T)$ denote the height of the droplet at location 0 at time $t$ in the PNG model. Then*

$$\mathbb{P}\{h(T) \le n\} = \mathbb{P}\{L(\frac{T}{\sqrt{2}}) \le n\}, \tag{3.129}$$

*where $L(t)$ is the Poissonized version of the length of the longest increasing subsequence given in Subsection* 2.1.

**Exercise 3.53.** Consider the PNG model with the droplet initial condition *on the half line* $\mathbb{R}_+$. When an island meets the edge $x = 0$, it stops expanding. Show that the height $h(0, T)$ equals the length of the longest increasing subsequence of a random *involution* of size $2\mathcal{N}$ with no fixed points where $\mathcal{N}$ is a Poisson random variable with parameter $\frac{T^2}{4}$. The asymptotic behavior of this model as $T \to \infty$, is analyzed in [**BR01a**] (see also [**SI04**]).

**Remark 3.54.** The PNG model can also interpreted as the following particle–antiparticle model on $\mathbb{R}$. A pair with a particle and an antiparticle is created at a random position and at a random time, given a Poisson process of rate 1 in the space-time plane. Upon creation, the particle travels to the right with constant speed 1 and the antiparticle travels to the left with constant speed 1. When a particle and an antiparticle collide, they annihilate each other. Further, assume that all the creations of pairs of a particle and an antiparticle occur in the forward light-cone $\{(x, t) \in \mathbb{R} \times \mathbb{R}_+ : |x| < t\}$. If we interpret the right edge of an island as a particle and the left edge of an island as an antiparticle in the PNG model with the droplet condition, then the birth of an island corresponds to the creation of a pair with a particle and an antiparticle. The collision of two islands corresponds to the collision, and the annihilation that follows, of a particle and an antiparticle. It is easy to see that the total number of particles and antiparticles that have crossed the location $x = 0$ by time $T$ equals the height $h(0, T)$ of the PNG model.

Chapter 4

# Bounds on the Expected Value of $\ell_N$

Let $\ell_N$ be the length of a longest increasing subsequence of a random permutation in $S_N$, as before. Ulam's problem, introduced in Chapter 1, is to prove that

$$\lim_{N\to\infty} \frac{\mathbb{E}[\ell_N]}{\sqrt{N}} = c \tag{4.1}$$

exists and to evaluate the constant $c$. In this chapter, we show that $c$ exists and obtain certain bounds for $c$. The first two sections are elementary while the third section uses a property of the Robinson-Schensted correspondence between permutations and Young tableaux discussed in Chapter 3. In an aside, in the last section of this chapter we consider the probability distribution induced on the conjugacy classes of $S_N$ by the uniform measure on $S_N$.

## 4.1. Lower Bound

The following basic deterministic result is due to Erdős and Szekeres [**ES35**] in 1935.

**Proposition 4.1.** *Let $c$ and $d$ be positive integers. Every sequence of $N = cd + 1$ distinct real numbers $s = a_1a_2\cdots a_N$ contains an increasing subsequence of length greater than or equal to $c+1$ or a decreasing subsequence of length at least $d+1$. As a consequence, for any $N$, $s$ contains an increasing or decreasing subsequence of length greater than or equal to $\sqrt{N-1}$.*

**Proof.** For any $i$, $1 \le i \le N$, let $\ell_i$ denote the length of the longest increasing subsequence in $s$ ending in $a_i$, and let $d_i$ denote the length of the longest decreasing subsequence in $s$ ending in $a_i$.

Suppose, by contradiction, that $\ell_i \le c$ and $d_i \le d$ for all $i$. Then the $N$ pairs $\{(\ell_i, d_i) : i = 1, \dots, N\}$ lie in a rectangle of size $cd < N$. So we must have $(\ell_{i_1}, d_{i_1}) = (\ell_{i_2}, d_{i_2})$ for some $1 \le i_1 < i_2 \le N$. If $a_{i_1} > a_{i_2}$, then $d_{i_1} < d_{i_2}$; on the other hand, if $a_{i_1} < a_{i_2}$, then $\ell_{i_1} < \ell_{i_2}$. This is a contradiction. Hence either $\ell_i > c$ or $d_i > d$, and because we are dealing with integers, $\ell_i \ge c + 1$ or $d_i \ge d + 1$.

Now let $N' = [\sqrt{N-1}]^2 + 1 \le N$. Then $s' = a_1 a_2 \cdots a_{N'}$ has either an increasing or a decreasing subsequence of length $[\sqrt{N-1}] + 1 = \sqrt{N-1} + \delta$ for some $0 < \delta \le 1$. But $\sqrt{N-1} + \delta > \sqrt{N-1}$, and so we are done. □

**Remark 4.2.** The roles of $c$ and $d$ in Proposition 4.1 can of course be switched.

Now as Haar measure (i.e., uniform measure) is invariant under reversal, $\pi = a_1 \cdots a_N \mapsto \pi^r = a_N \cdots a_1$, we must have for any $N$,

$$\mathbb{E}[\ell_N] = \frac{1}{2}\,\mathbb{E}[\ell_N(\pi) + \ell_N(\pi^r)] = \frac{1}{2}\,\mathbb{E}[\ell_N + d_N] \ge \frac{1}{2}\sqrt{N-1}. \tag{4.2}$$

Hence

**Corollary 4.3.**

$$\liminf_{N\to\infty} \frac{\mathbb{E}[\ell_N]}{\sqrt{N}} \ge \frac{1}{2}. \tag{4.3}$$

## 4.2. Existence of $c$

In this subsection, we prove that the limit in (4.1) exists. This result is due to Hammersley [**Ham72**].

Recall the Poissonized version $L(t)$ of $\ell_N$ discussed in Section 2.1: $L(t)$ is the length of the longest up/right path from $(0, 0)$ to $(t, t)$ constructed from points generated by a rate 1 point process in the first quadrant of $\mathbb{R}^2$. The basic relation between $L(t)$ and $\ell_N$ is given in (2.3),

$$\mathbb{P}\{L(t) \le n\} = \sum_{N=0}^{\infty} e^{-t^2} \frac{(t^2)^N}{N!}\, \mathbb{P}\{\ell_N \le n\}. \tag{4.4}$$

Here, we set $\ell_0 \equiv 0$. Consider the square $S = [0, t+s] \times [0, t+s]$ and also the squares $S_1 = [0, t] \times [0, t]$ and $S_2 = [t, t+s] \times [t, t+s]$. Given a realization of the Poisson process in the plane, it is clear that the length of the longest up/right path in $S$ is greater than or equal to the sum of the lengths of the longest up/right paths in $S_1$ and $S_2$. Figure 4.1 illustrates this for a specific example.

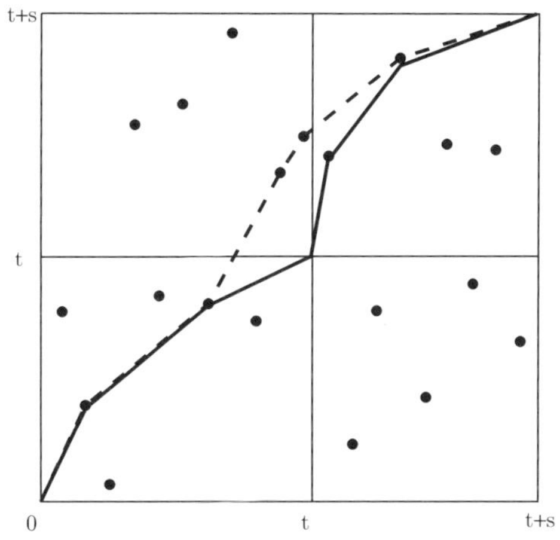


**Figure 4.1.** $L(t+s) \geq L(t) + L(s)$

As the measure induced on the set of point configurations under this Poisson process is invariant under translation, the following superadditivity relation holds:

$$\mathbb{E}[L(t+s)] \geq \mathbb{E}[L(t)] + \mathbb{E}[L(s)]. \tag{4.5}$$

Note that $\ell_N$ itself does not satisfy such a superadditivity relation.

The following result is standard and the proof is left as an exercise.

**Lemma 4.4** (Subadditivity Theorem). *We have the following:*

(a) *Suppose that* $\{a_n\}_{n\geq 0}$ *satisfies* $a_{n+m} \leq a_n + a_m$ *for all* $n, m \geq 0$. *Then*

$$\lim_{n\to\infty} \frac{a_n}{n} = c \tag{4.6}$$

*exists for some* $-\infty \leq c < \infty$ *and* $c = \inf_n \frac{a_n}{n}$.

(b) *Suppose that* $t \mapsto b_t$ *is a real-valued function of a parameter* $t \geq 0$ *such that* $b_{t+s} \leq b_t + b_s$ *for all* $t, s \geq 0$. *If, in addition,*

$$\sup_{0\leq t<\epsilon} |b_t| < \infty \tag{4.7}$$

*for some* $\epsilon > 0$, *then*

$$\lim_{t\to\infty} \frac{b_t}{t} = c \tag{4.8}$$

*exists for some* $-\infty \leq c < \infty$.

**Remark 4.5.** The example $a_n = -n^2$ shows that $c = -\infty$ can be attained.

**Corollary 4.6.** *There is a constant* $0 \le c \le \infty$ *such that*

$$\lim_{t\to\infty} \frac{\mathbb{E}[L(t)]}{t} = c. \tag{4.9}$$

**Proof.** In order to apply Lemma 4.4 to $b_t = -\mathbb{E}[L(t)]$, we need to check that $\mathbb{E}[L(t)]$ is uniformly bounded for, say, $0 \le t \le 1$. Noting that

$$\mathbb{E}[L(t)] \le \mathbb{E}[\# \text{ of points in } [0,t]\times[0,t]] = \sum_{k=0}^{\infty} k \cdot \frac{e^{-t^2} t^{2k}}{k!} = t^2, \tag{4.10}$$

is uniformly bounded for $0 \le t \le 1$ and $\mathbb{E}[L(t)] \ge 0$ completes the proof. □

Corollary 4.6 together with the following de-Poissonization lemma will imply that $\frac{\mathbb{E}[\ell_N]}{\sqrt{N}} \to c$ as $N \to \infty$.

**Lemma 4.7.** *For any* $0 < \epsilon < 1$ *and* $N \ge 1$,

$$\mathbb{E}[\ell_N] - \mathbb{E}[L(\sqrt{(1+\epsilon)N})] \le N e^{-\frac{1}{4}\epsilon^2 N} \tag{4.11}$$

*and*

$$\mathbb{E}[L(\sqrt{(1-\epsilon)N})] - \mathbb{E}[\ell_N] \le \frac{1}{\epsilon}\sqrt{\frac{N}{2\pi}} e^{-\frac{1}{2}\epsilon^2 N}. \tag{4.12}$$

**Proof.** Since

$$\mathbb{E}[L(t)] = \sum_{k=0}^{\infty} \mathbb{P}\{\mathcal{N} = k\}\, \mathbb{E}[L(t)|\mathcal{N} = k] = \sum_{k=0}^{\infty} \frac{e^{-t^2}(t^2)^k}{k!} \mathbb{E}[\ell_k], \tag{4.13}$$

$\phi(t) \equiv \mathbb{E}[L(t)]$ is the Poissonization of the sequence $a_N \equiv \mathbb{E}[\ell_N]$, $N = 0, 1, \dots$, as defined in (2.7). To apply the de-Poissonization Lemma 2.3, $a_N$ should satisfy the monotonicity $a_{N+1} \le a_N$ and the bound $0 \le a_N \le 1$. Corollary 2.9 implies the monotonicity relation

$$a_N \le a_{N+1}. \tag{4.14}$$

In addition, the basic bound

$$0 \le a_N \le N \tag{4.15}$$

is obvious. It is easy to check that the proof of Lemma 2.3 can be adapted in this case to yield the desired estimates (see Remark 2.4). Note that the direction of the inequalities (2.9) and (2.10) changed because of the reverse monotonicity, and the estimates are multiplied by $N$ due to the bound (4.15). □

We now obtain the following result.

**Proposition 4.8.** *There is* $0 < c \le \infty$ *such that*

$$\lim_{N\to\infty} \frac{\mathbb{E}[\ell_N]}{\sqrt{N}} = \lim_{t\to\infty} \frac{\mathbb{E}[L(t)]}{t} = c. \tag{4.16}$$

**Proof.** Fix $0 < \epsilon < 1$. From Lemma 4.7 and Corollary 4.6,

$$c\sqrt{1-\epsilon} \le \liminf_{N\to\infty} \frac{\mathbb{E}[\ell_N]}{\sqrt{N}} \le \limsup_{N\to\infty} \frac{\mathbb{E}[\ell_N]}{\sqrt{N}} \le c\sqrt{1+\epsilon}, \tag{4.17}$$

where $c \in [0, \infty]$ is the limit in Corollary 4.6. Since $0 < \epsilon < 1$ is arbitrary, this proves the result for some $0 \le c \le \infty$. But then Corollary 4.3 shows that $c \ge \frac{1}{2}$. □

The next result shows that $c < \infty$.

**Proposition 4.9.**

$$\lim_{N\to\infty} \frac{\mathbb{E}[\ell_N]}{\sqrt{N}} \le e. \tag{4.18}$$

**Proof.** We first prove that

$$\mathbb{P}\{\ell_N \ge n\} \le \binom{N}{n}\frac{1}{n!} \equiv m_n. \tag{4.19}$$

For $\pi \in S_N$, define

$$\begin{aligned} \nu_n(\pi) &= \#\text{ of increasing subsequences of } \pi \text{ of length } n \\ &= \#\{1 \le j_1 < \cdots < j_n \le N : \pi(j_1) < \cdots < \pi(j_n)\}. \end{aligned} \tag{4.20}$$

Then clearly $\ell_N(\pi) \ge n$ if and only if $\nu_n(\pi) \ge 1$. Hence,

$$\mathbb{P}\{\ell_N \ge n\} = \mathbb{P}\{\nu_n \ge 1\} \le \sum_{k=1}^{N} k \cdot \mathbb{P}\{\nu_n = k\} = \mathbb{E}[\nu_n]. \tag{4.21}$$

Write

$$\nu_n = \sum_{1\le j_1<\cdots<j_n\le N} 1_{A_{j_1,\ldots,j_n}}, \tag{4.22}$$

where

$$A_{j_1,\ldots,j_n} \equiv \{\sigma \in S_N : \sigma(j_1) < \cdots < \sigma(j_n)\}. \tag{4.23}$$

Since $|A_{j_1,\ldots,j_n}| = \binom{N}{n}(N-n)!$, we have

$$\begin{aligned} \mathbb{E}[\nu_n] &= \sum_{1\le j_1<\cdots<j_n\le N} \mathbb{P}\{A_{j_1,\ldots,j_n}\} = \sum_{1\le j_1<\cdots<j_n\le N} \frac{\binom{N}{n}(N-n)!}{N!} \\ &= \binom{N}{n}\frac{1}{n!}, \end{aligned} \tag{4.24}$$

which leads to (4.19).

Let $\lceil x \rceil$ denote the smallest integer that is greater than or equal to $x$. Clearly, $x \leq \lceil x \rceil < x+1$. For $n = \lceil e\sqrt{N} \rceil + k$ with any $k \geq 0$,

$$\frac{m_{n+1}}{m_n} = \frac{N-n}{(n+1)^2} = \frac{N - \lceil e\sqrt{N} \rceil - k}{(\lceil e\sqrt{N} \rceil + k + 1)^2} \leq \frac{N}{(e\sqrt{N})^2} = e^{-2}. \tag{4.25}$$

This implies that $\frac{m_{\lceil e\sqrt{N} \rceil + k}}{m_{\lceil e\sqrt{N} \rceil}} \leq e^{-2k}$ for all $k \geq 0$. Therefore, we obtain from (4.19)

$$\mathbb{P}\{\ell_N \geq \lceil e\sqrt{N} \rceil + k\} \leq e^{-2k} m_{\lceil e\sqrt{N} \rceil}, \qquad k \geq 0. \tag{4.26}$$

Using this, we obtain

$$\begin{aligned} \mathbb{E}[\ell_N] = \sum_{j=1}^{N} \mathbb{P}\{\ell_N \geq j\} &= \sum_{j=1}^{\lceil e\sqrt{N} \rceil - 1} \mathbb{P}\{\ell_N \geq j\} + \sum_{j=\lceil e\sqrt{N} \rceil}^{N} \mathbb{P}\{\ell_N \geq j\} \\ &\leq \sum_{j=1}^{\lceil e\sqrt{N} \rceil - 1} 1 + \sum_{k=0}^{N - \lceil e\sqrt{N} \rceil} e^{-2k} m_{\lceil e\sqrt{N} \rceil} \\ &\leq e\sqrt{N} + \frac{m_{\lceil e\sqrt{N} \rceil}}{1 - e^{-2}}. \end{aligned} \tag{4.27}$$

Using Stirling's formula, it is easy to show that

$$\lim_{N \to \infty} \frac{m_{\lceil e\sqrt{N} \rceil}}{\sqrt{N}} = 0. \tag{4.28}$$

Hence, (4.18) follows from (4.27). □

**Remark 4.10.** The above calculations tell us that $\frac{\mathbb{E}[L(t))]}{t}$ and $\frac{\mathbb{E}[\ell_N]}{\sqrt{N}}$ converge to the same constant $c$, $\frac{1}{2} \leq c \leq e$. But these calculations do not indicate whether $\frac{\ell_N}{\sqrt{N}}$ is concentrated near its mean or spread out. There is a stronger version of the subadditivity theorem which tells us that $\frac{\ell_N}{\sqrt{N}}$ is concentrated near its mean, for example see [**Kin73**] and [**Dur96**, Example 7.2, pp. 327–329]. In [**Kin73**], Kingman also improved the lower bound on $c$ showing that $c \geq \sqrt{\frac{8}{\pi}} > 1.59$.

## 4.3. Young Diagrams in a Markov Chain and an Optimal Upper Bound

Vershik and Kerov [**VK85**] gave a short and elegant proof of the following result.

**Proposition 4.11.** *For $N \geq 1$,*

$$\frac{\mathbb{E}[\ell_N]}{\sqrt{N}} \leq 2. \tag{4.29}$$

**Remark 4.12.** In view of the discussion in Chapter 1 where we stated that $\frac{\mathbb{E}[\ell_N]}{\sqrt{N}} \to 2$ as $N \to \infty$, this upper bound is optimal.

Let $\mathcal{Y} = \bigcup_{N=0}^{\infty} \mathcal{Y}_{\mathrm{N}}$ denote the set of the Young diagrams or, equivalently, partitions. The main idea in [**VK85**] is to consider a discrete-time Markov chain $\{\lambda_N\}_{N=0}^{\infty}$ with the state space $\mathcal{Y}$ such that $\lambda_N \in \mathcal{Y}_{\mathrm{N}}$ for each $N \geq 0$ and the marginal distribution at time $N$ is given by Plancherel measure,

$$\mathbb{P}\{\lambda_N = \mu\} = \frac{F_\mu^2}{N!}, \qquad \mu \in \mathcal{Y}_{\mathrm{N}}\,. \tag{4.30}$$

Recall that $F_\mu$ denotes the number of standard Young tableaux of shape $\mu$ (see Section 3.2). By convention, $F_\emptyset = 1$. For $\mu \in \mathcal{Y}$, let $|\mu|$ denote the number of boxes in $\mu$.

Suppose $\lambda$ and $\Lambda$ are Young diagrams of size $N-1$ and $N$, respectively. We say that $\lambda$ is an *ancestor* of $\Lambda$, or alternatively $\Lambda$ is a *descendent* of $\lambda$, if $\Lambda$ can be obtained from $\lambda$ by the addition of a single box. We denote the ancestor/descendent relation by $\lambda \uparrow \Lambda$. For example,

(4.31) 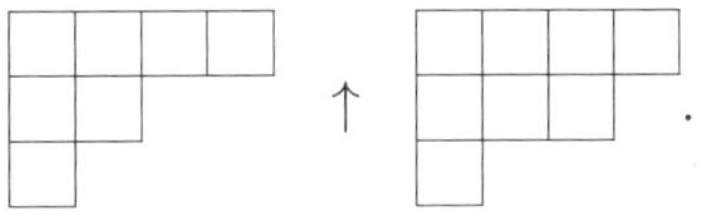

This relation gives rise to a natural directed graph structure on $\mathcal{Y}$, which we display graphically in Figure 4.2. Here the arrowed line corresponds to the relation $\lambda \uparrow \Lambda$. The following result is easily verified.

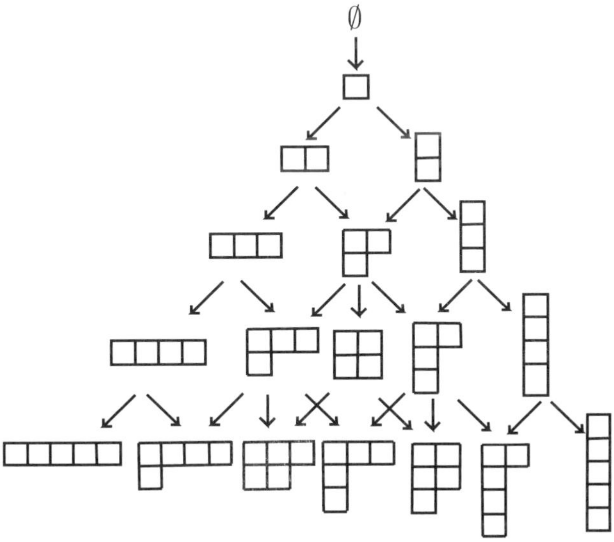

**Figure 4.2.** A part of the directed graph structure of $\mathcal{Y}$

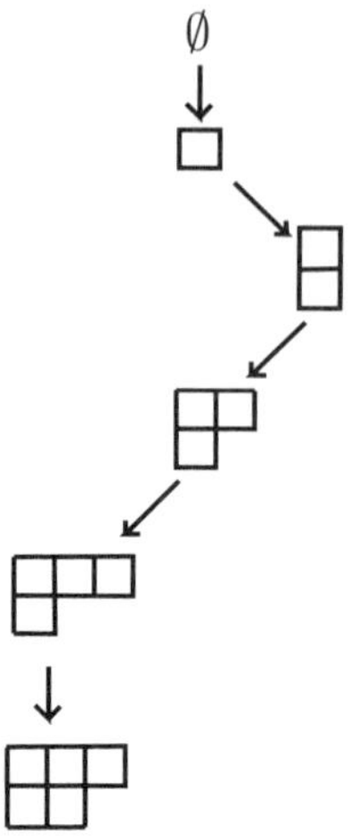

**Figure 4.3.** A part of the directed graph structure of $\mathcal{Y}$

**Lemma 4.13.** *For a given* $\lambda \in \mathcal{Y}$*, the set of SYT's* $P$ *with* $\operatorname{sh} P = \lambda$ *is in one-to-one correspondence with the set of paths on the directed graph from* $\emptyset$ *to* $\lambda$*.*

For example, the path in Figure 4.3 corresponds to the SYT

$$\begin{array}{|c|c|c|}\hline 1 & 3 & 4 \\ \hline 2 & 5 \\ \cline{1-2}\end{array}\,. \tag{4.32}$$

For $\lambda, \Lambda \in \mathcal{Y}$, define the *transition probability*

$$p_{\lambda\Lambda} = \begin{cases} \frac{F_\Lambda}{|\Lambda| F_\lambda} & \text{if } \lambda \uparrow \Lambda \\ 0 & \text{otherwise.} \end{cases} \tag{4.33}$$

This defines a bona fide transition probability with state space $\mathcal{Y}$ by the following lemma.

**Lemma 4.14.** *For* $\lambda \in \mathcal{Y}$*,*

$$\sum_{\mu \in \mathcal{Y}} p_{\lambda\mu} = 1. \tag{4.34}$$

**Proof.** First, note that

$$\sum_{\kappa : \kappa \uparrow \mu} F_\kappa = F_\mu, \qquad \mu \in \mathcal{Y}, \tag{4.35}$$

which follows immediately from Lemma 4.13 and the directed graph structure.

We prove (4.34) by induction in $|\lambda|$. If $|\lambda| = 0$, then $\lambda = \emptyset$, and (4.34) is obvious.

Assume that (4.34) is true if $|\lambda| = n-1$. Suppose $|\lambda| = n$. By definition,

$$
\begin{aligned}
\sum_{\mu} p_{\lambda\mu} &= \sum_{\mu\in\mathcal{Y}_{n+1}} p_{\lambda\mu} 1_{\lambda\uparrow\mu} = \sum_{\mu\in\mathcal{Y}_{n+1}} \frac{F_\mu}{(n+1)F_\lambda} 1_{\lambda\uparrow\mu} \\
&= \frac{1}{(n+1)F_\lambda} \sum_{\mu\in\mathcal{Y}_{n+1}} 1_{\lambda\uparrow\mu} \sum_{\kappa:\kappa\uparrow\mu} F_\kappa \qquad \text{by (4.35)} \\
&= \frac{1}{(n+1)F_\lambda} \sum_{\kappa\in\mathcal{Y}_n} F_\kappa \sum_{\mu\in\mathcal{Y}_{n+1}} 1_{\lambda\uparrow\mu} 1_{\kappa\uparrow\mu},
\end{aligned}
\tag{4.36}
$$

where $1_A$ denotes the indicator random variable of $A$. Considering the two cases $\kappa = \lambda$ and $\kappa \neq \lambda$ separately, the last double sum equals

$$
F_\lambda \sum_{\mu\in\mathcal{Y}_{n+1}} 1_{\lambda\uparrow\mu} + \sum_{\kappa\in\mathcal{Y}_n\setminus\{\lambda\}} F_\kappa \sum_{\mu\in\mathcal{Y}_{n+1}} 1_{\lambda\uparrow\mu} 1_{\kappa\uparrow\mu}. \tag{4.37}
$$

For $\kappa \in \mathcal{Y}_n \setminus\{\lambda\}$, the number of common descendent of $\lambda$ and $\kappa$ is either 0 or 1 (see Figure 4.2). Thus, the sum $\sum_{\mu\in\mathcal{Y}_{n+1}} 1_{\lambda\uparrow\mu} 1_{\kappa\uparrow\mu}$ is either 0 or 1. Let $D_n(\lambda)$ denote the set of $\kappa \in \mathcal{Y}_n \setminus\{\lambda\}$ such that $\kappa$ and $\lambda$ have a common descendent. Thus

$$
\sum_{\mu} p_{\lambda\mu} = \frac{1}{(n+1)F_\lambda}\left\{F_\lambda \cdot \#\{\text{descendents of } \lambda\} + \sum_{\kappa\in D_n(\lambda)} F_\kappa\right\}. \tag{4.38}
$$

Observe that $\kappa \in D_n(\lambda)$ if and only if $\kappa$ and $\lambda$ have a common ancestor, given by $\kappa \cap \lambda \in \mathcal{Y}_{n-1}$, and $\kappa \neq \lambda$. Thus,

$$
\sum_{\kappa\in D_n(\lambda)} F_\kappa = \sum_{\nu\in\mathcal{Y}_{n-1}} 1_{\nu\uparrow\lambda} \sum_{\kappa\in\mathcal{Y}_n\setminus\{\lambda\}} 1_{\nu\uparrow\kappa} F_\kappa. \tag{4.39}
$$

Now, by the induction hypothesis, for $\nu \in \mathcal{Y}_{n-1}$,

$$
\sum_{\kappa\in\mathcal{Y}_n} 1_{\nu\uparrow\kappa} F_\kappa = n F_\nu. \tag{4.40}
$$

Hence,

$$
\sum_{\nu\in\mathcal{Y}_{n-1}} 1_{\nu\uparrow\lambda} \sum_{\kappa\in\mathcal{Y}_n} 1_{\nu\uparrow\kappa} F_\kappa = n \sum_{\nu\in\mathcal{Y}_{n-1}} 1_{\nu\uparrow\lambda} F_\nu = n F_\lambda, \tag{4.41}
$$

using (4.35) for the second equality. This implies that

$$
\sum_{\kappa\in D_n(\lambda)} F_\kappa = n F_\lambda - \sum_{\nu\in\mathcal{Y}_{n-1}} 1_{\nu\uparrow\lambda} F_\lambda = n F_\lambda - \#\{\text{ancestors of } \lambda\} F_\lambda. \tag{4.42}
$$

Inserting this relation into (4.38), we obtain

$$
\sum_{\mu} p_{\mu\lambda} = \frac{n - \#\{\text{ancestors of } \lambda\} + \#\{\text{descendents of } \lambda\}}{n+1}. \tag{4.43}
$$

However, in the terminology of Chapter 3, it is clear that

$$\#\{\text{descendents of } \lambda\} = \#\{\text{outer corners of } \lambda\} \tag{4.44}$$

and

$$\#\{\text{ancestors of } \lambda\} = \#\{\text{inner corners of } \lambda\} \tag{4.45}$$

and, by the geometry

$$\#\{\text{outer corners of } \lambda\} - \#\{\text{inner corners of } \lambda\} = 1. \tag{4.46}$$

The proof of Lemma 4.14 is now complete. □

**Remark 4.15.** An alternative and very direct proof of (4.34) can be given by relating Young diagrams to the Markov moment problem via the hook formula in Theorem 3.42; see [**Ker93**].

The following lemma relates the above Markov chain with the Plancherel measure.

**Lemma 4.16.** *Let* $\{\lambda_N\}_{N=0}^{\infty}$ *be the Markov chain with transition probabilities given by* (4.33) *and initial distribution given by* $\mathbb{P}\{\lambda_0 = \emptyset\} = 1$. *Then the marginal distribution at time* $N$ *is given by the Plancherel measure* (4.30),

$$\mathbb{P}\{\lambda_N = \lambda\} = \frac{F_\lambda^2}{N!}, \qquad \lambda \in \mathcal{Y}_N, \tag{4.47}$$

*and* $\mathbb{P}\{\lambda_N = \lambda\} = 0$ *for* $\lambda \notin \mathcal{Y}_N$.

**Proof.** For $\lambda \in \mathcal{Y}_N$,

$$\mathbb{P}\{\lambda_N = \lambda\} = \sum_{(\lambda_1,\ldots,\lambda_{N-1})\in\Pi_N(\lambda)} p_{\emptyset\lambda_1} p_{\lambda_1\lambda_2} \cdots p_{\lambda_{N-1}\lambda}, \tag{4.48}$$

where $\Pi_N(\lambda)$ is the set of $(\lambda_1, \ldots, \lambda_{N-1})$ such that $\emptyset \uparrow \lambda_1 \uparrow \lambda_2 \uparrow \cdots \uparrow \lambda_{N-1} \uparrow \lambda$. Note that $|\Pi_N(\lambda)| = F_\lambda$ by Lemma 4.13. Hence,

$$\begin{aligned} \mathbb{P}\{\lambda_N = \lambda\} &= \sum_{(\lambda_1,\ldots,\lambda_{N-1})\in\Pi_N(\lambda)} \frac{F_{\lambda_1}}{F_\emptyset} \frac{F_{\lambda_2}}{2F_{\lambda_1}} \cdots \frac{F_\lambda}{NF_{\lambda_{N-1}}} \\ &= \frac{F_\lambda}{N!} |\Pi_N(\lambda)| \\ &= \frac{F_\lambda^2}{N!}. \end{aligned} \tag{4.49}$$

On the other hand, $\lambda_N \in \mathcal{Y}_N$ implies that $\mathbb{P}\{\lambda_N = \lambda\} = 0$ if $\lambda \notin \mathcal{Y}_N$. □

Following [**VK85**], we now prove (4.29).

**Proof of Proposition 4.11.** For $\lambda \in \mathcal{Y}$, let $\hat{\lambda}$ denote the unique Young diagram of size $|\lambda| + 1$ obtained from $\lambda$ by the addition of a box in the first row. Consider the Markov chain $\{\lambda_N\}_{N=0}^{\infty}$ as in Lemma 4.16. For any $k$, let $\chi_k$ be the indicator function of the event that $\{\lambda_k = \hat{\lambda}_{k-1}\}$. Hence

$$\begin{aligned} \mathbb{E}[\chi_k] &= \mathbb{P}\{\lambda_k = \hat{\lambda}_{k-1}\} \\ &= \sum_{\lambda \in \mathcal{Y}_{k-1}} \mathbb{P}\{\lambda_k = \hat{\lambda}_{k-1} | \lambda_{k-1} = \lambda\} \, \mathbb{P}\{\lambda_{k-1} = \lambda\} \\ &= \sum_{\lambda \in \mathcal{Y}_{k-1}} p_{\lambda\hat{\lambda}} \, \mathbb{P}\{\lambda_{k-1} = \lambda\}. \end{aligned} \tag{4.50}$$

By using the Schwarz inequality, (4.30), and (4.33), we obtain

$$\begin{aligned} \mathbb{E}[\chi_k] &\leq \left[ \sum_{\lambda \in \mathcal{Y}_{k-1}} p_{\lambda\hat{\lambda}}^2 \, \mathbb{P}\{\lambda_{k-1} = \lambda\} \right]^{1/2} \left[ \sum_{\lambda \in \mathcal{Y}_{k-1}} \mathbb{P}\{\lambda_{k-1} = \lambda\} \right]^{1/2} \\ &= \left[ \sum_{\lambda \in \mathcal{Y}_{k-1}} \left( \frac{F_{\hat{\lambda}}}{k F_{\lambda}} \right)^2 \frac{F_{\lambda}^2}{(k-1)!} \right]^{1/2} \cdot 1 \\ &= \frac{1}{\sqrt{k}} \left[ \sum_{\lambda \in \mathcal{Y}_{k-1}} \frac{1}{k!} F_{\hat{\lambda}}^2 \right]^{1/2} = \frac{1}{\sqrt{k}} \left[ \sum_{\lambda \in \mathcal{Y}_{k-1}} \mathbb{P}\{\lambda_k = \hat{\lambda}\} \right]^{1/2} \\ &< \frac{1}{\sqrt{k}}. \end{aligned} \tag{4.51}$$

Given a chain $\{\lambda_i\}_{i=0}^{\infty}$, for a fixed $N$,

$$\#\{\text{boxes in the first row of } \lambda_N\} = \sum_{k=1}^{N} \chi_k(\{\lambda_i\}_{i=0}^{\infty}). \tag{4.52}$$

Hence, from (4.30) and (4.51), we have

$$\begin{aligned} &\mathbb{E}_N^{\text{Plan}} \left[ \#\{\text{boxes in the first row of } \lambda_N\} \right] \\ &= \mathbb{E}\left[ \sum_{k=1}^{N} \chi_k \right] < \sum_{k=1}^{N} \frac{1}{\sqrt{k}} < 2\sqrt{N}, \end{aligned} \tag{4.53}$$

where $\mathbb{E}_N^{\text{Plan}}$ denotes the expectation with respect to the Plancherel measure on $\mathcal{Y}_{\mathrm{N}}$. The bound (4.29) now follows by the Robinson-Schensted correspondence. □

## 4.4. Asymptotics of the Conjugacy Classes of the Symmetric Group

In this section, we prove a basic limit theorem concerning the asymptotic behavior of conjugacy classes of the symmetric group $S_n$ equipped with the uniform measure as $n \to \infty$. In particular, we derive the distribution for the normalized length of the largest cycle in the cycle decomposition of a random permutation. This is another way (cf. Chapter 10 for yet other ways) of placing a measure on Young diagrams: one simply pushes the uniform measure on the symmetric group $S_n$ forward to a measure on cycle types which, via the one-to-one correspondence with partitions of $n$, induces a measure on Young diagrams. The results below are due to Vershik and Šmidt, and we follow their paper [**VŠ77**].

Let $\Delta_n \equiv \{(x_1, \ldots, x_n) \in [0,1]^n : \sum_{i=1}^n x_i \le 1\}$. Consider the family of mappings $T_n : [0,1]^n \to \Delta_n$ defined by

$$T_n(y_1, \ldots, y_n) = \left( y_1, y_2(1-y_1), \ldots, y_n \prod_{j=1}^{n-1} (1-y_j) \right). \tag{4.54}$$

These mappings have right inverses, which we denote by $T_n^{-1} : \Delta_n \to [0,1]^n$, and are defined as follows. For $x = (x_1, \ldots, x_n) \in \Delta_n$,

$$T_n^{-1}(x) \equiv \left( x_1, \frac{x_2}{1-x_1}, \ldots, \frac{x_n}{1-x_1-\cdots-x_{n-1}} \right) \tag{4.55}$$

if $x_1 + \cdots + x_{n-1} < 1$, and

$$T_n^{-1}(x) \equiv \left( x_1, \frac{x_2}{1-x_1}, \ldots, \frac{x_k}{1-x_1-\cdots-x_{k-1}}, 1, \ldots, 1 \right) \tag{4.56}$$

if $x_1 + \cdots + x_{k-1} < 1$, $x_1 + \cdots + x_k = 1$ for some $k < n$. One easily checks that $T_n \circ T_n^{-1} = \mathrm{id}_{\Delta_n}$, $T_n^{-1} \circ T_n$ is the identity on $[0,1)^{n-1} \times [0,1]$, and $T_n^{-1}$ is measurable.

Let $\mathcal{G}_n$ be the uniform probability measure on the symmetric group $S_n$. For $g \in S_n$, we define $x_1(g)$ to be the normalized length of the cycle containing 1 in the cycle decomposition of $g$ (by normalized length, we mean the length divided by $n$); we define $x_2(g)$ to be the normalized length of the cycle containing the smallest element not contained in the cycle which contains 1, and $x_i(g)$, $i \in \{3, \ldots, n\}$, are defined similarly with the convention that $x_i(g) = 0$ if $i$ is greater than the number of cycles in the cycle decomposition of $g$. Note that $C^n(g) \equiv (x_1(g), \ldots, x_n(g)) \in \Delta_n$. Let $\mu_n = C_*^n \mathcal{G}_n$ be the pushforward of the measure $\mathcal{G}_n$ by the mapping $C^n : S_n \to \Delta_n$.

For integers $r_1, \ldots, r_m \ge 1$ which satisfy $r_1 + \cdots + r_m \le n$, it is easy to see that

$$\begin{aligned} &\mathcal{G}_n \left\{ g \in S_n : x_1(g) = \frac{r_1}{n}, \ldots, x_m(g) = \frac{r_m}{n} \right\} \\ &\qquad = \frac{1}{n(n-r_1)\cdots(n-r_1-\cdots-r_{m-1})}. \end{aligned} \tag{4.57}$$

Indeed, the number of permutations $g \in S_n$ for which $x_1(g) = \frac{r_1}{n}$, $x_2(g) = \frac{r_2}{n}, \ldots, x_m(g) = \frac{r_m}{n}$ is given by

$$\begin{aligned}
&[(n-1)(n-2)\cdots(n-r_1+1)] \\
&\times [(n-r_1-1)\cdots(n-r_1-r_2+1)] \\
&\times \cdots \times [(n-(r_1+\cdots+r_{m-1})-1)\cdots(n-(r_1+\cdots+r_m)+1)] \\
&\times (n-(r_1+\cdots+r_m))!.
\end{aligned}$$

Dividing by $n!$ we arrive at equation (4.57). The right-hand side of (4.57) does not depend on the last constraint $r_m$. This fact will be used in an essential way in the analysis that follows. In particular, if $m = 1$, the probability that $x_1(g) = r$ is $\frac{1}{n}$. Thus, all values of $r$ are equally likely. In light of this fact, it is natural to expect that, in the scaling limit as $n \to \infty$, the infinite dimensional Lebesgue measure on $[0,1]^{\mathbb{N}}$ will play a key role.

**Exercise 4.17.** Show that for any fixed $m \in \mathbb{N}$, the probability that a random permutation picked from $S_n$ has less than $m$ cycles tends to 0 as $n \to \infty$. In particular, this probability is $O\left(\frac{(\log(n))^{m-1}}{n}\right)$.

Let $P_m^n$ be the projection of $[0,1]^n$ onto its first $m$ coordinates. First, we show that

$$\lim_{n\to\infty} (P_m^n \circ T_n^{-1})_* \mu_n([a_1,b_1] \times \cdots \times [a_m,b_m]) = \prod_{j=1}^{m} (b_j - a_j). \tag{4.58}$$

Let $a_i, b_i \in [0,1]$, $1 \le i \le m$, and suppose that $a_i \le b_i$ for all $i$. Define $R_n(a_1, b_1, \ldots, a_m, b_m)$ to be the set of $(r_1, \ldots, r_m) \in \mathbb{N}^m$ satisfying

$$\begin{aligned}
&r_1 \in [na_1, nb_1], \\
&r_2 \in [(n-r_1)a_2, (n-r_1)b_2], \ldots, \\
&r_m \in \left[(n - \sum_{i=1}^{m-1} r_i)a_m, (n - \sum_{i=1}^{m-1} r_i)b_m\right], \\
&\sum_{i=1}^{m} r_i \le n.
\end{aligned} \tag{4.59}$$

Now,

(4.60)

$$\begin{aligned}
&(P^n_m \circ T_n^{-1})_* \mu_n([a_1,b_1]\times\cdots\times[a_m,b_m]) \\
&\quad = \mathcal{G}_n\{g\in S_n : P^n_m\circ T_n^{-1}(C^n(g)) \in [a_1,b_1]\times\cdots\times[a_m,b_m]\} \\
&\quad = \mathcal{G}_n\,\{g \in S_n : g \text{ has at least } m \text{ cycles}, \\
&\qquad\quad P^n_m\circ T_n^{-1}(C^n(g)) \in [a_1,b_1]\times\cdots\times[a_m,b_m]\} + O\Big(\frac{(\log n)^{m-1}}{n}\Big) \\
&\quad = \sum_{R_n(a_1,\ldots,b_m)} \frac{1}{n(n-r_1)\cdots(n-r_1-\cdots-r_{m-1})} + O\Big(\frac{(\log n)^{m-1}}{n}\Big),
\end{aligned}$$

where the estimate in the third line uses Exercise 4.17. Since the summand does not depend on $r_m$, we can rewrite the right-hand side of the last line as

$$\begin{aligned}
&\sum_{R(a_1,\ldots,b_{m-1})} \frac{(b_m-a_m)(n-r_1-\cdots-r_{m-1})+O(1)}{n(n-r_1)\cdots(n-r_1-\cdots-r_{m-1})} \\
&\qquad = (b_m-a_m)\sum_{R(a_1,\ldots,b_{m-1})}\frac{1}{n(n-r_1)\cdots(n-r_1-\cdots-r_{m-2})} \\
&\qquad\quad + O\left(\frac{(\log n)^{m-1}}{n}\right),
\end{aligned}$$

where the error bound is obtained by estimating Riemann sums and noting that $r_1+\cdots+r_{m-1}\le n-1$ (since $r_m\ge 1$). The remaining sum is of the same form as the sum in the right-hand side of (4.60) except that there is one less index. Thus (4.58) is proved.

It is convenient to imbed the finite sequence valued random variables $A_n(g) = (x_1(g),\ldots,x_n(g))$ into one fixed infinite dimensional sequence space in which convergence to a natural infinite sequence valued process can be proven. Consider the space $Q=[0,1]^{\mathbb{N}}$ endowed with the product topology. $Q$ is complete, compact, separable, and metrizable: For example, the metric $d(x,y) = \sum_{i=1}^\infty \frac{1}{2^i}|x_i-y_i|$ on $Q$ induces the product topology. Let $\Delta = \{\{x\}\in Q : \sum_{j=1}^\infty x_i \le 1\}$, and define the continuous map $T: Q\to\Delta$ by

$$T(y) = \Big(y_1, y_2(1-y_1),\ldots, y_j\prod_{l=1}^{j-1}(1-y_j),\ldots\Big). \tag{4.61}$$

$T$ has a measurable right inverse $T^{-1}:\Delta\to Q$ defined in a similar manner to (4.55) and (4.56):

$$T^{-1}(x) = \Big(x_1, \frac{x_2}{1-x_1},\ldots,\frac{x_k}{1-x_1-\cdots-x_{k-1}},\ldots\Big) \tag{4.62}$$

if $x_1 + \cdots + x_q < 1$ for all $q \geq 1$, and

$$(4.63)\quad T^{-1}(x) = \left(x_1, \frac{x_2}{1-x_2}, \ldots, \frac{x_k}{1-x_1-\cdots-x_{k-1}}, 1, 1, \ldots\right)$$

if $x_1 + \cdots + x_{k-1} < 1$, $x_1 + \cdots + x_k = 1$ for some $k \geq 1$.

Let $\mathcal{L}$ be infinite dimensional Lebesgue measure on $Q$, and let $\mu = T_*\mathcal{L}$ be the pushforward of this measure onto $\Delta$. Define $X_n : S_n \to \Delta$ to be the $\Delta$-valued process

$$(4.64)\qquad X_n(g) = (x_1(g), \ldots, x_n(g), 0, \ldots).$$

We denote by $\mu_n$ the measure induced on $\Delta$ by $X_n$. We have already shown that the finite dimensional distributions of the $Q$-valued random variables $T^{-1}(X_n)$ converge to Lebesgue measure on finite dimensional cubes, $[0,1]^m$. Since all spaces are compact, tightness of the sequence of measures $T_*^{-1}\mu_n$ is automatic and the convergence of the finite dimensional distributions is sufficient to prove the weak convergence of $T^{-1}(X_n)$ to $\mathcal{L}$ on $Q$. Furthermore, the continuity of $T$ implies the weak convergence of $\mu_n$ to $\mu$ [**Bil99**].

We need to establish one last technical detail, viz., $\max_i X_n^{(i)} \to \max_i X^{(i)}$ in distribution. Here, $X$ is the $\Delta$-valued random variable with distribution $\mu$, and the superscripts denote the coordinates of these random variables. The convergence in distribution does not follow immediately from the above discussion because the maximum is not a continuous function on $Q$ or $\Delta$. We delay proving this last point until the end of this section, in order to give priority to the next calculation, which is the heart of the matter.

Using the facts established above, we can derive a functional equation for the asymptotic distribution of the normalized cycle length. Denote by $f(a) = \mu\{\max_{i\geq 1} x_i < a\}$ the distribution function of the maximal cycle length and observe that

$$(4.65)\quad \begin{aligned} f(a) &= \mathcal{L}\{y_1 < a,\ \ y_i \prod_{j=1}^{i-1}(1-y_j) < a,\ \ i = 2, 3, \ldots\} \\ &= \int_0^a \mathcal{L}\{y_1 = b,\ \ y_i \prod_{j=1}^{i-1}(1-y_j) < a,\ \ i = 2, 3, \ldots\}db \\ &= \int_0^a \mathcal{L}\{y_2 < \frac{a}{1-b},\ \ y_i \prod_{j=2}^{i-1}(1-y_j) < \frac{a}{1-b},\ \ i = 3, 4, \ldots\}db \\ &= \int_0^a f\left(\frac{a}{1-b}\right)db. \end{aligned}$$

Changing variables, we obtain

$$\frac{f(a)}{a} = \int_a^{\frac{a}{1-a}} \frac{f(b)}{b^2} db. \tag{4.66}$$

Differentiating both sides with respect to $a$ and denoting the density of $f$ by $p$ (note that (4.66) shows a priori that $f(a)$ is absolutely continuous), we find that

$$xp(x) = \int_0^{\frac{x}{1-x}} p(y)dy = 1 - \int_{\frac{x}{1-x}}^{\infty} p(y)dy. \tag{4.67}$$

In the last equality we used the fact that $p$ is supported on the positive axis. The fact that $p$ is actually supported on $[0,1]$ allows us to iteratively compute the density explicitly. If one knows $p$ in the interval $[\frac{1}{m}, \infty]$, then one can compute it in the interval $[\frac{1}{m+1}, \frac{1}{m}]$. We summarize by writing the expression for the functional equation of Vershik and Šmidt [**VŠ77**]:

$$p(x) = \frac{1}{x}\left(1 - \int_{\frac{x}{1-x}}^{\infty} p(y)dy\right). \tag{4.68}$$

We record the following basic facts about $p$: $p(x) = 0$ for $x \notin [0,1]$, $p(x) = \frac{1}{x}$ for $x \in [\frac{1}{2}, 1]$, $p(x) = \frac{1}{x}(1 + \log x - \log(1-x))$ for $x \in [\frac{1}{3}, \frac{1}{2}]$, and so on. Many more interesting properties of $p(x)$ can be found in [**VŠ77**].

Finally, we address the remaining technical point mentioned above. We recall the Skorohod representation theorem (see, for example, [**Dur96**]). Suppose $\Omega$ is a probability space, $\mathcal{Y}$ is a Polish space (a complete separable metric space), $X : \Omega \to \mathcal{Y}$ is a $\mathcal{Y}$-valued random variable, and $\{\Omega_n\}$ is a sequence of probability spaces each supporting a $\mathcal{Y}$-valued random variable $X_n$. If $X_n \to X$ in distribution, the Skorohod representation theorem asserts that there is a single probability space $(\tilde{\Omega}, \mathcal{F}, \mathbb{P})$ and a sequence of $\mathcal{Y}$-valued random variables $\{Y_n\}$ defined on $\tilde{\Omega}$ and a single random variable $Y : \tilde{\Omega} \to \mathcal{Y}$ such that $Y_n \to Y$ $\mathbb{P}$-almost surely and $\{Y_n\}, Y$ are distributed as $\{X_n\}, X$.

The Skorohod representation theorem applied to $\mathcal{Y} = \Delta$ allows us to compare $\mu_n$ and $\mu$ in terms of the almost sure convergence of a sequence of random variables on a new probability space. We note a basic fact about the measure $\mu$: $\mu\{\sum x_i = 1\} = 1$.

**Exercise 4.18.** Prove this fact. *Hint*: To show that $\mu\{\sum x_i = 1\} = 1$, we need only show that $y_1 + \sum_{i=2}^{\infty} y_i \prod_{j=1}^{i-1}(1-y_i) \le 1$ almost surely and that $\mathbb{E}\left[y_1 + \sum_{i=2}^{\infty} y_i \prod_{j=1}^{i-1}(1-y_i)\right] = 1$ where the $\{y_i\}$ are independent identically distributed random variables, uniform on $[0,1]$.

Combined with the fact that $\mu\{\sum x_i \le 1\} = 1$, almost sure convergence of the sequence of random variables on the new probability space implies the almost sure convergence of $\max x_i$. This, in turn, implies convergence in distribution of the maximum of the original random variables and completes the proof of the following basic result in [**VŠ77**].

**Proposition 4.19.** *Let $x_{\max}(N)$ be the maximum cycle length of a random permutation $\pi \in S_N$ (with respect to uniform measure). Then*

$$\lim_{N\to\infty} \mathbb{P}\left\{\frac{x_{\max}(N)}{N} \le x\right\} = \int_0^x p(y)dy, \tag{4.69}$$

*where $p(y)$ is the density function satisfying the functional equation* (4.68).

We see that the measure, $\mathbb{P}_N^{\text{conj}}$, induced on the Young diagrams by the pushforward of uniform measure onto the conjugacy classes (which naturally give rise to partitions of $N$) is very different from Plancherel measure. In particular, let $\lambda_1(N)$ be the number of boxes in the first row of a Young diagram of size $N$ and $\mathbb{P}_N^{\text{Plan}}$ be Plancherel measure on Young diagrams of size $N$. Then, as we have seen (see (1.26)),

$$\lim_{N\to\infty} \mathbb{P}_N^{\text{Plan}}\left\{\frac{\lambda_1(N) - 2\sqrt{N}}{N^{1/6}} \le x\right\} = F(x) \tag{4.70}$$

and

$$\lim_{N\to\infty} \mathbb{P}_N^{\text{conj}}\left\{\frac{\lambda_1(N)}{N} \le x\right\} = \int_0^x p(y)dy. \tag{4.71}$$

Another important qualitative difference is that, unlike the case of Plancherel measure, $\mathbb{E}_{\mathbb{P}_N^{\text{conj}}}[\lambda_1(N)]$ grows as $N\int_0^1 xp(x)dx$. There are important qualitative differences between the shapes of $\mathbb{P}_N^{\text{Plan}}$-random Young diagrams and $\mathbb{P}_N^{\text{conj}}$-random Young diagrams. In the case of Plancherel measure, the distribution of the length of the first row is the same as that for the number of rows, so $\mathbb{E}_N^{\text{Plan}}[\lambda_1(N)] = \mathbb{E}_N^{\text{Plan}}[\#\text{ of rows }] \sim 2\sqrt{N}$ as $N \to \infty$. We now show that $\mathbb{E}_N^{\text{conj}}[\#\text{of rows}] \sim \log(N)$. Indeed, let $r_N = r_N(\lambda)$ be the number of rows corresponding to a Young diagram $\lambda$ of size $N$. As $r_N(\lambda)$ equals the number of cycles for permutations corresponding to $\lambda$, we have, using the explicit form of (4.57),

$$\begin{aligned}
&\mathbb{E}_N^{\text{conj}}[r_N] \\
&= \sum_{n=1}^{N} \mathbb{P}_N^{\text{conj}}\{\text{cycle containing 1 has length } n\}\left(\mathbb{E}_{N-n}^{\text{conj}}[r_{N-n}] + 1\right) \\
&= \frac{1}{N} + \sum_{n=1}^{N-1} \frac{1}{N}\left(\mathbb{E}_{N-n}^{\text{conj}}[r_{N-n}] + 1\right) = 1 + \frac{1}{N}\sum_{n=1}^{N-1} \mathbb{E}_n^{\text{conj}}[r_n].
\end{aligned} \tag{4.72}$$

By induction (to be done as an exercise), we find that

$$\mathbb{E}_N^{\mathrm{conj}}[r_N] = \sum_{k=1}^{N} \frac{1}{k} \sim \log(N) \text{ as } N \to \infty. \tag{4.73}$$

Thus, $\mathbb{P}_N^{\mathrm{Plan}}$-random Young diagrams are as deep as they are long, whereas $\mathbb{P}_N^{\mathrm{conj}}$-random Young diagrams are shallow and long.

**Remark 4.20.** The distribution $f(a) = \mu\{\max_{i\geq 1} x_i < a\}$ has an interesting relation to number theory. Indeed, $f(a) = \rho\left(\frac{1}{a}\right), a > 0$, where $\rho(x)$ is the Dickman distribution [**Dic30**]. In [**Dic30**], Dickman proved the following result. Let $\#(n)$ denote the largest prime factor of $n$. For any positive integer $N$ assume that $n \in \{1, 2, \ldots, N\}$ is uniformly distributed. Then, for $a > 0$,

$$\lim_{N\to\infty} \mathbb{P}\left\{n \in \{1, 2, \ldots, N\} : \#(n) \leq n^{1/a}\right\} = \rho(a). \tag{4.74}$$

We note that the Dickman distribution is also precisely the distribution of the largest coordinate in the Poisson-Dirichlet process [**Bil99**].

Chapter 5

# Orthogonal Polynomials, Riemann-Hilbert Problems, and Toeplitz Matrices

In this chapter, we begin our study of orthogonal polynomials (OP's). We are primarily interested in two kinds of orthogonal polynomials, those that are orthogonal with respect to a weight on the line and those that are orthogonal with respect to a weight on the circle. We also give an introduction to Riemann-Hilbert problems and discuss some properties of Toeplitz matrices. The material in some sections follows the presentation in [**Dei99b**], and in such cases, we provide few proofs.

## 5.1. Orthogonal Polynomials on the Real Line (OPRL)

Let $d\mu(x)$ be a (positive) measure on the line. We always assume that $d\mu$ has finite moments of all orders:

$$\int_{\mathbb{R}} |x|^m d\mu(x) < \infty, \qquad m = 0, 1, 2, \ldots. \tag{5.1}$$

Applying the Gram-Schmidt process in $L^2(\mathbb{R}, d\mu)$ to the monomials $1$, $x$, $x^2$, $\ldots$, we obtain the *monic orthogonal polynomials*

$$\pi_n(x) = x^n + \cdots, \qquad n = 0, 1, 2, \ldots, \tag{5.2}$$

satisfying

$$\int_{\mathbb{R}} \pi_n(x)\pi_m(x)d\mu(x) = 0, \qquad \text{if } n \neq m. \tag{5.3}$$

The $\pi_n$'s are unique. Let

$$p_n(x) = \gamma_n \pi_n(x) = \gamma_n x^n + \cdots, \qquad \gamma_n > 0 \tag{5.4}$$

be the associated *orthonormal polynomials*

$$\int_{\mathbb{R}} p_n(x)p_m(x)d\mu(x) = \delta_{nm}. \tag{5.5}$$

In all that follows, unless stated otherwise, we will assume that our measures are supported on an infinite set of points. This will imply, in particular, that the set of orthogonal polynomials, $p_0, p_1, p_2, \ldots$, is infinite. Some standard references for *orthogonal polynomials* (OP's) are [**Sze75, Dei99b, KS, Sim05a, Sim05b**].

We note two formulae for the $p_n$'s. Let

$$\begin{aligned} H_n &= \det\left(\int s^{i+j}d\mu(s)\right)_{0\le i,j\le n} \\ &= \det\begin{pmatrix} \int s^0 d\mu(s) & \int s^1 d\mu(s) & \cdots & \int s^n d\mu(s) \\ \int s^1 d\mu(s) & \int s^2 d\mu(s) & \cdots & \int s^{n+1} d\mu(s) \\ \vdots & \vdots & \ddots & \vdots \\ \int s^n d\mu(s) & \int s^{n+1} d\mu(s) & \cdots & \int s^{2n} d\mu(s) \end{pmatrix} \end{aligned} \tag{5.6}$$

be the *Hankel determinant* associated with $d\mu$. $H_n$ can be expressed as a multi-integral

$$H_n = \frac{1}{(n+1)!}\int\cdots\int \prod_{0\le i<j\le n}(s_i - s_j)^2 d\mu(s_0)\cdots d\mu(s_n), \tag{5.7}$$

which is called the *Heine formula* for the Hankel determinant $H_n$. From (5.7) we see that

$$H_n > 0. \tag{5.8}$$

**Exercise 5.1.** Prove (5.7) using Andréief's determinantal identity (see Lemma 6.2 below), and the fact that $\prod_{0\le i<j\le n}(s_j - s_i)$ equals the Vandermonde determinant

$$\begin{aligned} V(s_0, s_1, \ldots, s_n) &= \det\begin{pmatrix} 1 & s_0 & \cdots & s_0^{n-1} & s_0^n \\ \vdots & \vdots & \ddots & \vdots & \vdots \\ 1 & s_n & \cdots & s_n^{n-1} & s_n^n \end{pmatrix} \\ &= \det(s_j^i)_{0\le i,j\le n} = \det(s_i^j)_{0\le i,j\le n}. \end{aligned} \tag{5.9}$$

We sometimes also use the notation $\Delta(s_0, \ldots, s_n)$ for $V(s_0, \ldots, s_n)$.

Set

$$H_n(x) \equiv \det \begin{pmatrix} \int s^0 d\mu & \int s^1 d\mu & \cdots & \int s^n d\mu \\ \int s^1 d\mu & \int s^2 d\mu & \cdots & \int s^{n+1} d\mu \\ \vdots & \vdots & \ddots & \vdots \\ \int s^{n-1} d\mu & \int s^n d\mu & \cdots & \int s^{2n-1} d\mu \\ 1 & x & \cdots & x^n \end{pmatrix}. \tag{5.10}$$

A simple calculation shows that $H_n(x)$'s satisfy the orthogonality conditions (5.3) and

$$p_n(x) = \frac{1}{\sqrt{H_{n-1}H_n}} H_n(x) = \sqrt{\frac{H_{n-1}}{H_n}}(x^n + \cdots). \tag{5.11}$$

Thus, the leading coefficient of $p_n(x)$ is

$$\gamma_n = \sqrt{\frac{H_{n-1}}{H_n}}. \tag{5.12}$$

The function $H_n(x)$ can also be written as a multi-integral as follows (to be done as an exercise)

$$\begin{aligned} p_n(x) &= \frac{1}{n!\sqrt{H_{n-1}H_n}} \int \cdots \int \prod_{i=0}^{n-1}(x - s_i) \\ &\times \prod_{0 \le i < j \le n-1} (s_i - s_j)^2 d\mu(s_0) \cdots d\mu(s_{n-1}). \end{aligned} \tag{5.13}$$

**Remark 5.2.** In the language of statistical mechanics, formula (5.7) is a *partition function* representation of $H_n$. Indeed, if $d\mu(x) = e^{-Q(x)}dx$ for some polynomial $Q(x) = a_{2m}x^{2m} + \cdots$, where $a_{2m} > 0$, we can write (5.7) as

$$\begin{aligned} H_n &= \frac{1}{(n+1)!} \int_{\mathbb{R}} \cdots \int_{\mathbb{R}} e^{-\sum_i Q(x_i)} e^{\sum_{i \neq j} \log|x_i - x_j|} dx_0 \cdots dx_n \\ &= \frac{1}{(n+1)!} \int \cdots \int e^{-P(x_0, \ldots, x_n)} dx_0 \cdots dx_n, \end{aligned} \tag{5.14}$$

where $P(x_0, \ldots, x_n)$ is the *Hamiltonian*

$$-\sum_{i \neq j} \log|x_i - x_j| + \sum_{i=1}^{n} Q(x_i). \tag{5.15}$$

We expect that, as $n \to \infty$, the leading order contribution to $H_n$ comes from the critical point(s) of $P$. Precise asymptotic information to all orders for $H_n, p_n, \pi_n, \gamma_n, \ldots$, however, is obtained from the Riemann-Hilbert representation of the orthogonal polynomials (see [**Dei99b**]).

For a given set of orthogonal polynomials $\{p_j\}_{j=0}^{\infty}$ with respect to a measure on the real line, there exits a sequence of numbers $\{a_i, b_i\}_{i=0}^{\infty}$, $a_i, b_i \in \mathbb{R}$, $b_j > 0$, such that the *three-term recurrence relation* holds (to be done as an exercise):

$$\begin{cases} b_{j-1}p_{j-1}(x) + a_j p_j(x) + b_j p_{j+1}(x) = x p_j(x), & j \geq 1, \\ a_0 p_0(x) + b_0 p_1(x) = x p_0, & j = 0. \end{cases} \tag{5.16}$$

Thus, associated with orthonormal polynomials, there is a *Jacobi operator*

$$J = \begin{pmatrix} a_0 & b_0 & & & 0 \\ b_0 & a_1 & b_1 & & \\ & b_1 & a_2 & b_2 & \\ & & b_2 & \ddots & \\ 0 & & & & \end{pmatrix} = J^T, \tag{5.17}$$

for which

$$J \begin{pmatrix} p_0(x) \\ p_1(x) \\ p_2(x) \\ \vdots \end{pmatrix} = x \begin{pmatrix} p_0(x) \\ p_1(x) \\ p_2(x) \\ \vdots \end{pmatrix}. \tag{5.18}$$

See [**Dei99b**] for the inverse problem of how to reconstruct the measure $d\mu$, given the Jacobi operator $J$.

From (5.16), we find

$$b_n = \frac{\gamma_n}{\gamma_{n+1}} = \sqrt{\frac{H_{n-1}H_{n+1}}{H_n^2}}. \tag{5.19}$$

There is an analogous formulae for $a_n$ (to be done as an exercise).

From (5.16), we have for any $n$,

$$\begin{pmatrix} a_0 - x & b_0 & & & 0 \\ b_0 & a_1 - x & b_1 & & \\ & \ddots & \ddots & \ddots & \\ & & b_{n-2} & a_{n-1} - x & b_{n-1} \\ 0 & & & b_{n-1} & a_n - x \end{pmatrix} \begin{pmatrix} p_0(x) \\ p_1(x) \\ \vdots \\ p_{n-1}(x) \\ p_n(x) \end{pmatrix} \tag{5.20}$$

$$= \begin{pmatrix} 0 \\ 0 \\ \vdots \\ 0 \\ -b_n p_{n+1}(x) \end{pmatrix}.$$

Hence, $p_{n+1}(\hat{x}) = 0$ if and only if $\hat{x}$ is an eigenvalue of the real symmetric matrix $J$. In particular the zeros of $p_{n+1}$ are all real. We leave it as an exercise to show that all zeros are distinct, and the zeros of $p_{n+1}$ and the zeros of $p_n$ interlace.

An important consequence of the three-term recurrence relation (5.16) is the *Christoffel-Darboux formula*

$$\sum_{i=0}^{n-1} p_i(x)p_i(y) = \frac{\gamma_{n-1}}{\gamma_n}\frac{p_n(x)p_{n-1}(y) - p_{n-1}(x)p_n(y)}{x-y}, \tag{5.21}$$

whose proof is left as an exercise.

## 5.2. Some Classical Orthogonal Polynomials

The following examples are from [**Sze75**, §2.4]. See also [**KS**] for a list of special orthogonal polynomials and their properties.

Here, $d\mu = w(x)dx$ is supported on an interval $[a,b]$ where $-\infty \le a < b \le \infty$. In each case, the polynomials below are proportional to the orthonormal polynomials with respect to the given weights.

- Jacobi polynomials $p_n^{(\alpha,\beta)}$: for $\alpha > -1$, $\beta > -1$,

$$w = (1-x)^\alpha(1+x)^\beta, \qquad a = -1, \quad b = 1. \tag{5.22}$$

- Laguerre polynomials $L_n^{(\alpha)}(x)$: for $\alpha > -1$,

$$w(x) = e^{-x}x^\alpha, \qquad a = 0, \quad b = \infty. \tag{5.23}$$

- Hermite polynomials $H_n(x)$:

$$w(x) = e^{-x^2}, \qquad a = -\infty, \quad b = \infty. \tag{5.24}$$

When $\alpha = \beta$, Jacobi polynomials are called Gegenbauer polynomials or ultraspherical polynomials. In particular, ultraspherical polynomials become

- Tchebichef (Chebychev) polynomials of the first kind, $T_n(x) = \cos n\theta$, $x = \cos\theta$, when $\alpha = \beta = -\frac{1}{2}$,
- Tchebichef (Chebychev) polynomials of the second kind, $U_n(x) = \frac{\sin(n+1)\theta}{\sin\theta}$, $x = \cos\theta$, when $\alpha = \beta = +\frac{1}{2}$,
- Legendre polynomials, or spherical polynomials, $P_n(x)$ when $\alpha = \beta = 0$.

Some classical orthogonal polynomials with respect to discrete measures (sometimes referred to as discrete orthogonal polynomials) can also be found in [**Sze75**, §2.8].

- Poisson-Charlier polynomials:

$$d\mu(x) = \sum_{j=0}^{\infty} \frac{e^{-a}a^j}{j!}\delta(x-j), \qquad a > 0. \tag{5.25}$$

- Meixner polynomials:

$$d\mu(x) = \sum_{j=0}^{\infty} \binom{j+k-1}{j} q^j \delta(x-j). \tag{5.26}$$

- Krawtchouk polynomials: $0 < p, q < 1$, $p + q = 1$,

$$d\mu(x) = \sum_{j=0}^{N} \binom{N}{j} p^j q^{N-j} \delta(x-j). \tag{5.27}$$

- Hahn polynomials: $\alpha > -1, \beta > -1$,

$$d\mu(x) = \sum_{j=0}^{N} \binom{\alpha+j}{j} \binom{\beta+N-j}{N-j} \delta(x-j). \tag{5.28}$$

Quite remarkably, these classical discrete polynomials show up in statistical models of the type we consider in Chapter 12. Indeed, the *Poisson-Charlier* polynomials are related to the length of weakly increasing subsequences in random words [**TW01**], and the *Meixner* polynomials are related to the two-dimensional directed last passage percolation model (see [**Joh01a**] and Subsection 9.3.1 below). The *Krawtchouk* polynomials arise in connection with zig-zag paths in random domino tilings of the Aztec diamond (see Section 12.2 below), in a certain simplified directed first-passage percolation model introduced by Seppäläinen (see [**Joh02**]), and also in non-intersecting random walks [**KOR02**], which are also related to a model (see [**BBDS06**]) for the intriguing random matrix phenomena, observed by Krbálek and Šeba [**KŠ00**], in the operation of the bus transportation system in Cuernavaca, Mexico. The Hahn polynomials arise in the *abc*-hexagon tiling problem [**Joh02**].

## 5.3. The Riemann-Hilbert Problem (RHP) for Orthogonal Polynomials

We now show how to rephrase the orthogonal polynomial problem as a *Riemann-Hilbert problem* (*RHP*), following Fokas, Its, and Kitaev [**FIK91**]. We restrict our attention here to measures $d\mu(z)$ on $\mathbb{R}$ which are absolutely continuous with respect to $dz$, i.e., $d\mu = w(z)dz$, $w(z) \geq 0$.

We first provide some general information on Riemann-Hilbert problems. Basic references are [**CG81, LS87**]; see also [**Dei99b**] and [**DZ03**]. Let $\Sigma$ be an oriented contour in the complex plane $\mathbb{C}$. By convention, as we follow the contour in the direction of the arrows, the $(\pm)$-sides of $\Sigma$ lie to the left/right, as indicated in Figure 5.1. Let $v$ be a map from $\Sigma$ to $G\ell(k)$, the set of $k \times k$ invertible complex-valued matrices, such that $v, v^{-1} \in L^\infty(\Sigma)$: $v$ is called a *jump matrix*. We say that an $\ell \times k$ matrix valued function

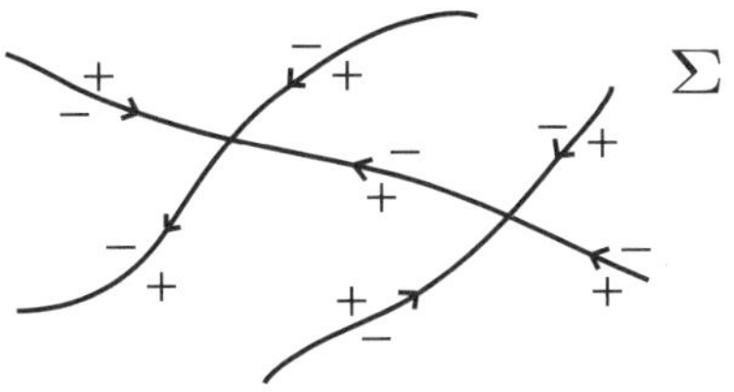


**Figure 5.1.** Contour Σ

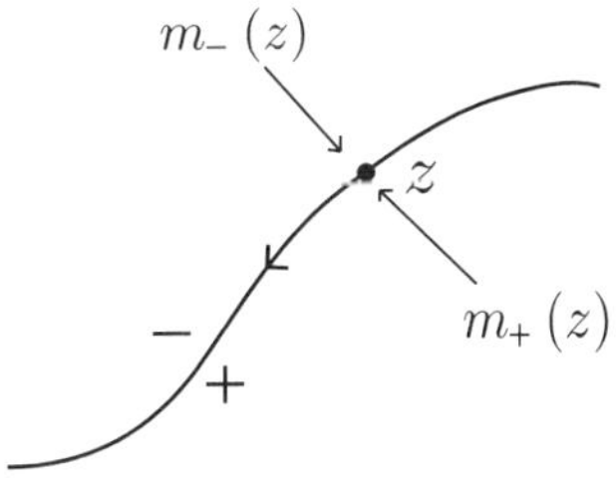


**Figure 5.2.** $m_{\pm}(z)$

$m(z)$, $z \in \mathbb{C} \setminus \Sigma$, *solves* the RHP $(\Sigma, v)$ if

- $m(z)$ is analytic in $\mathbb{C}\backslash\Sigma$,
- $m_+(z) = m_-(z)v(z)$ for $z \in \Sigma$ where (see Figure 5.2)

$$m_{\pm}(z) = \lim_{\substack{z' \to z, \\ z' \in (\pm)\text{-side of } \Sigma}} m(z'). \tag{5.29}$$

If, in addition, $\ell = k$ and

- $m(z) \to I$ as $z \to \infty$, where $I$ is the $k \times k$ identity matrix,

then we say that $m$ solves the *normalized* RHP $(\Sigma, v)$.

The precise sense in which the boundary values are attained, as well as the nature of the limit as $z \to \infty$ in the normalized situation, is a very important technical issue, which we will set aside for the moment, but to which we will return at a later point (see Section 5.5).

Let $w(z) \geq 0$ be a weight on $\mathbb{R}$ for which the associated measure $d\mu(x) = w(z)dz$ has finite moments as before. Let $\Sigma = \mathbb{R}$ be oriented from left to

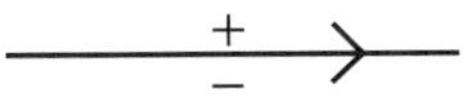


**Figure 5.3.** Contour $\Sigma = \mathbb{R}$

right (see Figure 5.3) and consider the jump matrix

$$v(z) = \begin{pmatrix} 1 & w(z) \\ 0 & 1 \end{pmatrix}, \qquad z \in \mathbb{R}. \tag{5.30}$$

To simplify the presentation that follows, we will assume that, in addition, $w \in \mathcal{S}(\mathbb{R})$, the Schwartz space. In particular $w \in L^\infty(\mathbb{R})$ and hence $v, v^{-1} \in L^\infty(\mathbb{R})$. Fix an integer $n \geq 0$ and consider the following RHP for a $2 \times 2$ matrix valued function $Y(z)$.

**Riemann-Hilbert Problem 5.3** (RHP for OP's on the real line [**FIK91**]). *Fix an integer $n \geq 0$, and let a $2\times 2$ matrix valued function $Y(z)$ satisfy the following conditions:*

- *$Y(z)$ analytic in $\mathbb{C} \setminus \mathbb{R}$,*
- *$Y_+(z) = Y_-(z)v(z)$ for $z \in \mathbb{R}$ where $v$ is given in* (5.30),

*and*

- $Y(z)\begin{pmatrix} z^{-n} & 0 \\ 0 & z^n \end{pmatrix} \to \begin{pmatrix} 1 & 0 \\ 0 & 1 \end{pmatrix} = I$ *as $z \to \infty$.*

*To specify the boundary behavior precisely in* **this** *case (cf. Section* 5.5*), we require that*

- *$Y(z)$ is continuous in $\overline{\mathbb{C}}_+ = \{\operatorname{Im} z \geq 0\}$ and in $\overline{\mathbb{C}}_- = \{\operatorname{Im} z \leq 0\}$ so that the limits $Y_\pm(z)$ exist pointwise, and*
- $Y(z)\begin{pmatrix} z^{-n} & 0 \\ 0 & z^n \end{pmatrix} \to I$ *pointwise as $z \to \infty$ in $\overline{\mathbb{C}}_\pm$, respectively.*

With these requirements, the solution of the RHP is unique.

**Lemma 5.4.** *The solution $Y$ to RHP* 5.3*, if it exists, is unique.*

**Remark 5.5.** The proof below applies to any RHP for which the jump matrix satisfies $\det v(z) \equiv 1$ and for which $\det Y(z) \to 1$ as $z \to \infty$.

**Proof.** Let $Y(z)$ be a solution to the RHP. Clearly, $\det Y(z)$ is analytic in $\mathbb{C}_\pm$ and continuous in $\overline{\mathbb{C}}_\pm$. Also, since $\det v(z) = 1$ for all $z \in \Sigma$,

$$(\det Y)_+ = (\det Y)_- \det v = (\det Y)_- \qquad \text{on } \Sigma. \tag{5.31}$$

Hence, by Morera's theorem, $\det Y(z)$ is analytic in $\mathbb{C}$. On the other hand, as $Y(z)\left(\begin{smallmatrix} z^{-n} & 0 \\ 0 & z^n \end{smallmatrix}\right) \to I$, we have $\det Y(z) \to 1$ as $z \to \infty$. So, $\det Y(z) \equiv 1$ for all $z \in \mathbb{C}$ by Liouville's theorem. Thus, since $Y(z)$ is a $2 \times 2$ matrix, $Y(z)^{-1}$ exists and is analytic for all $z \in \mathbb{C}_\pm$ and is continuous in $\overline{\mathbb{C}}_\pm$.

Suppose that $U(z)$ is another solution to the RHP. Set a new matrix $T(z) \equiv U(z)Y(z)^{-1}$. Then $T(z)$ is analytic in $\mathbb{C}_\pm$, continuous in $\overline{\mathbb{C}}_\pm$, and for $z \in \mathbb{R}$, $T_+(z) = U_+(z)Y_+(z)^{-1} = (U_-(z)v(z))(Y_{-1}(z)v(z))^{-1} = U_-(z)Y_-(z)^{-1} = T_-(z)$. Hence again by Morera's theorem, we see that

$T(z)$ is analytic in all of $\mathbb{C}$. But as $U(z)\left(\begin{smallmatrix} z^{-n} & 0 \\ 0 & z^n \end{smallmatrix}\right), Y(z)\left(\begin{smallmatrix} z^{-n} & 0 \\ 0 & z^n \end{smallmatrix}\right) \to I$ as $z \to \infty$, we conclude that $T(z) \equiv I$, i.e., $U(z) = Y(z)$. So the solution is unique, if it exists. □

Now we construct the solution $Y$ explicitly. Suppose first that the solution $Y$ exists, and write

$$Y(z) = \begin{pmatrix} Y_{11}(z) & Y_{12}(z) \\ Y_{21}(z) & Y_{22}(z) \end{pmatrix}. \tag{5.32}$$

Consider the first row of (5.32). On $\mathbb{R}$, the jump condition $Y_+ = Y_- v$ implies that

$$\begin{pmatrix} Y_{11} & Y_{12} \end{pmatrix}_+ = \begin{pmatrix} Y_{11} & Y_{12} \end{pmatrix}_- \begin{pmatrix} 1 & w \\ 0 & 1 \end{pmatrix}. \tag{5.33}$$

Hence $Y_{11+} = Y_{11-}$. As before we conclude that $Y_{11}(z)$ is analytic in $\mathbb{C}$. But from

$$\begin{pmatrix} Y_{11} & Y_{12} \end{pmatrix} \begin{pmatrix} z^{-n} & 0 \\ 0 & z^n \end{pmatrix} = \begin{pmatrix} Y_{11} z^{-n} & Y_{12} z^n \end{pmatrix} \to \begin{pmatrix} 1 & 0 \end{pmatrix}, \quad z \to \infty, \tag{5.34}$$

we conclude that $Y_{11}(z)$ is a monic polynomial of degree $n$,

$$Y_{11}(z) = z^n + \cdots . \tag{5.35}$$

Now (5.33) implies that

$$Y_{12+} = Y_{12-} + Y_{11-} w, \tag{5.36}$$

i.e.,

$$Y_{12+} = Y_{12-} + Y_{11} w, \qquad \text{where } Y_{11} = Y_{11+} = Y_{11-}. \tag{5.37}$$

This is an example of an additive RHP which can always, as we will show, be solved by the formula

$$Y_{12}(z) = (\mathrm{C}(Y_{11} w))(z), \tag{5.38}$$

where C denotes the *Cauchy operator*

$$(\mathrm{C}\, h)(z) \equiv \frac{1}{2\pi i} \int_\Sigma \frac{h(s)}{s - z} ds. \tag{5.39}$$

We will discuss the mapping and analyticity properties of the Cauchy operator below (see Section 5.5). In this section we just note that for $\Sigma = \mathbb{R}$, if $h \in \mathcal{S}(\mathbb{R})$ (actually $h \in H^1(\mathbb{R}) = \{h : h, h' \in L^2(\mathbb{R})\}$, the first Sobolev space, will do),

$$\begin{cases} (\mathrm{C}\, h)(z) \text{ is analytic in } \mathbb{C} \backslash \mathbb{R}, \\ (\mathrm{C}\, h)(z) \text{ is continuous in } \overline{\mathbb{C}}_\pm, \\ (\mathrm{C}\, h)(z) \to 0 \text{ as } z \to \infty \text{ in } \overline{\mathbb{C}}_\pm. \end{cases} \tag{5.40}$$

**Exercise 5.6.** Prove these facts; see, e.g., [**Dur70**], [**BK97**].

Most importantly, for $z \in \mathbb{R}$,

$$\begin{aligned}(5.41)\qquad &(\mathrm{C}\,h)_+(z) - (\mathrm{C}\,h)_-(z)\\ &= \lim_{\epsilon\downarrow 0}\left\{\int_{\mathbb{R}} \frac{h(s)}{s-z-i\epsilon}\frac{ds}{2\pi i} - \int_{\mathbb{R}} \frac{h(s)}{s-z+i\epsilon}\frac{ds}{2\pi i}\right\}\\ &= \lim_{\epsilon\downarrow 0}\frac{1}{\pi}\int_{\mathbb{R}} h(s)\frac{\epsilon}{(s-z)^2+\epsilon^2}ds\\ &= h(z)\end{aligned}$$

by the familiar property of the Poisson kernel (this can be done as an exercise). Thus, defining

$$(5.42)\qquad \mathrm{C}_{\pm}\,h(z) \equiv \lim_{\substack{z'\to z\\ z'\in(\pm)\text{-side of }\Sigma}} (\mathrm{C}\,h)(z'),$$

we have

$$(5.43)\qquad \mathrm{C}_+ - \mathrm{C}_- = 1.$$

This basic relation holds for all *reasonable* contours $\Sigma$.

Referring to (5.38), let

$$(5.44)\qquad H(z) \equiv (\mathrm{C}(Y_{11}w))(z).$$

We will show that $H = Y_{12}$. By (5.34), $Y_{12}(z) = o(z^{-n}) \to 0$ as $z \to \infty$. As $Y_{11}w \in \mathcal{S}(\mathbb{R})$ (recall that $Y_{11}$ is a polynomial), $H(z)$ is analytic in $\mathbb{C}\backslash\mathbb{R}$, continuous in $\overline{\mathbb{C}}_{\pm}$, and satisfies

$$(5.45)\qquad H_+ - H_- = Y_{11}w.$$

By (5.40) $H(z) \to 0$ as $z \to \infty$ in $\mathbb{C}_{\pm}$. Since $(Y_{12})_+ - (Y_{12})_- = Y_{11}w$, $(H - Y_{12})_+ = (H - Y_{12})_-$. Again by Morera's theorem, $H - Y_{12}$ is entire. As $H(z) - Y_{12}(z) \to 0$ as $z \to \infty$, it follows by Liouville's theorem that

$$(5.46)\qquad H(z) - Y_{12}(z) \equiv 0.$$

Therefore we obtain

$$(5.47)\qquad Y_{12}(z) = \int_{\mathbb{R}} \frac{Y_{11}(s)w(s)}{s-z}\frac{ds}{2\pi i}.$$

Note that in proving (5.47), we only used the fact that $Y_{12} = o(1)$ as $z \to \infty$, not the fact that $Y_{12} = o(z^{-n})$. The later condition implies $n$ *moment conditions* on $Y_{11}$ as follows. Note that

$$\begin{aligned}(5.48)\qquad \frac{1}{z-s} &= \frac{1}{z} + \frac{s}{z}\frac{1}{z-s} = \frac{1}{z} + \frac{s}{z^2} + \frac{s^2}{z^2}\frac{1}{z-s}\\ &= \frac{1}{z} + \frac{s}{z^2} + \cdots + \frac{s^k}{z^{k+1}} + \frac{s^{k+1}}{z^{k+1}}\frac{1}{z-s}\end{aligned}$$

for any integer $k \geq 0$. Inserting this relation into (5.47), we find

(5.49)
$$\begin{aligned}-Y_{12}(z) =&\frac{1}{z}\int Y_{11}(s)w(s)\frac{ds}{2\pi i} + \frac{1}{z^2}\int sY_{11}(s)w(s)\frac{ds}{2\pi i}\\ &+\cdots+\frac{1}{z^n}\int s^{n-1}Y_{11}(s)w(s)\frac{ds}{2\pi i} + \frac{1}{z^n}\int \frac{s^nY_{11}(s)w(s)}{s-z}\frac{ds}{2\pi i}.\end{aligned}$$

As $s^nY_{11}w \in \mathcal{S}(\mathbb{R})$ it follows from (5.40) that $\int \frac{s^nY_{11}(z)w}{s-z}\frac{ds}{2\pi i} \to 0$ as $z \to \infty$ in $\overline{\mathbb{C}}_{\pm}$. Thus the condition $Y_{12} = o(z^{-n})$ yields

$$\int s^jY_{11}(s)w(s)\frac{ds}{2\pi i} = 0, \qquad j = 0, 1, \ldots, n-1. \tag{5.50}$$

Therefore, $Y_{11}(s)$ is a monic polynomial of degree $n$ which is orthogonal to $1$, $s$, $s^2, \ldots, s^{n-1}$ with respect to the measure $w(s)ds$. Hence

$$Y_{11}(z) = \pi_n(z), \tag{5.51}$$

the $n$th monic orthogonal polynomial with respect to $w$.

Analyzing the second row $(Y_{21}, Y_{22})$ in a similar way, we find that $Y_{21}(z)$ is a polynomial of degree at most $n-1$, $Y_{21}(z) = \alpha z^{n-1} + \cdots$. We also find that

$$Y_{22}(z) = \int \frac{Y_{21}(s)w(s)}{s-z}\frac{ds}{2\pi i}. \tag{5.52}$$

The asymptotic condition on $Y(z)$ yields

$$Y_{22}(z)z^n \to 1, \tag{5.53}$$

which implies that

$$\int s^kY_{21}(s)w(s)ds = 0, \qquad k = 0, 1, 2, \ldots, n-2, \tag{5.54}$$

and

$$-\int s^{n-1}Y_{21}(s)w(s)\frac{ds}{2\pi i} = 1. \tag{5.55}$$

From (5.54) we conclude that $Y_{21}(z) = \alpha\pi_{n-1}(z)$, and (5.55) becomes

$$\begin{aligned}1 &= -\alpha\int s^{n-1}\pi_{n-1}w\frac{ds}{2\pi i} = -\alpha\int \pi_{n-1}(s)^2w(s)\frac{ds}{2\pi i}\\ &= -\frac{\alpha}{\gamma_{n-1}^2}\int p_{n-1}(s)^2w(s)\frac{ds}{2\pi i} = -\frac{\alpha}{2\pi i\gamma_{n-1}^2},\end{aligned} \tag{5.56}$$

where $\gamma_k$ is the leading coefficient of the orthonormal polynomial $p_k$; see (5.4). Thus,

$$\alpha = -2\pi i\gamma_{n-1}^2. \tag{5.57}$$

We conclude that if the above RHP has a solution $Y(z)$, then it is necessarily given by (5.58) below. Conversely, it is easy to check directly that the right-hand side of (5.58) gives a solution of the RHP with the right analyticity and continuity properties in $\mathbb{C}_\pm$ and $\overline{\mathbb{C}}_\pm$, respectively. Hence, we obtain the following.

**Theorem 5.7.** *The solution $Y(z)$ to the Riemann-Hilbert Problem 5.3 exists, is unique, and is given by*

$$Y(z) = \begin{pmatrix} \pi_n(z) & (\mathrm{C}(\pi_n w))(z) \\ -2\pi i \gamma_{n-1}^2 \pi_{n-1}(z) & -2\pi i \gamma_{n-1}^2 (\mathrm{C}(\pi_{n-1} w))(z) \end{pmatrix}. \tag{5.58}$$

*Here,* C *denotes the Cauchy operator defined in* (5.39).

**Exercise 5.8.** If we write for $Y = Y^{(n)}$,

$$Y^{(n)}(z) \begin{pmatrix} z^{-n} & 0 \\ 0 & z^n \end{pmatrix} = I + \frac{Y_1^{(n)}}{z} + \frac{Y_2^{(n)}}{z^2} + \cdots \tag{5.59}$$

as $z \to \infty$, we find

$$\begin{aligned} \gamma_{n-1}^2 &= -\frac{1}{2\pi i}(Y_1^{(n)})_{12}, \\ a_n &= (Y_1^{(n)})_{11} - (Y_1^{(n+1)})_{11}, \\ b_{n-1}^2 &= (Y_1^{(n)})_{12}(Y_1^{(n)})_{21}. \end{aligned} \tag{5.60}$$

Thus, all the quantities of interest in the theory of OP's can be expressed directly in terms of the solution of an associated RHP. This means in particular, that if we want to study the asymptotics of $\pi_n(z)$, for example, we may study instead the asymptotics of $Y_{11}(z)$, the (11)-entry of the solution of the RHP as $n \to \infty$. As we will see, this reformulation of the orthogonal polynomial problem for the weight $w$, as an RHP, turns out to be extremely useful.

**Exercise 5.9.** Find an appropriate RHP formulation for the *discrete* orthogonal polynomials (i.e., OP's with respect to a discrete measure). See, for example, [**BKMM07**].

## 5.4. Orthogonal Polynomials on the Unit Circle (OPUC) and Toeplitz Matrices

We now consider polynomials $\{p_n\}$ that are orthogonal with respect to a (positive) measure $d\mu(\theta)$ on the unit circle $\{z \in \mathbb{C} : |z| = 1\}$. Write

$$p_n(z) = \kappa_n z^n + \cdots = \kappa_n \pi_n(z), \quad \kappa_n > 0, \quad n = 0, 1, 2, \ldots, \tag{5.61}$$

where

$$\int_{-\pi}^{\pi} \overline{p_n(e^{i\theta})} p_m(e^{i\theta}) \frac{d\mu(\theta)}{2\pi} = \delta_{n,m}, \quad n, m \geq 0. \tag{5.62}$$

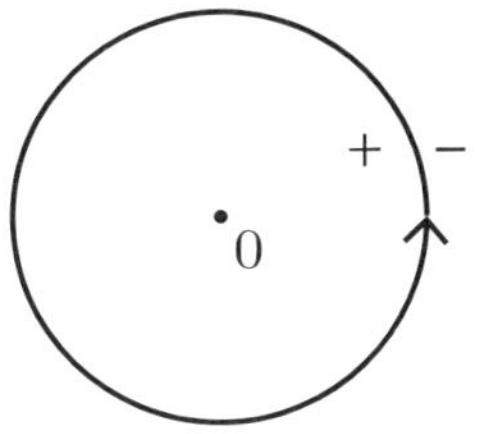


**Figure 5.4.** Contour $\Sigma = \{z \in \mathbb{C} : |z| = 1\}$

Note the complex conjugation on $p_n$. Let

$$\pi_n(z) = \kappa_n^{-1} p_n(z) = z^n + \cdots \tag{5.63}$$

be the associated monic orthogonal polynomials. References for orthogonal polynomials on the unit circle are [**Sze75, Sim05a, Sim05b**].

There are analogues (find them!—see Exercise 5.14 below) of all the preceding formulae for OP's with respect to $d\mu(x)$ on $\mathbb{R}$ for the case of the circle. We are particularly interested in the RHP formulation for these OP's. Let $\Sigma = \{z \in \mathbb{C} : |z| = 1\}$, oriented counterclockwise; see Figure 5.4. We only consider measures $d\mu(\theta) = w(e^{i\theta})\frac{d\theta}{2\pi}$ which are absolutely continuous with respect to Lebesgue measure on the circle.

**Riemann-Hilbert Problem 5.10** (RHP for OP's on the unit circle [**BDJ99**]). *Fix an integer $n \geq 0$. Find $Y(z)$ satisfying the following conditions:*

- *$Y(z)$ is analytic in $\mathbb{C}\backslash\Sigma$;*
- *$Y_+(z) = Y_-(z)v(z)$, $z \in \Sigma$, where $v(z) = \begin{pmatrix} 1 & z^{-n}w(z) \\ 0 & 1 \end{pmatrix}$;*
- *$Y(z)\begin{pmatrix} z^{-n} & 0 \\ 0 & z^n \end{pmatrix} \to I$ as $z \to \infty$.*

Note that the (12)-entry of the jump matrix is now $z^{-n}w$. Under suitable regularity assumptions (for example, $w(z) \in H^1(\Sigma, d\theta)$ is enough) one can prove the following result as before (this can be done as an exercise).

For any polynomial $h$ we define the *reverse polynomial* for $h$ by

$$h^*(z) \equiv z^n \overline{h(1/\bar{z})} = z^n \bar{h}(1/z). \tag{5.64}$$

Clearly, $h^*$ is a polynomial of the same degree as $h$.

**Theorem 5.11** (see, e.g., [**BDJ99**]). *RHP* 5.10 *above has a unique solution given by*

$$Y(z) = \begin{pmatrix} \pi_n(z) & (C\,f)(z) \\ -\kappa_{n-1}^2 \pi_{n-1}^*(z) & -\kappa_{n-1}^2 (C\,g)(z) \end{pmatrix}, \tag{5.65}$$

*where*

$$f(z) = z^{-n}\pi_n(z)w(z), \quad g(z) = z^{-n}\pi^*_{n-1}(z)w(z). \tag{5.66}$$

The following symmetry properties for $Y(z)$ follow immediately from the RHP. Let

$$\sigma_3 = \begin{pmatrix} 1 & 0 \\ 0 & -1 \end{pmatrix} \tag{5.67}$$

be the third Pauli matrix.

**Proposition 5.12** (Symmetry). *Let* $Y = Y(z;n)$ *be as above. Then* $\det Y(z) = 1$ *for all* $z$ *and*

$$Y(z) = \sigma_3 Y(0)^{-1} Y(z^{-1}) \begin{pmatrix} z^n & 0 \\ 0 & z^{-n} \end{pmatrix} \sigma_3, \tag{5.68}$$

$$Y(z) = \overline{Y(\bar{z})}. \tag{5.69}$$

**Proof.** As the jump matrix $v(z)$ for $Y$ has determinant 1, it follows that $\det Y_+(z) = \det Y_-(z)$ for $z \in \Sigma$, and so $\det Y(z)$ is entire. But $\det Y(z) = \det\left(Y(z)\left(\begin{smallmatrix} z^{-n} & 0 \\ 0 & z^n \end{smallmatrix}\right)\right) \to \det(I) = 1$ as $z \to \infty$. By Liouville's theorem we conclude that $\det Y(z) = 1$ for all $z$ as before.

Set

$$H(z) \equiv Y(z^{-1}) \begin{pmatrix} z^n & 0 \\ 0 & z^{-n} \end{pmatrix} \sigma_3, \qquad z \in \mathbb{C} \setminus (\Sigma \cup \{0\}). \tag{5.70}$$

Then a direct calculation shows that $H(z)$ has the same jump across $\Sigma$ as $Y$, $H_+(z) = H_-(z)\left(\begin{smallmatrix} 1 & z^{-n}w(z) \\ 0 & 1 \end{smallmatrix}\right)$. As $z \to 0$, $H(z) \to \sigma_3$, so $H(z)$ has a removable singularity at $z = 0$. On the other hand, as $z \to \infty$, $H(z)\left(\begin{smallmatrix} z^{-n} & 0 \\ 0 & z^n \end{smallmatrix}\right) \to Y(0)\sigma_3$. Thus, $\sigma_3 Y(0)^{-1} H(z)$ satisfies the same RHP as $Y(z)$. Relation (5.68) then follows by uniqueness for the RHP. The proof of (5.69) is similar. □

Letting $z \to 0$ in (5.68), we obtain the relation

$$Y(0) = \sigma_3 Y(0)^{-1} \sigma_3. \tag{5.71}$$

This implies that $Y_{22}(0) = Y_{11}(0) = \pi_n(0)$. On the other hand, since $\pi^*_{n-1}(0) = 1$, we have

$$Y_{21}(0) = -\kappa^2_{n-1}. \tag{5.72}$$

Also (cf. (5.47))

$$\begin{aligned} Y_{12}(0) &= \frac{1}{2\pi i}\int_\Sigma \frac{\pi_n(s)s^{-n}w(s)}{s}\,ds = \int_{-\pi}^{\pi} \pi_n(e^{i\theta})\overline{e^{in\theta}}w(e^{i\theta})\frac{d\theta}{2\pi} \\ &= \frac{1}{\kappa_n^2}\int_{-\pi}^{\pi} p_n(e^{i\theta})\overline{p_n(e^{i\theta})}w(e^{i\theta})\frac{d\theta}{2\pi} = \frac{1}{\kappa_n^2}. \end{aligned} \tag{5.73}$$

Therefore, we obtain:

**Corollary 5.13.** *The following relations hold for $Y(0)$, $n \geq 0$:*

$$Y(0) = \begin{pmatrix} \pi_n(0) & \kappa_n^{-2} \\ -\kappa_{n-1}^2 & \pi_n(0) \end{pmatrix}, \qquad \kappa_{-1} \equiv 0, \tag{5.74}$$

*and*

$$\det Y(0) = \pi_n(0)^2 + \left(\frac{\kappa_{n-1}}{\kappa_n}\right)^2 = 1. \tag{5.75}$$

*Moreover, $\pi_n(z)$ has real coefficients. Thus, $\pi_n(0)$ is real and*

$$|\pi_n(0)| < 1, \qquad n \geq 1. \tag{5.76}$$

Now, set

$$y(z) = (y_1(z), y_2(z)) \equiv \left(1, \kappa_{n-1}^{-2}\pi_n(0)\right) Y(z) \begin{pmatrix} z^{-1} & 0 \\ 0 & 1 \end{pmatrix}, \qquad n \geq 1. \tag{5.77}$$

Direct calculation using (5.74) shows that $y(z)$ is analytic in $\mathbb{C}\setminus\Sigma$, including $z = 0$, and that $y_+(z) = y_-(z) \left(\begin{smallmatrix} 1 & z^{-n+1}w(z) \\ 0 & 1 \end{smallmatrix}\right)$ for $z \in \Sigma$. As $z \to \infty$,

$$\begin{aligned} y(z) \begin{pmatrix} z^{-n+1} & 0 \\ 0 & z^{n-1} \end{pmatrix} &= \left(1, \kappa_{n-1}^{-2}\pi_n(0)\right) Y(z) \begin{pmatrix} z^{-n} & 0 \\ 0 & z^n \end{pmatrix} \begin{pmatrix} 1 & 0 \\ 0 & z^{-1} \end{pmatrix} \\ &\to (1, 0). \end{aligned} \tag{5.78}$$

By uniqueness for the RHP for $\pi_{n-1}(z)$, it follows that

$$y(z) = (1, 0) Y(z; n-1),$$

hence $y_1(z) = \pi_{n-1}(z)$. Using (5.77), this implies the well-known *Szegő recurrence relation*,

$$z\pi_{n-1}(z) = \pi_n(z) + \bar{\alpha}_{n-1}\pi_{n-1}^*(z) \tag{5.79}$$

where

$$\alpha_{n-1} \equiv -\overline{\pi_n(0)}, \qquad n \geq 1, \tag{5.80}$$

is the *Verblunsky coefficient* for the polynomials. The Verblunsky coefficients play a fundamental role in the theory of OPUC's; see [**Sim05a, Sim05b**]. In particular, for measures $d\mu$ on $\Sigma$, the map $d\mu \mapsto (\alpha_0, \alpha_1, \alpha_2, \dots)$ is a bijection onto $D \times D \times D \times \cdots$, where $D = \{|z| < 1\}$ is the unit disk, and so the $\alpha_j$'s parametrize the measures $d\mu$. Note that in our situation, all the Verblunsky coefficients are real,

$$\alpha_n = -\pi_{n+1}(0) \in \mathbb{R}, \qquad n \geq 0. \tag{5.81}$$

It is also useful to define

$$\alpha_{-1} \equiv -\pi_0(0) = -1. \tag{5.82}$$

For an integrable function $a(z)$ on the unit circle, let

$$a_k = \int_{|z|=1} z^{-k} a(z) \frac{dz}{2\pi i z} = \frac{1}{2\pi} \int_{-\pi}^{\pi} e^{-ik\theta} a(e^{i\theta}) d\theta, \qquad k \in \mathbb{Z}, \tag{5.83}$$

be the $k$th Fourier coefficient of $a$. Define the $(n+1) \times (n+1)$ matrix

$$T_n(a) = (a_{j-k})_{0 \le j,k \le n} = \begin{pmatrix} a_0 & a_{-1} & \cdots & a_{-n+1} & a_{-n} \\ a_1 & a_0 & \cdots & a_{-n+2} & a_{-n+1} \\ \vdots & & \ddots & & \vdots \\ a_{n-1} & a_{n-2} & \cdots & a_0 & a_{-1} \\ a_n & a_{n-1} & \cdots & a_1 & a_0 \end{pmatrix}. \tag{5.84}$$

This matrix is called the *Toeplitz matrix* for the function $a$, and $a$ is referred to as the *symbol* of $T_n(a)$. Define the corresponding *Toeplitz determinant* by

$$D_n(a) = \det(T_n(a)). \tag{5.85}$$

**Exercise 5.14.**

(a) Show that

$$D_n(a) = \frac{1}{(n+1)!} \int_{|z_0|=1} \cdots \int_{|z_n|=1} \prod_{0 \le j < k \le n} |z_k - z_j|^2 \prod_{j=0}^{n} w(z_j) \frac{dz_j}{2\pi i z_j}. \tag{5.86}$$

(b) Show that (cf. (5.11) and (5.13)) for a real nonnegative function $a$, the OPUC $p_n(z)$ with respect to the measure $a(e^{i\theta})\frac{d\theta}{2\pi}$ is given by

$$p_n(z) = \frac{1}{\sqrt{D_n D_{n-1}}} \det \begin{pmatrix} a_0 & a_{-1} & \cdots & a_{-n+1} & a_{-n} \\ a_1 & a_0 & \cdots & a_{-n+2} & a_{-n+1} \\ \vdots & \vdots & \ddots & \vdots & \vdots \\ a_{n-1} & a_{n-2} & \cdots & a_0 & a_{-1} \\ 1 & z & \cdots & z^{n-1} & z^n \end{pmatrix} \tag{5.87}$$

for $n \ge 1$, and also is given by

$$\begin{aligned} p_n(z) = {} & \frac{1}{n!\sqrt{D_{n-1} D_n}} \int_{|z_0|=1} \cdots \int_{|z_{n-1}|=1} \prod_{j=0}^{n-1} (z - z_j) \\ & \cdot \prod_{0 \le j < k \le n-1} |z_j - z_k|^2 \prod_{j=0}^{n-1} w(z_j) \frac{dz_j}{2\pi i z_j} \end{aligned} \tag{5.88}$$

for $n \ge 1$. Show, in addition, that

$$\kappa_n = \sqrt{\frac{D_{n-1}}{D_n}}, \qquad n \ge 0. \tag{5.89}$$

(Here, $D_{-1} \equiv 1$.)

(c) Prove the Szegő recurrence relations

$$
\begin{aligned}
z\pi_n(z) &= \pi_{n+1}(z) - \pi_{n+1}(0)\pi_n^*(z), \\
z\pi_n(z) &= \frac{\kappa_{n+1}^2}{\kappa_n^2}\big(\pi_{n+1}(z) - \pi_{n+1}(0)\pi_{n+1}^*(z)\big).
\end{aligned} \tag{5.90}
$$

(d) Show that all the zeros of $p_n(z)$ lie strictly inside $D = \{|z| < 1\}$. For example, for uniform measure $\frac{d\theta}{2\pi}$, $p_n(z) = z^n$, $n = 0, 1, \ldots$, and hence $p_n(z) = 0$ implies that $z = 0$, which certainly lies in $D$.

In Sections 5.7 and 5.8, we will prove two fundamental results on Toeplitz matrices which will be used in later chapters. The first result (see Section 5.7) is the strong Szegő limit theorem which concerns the asymptotics of Toeplitz determinants. The second result (see Section 5.8) concerns the inverses of Toeplitz matrices, when the size of the matrices tends to infinity. These results are well known in the literature and there are many proofs, from different points of view, of the strong Szegő limit theorem. In this text, we provide proofs which are based on so-called *integrable operators* and an associated RHP. Integrable operators are introduced in Section 5.6. The precise notion of a solution to an RHP and some basic results for RHP's are discussed in Section 5.5.

## 5.5. RHP: Precise Description

As indicated in Section 5.3, the notion of the solution of a RHP must be made precise. In particular, one must specify in what sense the limits, $m_\pm(z)$ in (5.29) are attained. In situations where, for example, $\Sigma$ is a simple closed contour and the jump matrix is smooth, it is appropriate to require that $m(z)$ is continuous up to the boundary in each component of $\mathbb{C} \setminus \Sigma$, and hence $m_\pm(z)$ may be taken to be pointwise limits in the specification of the RHP. This is the case, for example, in the RHP's discussed in RHP 5.3 when $\Sigma = \mathbb{R}$ and RHP 5.10 when $\Sigma = \{|z| = 1\}$ above. But, if the jump matrix $v$ of a RHP $(\Sigma, v)$ is no longer smooth, then the solution $m(z)$ of the RHP may not have pointwise limits $m_\pm(z)$ for all $z \in \Sigma$.

For example, let $\Sigma = \mathbb{R}$ and let

$$
v(z) = \begin{cases} i, & -1 < z < 1, \\ 1, & z < -1, \quad z > 1. \end{cases} \tag{5.91}
$$

Then the function $m(z) = \left(\frac{z-1}{z+1}\right)^{1/4}$ (defined with branch cut $[-1,1]$) satisfies

- $m(z)$ is analytic in $\mathbb{C} \setminus [-1,1]$,
- $m_+(z) = m_-(z)v(z)$, $z \in \mathbb{R} \setminus \{-1,1\}$,
- $m(z) \to 1$ as $z \to \infty$,

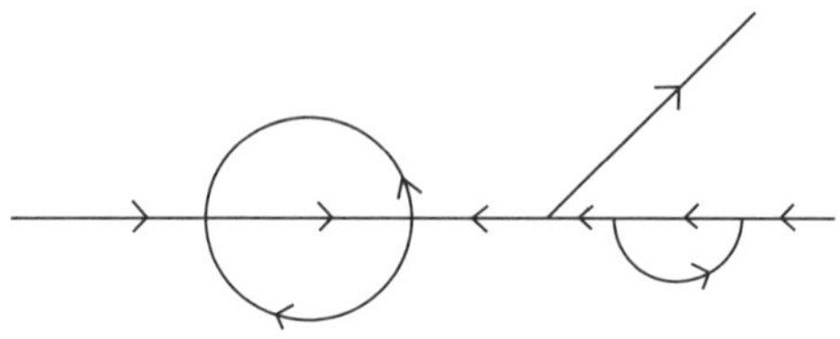

**Figure 5.5.** Contour $\Sigma$

but the limits $m_\pm(z)$ do not exist at $z = -1$. On the other hand, the function $m(z) = \left(\frac{z+1}{z-1}\right)^{1/4}$ solves the RHP with $v(z)$ replaced by $v(z)^{-1}$ and does not have a limit at $z = 1$. For such problems, the sense in which $m(z)$ solves the RHP $(\Sigma, v)$ must be specified in an appropriate generalized sense. RHP's with nonsmooth jump matrices occur, for example, in the asymptotic analysis of RHP's. In such cases, the original RHP's, which are typically smooth, are transformed asymptotically to RHP's with jump matrices that are only piecewise smooth.

We now describe some of the basic properties of the Cauchy operator (5.39),

$$(5.92) \qquad (C\,h)(z) = (C_\Sigma\,h)(z) \equiv \frac{1}{2\pi i}\int_\Sigma \frac{h(s)}{s-z}ds, \qquad z \in \mathbb{C}\backslash\Sigma,$$

for an oriented contour $\Sigma$ in the plane, e.g., see Figure 5.1. The properties and estimates below are true for a very general class of contours, which includes contours that are a finite unions of smooth curves in $\overline{\mathbb{C}}$. In particular, the results are true for contours which are finite unions of line segments and arcs of a circle (see Figure 5.5) which is all that is needed for the problems considered in this text. (See, e.g., [**Gol69**], [**Dur70**], [**BK97**], [**Gar07**], and [**DZ**] for further properties of Cauchy operators.)

**Remark 5.15** (Remarks on curves)**.** The Riemann sphere $\overline{\mathbb{C}}$ has a natural topological and analytical structure induced by the coordinate patches $\{|z| < c\}$, $\{|\zeta| < c' : \zeta = z^{-1}\}$ with $c, c' > 0$. By a curve (more precisely, a simple curve) $\Sigma$ in $\mathbb{C}$, we mean a closed set that is the homeomorphic image of an interval $[a, b]$, $-\infty < a < b < \infty$, or the unit circle $T = \{|z| = 1\}$. We say that the homeomorphism provides a *parametrization* for $\Sigma$. The curve is *smooth* if there is a homeomorphism, $\varphi$ say, which is a smooth map from $[a, b]$ or $T$ into (the manifold) $\overline{\mathbb{C}}$ with a smooth inverse. (Exercise: Show that $\Sigma = \mathbb{R}$, or more precisely $\Sigma = \overline{\mathbb{R}} = \mathbb{R} \cup \{\infty\}$, is a smooth curve in $\overline{\mathbb{C}}$.) If $\Sigma$ is smooth, then, a fortiori, it is locally rectifiable, and hence carries a natural measure, viz., arc-length measure, which we denote by $|dz|$, or sometimes $ds$, and an associated $\sigma$-algebra $\mathcal{A}$, making $\Sigma$ into a measure space $(\Sigma, \mathcal{A}, |dz|)$. $L^p(\Sigma)$ spaces are defined in the standard way, consisting of measurable functions $f$ for which the *norm* $\|f\|_p \equiv \left(\int_\Sigma |f(z)|^p|dz|\right)^{1/p}$, $1 \le p < \infty$, is finite: For $L^\infty(\Sigma)$,

$$\|f\|_\infty \equiv \text{ess sup}_{z\in\Sigma}|f(z)| \equiv \inf\{t > 0 : \text{meas}\{|f(z)| > t\} = 0\}.$$

If $\varphi$ and $\tilde{\varphi}$ are two smooth parametrizations for $\Sigma$, then $\tilde{\varphi}^{-1} \circ \varphi$ is either strictly increasing or strictly decreasing. In the former case we say that $\varphi$ and $\tilde{\varphi}$ induce the *same orientation* on $\Sigma$; in the latter case, they induce the *opposite orientation.* If $f \in L^1(\Sigma)$, then the *line integral* $\int f(z)dz$ is well defined, up to orientation, by

$$\int_\Sigma f(z)dz \equiv \int f(\varphi(t))\varphi'(t)dt \tag{5.93}$$

for any smooth parametrization $\varphi$ for $\Sigma$; if $\varphi$ on the right-hand side of (5.93) is replaced by $\tilde{\varphi}$ which has the opposite orientation, then the sign of the integral flips.

Sobolev space $H_k^p(\Sigma)$ on $\Sigma$ are defined in a natural way. Indeed, if $\Sigma$ is a smooth contour and $a$ and $b$ are finite points in $\Sigma$, we let $[a,b]$ denote the sub-arc of $\Sigma$ connecting $a$ and $b$. We say a function $f : \Sigma \to \mathbb{C}$ is *absolutely continuous* on a sub-arc $[a,b] \subset \Sigma$ if for any $\epsilon > 0$, there exists $\delta > 0$ such that

$$\sum_{j=1}^n |f(b_j) - f(a_j)| < \epsilon \tag{5.94}$$

whenever

$$\sum_{j=1}^n |b_j - a_j| < \delta \tag{5.95}$$

for any collection of disjoint sub-arcs $[a_1, b_1], \ldots, [a_n, b_n]$ in $[a,b]$. If $f$ is absolutely continuous on a sub-arc, then the derivative

$$f'(z) = \lim_{z'\to z, z'\in\Sigma} \frac{f(z') - f(z)}{z' - z}$$

exists for almost every $z \in [a,b]$. Moreover, $f' \in L^1([a,b])$ and the fundamental theorem of calculus holds for all sub-arcs $[a',b'] \subset [a,b]$:

$$\int_{[a',b']} f'(z)dz = f(b') - f(a'). \tag{5.96}$$

The space $H_1^p(\Sigma)$, $1 \le p \le \infty$, for example, is the space of all functions $f \in L^p(\Sigma)$ which are absolutely continuous on all sub-arcs in $\Sigma$, and such that $f' \in L^p(\Sigma)$, with norm $\|f\|_{H^1} = \|f\|_p + \|f'\|_p$. Higher order Sobolev spaces are defined similarly.

Henceforth, we will assume without further comment that the contour $\Sigma$ is a finite union $\Sigma = \bigcup_j \Sigma_j$ of smooth curves in $\mathbb{C}$ with only a finite number of points of self-intersection. The latter restriction ensures that $\overline{\mathbb{C}} \setminus \Sigma$ has only a finite number of components. For any curve $\Sigma_j \subset \Sigma$, the orientation induced by a parametrization of $\Sigma_j$ is indicated by an arrow (see, e.g., Figure 5.5). For such contours $\Sigma = \bigcup_j \Sigma_j$, and also, as noted above, for a much broader class of contours, the following results for $C_\Sigma$ hold true. Let $h \in L^p(\Sigma, |dz|)$ for some $1 \le p < \infty$. Then

$$(5.97) \qquad (\mathrm{C}_\pm h)(z) \equiv \lim_{\substack{z' \to z \\ z' \in (\pm)-\text{side of } \Sigma}} (\mathrm{C} h)(z')$$

exists as a nontangential limit for almost every $z \in \Sigma$. *Nontangential* means $z' \to z$ in any fixed (truncated) cone based at $z \in \Sigma$ and lying entirely in $\mathbb{C} \backslash \Sigma$. Furthermore, for $1 < p < \infty$, there is a constant $c_p > 0$, depending only on $\Sigma$ and $p$, such that

$$(5.98) \qquad \| \mathrm{C}_\pm h \|_{L^p(\Sigma)} \le c_p \|h\|_{L^p(\Sigma)}$$

for all $h \in L^p(\Sigma)$. Moreover,

$$(5.99) \qquad \mathrm{C}_\pm h = \pm \frac{1}{2} h + \frac{i}{2} \mathrm{H} h,$$

where

$$(5.100) \qquad (\mathrm{H} h)(z) \equiv \lim_{\epsilon \to 0} \frac{1}{\pi} \int_{\{s \in \Sigma : |s-z| > \epsilon\}} \frac{h(s)}{z-s} ds, \qquad z \in \Sigma,$$

is the *Hilbert transform*. The limit (5.100) exists a.e. on $\Sigma$ and also in $L^p(\Sigma, |dz|)$, and we have

$$(5.101) \qquad \| \mathrm{H} h \|_{L^p(\Sigma)} \le c'_p \|h\|_{L^p(\Sigma)}, \quad 1 < p < \infty,$$

for some constant $c'_p$. Clearly,

$$(5.102) \qquad \mathrm{C}_+ - \mathrm{C}_- = 1, \qquad \mathrm{C}_+ + \mathrm{C}_- = i\,\mathrm{H},$$

where 1 denotes the identity operator. Moreover, if $\gamma$ is a smooth open segment lying in $\Sigma$ (see Figure 5.6) and $h \in H^1_p(\gamma) = \{f : f, \frac{\partial f}{\partial z} \in L^p(\gamma, |dz|)\}$ for $1 < p < \infty$, then $(\mathrm{C} h)(z)$ is continuous up to $\gamma$. Thus, the limit $(\mathrm{C}_\pm h)(z)$ for $h \in H^1_p(\gamma)$ exists for all $z \in \gamma$ without the nontangential restriction.

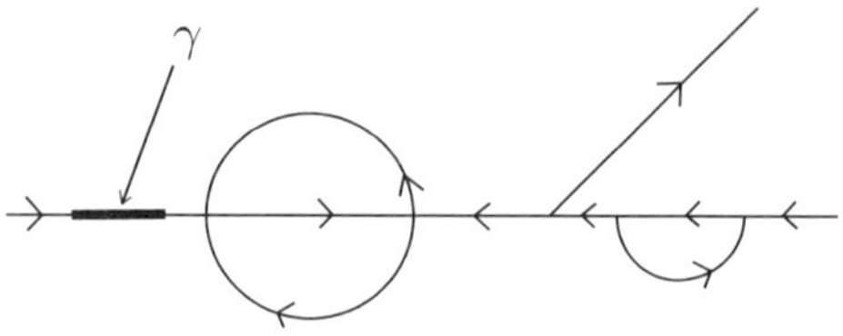


**Figure 5.6.** Open segment $\gamma \in \Sigma$

Furthermore, for $h \in H_p^1(\Sigma)$, $(\mathrm{C}\, h)(z)$ is bounded in $\mathbb{C}\backslash\Sigma$ and tends to zero uniformly as $z \to \infty$. Also, it follows from the above comments that if $\Sigma_1$ and $\Sigma_2$ are two contours in $\mathbb{C}$, then for $1 < p < \infty$

$$\|\, \mathrm{C}_\pm\, h\|_{L^p(\Sigma_2)} \leq c_p \|h\|_{L^p(\Sigma_1)} \tag{5.103}$$

for $h \in L^p(\Sigma_1)$. For example, if $\Gamma_1 = \mathbb{R}$, $\Gamma_2 = z_0 + \mathbb{R}_+ e^{i\theta}$ for $z_0 \in \mathbb{R}$, $0 < \theta < \pi$, then

$$\|\, \mathrm{C}_\pm\, h\|_{L^p(z_0+\mathbb{R}_+e^{i\theta})} \leq c_p \|h\|_{L^p(\mathbb{R})} \tag{5.104}$$

for $h \in L^p(\mathbb{R})$. Moreover, it is not hard to see that the constant $c_p$ can be chosen independently of $z_0 \in \mathbb{R}$ and $0 < \theta < \pi$ (this can be done as an exercise; cf. [**BDT88**]).

Let $1 < p < \infty$ and consider a pair $(f_+, f_-)$ of functions in $L^p(\Sigma)$. We say (cf. [**DZ03**]) that $f_\pm \in \partial\, \mathrm{C}_p$ if there exists a (unique) function $h \in L^p(\Sigma)$ such that

$$f_\pm(z) = (\mathrm{C}_\pm\, h)(z). \tag{5.105}$$

In turn we will call $f(z) \equiv \mathrm{C}\, h(z)$, $z \in \mathbb{C}\backslash\Sigma$, the *extension* of $f_\pm$ off $\Sigma$.

**Riemann-Hilbert Problem 5.16** (RHP; precise sense)**.** *Fix $1 < p < \infty$. Let $v$ be a $k \times k$ jump matrix on $\Sigma$. We say that a $k \times k$ matrix $m_\pm$ solves the* normalized RHP $(\Sigma, v)$ *if $m_\pm - I \in \partial\, \mathrm{C}_p$ and*

$$m_+(z) = m_-(z)v(z), \quad a.e.\ z \in \Sigma. \tag{5.106}$$

It turns out that the solution of an RHP is equivalent to the solution of a singular integral equation on $\Sigma$. Indeed, let

$$v(z) = (v^-(z))^{-1} v^+(z) \quad \text{a.e. } z \in \Sigma \tag{5.107}$$

be *any pointwise* factorization of $v$ with $v^\pm(z)$ invertible for a.e. $z \in \Sigma$ and such that $v^\pm$, $(v^\pm)^{-1} \in L^\infty(\Sigma)$. Define $w^\pm \in L^\infty(\Sigma)$ via

$$v^+ = I + w^+, \qquad v^- = I - w^-. \tag{5.108}$$

Let $w = (w^+, w^-)$. For row vectors $h = (h_1, \ldots, h_k)$, define the singular integral operator

$$\mathrm{C}_w\, h \equiv \mathrm{C}_+(hw^-) + \mathrm{C}_-(hw^+). \tag{5.109}$$

Clearly, $\mathrm{C}_w$ is a bounded operator from $L^p(\Sigma)$ to $L^p(\Sigma)$ for $1 < p < \infty$.

Now, suppose that

$$1 - \mathrm{C}_w \text{ is invertible in } L^p(\Sigma), \tag{5.110}$$

and suppose that $w^\pm \in L^p(\Sigma)$. Let $\mu \in I + L^p(\Sigma)$ solve the equation

$$(1 - C_w)\mu = I. \tag{5.111}$$

Here $I$ indicates the function on $\Sigma$ which is constant and equal to the identity matrix. More precisely, if we write $\mu = I + b$, $b \in L^p(\Sigma)$, then

$$(1 - C_w)b = C_w I = C_+(w^-) + C_-(w^+) \in L^p(\Sigma). \tag{5.112}$$

Set

$$m_\pm \equiv I + C_\pm(\mu(w^+ + w^-)). \tag{5.113}$$

Clearly, $\mu(w^+ + w^-) = (w^+ + w^-) + b(w^+ + w^-) \in L^p(\Sigma)$, so $m_\pm - I \in \partial C_p$. Now,

$$\begin{aligned} m_+ &= I + C_+(\mu(w^+ + w^-)) = I + C_+(\mu w^+) + C_+(\mu w^-) \\ &= I + (1 + C_-)(\mu w^+) + C_+(\mu w^-) \qquad (\text{as } C_+ - C_- = 1) \\ &= I + \mu w^+ + C_w \mu \\ &= \mu w^+ + \mu \qquad (\text{as } (1 - C_w)\mu = I) \\ &= \mu v^+. \end{aligned} \tag{5.114}$$

Similarly

$$m_- = \mu v^-. \tag{5.115}$$

Thus, $m_+ = m_-(v^-)^{-1}v^+ = m_- v$. Hence $m_\pm$ solves the normalized RHP $(\Sigma, v)$.

In summary, we have the following result.

**Theorem 5.17.** *Consider the normalized RHP $(\Sigma, v)$ as in RHP 5.16. Define the operator $C_w$ as in (5.109):*

$$C_w h \equiv C_+(hw^-) + C_-(hw^+), \tag{5.116}$$

*where $w = (w^+, w^-)$ and $w_\pm$ is defined through (5.108) and (5.107). If $1 - C_w$ is invertible in $L^p(\Sigma)$, then*

$$m_\pm \equiv I + C_\pm(\mu(w^+ + w^-)), \tag{5.117}$$

*where*

$$\mu \equiv (1 - C_w)^{-1} I = I + (1 - C_w)^{-1}(C_+(w^+) + C_-(w^+)) \tag{5.118}$$

*solves the normalized RHP $(\Sigma, v)$. In this case,*

$$\begin{aligned} m(z) &= I + C(\mu(w^+ + w^-))(z) \\ &= I + \frac{1}{2\pi i}\int_\Sigma \frac{\mu(s)(w^+(s) + w^-(s))}{s - z} ds \\ &= I + \frac{1}{2\pi i}\int_\Sigma \frac{m_+(s)(I - v^{-1}(s))}{s - z} ds. \end{aligned} \tag{5.119}$$

The last equality in (5.119) follows from (5.114).

In this way the RHP is turned into a problem about the singular integral operator $C_w$. If $w^{\pm}$ are small enough so that $C_w$ is of norm $< 1$, then $(1 - C_w)^{-1}$ exists via the Neumann series. In most cases, however, we must infer the existence of $(1 - C_w)^{-1}$ using the theory of Fredholm operators (see, e.g., [**DKM**$^{+}$**99b**]).

In the case that $v(z)$ is a (nonvanishing) scalar function, the RHP $(\Sigma, v)$ has a solution $m(z)$ given by the *Plemelj formula*

$$(5.120) \qquad m(z) = e^{C(\log v)(z)} = e^{\int_\Sigma \frac{\log v(s)}{s-z} \frac{ds}{2\pi i}}, \qquad z \in \mathbb{C} \setminus \Sigma,$$

provided that $\log v(s)$ is sufficiently smooth. Indeed by (5.99), for $z \in \Sigma$,

$$(5.121) \qquad m_+(z) = e^{C_+(\log v)(z)} = e^{C_-(\log v)(z) + \log v(z)} = m_-(z)v(z),$$

and clearly $m(z) \to 1$ as $z \to \infty$.

In particular, if $\Sigma$ is a smooth Jordan curve in $\mathbb{C}$ and $v(s)$ is a smooth function on $\Sigma$ with zero winding, then $\log v(s)$ is smooth and (5.120) is indeed a solution of $(\Sigma, v)$. Moreover, it is the unique solution and $m(z) \neq 0$ in $\mathbb{C} \setminus \Sigma$, and $m_\pm(z) \neq 0$ in $\Sigma$. The same is true, for example, if $\Sigma = \mathbb{R}$, $v(z)$ is nonvanishing, smooth, and without winding on $\Sigma$, and $\lim_{z\to\pm\infty} v(z) = 1$ (this can be done as an exercise).

If $v(z)$ has nonzero winding, then the solution $m(z)$ of the RHP $(\Sigma, v)$ may fail to exist, or it may exist but fail to be unique. For example, if $\Sigma$ is the unit circle and $v(z) = z$, then

$$(5.122) \qquad m(z) = \begin{cases} z + a, & |z| < 1, \\ 1 + \frac{a}{z}, & |z| > 1. \end{cases}$$

is a solution of $(\Sigma, v(z) = z)$, for any constant $a$, and so the solution of $(\Sigma, v)$ is not unique. On the other hand, if $v(z) = \frac{z+i}{z-i}$ on $\mathbb{R}$, oriented from $-\infty$ to $+\infty$, then the RHP $(\Sigma, v(z) = \frac{z+i}{z-i})$ cannot have a solution. For if a solution $m(z)$ did exist, then

$$(5.123) \qquad h(z) \equiv \begin{cases} m(z)(z-i), & \mathrm{Im}(z) > 0, \\ m(z)(z+i), & \mathrm{Im}(z) < 0, \end{cases}$$

is an entire function such that $h(z) = z + O(1)$ as $z \to \infty$. Hence $h(z) = z + b$ for some constant $b$. Thus

$$(5.124) \qquad m(z) \equiv \begin{cases} \frac{z+b}{z-i}, & \mathrm{Im}(z) > 0, \\ \frac{z+b}{z+i}, & \mathrm{Im}(z) < 0, \end{cases}$$

and there is no way to choose $b$ such that $m(z)$ is analytic in $\mathbb{C} \setminus \Sigma$. If $v(z) = \frac{z-i}{z+i}$, then we find

$$m(z) \equiv \begin{cases} \frac{z+c}{z+i}, & \mathrm{Im}(z) > 0, \\ \frac{z+c}{z-i}, & \mathrm{Im}(z) < 0, \end{cases} \tag{5.125}$$

and so $(\Sigma, v)$ has multiple solutions, one for each choice of $c$.

Note that for $v(z) = z$ on the unit circle $\Sigma$, $v(z)$ has nonzero winding and $\log v(z) = \log(e^{i\theta}) = i\theta$ is not smooth on $\Sigma$. On the other hand, $v(z) = \frac{z+i}{z-i}$ and $v(z) = \frac{z-i}{z+i}$ have nonzero winding on $\mathbb{R}$. Hence $\frac{\log v(s)}{s-z}$ is not integrable on $\mathbb{R}$, and so the integral in (5.120) does not converge.

## 5.6. Integrable Operators

In this section we consider a distinguished class of operators, the so-called *integrable operators*, on an oriented contour $\Sigma$ in $\mathbb{C}$. Integrable operators are special integral operators, and surprisingly many of the operators that arise in mathematics and mathematical physics are in fact integrable operators. In the next two sections, we use a particular integrable operator that is naturally associated to Toeplitz matrices. Another application of integrable operators can be found in Section 6.5 where we derive a differential equation for the Tracy-Widom distribution.

**Definition 5.18.** An operator K acting in $L^2(\Sigma) = L^2(\Sigma, |dz|)$ is said to be *integrable* if it has a kernel of the form

$$\mathrm{K}(z, z') = \frac{\Sigma_{j=1}^N f_j(z) g_j(z')}{z - z'}, \qquad z, z' \in \Sigma, \tag{5.126}$$

for some functions $f_i, g_j \in L^\infty(\Sigma)$, $1 \le i, j \le N < \infty$. Explicitly,

$$\mathrm{K}\, h(z) = \fint_\Sigma \mathrm{K}(z, z') h(z') dz' = \lim_{\epsilon \downarrow 0} \int_{\{z' \in \Sigma : |z' - z| > \epsilon\}} \mathrm{K}(z, z') h(z') dz'. \tag{5.127}$$

Examples of such operators have been considered for many years. Elements of the general theory of such operators appeared in the work of Sakhnovich in the late 1960s, and the full theory was developed by Its, Izergin, Korepin, and Slavnov [**IIKS90**] in the early 1990s. Integrable operators have many remarkable properties (see for example [**Dei99a**]):

- the space of integrable operators is an algebra, i.e., if $\mathrm{K}_1$, $\mathrm{K}_2$ are integrable and $\lambda_1, \lambda_2$ are constants, then $\lambda_1 \mathrm{K}_1 + \lambda_2 \mathrm{K}_2$ is integrable and so is $\mathrm{K}_1 \mathrm{K}_2$;
- if K is integrable and $(1 - \mathrm{K})^{-1}$ exists, then

$$\mathrm{R} = (1 - \mathrm{K})^{-1} - 1 \tag{5.128}$$

  is integrable.

**Exercise 5.19.** Show that the kernel of R is

$$\mathrm{R}(z,z') = \frac{\sum_{i=1}^{N} F_i(z)G_i(z')}{z-z'}, \tag{5.129}$$

where

$$F_i = (1-\mathrm{K})^{-1} f_i, \quad G_i = (1-\mathrm{K}^T)^{-1} g_i, \quad 1 \le i \le N. \tag{5.130}$$

Most remarkably, as we now show

- $(1-\mathrm{K})^{-1}$ can be computed in terms of the solution of a naturally associated RHP.

The following identity plays a key role. If $X_1$ and $X_2$ are Banach spaces, then the *resolvent formula*

$$\lambda(DE+\lambda)^{-1} + D(ED+\lambda)^{-1}E = 1 \tag{5.131}$$

holds for all bounded operator $D\colon X_1 \to X_2$, $E\colon X_2 \to X_1$, in the sense that if $-\lambda \neq 0$ lies in the resolvent set of $ED$, then $-\lambda$ also lies in the resolvent set of $DE$ and $(DE+\lambda)^{-1} = \frac{1}{\lambda}(1-D(ED+\lambda)^{-1}E)$. This formula has applications to an extremely wide variety of problems in mathematical physics; see, e.g., [**Dei78**]. In particular,

$$\mathrm{spec}(ED)\backslash\{0\} = \mathrm{spec}(DE)\backslash\{0\}. \tag{5.132}$$

**Exercise 5.20.** Prove (5.131).

Let K be an integrable operator on $\Sigma$, $\mathrm{K}(z,z') = \sum_{j=1}^{N} \frac{f_j(z)g_j(z')}{z-z'}$ for $z, z' \in \Sigma$. The action of K may be expressed as

$$(\mathrm{K}\,h)(z) = \pi \sum_{j=1}^{N} (\mathrm{H}(hg_j))(z) f_j(z), \quad z \in \Sigma, \tag{5.133}$$

where H is the Hilbert transform (5.100). Clearly, by the properties of H, K is a bounded operator on $L^p(\Sigma)$, $1 < p < \infty$, as $f_i, g_j \in L^\infty$. Define the column vectors

$$f = (f_1, \ldots, f_N)^T, \qquad g = (g_1, \ldots, g_N)^T. \tag{5.134}$$

Let $R_f$ denote the operator of right multiplication by the column vector $f$, taking row vector functions to scalar functions

$$R_f h(z) = h(z)f(z) = \sum_{i=1}^{N} h_i(z) f_i(z), \qquad h = (h_1, \ldots, h_N), \tag{5.135}$$

and let $R_{g^T}$ denote right multiplication by the row vector $g^T$, taking scalar functions to row vector functions

$$R_{g^T} h = h(z)g^T(z) = (h(z)g_1(z), \ldots, h(z)g_N(z)). \tag{5.136}$$

Clearly,

$$\mathrm{K}\,h = \pi R_f[\mathrm{H}(R_{g^T}h)] = (DE)h, \tag{5.137}$$

where $D \equiv R_f$, $E \equiv \pi H R_{g^T}$. Then $ED$ maps row vector functions to row vector functions and $(ED)u = \pi\,\mathrm{H}\,R_{g^T}R_f u = \pi\,\mathrm{H}(ufg^T)$. Using $\mathrm{C}_+ + \mathrm{C}_- = i\,\mathrm{H}$, we find that

$$(ED)u = \pi\,\mathrm{H}(ufg^T) = \mathrm{C}_+[u(-i\pi fg^T)] + \mathrm{C}_-[u(-i\pi fg^T)] = \mathrm{C}_w\,u, \tag{5.138}$$

where $\mathrm{C}_w$ is the operator defined in (5.116) and

$$w \equiv (w^+, w^-) = (-i\pi fg^T, -i\pi fg^T). \tag{5.139}$$

The operator in (5.138) arises in the solution of the normalized RHP $(\Sigma, v)$ with $v = (v^-)^{-1}v^+$ where

$$\begin{cases} v^- = I - w^- = I + i\pi fg^T, \\ v^+ = I + w^+ = I - i\pi fg^T. \end{cases} \tag{5.140}$$

Thus ,

$$v = (v^-)^{-1}v^+ = I - \frac{2i\pi}{1 + i\pi < g, f >} fg^T, \tag{5.141}$$

where $\langle g, f\rangle \equiv \sum_1^N g_i(z)f_i(z)$.

Note that due to (5.137), (5.138), and the resolvent formula (5.131), if $1-\mathrm{K}$ is invertible, so is $1-\mathrm{C}_w$. In this case, the normalized RHP $(\Sigma, v)$ has a (unique) solution given by (5.119). In situations where we do not know a priori that $1-\mathrm{K}$ is invertible, the uniqueness of the solution to the RHP, if it exists, can often be inferred by an argument similar to that in Lemma 5.4.

Let $F_i(z) = (1-\mathrm{K})^{-1}f_i(z)$ and $G_i(z) = (1-\mathrm{K}^T)^{-1}g_i(z)$ be the functions in (5.130), and set $F = (F_1, \ldots, F_N)^T$, $G = (G_1, \ldots, G_N)^T$. Then,

$$\begin{aligned} F &= (1-\mathrm{K})^{-1}f = (1-DE)^{-1}f = (1 + D(1-ED)^{-1}E)f \\ &= f + R_f(1-\mathrm{C}_w)^{-1}\pi\,\mathrm{H}(fg^T) \qquad \text{(by (5.138))} \\ &= f + R_f(1-\mathrm{C}_w)^{-1}(\mathrm{C}_w\,I) \\ &= \big(I + (1-\mathrm{C}_w)^{-1}(\mathrm{C}_w\,I)\big)f. \end{aligned} \tag{5.142}$$

Define $\mu \equiv (1-\mathrm{C}_w)^{-1}I$ as in (5.118). Then

$$F = \mu f = m_\pm(v_\pm)^{-1}f = m_\pm(I \mp i\pi fg^T)^{-1}f \tag{5.143}$$

using (5.114), (5.115), and (5.140). Therefore,

$$F = \frac{m_\pm f}{1 \mp i\pi\langle f, g\rangle}. \tag{5.144}$$

One also finds, with a little more calculation, that

$$G = \frac{(m_\pm^T)^{-1} g}{1 \mp i\pi\langle f, g\rangle}. \tag{5.145}$$

Summarizing, we have obtained the following result:

**Theorem 5.21** (Resolvent of an integrable operator). *For an integrable operator with kernel* (5.126), *set* $f \equiv (f_1, \ldots, f_N)^T$ *and* $g \equiv (g_1, \ldots, g_N)^T$. *Let* $m_\pm$ *be the solution to the normalized RHP* $(\Sigma, v)$, *where*

$$v(z) \equiv I - \frac{2\pi i}{1 + \pi i\langle g, f\rangle} f(z) g(z)^T \tag{5.146}$$

*as in* (5.141). *If* $(1 - \mathrm{K})^{-1}$ *exists, then the RHP* $(\Sigma, v)$ *has a unique solution. Moreover, the resolvent* $\mathrm{R} = (1 - \mathrm{K})^{-1} - 1$ *is also an integrable operator with kernel*

$$\mathrm{R}(z, z') = \sum_{i=1}^{N} \frac{F_i(z) G_i(z')}{z - z'}, \tag{5.147}$$

*where*

$$\begin{aligned} F &= (F_1, \ldots, F_N)^T = (1 - \mathrm{K})^{-1} f = \frac{m_\pm f}{1 \mp i\pi\langle f, g\rangle}, \\ G &= (G_1, \ldots, G_N)^T = (1 - \mathrm{K}^T)^{-1} g = \frac{(m_\pm^T)^{-1} f}{1 \mp i\pi\langle f, g\rangle}. \end{aligned} \tag{5.148}$$

We will be most interested in the case where $f_i(z)$, $g_i(z)$ are continuous on $\Sigma$ and

$$\sum_{i=1}^{N} f_i(z) g_i(z) = 0, \quad z \in \Sigma. \tag{5.149}$$

Under this condition, we have

$$F(z) = m_\pm f, \qquad G(z) = (m_\pm^T)^{-1} g, \tag{5.150}$$

where $m_\pm$ solve the normalized RHP $(\Sigma, I - 2\pi i f g^T)$.

## 5.7. The Strong Szegő Limit Theorem

The classical Strong Szegő Limit Theorem is:

**Theorem 5.22** (Strong Szegő Limit Theorem). *Let* $L(\theta)$ *be a real analytic function on the unit circle* $L(\theta + 2\pi) = L(\theta)$ *with Fourier coefficients* $L_k \equiv \int_{-\pi}^{\pi} e^{-ik\theta} L(\theta) \frac{d\theta}{2\pi}$. *Let*

$$\varphi(z) = e^{L(\theta)}, \qquad z = e^{i\theta}. \tag{5.151}$$

*Then* $\varphi(z)$ *can be analytically continued, and* $|\varphi(z)| \geq \mathrm{Re}(\varphi(z)) > 0$*, in* $\rho < |z| < \rho^{-1}$ *for some* $\rho \in (0,1)$ *and for any* $\rho_1 \in (\rho, 1)$*, the Toeplitz determinant* $D_n(\varphi) = \det(T_n(\varphi)) = \det\left(\varphi_{j-k}\right)_{j,k=0}^{n}$ *with symbol* $\varphi$ *satisfies*

$$\log D_n(\varphi) = (n+1)L_0 + \sum_{1}^{\infty} kL_k L_{-k} + O(\rho_1^n) \tag{5.152}$$

*as* $n \to \infty$.

**Remark 5.23.** The above result is stated for positive symbols. But, it can be checked that the same result holds for complex valued $\varphi$ as long as $\varphi$ is analytic and nonzero in some annulus $\rho < |z| < \rho^{-1}$ for some $\rho \in (0,1)$, and the winding number is zero. Indeed the proof given below applies to such symbols with minimal changes (see Remark 5.24 below). Note that for positive symbols, $L_{-k} = \overline{L_k}$.

Although this result is sufficient for our purposes in this book, the conditions on $L(\theta)$ can be weakened considerably; see [**GI71, Sim05a, Sim05b**]. Without the analyticity condition, the error term is, in general, of order $o(1)$. The first term $(n+1)L_0$ of (5.152) was obtained by Szegő in 1915 (called Szegő weak limit theorem) and the second term $\sum_1^\infty kL_kL_{-k}$ was obtained by Szegő 37 years later in 1952 in response to a question of Onsager concerning the Ising model. Szegő proved the result for $L \in C^{1+\epsilon}, \epsilon > 0$. There are many different proofs of the Strong Szegő Limit Theorem using tools from many areas of mathematics; see e.g. [**Sim05a, Sim05b**] for references.

Here we prove the theorem using the theory of integrable operators and their associated RHP's. We follow the proof given in [**Dei99a**]. One of the main ingredient of the proof is a steepest-descent analysis of the RHP. The steepest-descent method for RHP's was introduced by Deift and Zhou [**DZ93**] in 1993 and was was further extended to allow for genuinely nonlinear oscillations by Deift, Zhou, and Venakides [**DVZ97**]. The method is an analogue, in the noncommutative context of an RHP, of the classical steepest-descent/stationary-phase method for scalar integrals of the form

$$f(t) = \int_\Gamma h(z)e^{tg(z)}dz, \tag{5.153}$$

as $t \to \infty$, where $\Gamma$ is a contour in $\mathbb{C}$. See Appendix B for the basics of the classical steepest-descent method. The steepest-descent method for RHP's turns out to contain many new phenomena over and above the scalar case; see [**DKM$^+$99a, DKM$^+$99b**] for more information. We do not discuss the full power of the steepest-descent method for RHP's in this text. The asymptotic analysis of the RHP in this section is one of the most elementary examples of the method. For an extensive introduction to the steepest-descent method for RHP's, see for example [**Dei99b**].

The rest of this section is devoted to a proof of Theorem 5.22.

We first show that $D_n(\varphi)$ can be reformulated in terms of a Fredholm determinant of an integrable operator. Let $P_n = \{\sum_{j=0}^{n} a_j z^j : a_j \in \mathbb{C}\}$ denote the set of trigonometric polynomials of degree at most $n$. Define the operator $\tau_n \colon P_n \to P_n$ by

$$\tau_n z^k = \sum_{j=0}^{n} \varphi_{j-k} z^j, \qquad 0 \le k \le n. \tag{5.154}$$

Clearly, $\tau_n$ is conjugate to the Toeplitz matrix $T_n(\varphi) = (\varphi_{j-k})_{0\le j,k\le n}$, and hence $D_n(\varphi) = \det(\tau_n)$. Note that for any $k = 0, \dots, n$,

$$\begin{aligned} \tau_n z^k - z^k &= \sum_{j=0}^{n} \varphi_{j-k} z^j - \sum_{j=0}^{n} \delta_{j-k} z^j \\ &= \sum_{j=0}^{n} \left[ \oint \xi^{-j+k} \varphi(\xi) \frac{d\xi}{2\pi i \xi} \right] z^j - \sum_{j=0}^{n} \left[ \oint \xi^{-j+k} \frac{d\xi}{2\pi i \xi} \right] z^j \\ &= - \oint \mathrm{K}_n(z,\xi) \xi^k d\xi, \end{aligned} \tag{5.155}$$

where

$$\mathrm{K}_n(z,\xi) \equiv \sum_{j=0}^{n} \left(\frac{z}{\xi}\right)^j \frac{1-\varphi(\xi)}{2\pi i \xi} = \frac{z^{n+1}\xi^{-n-1} - 1}{z - \xi} \cdot \frac{1-\varphi(\xi)}{2\pi i}. \tag{5.156}$$

Let $\mathrm{K}_n : L^2(\Sigma, dz) \to L^2(\Sigma, dz)$ be the operator with kernel $\mathrm{K}_n(z,\xi)$. Clearly, $\mathrm{K}_n$ is of integrable type. We have shown in (5.155) that

$$(1 - \mathrm{K}_n) z^k = \sum_{j=0}^{n} \varphi_{j-k} z^j, \qquad 0 \le k \le n. \tag{5.157}$$

A direct calculation also shows that

$$(1 - \mathrm{K}_n) z^k = z^k + \sum_{j=0}^{n} \varphi_{j-k} z^j \qquad \text{for } k < 0 \text{ or } k > n. \tag{5.158}$$

As $\{z^k\}_{k=-\infty}^{\infty}$ is a basis for $L^2(\Sigma)$, it follows that

$$D_n(\varphi) = \det\left(\varphi_{j-k}\right)_{j,k=0}^{n} = \det(1 - \mathrm{K}_n). \tag{5.159}$$

Note that $\mathrm{K}_n$ is of finite rank, and hence trace class, so the Fredholm determinant $\det(1 - \mathrm{K}_n)$ is well defined. Also, as $\varphi(z) > 0$, $D_n(\varphi) > 0$ (see, e.g., (5.86)); hence, $\det(1 - \mathrm{K}_n) \ne 0$ and $(1 - \mathrm{K}_n)^{-1}$ exists. See Appendix A for the basics of trace class operators and Fredholm determinants.

We now express $D_n(\varphi)$ in terms of the solution of the RHP associated with the integrable operator $\mathrm{K}_n$. Using (5.159), we have the following elementary formula:

$$\begin{aligned}\log D_n(\varphi) = \log\det(1-\mathrm{K}_n) &= \int_0^1 \left[\frac{d}{dt}\log\det(1-t\,\mathrm{K}_n)\right]dt \\ &= \int_0^1 \left[\frac{d}{dt}\operatorname{tr}\log(1-t\,\mathrm{K}_n)\right]dt \\ &= -\int_0^1 \operatorname{tr}\left((1-t\,\mathrm{K}_n)^{-1}\,\mathrm{K}_n\right)dt.\end{aligned} \tag{5.160}$$

For $0 \le t \le 1$, set

$$\varphi_t(z) \equiv (1-t) + t\varphi(z), \qquad z \in \Sigma. \tag{5.161}$$

Clearly $\varphi_0 = 1$, $\varphi_1 = \varphi$, $\varphi_t$ is analytic, and $\mathrm{Re}(\varphi_t(z)) > 0$ in the same annulus $\rho < |z| < \rho^{-1}$. For each $t \in [0,1]$, let $\mathrm{K}_{t,n}$ be the operator with kernel

$$\mathrm{K}_{t,n}(z,\xi) = \frac{(\frac{z}{\xi})^{n+1}-1}{z-\xi}\cdot\frac{1-\varphi_t(\xi)}{2\pi i}. \tag{5.162}$$

Since $1-\varphi_t = t(1-\varphi)$,

$$\mathrm{K}_{t,n}(z,\xi) = t\,\mathrm{K}_n(z,\xi), \tag{5.163}$$

hence

$$\log D_n(\varphi) = -\int_0^1 \operatorname{tr}\left((1-\mathrm{K}_{t,n})^{-1}\,\mathrm{K}_{t,n}\right)\frac{dt}{t}. \tag{5.164}$$

Clearly, $\det(1-\mathrm{K}_{t,n}) = D_n(\varphi_t) > 0$ and hence $(1-\mathrm{K}_{t,n})^{-1}$ exists; together with (5.163) this justifies the calculation (5.160) leading to (5.164).

Observe that $\mathrm{K}_{t,n}$ is an integrable operator with kernel $\frac{\sum_{i=1}^2 f_{t,i}(z)g_{t,i}(\xi)}{z-\xi}$ on $L^2(\Sigma)$ where

$$f_t = \begin{pmatrix} f_{t,1} \\ f_{t,2}\end{pmatrix} \equiv \begin{pmatrix} z^{n+1} \\ -1\end{pmatrix}, \qquad g_t = \begin{pmatrix} g_{t,1} \\ g_{t,2}\end{pmatrix} \equiv \frac{1-\varphi_t(z)}{2\pi i}\begin{pmatrix} z^{-n-1} \\ 1\end{pmatrix}. \tag{5.165}$$

We have $\sum_{i=1}^n f_{t,i}(z)g_{t,i}(z) = 0$ for $z \in \Sigma$. From Theorem 5.21 in the previous section, associated to $\mathrm{K}_n$ we have the normalized RHP $(\Sigma, v_{t,n})$ where

$$v_{t,n} = I - 2\pi i f_t(z)g_t(z)^T = \begin{pmatrix} \varphi_t & (\varphi_t-1)z^{n+1} \\ z^{-(n+1)}(1-\varphi_t) & 2-\varphi_t. \end{pmatrix}. \tag{5.166}$$

Let $m_{t,\pm} = m_{t,n,\pm}$ be the solution to this normalized RHP $(\Sigma, v_{t,n})$, which exists, by Theorem 5.21, as $1-\mathrm{K}_{t,n}$ is invertible. From Theorem 5.21, we also have

$$((1-\mathrm{K}_{t,n})^{-1}\,\mathrm{K}_{t,n})(z,\xi)=\frac{\sum_{j=1}^{2}F_{t,j}(z)G_{t,j}(\xi)}{z-\xi}, \tag{5.167}$$

where

$$\begin{aligned} F_t&=\begin{pmatrix}F_{t,1}\\ F_{t,2}\end{pmatrix}\equiv m_{t,+}f_t=m_{t,+}\begin{pmatrix}z^{n+1}\\ -1\end{pmatrix},\\ G_t&=\begin{pmatrix}G_{t,1}\\ G_{t,2}\end{pmatrix}\equiv (m_{t,+}^T)^{-1}g_t=(m_{t,+}^T)^{-1}\begin{pmatrix}z^{-(n+1)}\\ 1\end{pmatrix}\frac{1-\varphi_t}{2\pi i}.\end{aligned} \tag{5.168}$$

From (5.164), (5.167), and (5.168),

$$\begin{aligned}\log D_n(\varphi)&=-\int_0^1\frac{dt}{t}\int_\Sigma\left[\sum_{j=1}^{2}F'_{t,j}(z)G_{t,j}(z)\right]dz\\ &=-\int_0^1\frac{dt}{t}\int_\Sigma G_t^T(z)F_t'(z)dz.\end{aligned} \tag{5.169}$$

In order to evaluate the integral in (5.169), we evaluate $m_{t,\pm}$ asymptotically as $n\to\infty$. This is done in the following way. First, factorize

$$v_{t,n}=\begin{pmatrix}1&0\\ z^{-(n+1)}(\varphi_t^{-1}-1)&1\end{pmatrix}\begin{pmatrix}\varphi_t&0\\ 0&\varphi_t^{-1}\end{pmatrix}\begin{pmatrix}1&(1-\varphi_t^{-1})z^{n+1}\\ 0&1\end{pmatrix}. \tag{5.170}$$

Then define a new contour $\Sigma$ consisting of three circles,

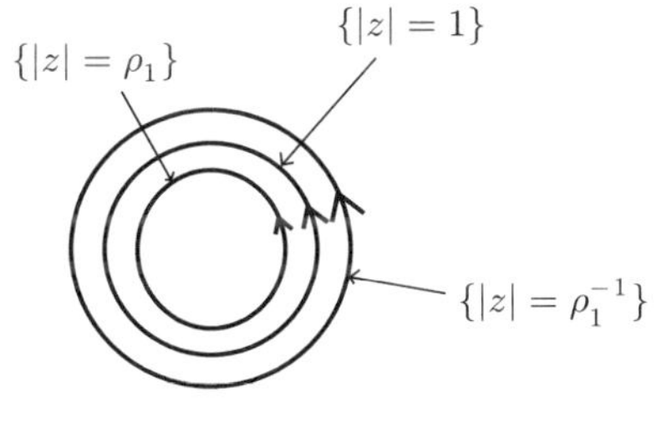


**Figure 5.7.** Contour $\Sigma^{(1)}$

$$\Sigma^{(1)}=\{|z|=\rho_1\}\cup\Sigma\cup\{|z|=\rho_1^{-1}\},\qquad \rho<\rho_1<1, \tag{5.171}$$

all oriented counterclockwise; see Figure 5.7. Define

- $m^{(1)}(z)\equiv m_t(z)$ for $|z|<\rho_1$ or $|z|>\rho_1^{-1}$,
- $m^{(1)}(z)\equiv m_t(z)\begin{pmatrix}1&(1-\varphi_t^{-1})z^{n+1}\\ 0&1\end{pmatrix}^{-1}$ for $\rho_1<|z|<1$,
- $m^{(1)}(z)\equiv m_t(z)\begin{pmatrix}1&0\\ z^{-(n+1)}(\varphi_t^{-1}-1)&1\end{pmatrix}$ for $1<|z|<\rho_1^{-1}$.

A direct calculation shows that $m_{\pm}^{(1)}(z)$ solves the new normalized RHP $(\Sigma^{(1)}, v_{t,n}^{(1)})$ where

- $v_{t,n}^{(1)}(z) = \begin{pmatrix} 1 & (1-\varphi_t^{-1})z^{n+1} \\ 0 & 1 \end{pmatrix}$ for $|z| = \rho_1$,
- $v_{t,n}^{(1)}(z) = \begin{pmatrix} \varphi_t & 0 \\ 0 & \varphi_t^{-1} \end{pmatrix}$ for $|z| = 1$,
- $v_{t,n}^{(1)}(z) = \begin{pmatrix} 1 & 0 \\ z^{-(n+1)}(\varphi_t^{-1} - 1) & 1 \end{pmatrix}$ for $|z| = \rho_1^{-1}$.

Note that $m^{(1)}(z)$ is indeed analytic in $\mathbb{C}\backslash\Sigma^{(1)}$ as $\varphi_t(z)^{-1}$ is analytic in the annulus $\{\rho_1 < |z| < \rho_1^{-1}\}$.

The new RHP $(\Sigma^{(1)}, v_{t,n}^{(1)})$ is clearly equivalent to the original RHP $(\Sigma, v_{t,n})$: the solution of the one problem can be obtained from the solution of the other by elementary algebraic operations, and vice versa. There is, however, a critical difference between the two problems: The exponential factors $z^{n+1}$ and $z^{-(n+1)}$ now live on contours in the complex plane where they are exponentially decreasing uniformly as $n \to \infty$. This implies that

$$\sup_{0\le t\le 1} \|v_{t,n}^{(1)} - I\|_{L^\infty(\Sigma^{(1)}\backslash\Sigma)} \le c\rho_1^n, \tag{5.172}$$

for some $c > 0$, and we expect that, as $n \to \infty$,

$$m_{\pm}^{(1)} \to m_{\pm}^{\infty}, \tag{5.173}$$

where $m_{\pm}^{\infty}$ solves the normalized RHP $(\Sigma, v_t^{\infty})$,

$$v_t^{\infty} = \begin{pmatrix} \varphi_t & 0 \\ 0 & \varphi_t^{-1} \end{pmatrix}, \qquad z \in \Sigma = \{|z| = 1\}. \tag{5.174}$$

As $v_t^{\infty}$ is diagonal, the RHP $(\Sigma, v_t^{\infty})$ can be solved explicitly, and we find (this can be done as an exercise)

$$m^{\infty}(z) = \begin{pmatrix} e^{q_t(z)} & 0 \\ 0 & e^{-q_t(z)} \end{pmatrix}, \tag{5.175}$$

where

$$q_t(z) = \frac{1}{2\pi i}\int_\Sigma \frac{\log \varphi_t(s)}{s-z}\,ds \tag{5.176}$$

is the solution of the scalar RHP

$$q_{t,+}(z) = q_{t,-}(z) + \log\varphi(z), \qquad z \in \Sigma. \tag{5.177}$$

It is a simple exercise to show that $m^{\infty}$ is continuous up to the boundary $\Sigma$, uniformly for $0 \le t \le 1$.

A deformation of contours as above to take advantage of the exponential decay of oscillatory factors in the jump matrices is the basic idea in the steepest-descent method for RHP's introduced in [**DZ93**]. In order to prove (5.173), set

$$N(z) \equiv m^{(1)}(z)(m^{\infty}(z))^{-1}. \tag{5.178}$$

We would like to show that $N(z) - I$ is "small". Clearly, $N(z)$ is analytic in $\mathbb{C}\backslash\Sigma^{(1)}$. For $z \in \Sigma$,

$$\begin{aligned} N_+(z) &= m_+^{(1)}(z)(m_+^{\infty}(z))^{-1} \\ &= m_-^{(1)}\begin{pmatrix} \varphi_t & 0 \\ 0 & \varphi_t^{-1}\end{pmatrix}\left[m_-^{\infty}\begin{pmatrix} \varphi_t & 0 \\ 0 & \varphi_t^{-1}\end{pmatrix}\right]^{-1} = m_-^{(1)}(m_-^{\infty})^{-1} \\ &= N_-(z). \end{aligned} \tag{5.179}$$

Thus, $N(z)$ is analytic across $\Sigma$. On the other hand, for $z \in \Sigma^{(1)}\backslash\Sigma$, using the fact that $m^{\infty}$ is analytic across $\Sigma^{(1)}\backslash\Sigma$,

$$\begin{aligned} N_+ &= m_+^{(1)}(m^{\infty})^{-1} \\ &= m_-^{(1)}v_{t,n}^{(1)}(m^{\infty})^{-1} = N_-m^{\infty}v_{t,n}^{(1)}(m^{\infty})^{-1} = N_-v_N, \end{aligned} \tag{5.180}$$

where

$$v_N \equiv m^{\infty}v_{t,n}^{(1)}(m^{\infty})^{-1} \tag{5.181}$$

for $z \in \Sigma^{(1)}\backslash\Sigma$. From the estimate (5.172) on $v_{t,n}^{(1)}$ and by noting that all entries of $m^{\infty}$ are uniformly bounded on $\Sigma^{(1)}\backslash\Sigma$, we find that

$$\sup_{0\le t\le 1}\|v_N - I\|_{L^{\infty}(\Sigma^{(1)}\backslash\Sigma)} = O(\rho_1^n). \tag{5.182}$$

The estimate (5.182) allows us to find the solution $N$ to the RHP $(\Sigma_N, v_N)$, $\Sigma_N \equiv \Sigma^{(1)}\setminus\Sigma$, using the general method discussed in Section 5.5, as follows. Set

$$(v_N^+, v_N^-) \equiv (v_N, I), \qquad (w_N^+, w_N^-) \equiv (v_N - I, 0). \tag{5.183}$$

Define the operator $\mathrm{C}_N\, h = \mathrm{C}_-(hw_N^+)$ (see (5.109) above) and consider the equation $(1-\mathrm{C}_N)\mu_N = I$ on $\Sigma^{(1)}\backslash\Sigma$. By (5.182), $\sup_{0\le t\le 1}\|w_N^+\|_{L^{\infty}(\Sigma^{(1)}\backslash\Sigma)} = O(\rho_1^n)$. Since $\mathrm{C}_-$ is a bounded operator, from (5.182) we find that

$$\|\,\mathrm{C}_N\,\|_{L^2(\Sigma_N)\to L^2(\Sigma_N)} \le C\|w_N^+\|_{L^{\infty}(\Sigma_N)} \le c\rho_1^n \tag{5.184}$$

for some $C, c > 0$, uniformly in $t \in [0,1]$. Hence, $(1-\mathrm{C}_N)^{-1}$ exists for all large enough $n$, and (see (5.118))

$$\mu_N \equiv (1-\mathrm{C}_N)^{-1}I = I + (1-\mathrm{C}_N)^{-1}(\mathrm{C}_N\, I) = I + O(\rho_1^n), \tag{5.185}$$

where the $O(\rho_1^n)$ term is uniform in $L^2(\Sigma_N)$ for $0 \le t \le 1$. Thus, for $z \in \mathbb{C}\backslash(\Sigma_N)$, and in particular for $z \in \Sigma$, we have from (5.119) that, uniformly for $0 \le t \le 1$,

$$N(z) = I + \frac{1}{2\pi i}\int_{\Sigma_N} \frac{\mu_N(s)w_N^+(s)}{s-z}ds = I + O(\rho_1^n) \tag{5.186}$$

where the $O(\rho_1^n)$ term is uniform in $z \in \Sigma$, together with any number of derivatives. Recalling the definition of $N(z)$, we see that for $z \in \Sigma$,

$$m_+^{(1)}(z) = N(z)m_+^\infty(z) = m_+^\infty(z) + O(\rho_1^n), \tag{5.187}$$

uniformly in $z \in \Sigma$ and $0 \le t \le 1$. This implies that on $\Sigma$,

$$\begin{aligned} m_{t,n+}(z) &= m_+^{(1)} \begin{pmatrix} 1 & (1-\varphi_t^{-1})z^{n+1} \\ 0 & 1 \end{pmatrix} \\ &= \begin{pmatrix} e^{q_{t+}} & 0 \\ 0 & e^{-q_{t+}} \end{pmatrix} \begin{pmatrix} 1 & (1-\varphi_t^{-1})z^{n+1} \\ 0 & 1 \end{pmatrix} + O(\rho_1^n). \end{aligned} \tag{5.188}$$

Inserting (5.188) into (5.169), we find

$$\begin{aligned} &\log D_n(\varphi) \\ &= \frac{1}{2\pi i}\int_\Sigma \int_0^1 \left(\frac{n+1}{z} + 2q_{t+}' - \varphi_t'\varphi_t^{-1}\right)\frac{\varphi_t - 1}{t\varphi_t} dtdz + O(\rho_1^n) \\ &= (n+1)L_0 + \frac{1}{2\pi i}\int_\Sigma \int_0^1 (2q_{t+}' - \varphi_t'\varphi_t^{-1})\frac{\varphi_t - 1}{t\varphi_t} dtdz + O(\rho_1^n) \end{aligned} \tag{5.189}$$

as

$$\frac{d}{dt}\log\varphi_t = \frac{\varphi_t - 1}{t\varphi_t}, \tag{5.190}$$

and hence

$$\frac{(n+1)}{2\pi i}\int_\Sigma \int_0^1 \frac{\varphi_t - 1}{t\varphi_t} dt\frac{dz}{z} = \frac{n+1}{2\pi i}\int_\Sigma \log\varphi(z)\frac{dz}{z} = (n+1)L_0. \tag{5.191}$$

From (5.177), $q_{t+}' = q_{t-}' + (\log\varphi_t)'$, hence $2q_{t+}' - \varphi_t'\varphi_t^{-1} = q_{t+}' + q_{t+}'$. This implies

$$\begin{aligned} &\log D_n(\varphi) \\ &= (n+1)L_0 + \frac{1}{2\pi i}\int_\Sigma \int_0^1 (q_{t+}' + q_{t-}')\frac{\varphi_t - 1}{t\varphi_t} dtdz + O(\rho_1^n). \end{aligned} \tag{5.192}$$

We now evaluate the double integral. First note that from (5.176),

$$q_t'(z) = \frac{1}{2\pi i}\int_\Sigma \frac{\log\varphi_t(s)}{(s-z)^2}ds. \tag{5.193}$$

Observe that, due to the analyticity, for $|z| < 1$, the contour can be deformed to $|s| = 1 + \epsilon$, and for $|z| > 1$ the contour can be deformed to $|s| = 1 - \epsilon$.

Then, for $\epsilon > 0$ small,

$$\begin{aligned}(5.194)\qquad &\int_{|z|=1-\epsilon}\int_0^1 q_t'(z)\frac{\varphi_t(z)-1}{t\varphi_t(z)}dtdz\\ &= \frac{1}{2\pi i}\int_{|z|=1-\epsilon}\int_{|s|=1+\epsilon}\int_0^1 \frac{\log\varphi_t(s)}{(s-z)^2}\frac{\varphi_t(z)-1}{t\varphi_t(z)}dtdsdz.\end{aligned}$$

Integrating by parts in the $t$-variable using (5.190), we find that (5.194) equals

$$\begin{aligned}(5.195)\qquad &\frac{1}{2\pi i}\int_{|z|=1-\epsilon}\int_{|s|=1+\epsilon}\frac{\log\varphi(s)\log\varphi(z)}{(s-z)^2}dsdz\\ &-\frac{1}{2\pi i}\int_{|z|=1-\epsilon}\int_{|s|=1+\epsilon}\int_0^1\frac{(\varphi_t(s)-1)\log\varphi_t(z)}{t\varphi_t(s)(s-z)^2}dtdsdz.\end{aligned}$$

It is an exercise to evaluate the first double integral and check that

$$(5.196)\qquad \frac{1}{2\pi i}\int_{|z|=1-\epsilon}\int_{|s|=1+\epsilon}\frac{\log\varphi(s)\log\varphi(z)}{(s-z)^2}dsdz = 2\pi i\sum_{k=1}^{\infty}kL_kL_{-k}.$$

On the other hand, by performing the $z$-integral first and using (5.193), the second double integral in (5.195) equals

$$(5.197)\qquad \int_{|s|=1+\epsilon}\int_0^1 q_t'(s)\frac{\varphi_t(s)-1}{t\varphi_t(s)}dtds.$$

Combining (5.194), (5.195), (5.196), and (5.197) and taking the limit $\epsilon\downarrow 0$, we find that

$$\begin{aligned}(5.198)\qquad &\int_\Sigma\int_0^1 q_{t+}'(z)\frac{\varphi_t(z)-1}{t\varphi_t(z)}dtdz\\ &= 2\pi i\sum_{k=1}^{\infty}kL_kL_{-k} - \int_\Sigma\int_0^1 q_{t-}'(s)\frac{\varphi_t(s)-1}{t\varphi_t(s)}dtds.\end{aligned}$$

Inserting this and (5.191) into (5.192), we finally obtain

$$(5.199)\qquad \log D_n(\varphi) = (n+1)L_0 + \sum_1^{\infty}kL_kL_{-k} + O(\rho_1^n),\qquad n\to\infty.$$

This proves Szegő's Strong Limit Theorem in the case that $\varphi(z) = e^{L(\theta)}$ is real analytic on $\Sigma$.

**Remark 5.24.** As described above, if $\varphi(e^{i\theta}) > 0$, the invertibility of $1-t\,\mathrm{K}_n$, for all $n\geq 0$, follows from the fact that $t\,\mathrm{K}_n = K_{t,n}$, and $\det(1-\mathrm{K}_{t,n}) = D_n(\varphi_t) > 0$. If $\varphi(e^{i\theta})$ is complex valued, this argument breaks down, but if $\varphi(e^{i\theta})$ has zero winding, then the invertibility of $(1-t\,\mathrm{K}_n)$, and hence the existence of the solution $m_{t,n}$ of the associated RHP, follows *for large $n$*, by the following argument:

(i) Note again as above that $(1-\mathrm{K}_{t,n})^{-1}$ exists if and only if $(1-\mathrm{C}_{w_{t,n}})^{-1}$ exists, where $\mathrm{C}_{w_{t,n}}$ is the operator (5.109) associated with the RHP $(\Sigma, v_{t,n})$.

(ii) Extend $\Sigma \to \Sigma^{(1)}$ as in (5.171), and extend $v_{t,n}$ to $\Sigma^{(1)}$ by setting $v_{t,n}^{(e)}(z) = v_{t,n}(z)$ for $z \in \Sigma$, and $v_{t,n}^{(e)}(z) = I$ for $z \in \Sigma^{(1)} \setminus \Sigma$. It is a simple calculation (this can be done as an exercise: cf. [**DZ03**]) to show that $(1-\mathrm{C}_{w_{t,n}})^{-1}$ exists if and only if the operator $(1-\mathrm{C}_{w_{t,n}^{(e)}})$ associated with the RHP $(\Sigma^{(1)}, v_{t,n}^{(e)})$ is invertible in $L^2(\Sigma^{(1)}, |dz|)$.

(iii) Using the theory of *inhomogeneous RHP's of type* 2 (IRHP2), it then follows that $(1-\mathrm{C}_{w_{t,n}^{(e)}})^{-1}$ exists if and only if the operator $1-\mathrm{C}_N$ above is invertible, which it is for large $n$, say $n \geq n_0$.

(iv) We conclude that $(1-\mathrm{K}_{t,n})^{-1}$ exists in $L^2(\Sigma, |dz|)$ for large $n$. Moreover, the estimates are uniform in $t \in [0,1]$, and we have the bound

$$\|(1-\mathrm{K}_{t,n})^{-1}\|_{L^2} \leq c, \qquad \text{for } t \in [0,1],\ n \geq n_0. \tag{5.200}$$

## 5.8. Inverses of Large Toeplitz Matrices

In this section, we discuss an approximate formula for the inverse of a large Toeplitz matrix, a result which is used in a later chapter. In order to obtain such a formula, we first need to discuss some properties of semi-infinite Toeplitz matrices.

Let $\{a_n\}_{-\infty}^{\infty}$ be a sequence of complex numbers. The semi-infinite Toeplitz matrix associated with $\{a_n\}_{-\infty}^{\infty}$ is given by

$$A = \begin{pmatrix} a_0 & a_{-1} & a_{-2} & \cdots \\ a_1 & a_0 & a_{-1} & \cdots \\ a_2 & a_1 & a_1 & \cdots \\ \vdots & \vdots & \vdots & \ddots \end{pmatrix}. \tag{5.201}$$

$A$ acts on sequences $c = \{c_k\}_{k=0}^{\infty}$ in the Hilbert space

$$\ell_2(\mathbb{N}_0) = \{\{c_k\}_{k=0}^{\infty} : \sum_{k=0}^{\infty} |c_k|^2 < \infty\}$$

in the natural way: $\{c_k\}_{k=0}^{\infty} \mapsto \{\sum_{k=0}^{\infty} a_{j-k}c_k\}_{j=0}^{\infty}$. One calls $A$ the *Toeplitz operator* associated with $\{a_n\}_{-\infty}^{\infty}$. The following results are standard; see, e.g., [**BS99**], and the proofs are left as exercises.

Let $\Sigma = \{z : |z| = 1\}$ be the unit circle as before. Set $\mathbb{N}_0 = \{0, 1, 2, \dots\}$.

**Proposition 5.25.** *The Toeplitz matrix $A$ defines a bounded operator on $\ell_2(\mathbb{N}_0)$ if and only if there is a bounded function $a(z)$, $z \in \Sigma$, whose nth*

*Fourier coefficient is equal to* $a_n$. *In this case,* $A$ *is denoted as* $T(a)$, *and we have*

$$\|T(a)\| = \|a\|_\infty, \tag{5.202}$$

*where the left-hand side is the operator norm of* $T(a)$ *on* $\ell_2(\mathbb{N}_0)$.

**Remark 5.26.** By Fourier theory, $\ell_2(\mathbb{Z})$ is unitarily equivalent to $L^2\left(\Sigma, \frac{d\theta}{2\pi}\right)$,

$$\{c_k\}_{-\infty}^{\infty} \mapsto c(z) = \sum_{k=-\infty}^{\infty} c_k z^k, \qquad \sum_{k=-\infty}^{\infty} |c_k|^2 = \int_{-\pi}^{\pi} |c(z)|^2 \frac{d\theta}{2\pi}.$$

Under this map $\ell_2(\mathbb{N}_0)$ maps onto the Hardy space

$$\mathbb{H}^{2+} = \left\{ c(z) : c(z) = \sum_{k=0}^{\infty} c_k z^k, \|c\|^2_{L^2(\Sigma, \frac{d\theta}{2\pi})} = \sum_{k=0}^{\infty} |c_k|^2 < \infty \right\}.$$

On $\mathbb{H}^{2+}$, $T(a)$ takes the form

$$c \in \mathbb{H}^{2+} \mapsto P^+(ac) \in \mathbb{H}^{2+}, \tag{5.203}$$

where $P^+$ denotes the orthogonal projection in $L^2$ onto $\mathbb{H}^{2+}$: $\sum_{k=-\infty}^{\infty} c_k z^k \mapsto \sum_{k=0}^{\infty} c_k z^k$.

*Wiener algebra* is the set

$$\mathrm{W} = \left\{ a(z) = \sum_{-\infty}^{\infty} a_n z^n : \sum_{-\infty}^{\infty} |a_n| < \infty \right\}. \tag{5.204}$$

W is an algebra under convolution: for $a = \sum a_n z^n, b = \sum b_n z^n$ where $a, b \in \mathrm{W}$, the convolution $a \circ b(z) \equiv \sum_{-\infty}^{\infty} c_n z^n \in W$, where $c_n = \sum_k a_{n-k} b_k$. If $a \in \mathrm{W}$ satisfies $a(z) \neq 0$, $z \in \Sigma$, and has zero winding number, then $a$ has a Riemann-Hilbert factorization (or more classically, a Wiener-Hopf factorization) as follows: since the winding number is 0, it follows by a nontrivial result of Wiener that one can write $a(z) = e^{b(z)}$, $z \in \Sigma$, for some $b(z) = \sum_{-\infty}^{\infty} b_n z^n \in \mathrm{W}$ (see, e.g. [**Kat04**]). Let

$$a_+(z) = e^{\sum_{n=0}^{\infty} b_n z^n}, \qquad a_-(z) = e^{\sum_{n=-\infty}^{-1} b_n z^n}. \tag{5.205}$$

Then clearly

(i) $a(z) = a_+(z) a_-(z)$, $z \in \Sigma$;

(ii) $a_+(z)$ (resp., $a_-(z)$) has an analytic continuation to $|z| < 1$ (resp., $|z| > 1$), and both are continuous up to the boundary;

(iii) $a_+(z) \neq 0$ for $|z| \leq 1$ and $a_-(z) \neq 0$ for $|z| \geq 1$.

In addition,

(iv) $a_+, a_- \in W$, again by Wiener's result.

**Remark 5.27.** If, in addition, $a(z)$ is analytic in a neighborhood of $\Sigma$, then the existence of the factorization is an elementary exercise, and one does not need Wiener's deeper result. Also, we have

(ii)′ for some $0 < \rho < 1$, $a_+(z)$ (resp. $a_-(z)$) has an analytic continuation to $|z| < \rho^{-1}$ (resp. $|z| > \rho$), and

(iii)′ $a_+(z) \neq 0$ for $|z| < \rho^{-1}$ and $a_-(z) \neq 0$ for $|z| > \rho$.

The following result is due to Krein. See, for example, Section 1.5 of [**BS99**].

**Proposition 5.28.** *Suppose $a \in \mathrm{W}$. Then $T(a)$ is invertible if and only if $a(z) \neq 0$ for all $z \in \Sigma$ and has zero winding number. In this case,*

$$T(a)^{-1} = T(a_+^{-1})T(a_-^{-1}). \tag{5.206}$$

Let $T_n(a)$ be the $(n+1) \times (n+1)$-dimensional Toeplitz matrix associated with symbol $a(z)$ as in (5.84). The following result shows that, as $n \to \infty$, $T_n(a)^{-1}$ is close to $T(a)^{-1}$ in a precise sense. This result is essentially due to Widom and is contained in (the proof of) Theorem 2.15 of [**BS99**].

**Theorem 5.29** ([**BS99**, Theorem 2.15])**.** *Suppose $a \in \mathrm{W}$ and has zero winding number. In addition, suppose that*

$$\sum_{-\infty}^{\infty} |n|^\alpha |a_n| < \infty \tag{5.207}$$

*for some $\alpha > 0$. Then there exists $n_0 \in \mathbb{N}_0$ such that $T_n(a)$ is invertible for all $n \geq n_0$ and a constant $C$, independent of $n$, such that*

$$\begin{aligned} &\left|(T_n(a))^{-1}_{j,k} - (T(a_+^{-1})T(a_-^{-1}))_{j,k}\right| \\ &\quad \leq C \min\left\{\frac{1}{(n+1-j)^\alpha}, \frac{1}{(n+1-k)^\alpha}\right\}, \qquad 0 \leq j,k \leq n, \end{aligned} \tag{5.208}$$

*for $n \geq n_0$.*

We prove this theorem for the case when $a(z)$ is analytic in an annulus $\{z : \rho < |z| < \rho^{-1}\}$ for some $0 < \rho < 1$ using Riemann-Hilbert techniques as in [**DÖ06**]. This result suffices for the application of the theorem in this text. The general case (5.207) is treated in [**DÖ06**] where more general so-called *Beurling weights* are also considered. For $a(z)$ analytic, we prove that there exists an integer $n_0 = n_0(a) > 0$ and constants $C, \eta > 0$ such that for all $n \geq n_0$

$$|(T_n(a))^{-1}_{j,k} - \left(T(a_+^{-1})T(a_-^{-1})\right)_{j,k}| \leq C \min\left\{e^{-\eta(n+1-j)}, e^{-\eta(n+1-k)}\right\} \tag{5.209}$$

uniformly for $0 \leq j,k \leq n$. Clearly, (5.209) implies (5.208) for any $\alpha > 0$.

**Proof of Theorem 5.29 when $a(z)$ is analytic.** We refer to the calculations of Section 5.7. We note that the Toeplitz matrix $T_n(a)$ is conjugate to the operator $\tau_n$ on the set of set of trigonometric polynomials of degree $\leq n$. From (5.154) and (5.157) we have

$$\tau_n z^k = (1 - \mathrm{K}_n) z^k = \sum_{j=0}^{n} a_{j-k} z^j, \quad 0 \leq k \leq n, \tag{5.210}$$

where $\mathrm{K}_n$ is the integrable operator on $L^2(\Sigma, |dz|)$ with kernel

$$\mathrm{K}_n(z,\xi) = \frac{z^{n+1}\xi^{-n-1} - 1}{z - \xi} \frac{1 - a(\xi)}{2\pi i}. \tag{5.211}$$

By Remark 5.24, $(1 - \mathrm{K}_n)^{-1}$ exists for $n$ large, and we have the bound $\|(1-\mathrm{K}_n)^{-1}\|_{L^2(\Sigma)} \leq c$ for $n \geq n_0$. Setting $\mathrm{R}_n \equiv (1-\mathrm{K}_n)^{-1} - 1 = (1-\mathrm{K}_n)^{-1}\mathrm{K}_n$, we find

$$\begin{aligned} (T_n(a))^{-1}_{j,k} &= \langle z^j, \tau_n^{-1} z^k \rangle = \langle z^j, (1 - \mathrm{K}_n)^{-1} z^k \rangle \\ &= \delta_{j,k} + \langle z^j, \mathrm{R}_n\, z^k \rangle \end{aligned} \tag{5.212}$$

where $\langle f, g \rangle = \int_{|z|=1} \overline{f(z)} g(z) \frac{dz}{2\pi i z}$. From (5.167) and (5.168), we have

$$\mathrm{R}_n(z,\xi) = \frac{F^T(z) G(\xi)}{z - \xi}, \qquad z, \xi \in \Sigma, \tag{5.213}$$

where

$$F(z) = m_+(z) \begin{pmatrix} z^{n+1} \\ -1 \end{pmatrix}, \quad G(\xi) = (m_+(\xi)^T)^{-1} \begin{pmatrix} \xi^{-n-1} \\ 1 \end{pmatrix} \frac{1 - a(\xi)}{2\pi i}. \tag{5.214}$$

Here $m(z)$ is the solution of the normalized RHP on $\Sigma$ satisfying

- $m(z)$ is analytic in $\mathbb{C} \setminus \Sigma$,
- $m_+(z) = m_-(z) v(z)$, $z \in \Sigma$,
- $m(z) \to I$ as $z \to \infty$,

where

$$\begin{aligned} v(z) =& I - 2\pi i \begin{pmatrix} z^{n+1} \\ -1 \end{pmatrix} \begin{pmatrix} z^{-n-1}, & 1 \end{pmatrix} \frac{1 - a(z)}{2\pi i} \\ =& \begin{pmatrix} a & (a-1) z^{n+1} \\ (1-a) z^{-n-1} & 2 - a \end{pmatrix}. \end{aligned} \tag{5.215}$$

(The notation here differs slightly from [**DÖ06**].)

As before, for any fixed $\rho_1$ satisfying $\rho < \rho_1 < 1$, let $\Sigma^{(1)} = \{|z| = \rho_1\} \cup \Sigma \cup \{|z| = \rho_1^{-1}\}$, oriented counterclockwise. Define $m^{(1)}$ as in the previous section (with the change that $\varphi_t = a$) and proceed as before. We note that the solution $q(z)$ to the scalar RHP (5.177) is related to the Wiener-Hopf factorization of $a = a_+ a_-$ in the following way

$$e^{q(z)} = \begin{cases} a_+(z), & |z| < 1, \\ a_-(z)^{-1}, & |z| > 1. \end{cases} \tag{5.216}$$

The matrix $N(z) \equiv m^{(1)}(z)(m^\infty(z))^{-1}$ in (5.178) obeys the estimate (5.186)

$$N(z) = 1 + e_n(z), \tag{5.217}$$

where

$$\|e_n\|_{L^\infty(\Sigma)} \leq c\rho_1^n, \quad n \to \infty, \tag{5.218}$$

for some constant $c > 0$. Using the relation $m^{(1)}(z) = N(z)m^\infty(z)$ and the explicit formula (5.175) for $m^\infty$, the solution $m(z)$ for $n$ sufficiently large and for $z \in \Sigma$, of the form (after the change (5.216))

$$\begin{aligned} m_+(z) &= N(z)m_+^\infty(z) \begin{pmatrix} 1 & (1 - a(z)^{-1})z^{n+1} \\ 0 & 1 \end{pmatrix} \\ &= N(z) \begin{pmatrix} a_+(z) & 0 \\ 0 & a_+(z)^{-1} \end{pmatrix} \begin{pmatrix} 1 & (1 - a(z)^{-1})z^{n+1} \\ 0 & 1 \end{pmatrix}. \end{aligned} \tag{5.219}$$

Inserting (5.219) into (5.214), we obtain

$$F(z) = N(z)\hat{F}(z), \qquad G(\xi) = (N(\xi)^T)^{-1}\hat{G}(\xi), \tag{5.220}$$

where

$$\hat{F}(z) = \begin{pmatrix} a_-(z)^{-1}z^{n+1} \\ -a_+(z)^{-1} \end{pmatrix}, \qquad \hat{G}(\xi) = \begin{pmatrix} a_+(\xi)^{-1}\xi^{-n-1} \\ a_-(\xi)^{-1} \end{pmatrix} \frac{1 - a(\xi)}{2\pi i}. \tag{5.221}$$

Recalling the definition of the Hilbert transform

$$(\mathrm{H}\, h)(z) = P.V.\frac{1}{\pi}\int_\Sigma \frac{h(s)}{z - s}ds, \qquad z \in \Sigma, \tag{5.222}$$

we obtain

$$\begin{aligned} \langle z^j, \mathrm{R}_n\, z^k \rangle &= \int_\Sigma z^{-j}\left[\int_\Sigma \frac{F^T(z)G(\xi)}{z - \xi}\xi^k d\xi\right]\frac{dz}{2\pi i z} \\ &= \pi \int_\Sigma F^T(z)z^{-i}\,\mathrm{H}(G(\diamond)\diamond^k)(z)\frac{dz}{2\pi i z} \\ &= \pi \left\langle \overline{F(\diamond)}\diamond^j, \mathrm{H}(G(\diamond)\diamond^k)\right\rangle. \end{aligned} \tag{5.223}$$

A standard computation (this can be done as an exercise) shows that for $c(z) = \sum_{k=-\infty}^\infty c_k z^k \in L^2\left(\Sigma, \frac{d\theta}{2\pi}\right)$, $\sum_{k=-\infty}^\infty |c_k|^2 < \infty$,

$$\mathrm{H}\left(\sum_{k=-\infty}^\infty c_k z^k\right) = -i\sum_{k=-\infty}^\infty \operatorname{sgn}(k)c_k z^k, \tag{5.224}$$

where $\operatorname{sgn}(k) = 1$ for $k \geq 0$ and $\operatorname{sgn}(k) = -1$ for $k < 0$. In particular, H is bounded on $L^2(\Sigma)$ with norm 1. It follows now from (5.223) and (5.217)-(5.221) that

$$\langle z^j, \mathrm{R}_n\, z^k \rangle = \pi \left\langle \overline{\hat{F}(\diamond)\diamond^j}, \mathrm{H}(\hat{G}(\diamond)\diamond^k) \right\rangle + d_n, \tag{5.225}$$

where

$$|d_n| \leq c' \rho_1^n \tag{5.226}$$

uniformly for $0 \leq j, k \leq n$ for some constant $c' > 0$.

Inserting (5.221) into the right-hand side of (5.225) and using (5.224), we find, after some tedious but elementary algebra,

$$\begin{aligned} \langle z^j, \mathrm{R}_n\, z^k \rangle = -\delta_{j,k} + &\sum_{m=\max\{j+k-n,0\}}^{\infty} (a_+^{-1})_{j-m}(a_-^{-1})_{m-k} \\ - &\sum_{m=-\infty}^{\min\{j+k-n,0\}-1} (a_+^{-1})_{j-m}(a_-^{-1})_{m-k} + d_n, \end{aligned} \tag{5.227}$$

where $(a_\pm^{-1})_k$ denote the Fourier coefficients of $a_\pm^{-1}$, respectively. So, from (5.212),

$$\begin{aligned} &(T_n(a))^{-1}_{jk} \\ &= \sum_{m=0}^{\infty} (a_+^{-1})_{j-m}(a_-^{-1})_{m-k} - \sum_{m=-\infty}^{j+k-n-1} (a_+^{-1})_{j-m}(a_-^{-1})_{m-k} + d_n \\ &= \left(T(a_+^{-1})T(a_-^{-1})\right)_{jk} - \sum_{m=n+1-j-k}^{\infty} (a_+^{-1})_{j+m}(a_-^{-1})_{-(m+k)} + d_n. \end{aligned} \tag{5.228}$$

But from (ii)$'$, (iii)$'$ in Remark 5.27, $a_+, a_-$ are analytic and nonzero in open neighborhoods of $\{|z| \leq 1\}$, $\{|z| \geq 1\}$, respectively. Hence, by standard Paley-Wiener arguments

$$|(a_+^{-1})_j|, |(a_-^{-1})_{-j}| \leq c e^{-\lambda j}, \quad j \geq 0,$$

for some constants $c, \lambda > 0$. Thus, for $0 \leq j, k \leq n$,

$$\begin{aligned} &\left| \sum_{m=n+1-j-k}^{\infty} (a_+^{-1})_{j+m}(a_-^{-1})_{-(m+k)} \right| \\ &\leq c^2 \sum_{m=n+1-j-k}^{\infty} e^{-\lambda(j+2m+k)} \\ &\leq C e^{-\lambda(n+1-j)} e^{-\lambda(n+1-k)} \\ &\leq C \min\{e^{-\lambda(n+1-j)}, e^{-\lambda(n+1-k)}\}. \end{aligned} \tag{5.229}$$

Together with (5.226), this proves the estimate (5.209). □

We complete this section with a corollary of the above theorem which we will use later. The matrix $T_n(a)^{-1}$ acts on $\ell_2(\{0,1,\ldots,n\})$ and can be extended to an operator on $\ell_2(\mathbb{N}_0)$ in a natural way: $T_n(a)^{-1}_{j,k} = \delta_{j-k}$ for $j > n$ or $k > n$. Let $Q_n$ denote the projection in $\ell_2(\mathbb{N}_0)$ onto $\ell_2(\{0,1,\ldots,n\})$.

**Corollary 5.30.** *Let $a$ be as in Theorem 5.29. In addition, assume that the Beurling weight $\alpha > 1$. Then*

$$Q_n T_n(a)^{-1} Q_n \to T(a)^{-1} \quad \textit{strongly} \tag{5.230}$$

*in $\ell_2(\mathbb{N}_0)$ as $n \to \infty$, i.e., for every $f \in \ell_2(\mathbb{N}_0)$,*

$$\|Q_n T_n(a)^{-1} Q_n f - T(a)^{-1} f\|_{\ell_2(\mathbb{N}_0)} \to 0$$

*as $n \to \infty$.*

**Proof.** Fix $f \in \ell_2(\mathbb{N}_0)$. Set $P_n = 1 - Q_n$. Note that

$$\begin{aligned} &\|Q_n T_n(a)^{-1} Q_n f - T(a)^{-1} f\| \\ &= \|Q_n(T_n(a)^{-1} - T(a)^{-1})Q_n f - Q_n T(a)^{-1} P_n f - P_n T(a)^{-1} f\| \\ &\le \|Q_n(T_n(a)^{-1} - T(a)^{-1})Q_n f\| + \|Q_n T(a)^{-1} P_n f\| \\ &\quad + \|P_n T(a)^{-1} f\|. \end{aligned} \tag{5.231}$$

As $P_n \to 0$ strongly, $\|Q_n T(a)^{-1} P_n f\| \to 0$ and $\|P_n T(a)^{-1} f\| \to 0$ as $n \to \infty$. In order to estimate the remaining term in (5.231), observe that

$$\begin{aligned} &\|Q_n(T_n(a)^{-1} - T(a)^{-1})Q_n f\| \\ &\le \|Q_n(T_n(a)^{-1} - T(a)^{-1})Q_{[n/2]} f\| \\ &\quad + \|Q_n(T_n(a)^{-1} - T(a)^{-1})(Q_n - Q_{[n/2]}) f\|. \end{aligned} \tag{5.232}$$

From Theorem 5.29,

$$\begin{aligned} &\|Q_n(T_n(a)^{-1} - T(a)^{-1})Q_{[n/2]} f\|^2 \\ &= \sum_{j=0}^{n} \left| \sum_{k=0}^{[n/2]} (T_n(a)^{-1}(j,k) - T(a)^{-1}(j,k)) f(k) \right|^2 \\ &\le C^2 \sum_{j=0}^{n} \left( \sum_{k=0}^{[n/2]} \min\left\{ \frac{1}{(n+1-j)^\alpha}, \frac{1}{(n+1-k)^\alpha} \right\} |f(k)| \right)^2 \\ &\le C^2 \sum_{j=0}^{n} \sum_{k=0}^{[n/2]} \min\left\{ \frac{1}{(n+1-j)^{2\alpha}}, \frac{1}{(n+1-k)^{2\alpha}} \right\} \|f\|^2 \\ &\le C^2 \frac{(n+1)([n/2]+1)}{(n+1-[n/2])^{2\alpha}} \|f\|^2 \to 0 \end{aligned} \tag{5.233}$$

as $n \to \infty$. On the other hand, using $Q_n Q_{[n/2]} = Q_{[n/2]}$,

$$\begin{aligned} &\|Q_n(T_n(a)^{-1} - T(a)^{-1})(Q_n - Q_{[n/2]})f\| \\ &\leq \|Q_n(T_n(a)^{-1} - T(a)^{-1})Q_n\| \|(Q_n - Q_{[n/2]})f\|. \end{aligned} \tag{5.234}$$

By (5.208),

$$\begin{aligned} &\|Q_n(T_n(a)^{-1} - T(a)^{-1})Q_n\|^2 \\ &\leq \sum_{0 \leq j,k \leq n} \left|T_n(a)^{-1}(j,k) - T(a)^{-1}(j,k)\right|^2 \\ &\leq C^2 \sum_{0 \leq j,k \leq n} \min\left\{\frac{1}{(n+1-j)^{2\alpha}}, \frac{1}{(n+1-k)^{2\alpha}}\right\} \\ &\leq 2C^2 \sum_{j=0}^{n} \frac{1}{(j+1)^{2\alpha-1}} < \infty. \end{aligned} \tag{5.235}$$

Since

$$\|(Q_n - Q_{[n/2]})f\|^2 \leq \|(1 - Q_{[n/2]})f\|^2 = \sum_{k=[n/2]+1}^{\infty} |f(k)|^2 \to 0, \tag{5.236}$$

it follows that $\|Q_n(T_n(a)^{-1} - T(a)^{-1})(Q_n - Q_{[n/2]})f\| \to 0$ as $n \to \infty$. This completes the proof. □

Chapter 6

# Random Matrix Theory

There are now several basic references for random matrix theory, see for example [**Meh04, Dei99b, Blo09, For10, AGZ10, BS10, ABDF11, Tao12**]. We will mostly follow [**Dei99b**]. However, we will not provide all the details: the interested reader should consult the above references.

## 6.1. Unitary Ensembles and the Eigenvalue Density Function

Let $\mathcal{M} = \mathcal{M}(n)$ denote the space of $n \times n$ Hermitian matrices $M = (M_{ij}) = M^*$. Note that $\mathcal{M} \cong \mathbb{R}^{n^2}$ as a real vector space. For a given function $F : \mathcal{M} \to \mathbb{R}$, $F(M) \to \infty$ as $\|M\| \to \infty$, we consider the probability distribution $P^{(n)}$ on $\mathcal{M}$ defined by

$$P^{(n)}(M)dM = c \cdot e^{-F(M)}dM = c \cdot e^{-F(M)} \prod_{i=1}^{n} dM_{ii} \prod_{i<j} dM_{ij}^R dM_{ij}^I, \tag{6.1}$$

where $M_{ij}^R$ and $M_{ij}^I$ denotes the real and imaginary parts of $M_{ij}$, respectively, and $c$ is a normalization constant determined by the condition

$$c \cdot \int_{\mathcal{M}} e^{-F(M)}dM = 1. \tag{6.2}$$

The role of $F$ is to transform Lebesgue measure on $\mathcal{M}$, which is not finite, into a probability measure on $\mathcal{M}$.

Following Wigner, we require all coordinate systems to be equivalent. This means that

$$e^{-F(\tilde{M})}d\tilde{M} = e^{-F(M)}dM, \tag{6.3}$$

where $\tilde{M} = UMU^*$ for any unitary matrix $U$. As $d\tilde{M} = dM$, it follows from (6.3) (see for example [**Dei99b**]) that $e^{-F(\tilde{M})} = e^{-F(M)}$, i.e., $e^{-F(UMU^*)} = e^{-F(M)}$ for all unitary $U$ and Hermitian $M$. Choosing $U$ to diagonalize $M$, we see that $F$ depends only on the eigenvalues of $M$. Of particular interest is the case

$$F(M) = \operatorname{tr} Q(M) \tag{6.4}$$

for some polynomial $Q(x) = q_{2j}x^{2j} + \cdots + q_0$, where $q_{2j} > 0$ to ensure normalizability.

Observe that if $Q(x) = ax^2 + bx$, $a > 0$,

$$e^{-\operatorname{tr} Q(M)} = \prod_{i=1}^{n} e^{-aM_{ii}^2 - bM_{ii}} \prod_{i<j} e^{-2a(M_{ij}^R)^2} \prod_{i<j} e^{-2a(M_{ij}^I)^2}. \tag{6.5}$$

Thus, $c \cdot e^{-\operatorname{tr} Q(M)} dM$ corresponds to a distribution in which all the $n^2$ variables $(M_{ii}, M_{ij}^R, M_{ij}^I)$ are statistically independent, i.e.,

$$c \cdot e^{-\operatorname{tr} Q(M)} dM = \prod_{i=1}^{n} f_i(M_{ii}) dM_{ii} \prod_{i<j} f_{ij}^R(M_{ij}^R) dM_{ij}^R \prod_{i<j} f_{ij}^I(M_{ij}^I) dM_{ij}^I \tag{6.6}$$

for suitable (Gaussian) functions $f_i$, $f_{ij}^R$, $f_{ij}^I$.

Conversely, if we insist that

(1) $\operatorname{const} e^{-F(M)} dM$ is invariant under the change of basis $M \to \tilde{M} = UMU^*$, $U$ unitary, and that
(2) the $n^2$ variables $\{M_{ii}, M_{ij}^R, M_{ij}^I\}$ are statistically independent,

then one can show (see [**Meh04**]) that $Q(M) = ax^2 + bx + c$ for some real constants $a > 0$, $b$, $c$. Shifting and rescaling, we obtain

$$P^{(n)}(M) = \operatorname{const} \cdot e^{-\operatorname{tr}(M^2)} dM. \tag{6.7}$$

The *Gaussian Unitary Ensemble* (GUE) is defined as the probability space $\mathcal{M} = \mathcal{M}_n$ with the probability distribution (6.7). When the *potential* $M^2$ in (6.7) is replaced by other suitable functions, $Q(M)$,

$$P^{(n)}(M) = \operatorname{const} \cdot e^{-\operatorname{tr} Q(M)} dM, \tag{6.8}$$

$\mathcal{M}$ endowed with this measure is called a *unitary ensemble* as the distribution is invariant under unitary conjugation. Many other matrix ensembles are of interest, in particular, the so-called Wigner ensembles in which the entries $M_{ij}$ of the matrices are independent subject only to the symmetry condition $M_{ij} = \overline{M_{ji}}$. However, we only consider unitary ensembles which are sufficient for our main purposes.

Given a unitary ensemble, we are interested principally in the distribution of the eigenvalues. We now derive the density function of the eigenvalues from the density function of the matrix (6.8). Every $n \times n$ Hermitian matrix $M$ has a spectral representation, $M = U\Xi U^*$, where $U$ is unitary and $\Xi = \operatorname{diag}(\xi_1, \ldots, \xi_n)$ is the diagonal matrix of the (real) eigenvalues of $M$. From now on we will assume that the eigenvalues are *ordered* $\xi_1 \geq \xi_2 \geq \cdots \geq \xi_n$. One can show (we leave this as an exercise) that generically the eigenvalues of $M$ are simple. More precisely, the set of Hermitian matrices with simple spectrum is open and dense in $\mathbb{R}^{n^2} \cong \mathcal{M}(n)$ and has full measure, i.e., the complement of the set has measure 0. Let

$$\mathcal{A} = \{M \in \mathcal{M} \colon M \text{ has simple spectrum}\} \subset \mathbb{R}^{n^2}. \tag{6.9}$$

If $M \in \mathcal{A}$ has distinct spectrum $\xi_1 > \xi_2 > \cdots > \xi_n$, the associated orthonormal eigenvectors are uniquely determined up to a phase. Thus, the spectral representation $M = U\Xi U^*$ determines a (well defined) bijective map from $\mathcal{A}$ onto the homogeneous manifold $\mathcal{U}(n)\backslash\mathbb{T}^n$

$$\varphi \colon M \mapsto \{\xi_1, \ldots, \xi_n, U \bmod \mathbb{T}^n\}, \tag{6.10}$$

where $\mathcal{U}(n)$ is the group of $n \times n$ unitary matrices and $\mathbb{T}^n$ denotes the $n$-torus $\{(e^{i\theta_1}, \ldots, e^{i\theta_n}) \colon 0 \leq \theta_j < 2\pi\}$. Thus the spectral representation gives rise to a smooth local parametrization of the matrices in $\mathcal{A}$ in terms of $n^2$ variables $\xi_1 > \cdots > \xi_n, p_1, \ldots, p_\ell$ where $\ell = n(n-1)$ is the dimension of the manifold $\mathcal{U}(n)\backslash\mathbb{T}$. A direct, but nontrivial computation, see for example [**Dei99b**], shows that the Jacobian for the transformation $\varphi$ is

$$\frac{\partial M}{\partial(\xi_1, \ldots, \xi_n, p_1, \ldots, p_\ell)} = G(p_1, \ldots, p_\ell) \prod_{i<j} (\xi_i - \xi_j)^2 \tag{6.11}$$

for some function $G$. Thus the Jacobian is a product of two factors, one of which depends only on the eigenvalues and the other depends only on the *angular parts* $p_1, \ldots, p_\ell$. We have the following basic result.

**Theorem 6.1** (Eigenvalue Density Function). *Let $f(M)$ be a bounded class function on $\mathcal{M}$, i.e., $f(M) = f(UMU^*)$ for all unitary matrix $U$. Then $f$ is a (symmetric) function of the eigenvalues of $M$, $f(M) = \tilde{f}(\xi_1, \ldots, \xi_n)$, and*

$$\begin{aligned} &\text{const} \cdot \int_{\mathcal{M}_n} f(M) e^{-\operatorname{tr} Q(M)} dM \\ &= \frac{1}{\hat{Z}_n} \int_{\xi_1 > \cdots > \xi_n} \tilde{f}(\xi_1, \ldots, \xi_n) e^{-\sum Q(\xi_i)} \prod_{i<j} (\xi_i - \xi_j)^2 d\xi_1 \cdots d\xi_n \\ &= \frac{1}{Z_n} \int_{\mathbb{R}^n} \tilde{f}(\xi_1, \ldots, \xi_n) e^{-\sum Q(\xi_i)} \prod_{i<j} (\xi_i - \xi_j)^2 d\xi_1 \cdots d\xi_n. \end{aligned} \tag{6.12}$$

*Here, $Z_n$ and $\hat{Z}_n$ are normalization constants and satisfy $Z_n = n!\hat{Z}_n$.*

Here we have used the symmetry of the integrand. Thus, the density function of the (symmetrized) eigenvalues is given by

$$P^{(n)}(\xi_1,\dots,\xi_n) = \frac{1}{Z_n} e^{-\sum Q(\xi_i)} \prod_{i<j} (\xi_i - \xi_j)^2 d\xi_1 \cdots d\xi_n, \tag{6.13}$$

where

$$Z_n = \int_{\mathbb{R}^n} e^{-\sum Q(\xi_i)} \prod_{i<j} (\xi_i - \xi_j)^2 d\xi_1 \cdots d\xi_n. \tag{6.14}$$

For the ordered eigenvalues $\xi_1 \geq \cdots \geq \xi_n$, the distribution $\hat{P}^{(n)}(\xi)d\xi$ is the same as $P^{(n)}(\xi)d\xi$ except the normalization constant is now

$$\hat{Z}_n = \int_{\xi_1 > \cdots > \xi_n} e^{-\sum Q(\xi_i)} \prod_{i<j} (\xi_i - \xi_j)^2 d\xi_1 \cdots d\xi_n = \frac{1}{n!} Z_n. \tag{6.15}$$

Note that square of the Vandermonde determinant $\prod_{i<j}(\xi_i - \xi_j)^2$ in the expression for the measure implies that there is *level repulsion*, i.e., the probability that eigenvalues are close is very small.

## 6.2. Andréief's Formula and the Computation of Basic Statisitcs

Let $h$ be a bounded measurable real valued function on $\mathbb{R}$. We compute

$$\begin{aligned} (*) &\equiv \mathbb{E}\Big[\prod_{j=1}^n (1 + h(\xi_j))\Big] \\ &= \int_{\xi_1 > \cdots > \xi_n} \hat{P}^{(n)}(\xi_1,\dots,\xi_n) \prod_{j=1}^n (1 + h(\xi_j)) d\xi_j. \end{aligned} \tag{6.16}$$

Since the integrand in (6.16) is a symmetric function of $\xi_1,\dots,\xi_n$, $(*)$ equals

$$\begin{aligned} &\int_{\mathbb{R}^n} P^{(n)}(\xi_1,\dots,\xi_n) \prod_{j=1}^n (1 + h(\xi_j)) d\xi_j \\ &= \frac{1}{Z_n} \int_{\mathbb{R}^n} \prod_{1 \leq i < j \leq n} (\xi_i - \xi_j)^2 \prod_{j=1}^n e^{-Q(\xi_j)} (1 + h(\xi_j)) d\xi_j. \end{aligned} \tag{6.17}$$

We will obtain an explicit formula for (6.17); other statistics such as gap probabilities and correlation functions then follow by specializing $h$ (see Section 6.3).

Using the Vandermonde determinant formula (5.9),

$$\prod_{1 \leq i < j \leq n} (\xi_j - \xi_i) = \det(\xi_j^{i-1})_{1 \leq i,j \leq n}, \tag{6.18}$$

we may write (6.17) as

$$(6.19)\qquad (*) = \frac{1}{Z_n}\int_{R^n} \det(f_i(\xi_j))\det(g_i(\xi_j))\prod_{j=1}^{n}(1+h(\xi_j))d\mu(\xi_j),$$

where $f_i(s) = g_i(s) = s^{i-1}$ and $d\mu(s) = e^{-Q(s)}ds$. Letting $h = 0$ in (6.19) we obtain a formula for $Z_n$:

$$(6.20)\qquad Z_n = \int_{R^n} \det(\xi_j^{i-1})_{1\le i,j\le n}\det(\xi_j^{i-1})_{1\le i,j\le n}\prod_{j=1}^{n} d\mu(\xi_j) > 0.$$

In the calculations that follow, up until (6.27), the specific form of the $f_i$'s and $g_i$'s is not important as long as the integral in (6.19) is nonzero and finite. The following basic determinantal identity is due to Andréief [**And83**] and will be used in several places in this book.

**Lemma 6.2** (Andréief's Formula). *Let $(X, \mu)$ be a measure space. For two sequences of square integrable functions $f_i(x)$, $g_i(x)$, $i = 1, 2, \ldots, n$,*

$$(6.21)\qquad \begin{aligned}&\int_X\cdots\int_X \det\left(f_i(x_j)\right)_{1\le i,j\le n}\det\left(g_i(x_j)\right)_{1\le i,j\le n}\prod_{k=1}^{n} d\mu(x_k)\\ &= n!\cdot\det\left(\int_X f_i(x)g_j(x)d\mu(x)\right)_{1\le i,j\le n}.\end{aligned}$$

**Proof of Lemma 6.2.** By the definition of $\det\left(g_i(x_j)\right)$, the left-hand side of (6.21) is equal to

$$\begin{aligned}
&\sum_{\pi\in S_n}\operatorname{sgn}\pi\int_X\cdots\int_X \det\left(f_i(x_j)\right)\prod_{j=1}^{n} g_{\pi(j)}(x_j)\prod_{k=1}^{n} d\mu(x_k)\\
&=\sum_{\pi\in S_n}\operatorname{sgn}\pi\int_X\cdots\int_X\\
&\times\det\begin{pmatrix} f_1(x_1)g_{\pi(1)}(x_1) & f_1(x_2)g_{\pi(2)}(x_2) & \cdots & f_1(x_n)g_{\pi(n)}(x_n)\\ f_2(x_1)g_{\pi(1)}(x_1) & f_2(x_2)g_{\pi(2)}(x_2) & \cdots & f_2(x_n)g_{\pi(n)}(x_n)\\ \vdots & \vdots & \ddots & \vdots\\ f_n(x_1)g_{\pi(1)}(x_1) & f_n(x_2)g_{\pi(2)}(x_2) & \cdots & f_n(x_n)g_{\pi(n)}(x_n)\end{pmatrix}\prod_{k=1}^{n} d\mu(x_k)\\
&=\sum_{\pi\in n}\operatorname{sgn}\pi\cdot\det\left(\int_X f_i(x)g_{\pi(j)}(x)d\mu(x)\right)_{1\le i,j\le n}\\
&=\sum_{\pi\in S_n}\det\left(\int_X f_i(x)g_j(x)d\mu(x)\right)\\
&= n!\cdot\det\left(\int_X f_i(x)g_j(x)d\mu(x)\right)
\end{aligned}$$

as desired. □

**Exercise 6.3.** Show that Andréief's formula generalizes the Cauchy-Binet formula for the determinant of the product $AB$ of two rectangular matrices $A$, $B$, where, say $A$ is $k \times m$ and $B$ is $m \times k$.

Applying Lemma 6.2 with $X = \mathbb{R}$, (6.19) becomes

$$(*) = \frac{1}{Z_n} \det \left( \int_{\mathbb{R}} f_i(s) g_j(s)(1 + h(s)) d\mu(s) \right)_{1 \le i,j \le n}, \tag{6.22}$$

which can be rewritten in matrix form

$$(*) = \frac{1}{Z_n} \det \big(A + BM_hC\big), \tag{6.23}$$

where the operators $B, C, M_h$, and $A$ are defined as follows. Let $L^2(X) = L^2(X, d\mu)$ be the set of a square-integrable function on $X$ with respect to the measure $\mu$, and let $\ell_2(\mathbb{N}^{(n)})$ be the set of square-summable functions on $\mathbb{N}^{(n)} = \{1, 2, \ldots, n\}$. The operators $B : L^2(X) \to \ell_2(\mathbb{N}^{(n)})$ and $C : \ell_2(\mathbb{N}^{(n)}) \to L^2(X)$ are defined by the kernels

$$B(i, s) \equiv f_i(s), \qquad C(s, j) \equiv g_j(s), \qquad s \in X, \quad i, j \in \mathbb{N}^{(n)}. \tag{6.24}$$

The operator $M_h : L^2(X) \to L^2(X)$ denotes multiplication by $h : (M_h\alpha)(s) = h(s)\alpha(s)$. Note that as $h$ is a bounded function $M_h$ is a bounded operator on $L^2(X)$. Finally, $A = BC$. Explicitly, $A : \ell_2(\mathbb{N}^{(n)}) \to \ell_2(\mathbb{N}^{(n)})$ has kernel

$$A(i, j) = \int_X B(i, s)C(s, j) d\mu(s) = \int_X f_i(s) g_j(s) d\mu(s). \tag{6.25}$$

Now, as $Z_n$ equals $(*)$ when $h \equiv 0$, $Z_n = \det(A)$. As $Z_n \neq 0$ by assumption, $A$ is invertible and

$$(*) = \det(1 + A^{-1}BM_hC) = \det(1 + CA^{-1}BM_h), \tag{6.26}$$

where the general identity $\det(1 + TS) = \det(1 + ST)$ (see (A.39)) is used in the last equality. The operator $CA^{-1}BM_h$ in (6.26) acts on $L^2(X)$.

A primary focus of random matrix theory is the statistics of the eigenvalues as the dimension $n$ of the matrices tends to infinity. Observe that $n$ is now a parameter in the kernel of the operator $CA^{-1}BM_h$ but the space $X$ is fixed. This should be contrasted with the original formula (6.19), or the matrix formula (6.22), where the space $\mathbb{R}^n$ of integration and the size of determinant depends on $n$. As opposed to the situation where the underlying space varies with $n$, the asymptotic analysis of (operator) determinants on a fixed space falls into a well developed functional analytic framework that is well understood; see for example [**Sim05c**] and Appendix A.

We now utilize the specific forms

$$f_i(s) = g_i(s) = s^{i-1}. \tag{6.27}$$

Observe that in (6.22), due to row and column operations,

$$(6.28)\qquad \begin{aligned}(*) &= \frac{1}{Z_n}\det\left(\int_X s^{i-1}s^{j-1}(1+h(s))d\mu(s)\right)_{1\le i,j\le n}\\ &= \frac{1}{Z_n}\det\left(\int_X \pi_{i-1}(s)\pi_{j-1}(s)(1+h(s))d\mu(s)\right)_{1\le i,j\le n}\end{aligned}$$

for any monomials $\pi_k(s) = s^k +$lower degree. Furthermore, if we let $p_k(x) = \sum_{q=0}^k \gamma_{kq}x^q$, $\gamma_{kk} = \gamma_k > 0$, $k = 0, 1, \dots$, be a polynomial of degree $k$ with a positive leading coefficient, then

$$(6.29)\qquad (*) = \frac{1}{\tilde{Z}_n}\det\left(\int_X p_{i-1}(s)p_{j-1}(s)(1+h(s))d\mu(s)\right)_{1\le i,j\le n},$$

where $\tilde{Z}_n = Z_n\prod_{k=1}^n \gamma_{k-1}$. Now, choose the $p_k(s)$'s to be the *orthonormal polynomials* with respect to the measure $d\mu$. Then, taking $f_i(s) = g_i(s) = p_{i-1}(s)$ in (6.22)–(6.25), we obtain

$$(6.30)\qquad A(i,j) = \int_{\mathbb{R}} B(i,s)C(s,j)d\mu(s) = \int_{\mathbb{R}} p_{i-1}(s)p_{j-1}(s)d\mu(s) = \delta_{i,j}.$$

Hence, for the case at hand, $A$ is the identity matrix. Therefore,

$$(6.31)\qquad (*) = \det(1 + CBM_h) = \det(1 + \tilde{K}M_h),$$

where the operator $\tilde{K} = CB : L^2(\mathbb{R}, d\mu) \to L^2(\mathbb{R}, d\mu)$ has the kernel

$$(6.32)\qquad \tilde{K}(u,v) = \sum_{i=1}^n p_{i-1}(u)p_{i-1}(v).$$

By the Christoffel-Darboux formula (5.21),

$$(6.33)\qquad \tilde{K}(u,v) = \frac{\gamma_{n-1}}{\gamma_n}\cdot\frac{p_n(u)p_{n-1}(v) - p_{n-1}(u)p_n(v)}{u-v}.$$

Mapping $L^2(\mathbb{R}, d\mu)$ with $d\mu(s) = e^{-Q(s)}ds$ isomorphically onto $L^2(\mathbb{R}, ds)$, we obtain the following fundamental result for the unitary ensemble.

**Theorem 6.4.** *Let $h : \mathbb{R} \to \mathbb{R}$ be a bounded measurable function. Let $\{p_k(s)\}_{k=0}^\infty$ be the orthonormal polynomials with respect to $d\mu(s) = e^{-Q(s)}ds$ on $\mathbb{R}$ and denote by $\gamma_k$ the leading coefficient of $p_k(s)$. Then*

$$(6.34)\qquad \begin{aligned}\mathbb{E}\Big[\prod_{j=1}^n(1+h(\xi_j))\Big] &\equiv \int_{\xi_1>\cdots>\xi_n} \hat{P}^{(n)}(\xi_1,\dots,\xi_n)\prod_{j=1}^n(1+h(\xi_j))d\xi_j\\ &= \det(1+KM_h),\end{aligned}$$

*where $K : L^2(\mathbb{R}) \to L^2(\mathbb{R})$ is defined by the kernel*

$$(6.35)\qquad K(u,v) = \frac{\gamma_{n-1}}{\gamma_n}\cdot\frac{p_n(u)p_{n-1}(v) - p_{n-1}(u)p_n(v)}{u-v}e^{-\frac{1}{2}Q(u)-\frac{1}{2}Q(v)},$$

*and $M_h : L^2(\mathbb{R}) \to L^2(\mathbb{R})$ denotes multiplication by $h$.*

**Remark 6.5.** It follows from (6.32) that $K$ is a (bounded) finite rank operator and hence $KM_h$ is trace class in $L^2(\mathbb{R})$ for any bounded function $h$.

## 6.3. Gap Probabilities and Correlation Functions

Theorem 6.4 implies formulae for the gap probabilities and correlation functions of the eigenvalues as special cases. The computations in this section do not rely on the precise form of the underlying measure $\mu$. In fact, the computations are valid for general determinantal point processes (see Chapter 11).

Recall that the eigenvalues are ordered as $\xi_1 \geq \xi_2 \cdots \geq \xi_n$.

**6.3.1. Gap probability.** For a subset $I \subset \mathbb{R}$, the *gap probability* is defined by

$$\begin{aligned}\mathbb{P}\{\text{no eigenvalues in } I\} &= \int_{\xi_1>\cdots>\xi_n, \xi_j \notin I} \hat{P}^{(n)}(\xi) d^n\xi \\ &= \int_{\xi_1>\cdots>\xi_n} \prod_{j=1}^{n} (1-\chi_I(\xi_j)) \cdot \hat{P}^{(n)}(\xi) d^n\xi \\ &= \mathbb{E}\Big[\prod_{j=1}^{n}(1-\chi_I(\xi_j)\Big],\end{aligned} \tag{6.36}$$

where $\chi_I$ is the indicator function for $I$. It follows by Theorem 6.4 that the gap probability equals $\det(1-KM_{\chi_I})$. Note that $M_{\chi_I} = M_{\chi_I}^2$ and hence $\det(1-KM_{\chi_I}) = \det(1-KM_{\chi_I}^2) = \det(1-M_{\chi_I}KM_{\chi_I})$. Thus

$$\mathbb{P}\{\text{no eigenvalues in } I\} = \det(1-K_I), \tag{6.37}$$

where $K_I$ is the restriction of $K$ to $L^2(I)$.

**6.3.2. Distribution function of the $k$th largest eigenvalue.** There is no eigenvalue in the interval $(x,\infty)$ if and only if the largest eigenvalue is less than or equal to $x$. Hence by taking $I=(x,\infty)$ in (6.37) we find that

$$\mathbb{P}\{\xi_1 \leq x\} = \det(1-K_x), \tag{6.38}$$

where $K_x$ is the restriction of $K$ to $L^2((x,\infty))$.

We now consider the distribution of the second largest eigenvalue. The event that $\xi_2 \leq x$ is the disjoint union of two events, $A = \{\xi_1 \leq x\}$ and $B = \{\xi_1 > x, \xi_2 \leq x\}$. In the second event $B$, note that $\xi_j \leq x$ for all

$j = 2, 3, \ldots, n$. Hence

$$\begin{aligned}\mathbb{P}\{\xi_2 \le x\} &= \mathbb{P}\{\xi_1 \le x\} + \mathbb{P}\{\xi_1 > x, \xi_2 \le x\} \\ &= \mathbb{P}\{\xi_1 \le x\} + \mathbb{E}\left[\chi_I(\xi_1)\prod_{j=2}^{n}(1-\chi_I(\xi_j))\right],\end{aligned} \tag{6.39}$$

where $I = (x, \infty)$. In order to compute the expectation, observe that

$$\begin{aligned}-\frac{d}{dt}\bigg|_{t=1} \mathbb{E}\left[\prod_{j=1}^{n}(1-t\chi_I(\xi_j))\right] &= \sum_{k=1}^{n}\mathbb{E}\left[\chi_I(\xi_k)\prod_{j=1, j\neq k}^{n}(1-\chi_I(\xi_j))\right] \\ &= \mathbb{E}\left[\chi_I(\xi_1)\prod_{j=2}^{n}(1-\chi_I(\xi_j))\right],\end{aligned} \tag{6.40}$$

since if one of $\xi_k$, $k \ge 2$ is larger than $x$, then $\xi_1$ should be larger than $x$. Using Theorem 6.4 for the function $h(s) = -t\chi_I(s)$, we obtain

$$\begin{aligned}\mathbb{P}\{\xi_2 \le x\} &= \mathbb{P}\{\xi_1 \le x\} + \mathbb{P}\{\xi_1 \ge x, \xi_2 \le x\} \\ &= \mathbb{P}\{\xi_1 \le x\} - \frac{d}{dt}\bigg|_{t=1} \det(1 - tK_x),\end{aligned} \tag{6.41}$$

where $K_x$ is the restriction of the operator $K$ in (6.35) to $L^2((x, \infty))$. Inductively, we obtain the following (we leave this as an exercise).

**Proposition 6.6.** *Let $K_x$ be as above. Then*

$$\begin{aligned}\mathbb{P}\{\xi_1 \le x\} &= \det(1 - K_x), \\ \mathbb{P}\{\xi_{k+1} \le x\} &= \mathbb{P}\{\xi_k \le x\} + (-1)^k \frac{d^k}{dt^k}\bigg|_{t=1} \det(1 - tK_x), \qquad k \ge 1.\end{aligned} \tag{6.42}$$

### 6.3.3. Joint distribution of first $k$ eigenvalues.

We now compute $\mathbb{P}\{\xi_1 \le x_1, \ldots, \xi_k \le x_k\}$ for $x_k \le \cdots \le x_1$. We illustrate the computation for the cases $k = 2$ and $k = 3$. When $k = 2$,

$$\begin{aligned}\mathbb{P}\{\xi_1 \le x_1, \xi_2 \le x_2\} &= \mathbb{P}\{\xi_1 \le x_2\} + \mathbb{P}\{x_2 < \xi_1 \le x_1, \xi_2 \le x_2\} \\ &= \mathbb{P}\{\xi_1 \le x_2\} + \mathbb{E}\left[\chi_{I_2}(\xi_1)\prod_{j=2}^{n}(1-\chi_{I_1\cup I_2}(\xi_j))\right],\end{aligned} \tag{6.43}$$

where $I_1 = (x_1, \infty)$ and $I_2 = (x_2, x_1]$. But

$$
\begin{aligned}
&-\frac{\partial}{\partial t_2}\bigg|_{t_2=1} \mathbb{E}\left[\prod_{j=1}^{n}(1-(\chi_{I_1}+t_2\chi_{I_2})(\xi_j))\right] \\
&\qquad = \sum_{k=1}^{n} \mathbb{E}\left[\chi_{I_2}(\xi_k) \prod_{j=1, j\neq k}^{n} (1-\chi_{I_1\cup I_2}(\xi_j))\right] \qquad (6.44)\\
&\qquad = \mathbb{E}\left[\chi_{I_2}(\xi_1)\prod_{j=2}^{n}(1-\chi_{I_1\cup I_2}(\xi_j))\right]
\end{aligned}
$$

since $\xi_1 \geq \xi_2 \geq \cdots$. Hence

$$
\begin{aligned}
&\mathbb{P}\{\xi_1 \leq x_1, \xi_2 \leq x_2\} \\
&\qquad = \mathbb{P}\{\xi_1 \leq x_2\} - \frac{\partial}{\partial t_2}\bigg|_{t_2=1} \mathbb{E}\left[\prod_{j=1}^{n}(1-(\chi_{I_1}+t_2\chi_{I_2})(\xi_j))\right] \qquad (6.45)\\
&\qquad = F_2(1,1) - \frac{\partial F_2}{\partial t_2}(1,1),
\end{aligned}
$$

where

$$
F_2(t_1,t_2) \equiv \mathbb{E}\left[\prod_{j=1}^{n}(1-(t_1\chi_{I_1}+t_2\chi_{I_2})(\xi_j))\right]. \qquad (6.46)
$$

When $k = 3$, let $x_3 \leq x_2 \leq x_1$. Setting $I_1 = (x_1, \infty)$, $I_2 = (x_2, x_1]$, and $I_3 = (x_3, x_2]$,

$$
\begin{aligned}
&\mathbb{P}\{\xi_1 \leq x_1, \xi_2 \leq x_2, \xi_3 \leq x_3\} \\
&= \mathbb{P}\{\xi_1 \leq x_3\} + \mathbb{P}\{x_3 < \xi_1 \leq x_2, \xi_2 \leq x_2, \xi_3 \leq x_3\} \\
&\qquad + \mathbb{P}\{x_2 \leq \xi_1 \leq x_1, \xi_2 \leq x_2, \xi_3 \leq x_3\} \qquad (6.47)\\
&= \mathbb{P}\{\xi_1 \leq x_3\} + \mathbb{P}\{\xi_1 \in I_3, \xi_2 \leq x_3\} + \mathbb{P}\{\xi_1 \in I_3, \xi_2 \in I_3, \xi_3 \leq x_3\} \\
&\qquad + \mathbb{P}\{\xi_1 \in I_2, \xi_2 \leq x_3\} + \mathbb{P}\{\xi_1 \in I_2, \xi_2 \in I_3, \xi_3 \leq x_3\}.
\end{aligned}
$$

If we set

$$
F_3(t_1,t_2,t_3) \equiv \mathbb{E}\left[\prod_{j=1}^{n}(1-(t_1\chi_{I_1}+t_2\chi_{I_2}+t_3\chi_{I_3})(\xi_j))\right], \qquad (6.48)
$$

a computation similar to (6.44) shows that

$$
\begin{aligned}
&\mathbb{P}\{\xi_1 \leq x_1, \xi_2 \leq x_2, \xi_3 \leq x_3\} \\
&\qquad = F_3(1,1,1) - \frac{\partial}{\partial t_3}F_3(1,1,1) + \frac{1}{2}\left(\frac{\partial^2}{\partial t_3^2}F_3\right)(1,1,1) \qquad (6.49)\\
&\qquad\qquad - \left(\frac{\partial}{\partial t_2}F_3\right)(1,1,1) + \left(\frac{\partial^2}{\partial t_2 \partial t_3}F_3\right)(1,1,1).
\end{aligned}
$$

We leave it as an exercise to prove the following result for general $k$.

**Proposition 6.7.** *For $k \in \mathbb{N}$, let*

$$\Lambda_k = \{n = (n_1, \ldots, n_k) \in \mathbb{N}_0^k : n_1 + \cdots + n_r \le r - 1, r = 1, 2, \ldots, k\}. \tag{6.50}$$

*Set $|n| = n_1 + \cdots + n_k$. Then, for $x_k \le \cdots \le x_1$,*

$$\begin{aligned} &\mathbb{P}\{\xi_1 \le x_1, \xi_2 \le x_2, \ldots, \xi_k \le x_k\} \\ &\quad = \sum_{n \in \Lambda_k} \frac{(-1)^{|n|}}{n_1! \cdots n_k!} \frac{\partial^{|n|}}{\partial^{n_1} t_1 \cdots \partial^{n_k} t_k} F_k(1, \ldots, 1), \end{aligned} \tag{6.51}$$

*where*

$$\begin{aligned} &F_k(t_1, \ldots, t_k) \\ &= \mathbb{E}\left[\prod_{j=1}^{n} (1 - (t_1 \chi_{(x_1,\infty)} + t_2 \chi_{(x_2,x_1]} + \cdots + t_k \chi_{(x_k,x_{k-1}]})(\xi_j))\right] \\ &= \det\left(1 - (t_1 K \chi_{(x_1,\infty)} + t_2 K \chi_{(x_2,x_1]} + \cdots + t_k K \chi_{(x_k,x_{k-1}]})\right). \end{aligned} \tag{6.52}$$

**6.3.4. Correlation functions.** Let $x_1, \ldots, x_k$ be distinct real numbers. Let $\Delta_j$, $j = 1, \ldots, k$, be small positive numbers such that the sets $I_j \equiv [x_j, x_j + \Delta_j]$ are disjoint. Consider

$$\mathbb{P}\{I_j \text{ contains exactly one eigenvalue}, j = 1, 2, \ldots, k\}. \tag{6.53}$$

Here it is not important which eigenvalue is contained in which interval. We are only interested in the event that there is exactly one eigenvalue in each of $I_j$'s.

Let $y_1 > \cdots > y_k$ be the ordered rearrangement of $x_1, \ldots, x_k$, and let $J_1, \ldots, J_k$ be the corresponding rearrangement of the intervals $I_1, \ldots, I_k$ so that $J_1$ lies to the right of $J_2$, etc. Since all $J_1, \ldots, J_k$ can possibly contain any of $\xi_{i_1}, \ldots, \xi_{i_k}$, where $i_1 < \cdots < i_k$, (6.53) equals

$$\sum_{i_1 < \cdots < i_k} \mathbb{E}\,[\prod_{j=1}^{k} \chi_{J_j}(\xi_{i_j})] = \mathbb{E}\,[\sum_{i_1 < \cdots < i_k} \prod_{j=1}^{k} \chi_{J_j}(\xi_{i_j})]. \tag{6.54}$$

Now since the term inside of $\mathbb{E}$ is a symmetric function of $\xi_1, \ldots, \xi_n$, (6.54) equals

$$\begin{aligned} &\int_{\mathbb{R}^n} \left[\sum_{i_1 < \cdots < i_k} \prod_{j=1}^{k} \chi_{J_j}(\xi_{i_j})\right] P^{(n)}(\xi) d^n\xi \\ &\quad = \sum_{i_1 < \cdots < i_k} \int_{\mathbb{R}^n} \prod_{j=1}^{k} \chi_{J_j}(\xi_{i_j}) P^{(n)}(\xi) d^n\xi \\ &\quad = \sum_{i_1 < \cdots < i_k} \int_{\mathbb{R}^n} \chi_{J_1}(\eta_1) \cdots \chi_{J_k}(\eta_k) P^{(n)}(\eta) d^n\eta. \end{aligned} \tag{6.55}$$

Here in the last line we changed variables so that $\xi_{i_1} \mapsto \eta_1, \dots, \xi_{i_k} \mapsto \eta_k$ and used the fact that $P^{(n)}(\xi)$ is symmetric in $\xi_1, \dots, \xi_n$. As the last integral is now the same for all $i_1 < \cdots < i_k$, (6.55) equals

$$\binom{n}{k} \int_{\mathbb{R}^n} \chi_{J_1}(\eta_1) \cdots \chi_{J_k}(\eta_k) P^{(n)}(\eta) d^n \eta. \tag{6.56}$$

Therefore, we obtain

$$\begin{aligned} &\lim_{\Delta_j \downarrow 0, j=1,\dots,k} \frac{\mathbb{P}\{I_j \text{ contains exactly one eigenvalue, } j = 1, 2, \dots, k\}}{\Delta_1 \cdots \Delta_k} \\ &= \binom{n}{k} \int_{\mathbb{R}^{n-k}} P^{(n)}(y_1, \dots, y_k, \eta_{k+1}, \dots, \eta_n) d\eta_{k+1} \cdots d\eta_n, \end{aligned} \tag{6.57}$$

where $y_1 > \cdots > y_k$ is the ordered rearrangement of $x_1, \dots, x_k$. Since the last integral is symmetric in $y_1, \dots, y_k$, it remains the same when we replace $y_j$'s by $x_j$'s.

The following calculations will show that the eigenvalues of matrices chosen from a unitary ensemble give rise to a determinantal point process in a sense made precise in Chapter 11. We explain there that the statistics of such processes are conveniently described in terms of *correlation functions*; see Definition 11.1. Here we describe correlation functions only in the context of random matrices of a unitary ensemble.

**Definition 6.8.** The $k$th correlation function is defined by

$$\begin{aligned} &R_k(x_1, \dots, x_k) \\ &\equiv \frac{n!}{(n-k)!} \int_{\mathbb{R}^{n-k}} P^{(n)}(x_1, \dots, x_k, \xi_{k+1}, \dots, \xi_n) d\xi_{k+1} \cdots d\xi_n. \end{aligned} \tag{6.58}$$

By the considerations above, we have

$$\begin{aligned} &\lim_{\Delta_1 \downarrow 0 \cdots \Delta_k \downarrow 0} \frac{\mathbb{P}\{[x_j, x_j + \Delta_j] \text{ contains an eigenvalue, } 1 \le j \le k\}}{\Delta_1 \cdots \Delta_k} \\ &= k! R_k(x_1, \dots, x_k). \end{aligned} \tag{6.59}$$

In the above limit, the probability that one of the intervals $[x_j, x_j + \Delta_j]$ contains more than one eigenvalue is negligible compared to $\Delta_1 \cdots \Delta_k$.

The correlation functions can be used to compute the following expectations.

**Lemma 6.9.** *For a subset $I$ of $\mathbb{R}$,*

(6.60)

$$\int_{I^k} R_k(x_1, \dots, x_k) dx_1 \cdots dx_k = \mathbb{E}[\# \textit{ of } k\textit{-tuples of eigenvalues in } I].$$

*In particular,*

$$\int_I R_1(x)dx = \mathbb{E}[\#\ \textit{of eigenvalues in } I]. \tag{6.61}$$

Here, the $k$-tuples are ordered. For example, for $k = 2$, if two eigenvalues $\xi_i$ and $\xi_j$ are in $I$, then both pairs $(\xi_i, \xi_j)$ and $(\xi_j, \xi_i)$ are counted.

**Proof.** From the definition of $R_k$,

$$\int_{I^k} R_k(x_1, \dots, x_k)dx_1 \cdots dx_k = \frac{n!}{(n-k)!} \int_{\mathbb{R}^n} \prod_{i=1}^k \chi_I(\xi_i) P^{(n)}(\xi) d^n\xi. \tag{6.62}$$

Let $T_k = \{(i_1, \dots, i_k) \in \{1, \dots, n\}^k : i_1, \dots, i_k \text{ are all distinct}\}$ be the set of $k$-tuples of distinct indices from $\{1, \dots, n\}$. Note that $|T_k| = \frac{n!}{(n-k)!}$. Using the symmetry of $P^{(n)}$, the above integral equals

$$\int_{\mathbb{R}^n} \sum_{(i_1,\dots,i_k)\in T_k} \prod_{j=1}^k \chi_I(\xi_{i_j}) P^{(n)}(\xi) d^n\xi = \mathbb{E}\left[ \sum_{(i_1,\dots,i_k)\in T_k} \prod_{j=1}^k \chi_I(\xi_{i_j}) \right]. \tag{6.63}$$

The expectation is precisely the right-hand side of (6.60). □

Finally, we evaluate the correlation functions in terms of $K$. For a parameter $t$ and a subset $I$ of $\mathbb{R}$, consider

$$\mathbb{E}\left[ \prod_{i=1}^n (1 - t\chi_I(\xi_i)) \right]. \tag{6.64}$$

By multiplying out the product and collecting the terms with the same powers of $t$, (6.64) equals

$$\begin{aligned} &\sum_{k=0}^n (-t)^k \, \mathbb{E}\left[ \sum_{i_1<\cdots<i_k} \prod_{j=1}^k \chi_I(\xi_{i_j}) \right] \\ &= \sum_{k=0}^n (-t)^k \sum_{i_1<\cdots<i_k} \int \prod_{j=1}^k \chi_I(\xi_{i_j}) P^{(n)}(\xi) d^n\xi. \end{aligned} \tag{6.65}$$

Using (6.62), we find that (6.65) equals

$$\begin{aligned} &\mathbb{E}\left[ \prod_{i=1}^n (1 - t\chi_I(\xi_i)) \right] \\ &= \sum_{k=0}^n (-t)^k \sum_{i_1<\cdots<i_k} \frac{(n-k)!}{n!} \int_{I^k} R_k(x_1, \dots, x_k)dx_1 \cdots dx_k \\ &= \sum_{k=0}^n \frac{(-t)^k}{k!} \int_{I^k} R_k(x_1, \dots, x_k)dx_1 \cdots dx_k. \end{aligned} \tag{6.66}$$

On the other hand, from Theorem 6.4, (6.64) equals

$$\det(1 - tK_I), \tag{6.67}$$

where $K_I$ is the restriction of $K$ to $L^2(I)$. Using the Fredholm expansion of a determinant (see (A.42)),

$$\begin{aligned} &\det(1 - tK_I) \\ &= \sum_{k=0}^{\infty} \frac{(-t)^k}{k!} \int_{\mathbb{R}^k} \det \begin{pmatrix} K_I(x_1, x_1) & \cdots & K_I(x_1, x_k) \\ \vdots & \ddots & \vdots \\ K_I(x_k, x_1) & \cdots & K_I(x_k, x_k) \end{pmatrix} dx_1 \cdots dx_k \\ &= \sum_{k=0}^{\infty} \frac{(-t)^k}{k!} \int_{I^k} \det \left(K(x_i, x_j)\right)_{1 \le i,j \le k} dx_1 \cdots dx_k. \end{aligned} \tag{6.68}$$

Note that the above series terminates at $k = n$ since $K$ is of rank $n$. Comparing (6.68) with (6.66), we find

$$\int_{I^k} R_k(x_1, \ldots, x_k) dx_1 \cdots dx_k = \int_{I^k} \det \left(K(x_i, x_j)\right)_{1 \le i,j \le k} dx_1 \cdots dx_k. \tag{6.69}$$

Since this holds for arbitrary subsets $I$ of $\mathbb{R}$, we obtain, using the symmetry of $R_k(x_1, \ldots, x_k)$ and $\det(K(x_i, x_j))_{1 \le i,j \le k}$, the following result (we leave this as an exercise).

**Proposition 6.10.** *For $k = 1, 2, \ldots, n$,*

$$R_k(x_1, \ldots, x_k) = \det \left(K(x_i, x_j)\right)_{1 \le i,j \le k}. \tag{6.70}$$

This result shows that the eigenvalues of a random matrix chosen from a unitary ensemble give rise to a determinantal point process in the sense of Chapter 11.

**Remark 6.11.** In [**Meh04**], the relation (6.70) is obtained in a different way using the so-called *integrating out* lemma. This lemma, which is due to Gaudin, shows how to *integrate out* variables not under consideration when computing certain statistics (see also [**Dei99b**]).

## 6.4. Scaling Limits and Universality

One of the principal results in RMT is that, in the appropriate scaling limits, statistical quantities such as (6.37), (6.38), and (6.70) have a *universal form*, independent of the particular choice of the normalizing function $F$ in (6.1). In other words, on the appropriate scale, the particular choice of normalizing function $F$ in (6.1) is irrelevant.

Consider the particular case of unitary ensembles with measures of the form

$$\frac{1}{Z_n} e^{-n \operatorname{Tr} Q(M)} dM \tag{6.71}$$

on the $n \times n$ Hermitian matrices where we assume that $e^{-nQ(x)}dx$ has finite moments, $\int |x|^m e^{-nQ(x)}dx < \infty$ for $m = 0, 1, 2, \ldots$: $Q(x) = q_{2m}x^{2m} + \cdots$, $q_{2m} > 0$ certainly satisfies this condition. We build the kernel $K = K^{(n)}$ in (6.35) out of orthogonal polynomials with respect to the *varying* weight $e^{-nQ(x)}dx$ as follows. Let $\{p_k^{(n)}\}_{k=0}^{\infty}$ be the orthonormal polynomials with respect to $e^{-nQ(x)}dx$. Then

$$K^{(n)}(x,y) = b_{n-1}^{(n)} \frac{p_n^{(n)}(x)p_{n-1}^{(n)}(y) - p_{n-1}^{(n)}(x)p_n^{(n)}(y)}{x-y} e^{-\frac{n}{2}Q(x)-\frac{n}{2}Q(y)}, \tag{6.72}$$

where $b_k^{(n)} = \frac{\gamma_k^{(n)}}{\gamma_k^{(n)}}$. One shows that for a very large class of $Q$'s (see [**Dei99b**]),

$$\begin{aligned} &\frac{1}{n}K^{(n)}(x,x) \\ &= \frac{1}{n} b_{n-1}^{(n)} \left( (p_n^{(n)})'(x) p_{n-1}^{(n)}(x) - p_n^{(n)}(x) (p_{n-1}^{(n)})'(x) \right) e^{-nQ(x)} \\ &\to w(x) \end{aligned} \tag{6.73}$$

as $n \to \infty$, where $d\mu_{eq}(x) = w(x)dx$ is the so-called *equilibrium measure* for $Q$. The convergence in (6.73) is in general weak, i.e., $\frac{1}{n}\int \phi(x)K^{(n)}(x,x)dx \to \int \phi(x)w(x)dx$ for all smooth functions $\phi(x)$ with compact support, but under appropriate additional assumptions (see [**DKM$^+$99b**]), the convergence is also pointwise. When $Q(x) = x^2$, the GUE case, the equilibrium measure is given by

$$w(x) = \frac{1}{\pi}\sqrt{2-x^2}\,\chi_{|x|<\sqrt{2}}dx, \tag{6.74}$$

the so-called *semi-circle law*. More generally, if $Q(x)$ is

- real analytic,
- $\frac{Q(x)}{\log|x|} \to +\infty$ as $|x| \to \infty$,

then the convergence in (6.73) is (weak and) pointwise and $d\mu_{eq}(x) = w(x)dx$ is absolutely continuous with respect to Lebesgue measure and is supported on a finite union of intervals. The equilibrium measure can be obtained by solving an energy minimization problem (see, e.g., [**Dei99b**]).

From (6.61) and (6.70), we see that as $n \to \infty$

$$\begin{aligned} \mathbb{E}[\#\text{ of eigenvalues in } (x, x+\Delta x)] &= \int_x^{x+\Delta x} K^{(n)}(t,t)dt \\ &\sim nw(x)\Delta x. \end{aligned} \tag{6.75}$$

Consider the eigenvalues in a neighborhood of a point $x$ in the support of $\mu_{eq}$, i.e., $w(x) > 0$, and scale the eigenvalues so that the expected number of eigenvalues per unit interval is finite. By convention we scale so that

$$\text{(6.76)} \qquad \frac{\text{expected number of eigenvalues in a small interval about } x}{\text{length of interval}} = 1.$$

As $\frac{nw(x)\Delta x}{\Delta x} \sim K^{(n)}(x,x)$, this means we should scale by $K^{(n)}(x,x)$. For $\theta > 0$, set

$$\text{(6.77)} \quad A_{x,n}(\theta) = \mathbb{P}\left\{\text{no eigenvalues in } \left(x - \frac{\theta}{K^{(n)}(x,x)}, x + \frac{\theta}{K^{(n)}(x,x)}\right)\right\}.$$

**Theorem 6.12** (Universality for the gap probability in the bulk). *Let $A_{x,n}(\theta)$ be as above. Then*

$$\text{(6.78)} \qquad \lim_{n\to\infty} A_{x,n}(\theta) = \det(1 - \mathbb{S}_\theta),$$

*where $\mathbb{S}_\theta$ is the operator with kernel*

$$\text{(6.79)} \qquad \mathbb{S}(\xi,\eta) = \frac{\sin \pi(\xi - \eta)}{\pi(\xi - \eta)}$$

*acting on $L^2((-\theta,\theta))$.*

In other words, apart from the choice of scaling, the precise form of the weight $e^{-nQ(x)}$ washes out from the asymptotic limit.

The proof of Theorem 6.12 proceeds as follows. After scaling in (6.37), the gap probability (6.77) takes the form

$$\text{(6.80)} \qquad A_{x,n}(\theta) = \det(1 - D_{x,n}),$$

where

$$\text{(6.81)} \quad D_{x,n}(\xi,\eta) = \frac{1}{K^{(n)}(x,x)} K^{(n)}\left(x + \frac{\xi}{K^{(n)}(x,x)}, x + \frac{\eta}{K^{(n)}(x,x)}\right)$$

acts on $L^2((-\theta,\theta))$. Since $K^{(n)}$ is given in terms of orthonormal polynomials via the Christoffel-Darboux formula (6.73), we see that the proof of (6.78) reduces to controlling the so-called *Plancherel-Rotach* asymptotics for the $p_j^{(n)}$'s. Plancherel and Rotach (1929) first computed this type of asymptotics for Hermite polynomials; see [**Sze75**, Theorem 8.22.9]. Indeed, one can prove (6.78) for the Gaussian case (GUE) using these classical estimates. If $Q(x) \neq x^2$, this type of asymptotics was proved more recently (1997–99) in [**DKM$^+$99a, DKM$^+$99b**]. The proof of [**DKM$^+$99a, DKM$^+$99b**] uses the steepest-descent method for RHP's. Here the method is more sophisticated than in Section 4.7 and requires the introduction of a so-called *g-function.* The $g$-function is a general tool for RHP's and was introduced in [**DVZ97**] in their analysis of the zero-dispersion limit for the Korteweg-de Vries equation. In [**BI99**], Bleher and Its also use RH methods to prove universality for quartic potentials, $Q(x) = x^4 - cx^2$. For another approach to universality; see [**PS97**]. Note that there are analogous results for the $k$-point correlation functions; see [**DKM$^+$99a, DKM$^+$99b**].

We now consider the gap probability at the edge of the support of the equilibrium measure. From (6.74), for GUE where the probability density is proportional to $e^{-n\operatorname{Tr} M^2}$, we see that in a small interval $(\sqrt{2}-\Delta, \sqrt{2})$ at the top of the support of the equilibrium measure, the expected number of eigenvalues is approximately

$$n\int_{\sqrt{2}-\Delta}^{\sqrt{2}} \sqrt{2-x^2}dx \approx 2^{3/4}n\int_{\sqrt{2}-\Delta}^{\sqrt{2}} \sqrt{\sqrt{2}-x}dx = \frac{2^{7/4}}{3}n\Delta^{3/2}, \tag{6.82}$$

so we must choose $\Delta \sim n^{-2/3}$ in order that the expected number of eigenvalues per unit interval is finite and nonzero. We anticipate a nontrivial limit for the distribution function of the largest eigenvalue $\xi_{\max}$ under the scaling

$$\begin{aligned} &\mathbb{P}\left\{\xi_{\max} \leq \sqrt{2} + \frac{t}{\sqrt{2}n^{2/3}}\right\} \\ &\quad = \mathbb{P}\left\{\text{no eigenvalues in } \left(\sqrt{2} + \frac{t}{\sqrt{2}n^{2/3}}, \infty\right)\right\}. \end{aligned} \tag{6.83}$$

This reduces to evaluating the kernel

$$\frac{1}{\sqrt{2}n^{2/3}}K^{(n)}\left(\sqrt{2} + \frac{\xi}{\sqrt{2}n^{2/3}}, \sqrt{2} + \frac{\eta}{\sqrt{2}n^{2/3}}\right), \tag{6.84}$$

and one shows using classical properties of the Hermite functions that this kernel converges to the so-called *Airy kernel*

$$\mathbb{A}(u,v) = \frac{\operatorname{Ai}(u)\operatorname{Ai}'(v) - \operatorname{Ai}'(u)\operatorname{Ai}(v)}{u-v}, \tag{6.85}$$

where $\operatorname{Ai}(u)$ is the Airy function (see (1.12)).

Thus we have:

**Theorem 6.13** (Edge asymptotics for GUE). *For GUE,*

$$\lim_{n\to\infty} \mathbb{P}\left\{(\xi_{\max} - \sqrt{2})\sqrt{2}n^{2/3} \leq t\right\} = \det(1 - \mathbb{A}_t), \tag{6.86}$$

*where* $\mathbb{A}_t$ *has kernel* $\mathbb{A}(u,v)$ *as in* (6.85) *above, and acts on* $L^2((t,\infty), dt)$.

**Exercise 6.14.** Show, using $\operatorname{Ai}''(s) = s\operatorname{Ai}(s)$, that

$$\mathbb{A}(u,v) = \int_0^\infty \operatorname{Ai}(u+s)\operatorname{Ai}(s+v)ds. \tag{6.87}$$

Then show that the operator with kernel $\operatorname{Ai}(u+s)$, $u > t$ and $s > 0$, is a Hilbert-Schmidt operator (use the asymptotics (1.12)), and conclude that $\mathbb{A}_t$ is a trace-class operator. See Appendix A.

**Lemma 6.15.** *For every $t \in \mathbb{R}$, the self-adjoint operator $\mathbb{A}_t$ satisfies $0 \le \mathbb{A}_t \le 1$, and so $\|\mathbb{A}_t\| \le 1$. Moreover, $1 - \mathbb{A}_t$ is invertible.*

**Proof.** Recall that

$$\mathrm{Ai}(x) = \frac{1}{2\pi} \int_{-\infty}^{\infty} e^{i(\frac{y^3}{3}+xy)} dy \tag{6.88}$$

and $\mathrm{Ai}(x)$ is real for real $x$. For $f \in L^2((t,\infty))$ with compact support,

$$\begin{aligned} 0 \le (f, \mathbb{A}_t f) &= \int_0^\infty \left| \int_t^\infty \mathrm{Ai}(s+u) f(u) du \right|^2 ds \\ &\le \int_{-\infty}^\infty \left| \int_{-\infty}^\infty \mathrm{Ai}(s+u) f_t(u) du \right|^2 ds, \end{aligned} \tag{6.89}$$

where $f_t(u) = \chi_{(t,\infty)}(u) f(u)$. Since

$$\int_{-\infty}^\infty \mathrm{Ai}(s+u) f_t(u) du = \frac{1}{\sqrt{2\pi}} \int_{-\infty}^\infty e^{isy} e^{i\frac{y^3}{3}} \hat{f}_t(-y) dy = \hat{g}(-s), \tag{6.90}$$

where $\hat{f}_t(y) = \frac{1}{\sqrt{2\pi}} \int e^{-iyu} f_t(u) du$ is the Fourier transform of $f_t$ and $g(y) = e^{i\frac{y^3}{3}} \hat{f}_t(-y)$, we obtain

$$\begin{aligned} 0 \le (f, \mathbb{A}_t f) \le \int_{-\infty}^\infty |\hat{g}(-s)|^2 \, ds &= \int_{-\infty}^\infty |g(y)|^2 \, dy \\ &= \int_{-\infty}^\infty |f_t(u)|^2 \, du = (f, f). \end{aligned} \tag{6.91}$$

Thus, $0 \le \mathbb{A}_t \le 1$. Since $\mathbb{A}_t$ is self-adjoint, $\|\mathbb{A}_t\| = \sup_{\|f\|=1} |(f, \mathbb{A}_t f)| \le 1$.

Since $\mathbb{A}_t$ is a compact operator, in order to show that $1 - \mathbb{A}_t$ is invertible, it is enough to show that there is no $f \in L^2((t,\infty))$, $f \not\equiv 0$, such that $\mathbb{A}_t f = f$. Suppose, on the contrary, that there is such an $f$. Then, for such $f$, we must have $(f, \mathbb{A}_t f) = \int_{-\infty}^\infty |\hat{g}(-s)|^2 ds$ in (6.91). On the other hand, (6.89) implies that $(f, \mathbb{A}_t f) = \int_0^\infty |\hat{g}(-s)|^2 ds$. Hence $\hat{g}(-s) = 0$ for $s < 0$, that is

$$\int_t^\infty \mathrm{Ai}(s+u) f(u) du = 0 \qquad \text{for } s < 0, \tag{6.92}$$

and it follows then by the analyticity properties of $\mathrm{Ai}(x)$ (we leave this as an exercise; see (6.88)) that (6.92) must hold for $s > 0$ as well, which implies $f \equiv 0$. This is a contradiction. Therefore, $1 - \mathbb{A}_t$ is invertible. □

The behavior described by Theorem 6.13 is universal (see (6.94) below) at any endpoint of the support of the equilibrium measure, as long as $d\mu_{eq}(x) = w(x)dx$ *vanishes like a square root* at the largest endpoint. It is possible for $w(x)$ to have different power-law behavior at an endpoint and then the result (6.86) will change. But again, there will be universality in the class of all $Q$'s which have the same power-law behavior at the endpoints of the equilibrium measure. Square root behavior at endpoints is generic among all $Q$'s in the sense that if we replace

$$e^{-nQ} \to e^{-n\alpha Q}, \qquad \alpha > 0, \tag{6.93}$$

then nonsquare root behavior can only occur for a discrete set of $\alpha$'s [**KM00**]. Hence, we have the following; see [**DG07**].

**Theorem 6.16** (Universality for the edge asymptotics). *Consider the unitary ensemble* (6.71). *For a generic potential $Q$ whose equilibrium measure vanishes like a square root at the leading edge $a$ of the equilibrium measure, there is a constant $b$ such that*

$$\lim_{n\to\infty} \mathbb{P}\left\{(\xi_{\max} - a)bn^{2/3} \le t\right\} = \det(1 - \mathbb{A}_t), \tag{6.94}$$

*where $\mathbb{P}$ is the unitary ensemble* (6.71).

The limit $\det(1 - \mathbb{A}_t)$ is just a different expression for the Tracy-Widom distribution $F(t)$ in (1.16). The equality of the two expressions is derived in the next section.

**Remark 6.17.** We have discussed several important scaling limits for unitary ensembles, Theorem 6.12, and Theorem 6.16. Such scaling limits for Wigner matrices, where the entries are assumed to be independent and invariance under unitary conjugation does not generally hold, have also been studied extensively. For Wigner matrices, [**Joh01b, EPR$^+$10, TV11, ERS$^+$10**] proved the bulk scaling limit analogue to (6.78) and [**SS98a, SS98b, Sos99, TV10, EKYY12, LY14**] proved the edge scaling limit analogue to (6.94). The techniques used to prove such limits for Wigner matrices are very different from those used for unitary ensembles and are beyond the scope of this text.

## 6.5. The Tracy-Widom Distribution Function

Quite remarkably, $\det(1-\mathbb{S}_\theta)$ and $\det(1-\mathbb{A}_t)$ in the previous section can be expressed in terms of special solutions of the Painlevé V and II equations, respectively. The first result is a famous calculation of Jimbo, Miwa, Mori, and Sato [**JMMS80**], and the second is a (now equally famous) calculation of Tracy and Widom [**TW94**].

The result of [**TW94**] is the following.

**Theorem 6.18.** *Let $u(s)$ be the (unique, global) solution of the Painlevé II equation (see* (1.13)*)*

$$u'' = su + 2u^3, \tag{6.95}$$

*satisfying*

$$u(s) \sim \mathrm{Ai}(s) \qquad \textit{as } s \to +\infty. \tag{6.96}$$

*Then*

$$F(t) \equiv e^{-\int_t^\infty (s-t)(u(s))^2 ds} = \det(1 - \mathbb{A}_t). \tag{6.97}$$

We give a proof of Theorem 6.18 using Riemann-Hilbert methods following the argument in [**BD02**]. For the original proof, see [**TW94**].

**Proof.** The proof consists of two parts.

**Part 1: Expressing the Airy kernel in terms of an RHP.** By translation,

$$\det(1 - \mathbb{A}_t) = \det(1 - \mathcal{A}_t), \tag{6.98}$$

where $\mathcal{A}_t$ acts on $L^2((0,\infty))$ and has kernel

$$\mathcal{A}_t(\xi,\eta) = \frac{\mathrm{Ai}(\xi+t)\,\mathrm{Ai}'(\eta+t) - \mathrm{Ai}'(\xi+t)\,\mathrm{Ai}(\eta+t)}{\xi-\eta}, \qquad \xi,\eta > 0. \tag{6.99}$$

Clearly, (6.97) is equivalent to showing

$$-\frac{d^2}{dt^2}\log\det(1-\mathcal{A}_t) = (u(t))^2 \tag{6.100}$$

since, on the one hand, $u(t)$ approaches 0 rapidly as $t \to +\infty$ (see (6.96) and (1.12)), and, on the other hand, $\mathcal{A}_t \to 0$ in trace norm (we leave this as an exercise). Note that $\det(1-\mathcal{A}_t) > 0$ by Lemma 6.15.

Now

$$\frac{d}{dt}\log\det(1-\mathcal{A}_t) = \frac{d}{dt}\operatorname{Tr}\log(1-\mathcal{A}_t) = -\operatorname{Tr}\big((1-\mathcal{A}_t)^{-1}\mathcal{A}_t'\big), \tag{6.101}$$

where $\mathcal{A}_t'$ denotes the derivative of the operator $\mathcal{A}_t$ with respect to $t$. From (6.99) and the differential equation $\mathrm{Ai}''(x) = x\,\mathrm{Ai}(x)$, we have

$$\mathcal{A}_t'(\xi,\eta) = -\,\mathrm{Ai}(\xi+t)\,\mathrm{Ai}(\eta+t). \tag{6.102}$$

Note, in particular, that $\mathcal{A}_t'(\xi,\eta)$ is a rank 1 operator and so

$$\frac{d}{dt}\log\det(1-\mathcal{A}_t) = \langle(1-\mathcal{A}_t)^{-1}f, f\rangle, \qquad f(x) \equiv \mathrm{Ai}(x+t), \tag{6.103}$$

where $\langle,\rangle$ denotes here the standard inner product on $L^2((0,\infty))$.

We now use the fact that $\mathcal{A}_t$ is an integrable operator as defined in Section 5.6. Setting

$$g(x) \equiv \mathrm{Ai}'(x+t), \tag{6.104}$$

we have

$$\mathcal{A}_t(\xi,\eta) = \frac{f(\xi)g(\eta) - g(\xi)f(\eta)}{\xi-\eta}. \tag{6.105}$$

From Theorem 5.21, if we set $R = (1-\mathcal{A}_t)^{-1}-1$, then $R$ is also an integrable operator with kernel

$$R(\xi,\eta) = \frac{F(\xi)G(\eta) - G(\xi)F(\eta)}{\xi - \eta}, \tag{6.106}$$

where

$$F(x) = \big((1-\mathcal{A}_t)^{-1}f\big)(x), \qquad G(x) = \big((1-\mathcal{A}_t)^{-1}g\big)(x). \tag{6.107}$$

Here we have used the fact that $\mathcal{A}_t$ is symmetric. In particular,

$$\frac{d}{dt}\log\det(1-\mathcal{A}_t) = \langle F, f\rangle. \tag{6.108}$$

By Theorem 5.21, the functions $F$ and $G$ can be expressed as

$$\begin{pmatrix} F(x) \\ G(x) \end{pmatrix} = m_+(x)\begin{pmatrix} f(x) \\ g(x) \end{pmatrix} \tag{6.109}$$

in terms of the solution $m$ to the normalized RHP $(\Sigma, v)$, where $\Sigma = (0,\infty)$, oriented from the left to the right, and the jump matrix $v$ is given by

$$\begin{aligned} v(z) &= I - 2\pi i \begin{pmatrix} f(z) \\ g(z) \end{pmatrix} (g(z), -f(z)) \\ &= \begin{pmatrix} 1 - 2\pi i f(z)g(z) & 2\pi i f(z)^2 \\ -2\pi i g(z)^2 & 1 + 2\pi i f(z)g(z) \end{pmatrix}. \end{aligned} \tag{6.110}$$

Note that

$$v(z)^{-1} = I + 2\pi i \begin{pmatrix} f(z) \\ g(z) \end{pmatrix} (g(z), -f(z)). \tag{6.111}$$

From (5.119), we have

$$m(z) = I + \frac{1}{2\pi i}\int_\Sigma \frac{m_+(s)(I - v^{-1}(s))}{s-z}ds. \tag{6.112}$$

Again by Theorem 5.21, the solution of the RHP $(\Sigma, v)$ exists as $1-\mathcal{A}_t$ is invertible (see Lemma 6.15). Note that $\det v(z) \equiv 1$ and hence $\det m(z) \equiv 1$ (see the Proof of Riemann-Hilbert Problem 5.3). Inserting the formula (6.111) for $v^{-1}$, we find

$$\begin{aligned} m(z) &= I - \int_0^\infty m_+(s)\begin{pmatrix} f(s) \\ g(s) \end{pmatrix} (g(s), -f(s))\frac{ds}{s-z} \\ &= I - \int_0^\infty \begin{pmatrix} F(s) \\ G(s) \end{pmatrix} (g(s), -f(s))\frac{ds}{s-z}. \end{aligned} \tag{6.113}$$

As $z\to\infty$, $m(z)$ has the expansion

$$m(z) = I + \frac{m_1}{z} + \frac{m_2}{z^2} + O(z^{-3}), \ z\to\infty, \tag{6.114}$$

for some constant matrices $m_1$ and $m_2$. Now,

$$m_1 = \int_0^\infty \begin{pmatrix} F(s) \\ G(s) \end{pmatrix} (g(s), -f(s))ds. \tag{6.115}$$

Hence, we see from (6.108) that

$$\frac{d}{dt} \log \det(1 - \mathcal{A}_t) = -(m_1)_{12}, \tag{6.116}$$

where

$$m_1 = \begin{pmatrix} (m_1)_{11} & (m_1)_{12} \\ (m_1)_{21} & (m_1)_{22} \end{pmatrix}. \tag{6.117}$$

**Part 2: Deriving a differential equation from the RHP.** Having expressed the derivative of $\log\det(1 - \mathcal{A}_t)$ in terms of the solution of the RHP $(\Sigma, v)$, we now show that the jump condition of the RHP implies that $(m_1)_{12}$ solves the Painlevé II equation. A key observation is the following: if $U(z)$ and $V(z)$ are entire and

$$B(z) \equiv \begin{pmatrix} f(z) & U(z) \\ g(z) & V(z) \end{pmatrix} \tag{6.118}$$

has determinant 1 for all $z \in \mathbb{C}$, the jump matrix (6.110) satisfies

$$B(z)^{-1} v(z) B(z) = \begin{pmatrix} 1 & 2\pi i \\ 0 & 1 \end{pmatrix} \equiv v_0, \qquad z \in \Sigma. \tag{6.119}$$

This can be checked directly. Such a matrix $B(z)$ can be obtained in the following way. Let $P(z)$ be the fundamental matrix solution of the linear differential equation

$$\frac{d}{dz} P(z) = \begin{pmatrix} 0 & 1 \\ z & 0 \end{pmatrix} P(z), \tag{6.120}$$

such that $\det P(z) = 1$ and the first column of $P(z)$ is $(\mathrm{Ai}(z), \mathrm{Ai}'(z))^t$. (As an exercise, show that such a solution exists.) Then

$$B(z) = P(z + t) \tag{6.121}$$

is entire, has determinant 1, and the first column is $(f, g)^t$.

With this choice of $B(z)$, set

$$M(z) = m(z) B(z), \qquad z \in \mathbb{C} \setminus \Sigma. \tag{6.122}$$

Then, from (6.119), $M$ satisfies the jump condition

$$M_+(z) = M_-(z) v_0, \qquad z \in \Sigma, \tag{6.123}$$

and the asymptotic condition

$$M(z) = \left(I + \frac{m_1}{z} + \frac{m_2}{z^2} + O(z^{-3})\right)P(z+t), \qquad z \to \infty. \tag{6.124}$$

Note that the jump matrix $v_0$ for $M$ is constant. Hence, by taking derivatives of (6.123), $\frac{\partial}{\partial t}M$ and $\frac{\partial}{\partial z}M$ satisfy the same jump condition as $M$. This implies that $\frac{\partial M}{\partial t}M^{-1}$ and $\frac{\partial M}{\partial z}M^{-1}$ have no jumps across the interval $(0,\infty)$, and therefore are analytic in $\mathbb{C}\setminus\{0\}$. Note that $M^{-1}$ exists as $\det m(z) = \det B(z) = 1$.

The functions $\frac{\partial M}{\partial t}M^{-1}$ and $\frac{\partial M}{\partial z}M^{-1}$ are determined once we know their behavior as $z \to 0$ and $z \to \infty$. First consider the limit $z \to 0$. By general theory (see Riemann-Hilbert Problem 5.16 in Section 5.5), $m_\pm(z) \in L^2(\mathbb{R})$: a fortiori, $m_\pm(z) \in L^2(0,\epsilon)$ for any $\epsilon > 0$. Since $P(z+t)$ is bounded near $z = 0$, $M(z)$ satisfies the same $L^2((0,\epsilon))$ condition. Note that

$$E(z) = \begin{pmatrix} 1 & -\log z \\ 0 & 1 \end{pmatrix}, \tag{6.125}$$

where $\log z$ is defined with branch cut $[0,\infty)$, satisfies the same jump condition (6.123) as $M(z)$. Thus, $M(z)E(z)^{-1}$ has no jump on $(0,\infty)$ and $M(z)E(z)^{-1}$ is analytic in $\mathbb{C}\setminus\{0\}$ with an isolated singularity at $z = 0$. But, as $E_\pm(z)^{-1}$ is in $L^2((0,\epsilon))$, it follows that $M_+E_+^{-1} \in L^1((0,\epsilon))$, hence the singularity at $z = 0$ is removable (we leave this as an exercise). So, we have $M(z)E(z)^{-1} = H(z)$ for some entire function $H(z)$. Clearly, $\det H(z) \equiv 1$. Thus,

$$\frac{\partial M}{\partial z}M^{-1} = \frac{\partial (HE)}{\partial z}(HE)^{-1} = \frac{\partial H}{\partial z}H^{-1} + H\begin{pmatrix} 0 & -\frac{1}{z} \\ 0 & 0 \end{pmatrix}H^{-1}. \tag{6.126}$$

The term $\frac{\partial H}{\partial z}H^{-1}$ is entire. The last term has a simple pole at $z = 0$. Setting

$$N \equiv \begin{pmatrix} ab & -a^2 \\ b^2 & -ab \end{pmatrix}, \qquad a(t) \equiv H_{11}(0), \quad b(t) \equiv H_{21}(0), \tag{6.127}$$

we find that

$$\frac{\partial M}{\partial z}M^{-1} - \frac{N}{z} \tag{6.128}$$

is entire.

Now we consider the limit $z \to \infty$. From (6.124),

$$\begin{aligned} \frac{\partial M}{\partial z}M^{-1} = &\left(I + \frac{m_1}{z} + \frac{m_2}{z^2} + O(z^{-3})\right)P'(z+t)P(z+t)^{-1} \\ &\times \left(I - \frac{m_1}{z} + \frac{m_1^2 - m_2}{z^2} + O(z^{-3})\right) + O(z^{-2}) \end{aligned} \tag{6.129}$$

as $z \to \infty$. Using the differential equation (6.120) for $P$, this implies that

$$\begin{aligned}\frac{\partial M}{\partial z}M^{-1}(z) =&zB_1 + B_2 + [m_1, B_1] \\ &+ \frac{1}{z}\Big([m_2, B_1] + [m_1, B_2] - [m_1, B_1]m_1\Big) + O(z^{-2}),\end{aligned} \tag{6.130}$$

where

$$B_1 \equiv \begin{pmatrix} 0 & 0 \\ 1 & 0 \end{pmatrix}, \qquad B_2 = B_2(t) \equiv \begin{pmatrix} 0 & 1 \\ t & 0 \end{pmatrix}, \tag{6.131}$$

and $[A, B] = AB - BA$ is the commutator. Therefore, by Liouville's theorem, we obtain

$$\frac{\partial M}{\partial z}M^{-1} - \frac{N}{z} =zB_1 + B_2 + [m_1, B_1]. \tag{6.132}$$

Comparing the term of order $1/z$, we also obtain the identity

$$-N + [m_2, B_1] + [m_1, B_2] - [m_1, B_1]m_1 = 0. \tag{6.133}$$

A similar argument applied to $\frac{\partial M}{\partial t}M^{-1}$ implies that

$$\frac{\partial M}{\partial t}M^{-1} =zB_1 + B_2 + [m_1, B_1] \tag{6.134}$$

and

$$m_1' + [m_2, B_1] + [m_1, B_2] - [m_1, B_1]m_1 = 0. \tag{6.135}$$

Here $m_1'$ is the derivative with respect to $t$. We leave the details as an exercise.

From (6.133) and (6.135), we find that

$$N = -m_1'. \tag{6.136}$$

The equations (6.132) and (6.134) are of Lax pair form $\frac{\partial M}{\partial z} = \tilde{A}M$ and $\frac{\partial M}{\partial t} = \tilde{B}M$ for suitable matrices $\tilde{A}$ and $\tilde{B}$. The compatibility condition $\frac{\partial^2 M}{\partial z \partial t} = \frac{\partial^2 M}{\partial t \partial z}$ implies that $\frac{\partial \tilde{A}}{\partial t} - \frac{\partial \tilde{B}}{\partial z} + [\tilde{A}, \tilde{B}] = 0$. Substituting the formulae for $\tilde{A}$ and $\tilde{B}$ into this condition, we obtain

$$zB_1' + \big(Q' - B_1 + [N, B_1]\big) + \frac{1}{z}\big(N' + [N, Q]\big) = 0, \tag{6.137}$$

where $Q = B_2 + [m_1, B_1]$. Using $B_1' = 0$, $B_2' = B_1$ and (6.136), (6.137) simplifies to the identity

$$-N' + [B_2, N] + [[m_1, B_1], N] = 0. \tag{6.138}$$

Note that since $\det m(z) = 1$ for all $z$, it follows from (6.114) that $\operatorname{Tr} m_1 = 0$. Set

$$m_1 = \begin{pmatrix} p(t) & q(t) \\ r(t) & -p(t) \end{pmatrix}. \tag{6.139}$$

Then, using the formula for $N$ in (6.127), we obtain

$$\begin{aligned} (ab)' &= b^2 + ta^2 - 2pa^2, \\ (a^2)' &= 2ab + 2qa^2, \\ (b^2)' &= 2tab - 4pab - 2qb^2. \end{aligned} \tag{6.140}$$

It is easy to check that two of the above equations imply the third. If $a(t) \neq 0$, the second equation implies that

$$b = a' - qa. \tag{6.141}$$

Inserting (6.141) into the third equation of (6.140), we find that if $b(t) \neq 0$,

$$a'' - (t - 2p + q' + q^2)a = 0. \tag{6.142}$$

On the other hand, the (12)-entry of the identity (6.135) (note that the (12)-entry of $[m_2, B_1]$ is 0) yields

$$q' + 2p - q^2 = 0. \tag{6.143}$$

Since the (12)-entries of the identity (6.136) implies that

$$q' = a^2, \tag{6.144}$$

we find from (6.142) and (6.143) that

$$a'' - (ta + 2a^3) = 0 \tag{6.145}$$

for $t$ such that $a(t) \neq 0$ and $b(t) \neq 0$. This is precisely the Painlevé II equation (6.95). One can show (we leave this as an exercise) that $a(t)$ and $b(t)$ are analytic in $t$ and not identically zero. Therefore, (6.145) holds for all $t$ by analytic continuation. From (6.116), (6.139), and (6.144), we have

$$-\frac{d^2}{dt^2} \log\det(1 - \mathcal{A}_t) = (m_1)'_{12} = q' = a^2, \tag{6.146}$$

where $a$ solves the Painlevé II equation (6.145).

It remains to check that $a(t)$ has the same asymptotic condition (6.96) as $u(t)$. Since $\log\det(1 - \mathcal{A}_t) = \operatorname{Tr}\log(1 - \mathcal{A}_t) \sim -\operatorname{Tr}(\mathcal{A}_t)$ as $t \to \infty$,

$$a^2(t) = -\frac{d^2}{dt^2} \log\det(1 - \mathcal{A}_t) \sim \frac{d^2}{dt^2} \operatorname{Tr} \mathcal{A}_t. \tag{6.147}$$

But, from (6.99), using $\operatorname{Ai}''(u) = u \operatorname{Ai}(u)$,

$$\operatorname{Tr} \mathcal{A}_t = \int_t^\infty [(\operatorname{Ai}'(u))^2 - u(\operatorname{Ai}(u))^2] du, \tag{6.148}$$

and hence

$$a^2(t) \sim \frac{d^2}{dt^2} \operatorname{Tr} \mathcal{A}_t = (\operatorname{Ai}(t))^2 \tag{6.149}$$

as $t \to +\infty$. Noting that the Painlevé II equation is invariant under the change $a \mapsto -a$, this proves Theorem 6.18. □

**Remark 6.19.** The above calculation in which a desired relation emerges (here, the relation is the compatibility of the mixed derivatives) is standard in the theory of integrable systems. The derivation of the Lax pair is standard in RH-theory and follows whenever the jump matrix is (or can be reduced to one that is) independent of the external parameters for the problem. This basic idea is again used in Section 7.3 below to obtain difference/differential equations associated to the longest increasing subsequence problem.

*Chapter 7*

# Toeplitz Determinant Formula

We now return to the discussion of the longest increasing subsequence problem. The goal of this and the next chapter is to prove determinantal formulae for the distribution function, $\mathbb{P}\{L(t) \le n\}$, where $L(t)$ denotes the length of the longest up/right path with nodes in the square with corners $(0,0)$ to $(t,t)$, in the case that the nodes are generated by a Poisson process (of parameter 1) in the first quadrant (see Section 2.1). Equivalently, $L(t)$ equals the length of the longest increasing subsequence of a random permutation of size $\mathcal{N}$ where $\mathcal{N}$ is a Poisson random variable with parameter $t^2$. Recall (2.3)

$$\mathbb{P}\{L(t) \le n\} = e^{-t^2} \sum_{N=0}^{\infty} \frac{t^{2N}}{N!} \mathbb{P}\{\ell_N \le n\}. \tag{7.1}$$

Here $\ell_0 \equiv 0$.

In this chapter we prove a Toeplitz determinant formula for $\mathbb{P}\{L(t) \le n\}$. We show in the next chapter that there is a different Fredholm determinant formula for the same quantity. The Toeplitz determinant formula was obtained by Gessel [**Ges90**] in 1990:

**Theorem 7.1** (Gessel's formula). *Let*

$$\varphi(z) = e^{t(z+z^{-1})}. \tag{7.2}$$

*Then for* $n = 0, 1, 2, \dots,$

$$\mathbb{P}\{L(t) \le n\} = e^{-t^2} D_{n-1}(\varphi), \tag{7.3}$$

*where* $D_{n-1}(\varphi) = \det(T_{n-1}(\varphi))$ *is the determinant of the* $n \times n$ *Toeplitz matrix with symbol* $\varphi$*. By convention, we set*

$$D_{-1}(\varphi) \equiv 1. \tag{7.4}$$

Recall that a Toeplitz matrix is

$$T_{n-1}(\varphi) = (\varphi_{j-k})_{0\le j,k\le n-1}\,, \tag{7.5}$$

where

$$\varphi_k \equiv \oint_{|z|=1} z^{-k}\varphi(z)\frac{dz}{2\pi i z} = \frac{1}{2\pi}\int_{-\pi}^{\pi} e^{-ik\theta}e^{2t\cos\theta}\,d\theta, \qquad k \in \mathbb{Z}, \tag{7.6}$$

is the $k$th Fourier coefficient of $\varphi$ (see (5.84) of Section 5.4).

Gessel proved a more general result of which Theorem 7.1 is a special case; see Section 10.4.

The Robinson-Schensted correspondence in Chapter 3 established a bijection between permutations $\{\pi\}_{\pi\in S_N}$ and pairs $\{(P(\pi), Q(\pi))\}$ of SYT's satisfying $\mathrm{sh}(P(\pi)) = \mathrm{sh}(Q(\pi))$, and moreover, the number of boxes in the first row of $\mathrm{sh}(P(\pi))$ equals $\ell_N(\pi)$. Hence

$$\begin{aligned}&\#\{\pi \in S_N : \ell_N(\pi) \le n\}\\ &= \#\{(P,Q) :\ P,\, Q \text{ are SYT's },\mathrm{sh}(P) = \mathrm{sh}(Q) \vdash N, (\mathrm{sh}(P))_1 \le n\}.\end{aligned} \tag{7.7}$$

Note that the transpose of an SYT is also an SYT and $(\mathrm{sh}(P^t))_1 = r(\mathrm{sh}(P))$. Thus

$$\begin{aligned}&\#\{\pi \in S_N : \ell_N(\pi) \le n\}\\ &= \#\{(P,Q) :\ P,\, Q \text{ are SYT's },\mathrm{sh}(P) = \mathrm{sh}(Q) \vdash N, r(\mathrm{sh}(P)) \le n\}\\ &= \sum_{\lambda\vdash N, r(\lambda)\le n} F_\lambda^2,\end{aligned} \tag{7.8}$$

where $F_\lambda$ is the number of SYT's with shape $\lambda$ (see (1.31)). Therefore (7.1) becomes

$$\begin{aligned}\mathbb{P}\{L(t) \le n\} &= e^{-t^2}\sum_{N=0}^{\infty}\frac{t^{2N}}{(N!)^2}\sum_{\lambda\vdash N, r(\lambda)\le n} F_\lambda^2\\ &= e^{-t^2}\sum_{r(\lambda)\le n}\frac{t^{2|\lambda|}}{(|\lambda|!)^2}F_\lambda^2,\end{aligned} \tag{7.9}$$

where the last sum is over all partitions (of any number) such that $r(\lambda) \le n$. Here, as before, for the empty partition $\lambda = \emptyset$, $F_\emptyset \equiv 1$ and $|\emptyset| \equiv 1$.

Various formulae for $F_\lambda$ were obtained in Section 3.2, and, a priori, any one of these formulae can be inserted into (7.9) to compute $\mathbb{P}(L(t) \le n)$. There are many proofs of Gessel's formula [**Ges90, Rai98,**

**GWW98, BDJ99**], see [**BDJ99**] for more historical information. We provide two of them. The first proof uses formula (3.67) for $F_\lambda$, and the second (shorter) proof uses the determinantal formula (3.62) of Frobenius and Young.

## 7.1. First Proof

The first proof is a direct calculation and is a modification of what appears in the Appendix in [**BDJ99**]. Since all partitions in the last sum in (7.9) have at most $n$ parts, a partition in the sum can be written as

$$\lambda = (\lambda_1, \lambda_2, \dots, \lambda_n), \qquad \lambda_1 \geq \cdots \geq \lambda_n \geq 0. \tag{7.10}$$

Note that some of $\lambda_j$'s are allowed to be 0. Then by (3.67) and Remark 3.32,

$$F_\lambda = \frac{|\lambda|!}{\ell_1! \cdots \ell_n!} \prod_{1 \leq i < j \leq n} (\ell_i - \ell_j), \tag{7.11}$$

where we set

$$\ell_j = \lambda_j + n - j, \qquad 1 \leq j \leq n. \tag{7.12}$$

Note that for $\lambda = (0, 0, \dots, 0)$, the right-hand side of (7.11) becomes

$$\frac{1}{(n-1)!(n-2)! \cdots 1!} \prod_{1 \leq i < j \leq n} (j - i) = 1. \tag{7.13}$$

Hence the formula (7.11) still holds for the empty partition $\lambda = \emptyset$ which is represented as $\lambda = (0, 0, \dots, 0)$ in (7.10). The condition $\lambda_1 \geq \cdots \geq \lambda_n$ implies that $\ell_1 > \ell_2 > \cdots > \ell_n \geq 0$ and

$$|\lambda| = \sum_{j=1}^{n} \ell_j - \frac{n(n-1)}{2}. \tag{7.14}$$

Inserting (7.11) into (7.9), we obtain

$$\mathbb{P}\{L(t) \leq n\} = e^{-t^2} t^{-n(n-1)} \sum_{\ell_1 > \ell_2 > \cdots > \ell_n \geq 0} \Delta(\ell)^2 \prod_{j=1}^{n} \frac{t^{2\ell_j}}{(\ell_j!)^2}, \tag{7.15}$$

where $\ell = (\ell_1, \dots, \ell_n)$, and

$$\Delta(\ell) = \prod_{1 \leq i < j \leq n} (\ell_j - \ell_i) = \det(\ell_j^{i-1})_{1 \leq i,j \leq n} \tag{7.16}$$

is the Vandermonde determinant (5.9). Note that the summand in (7.15) is a symmetric function in $\ell_1, \dots, \ell_n$. Also the summand is zero when two of

$\ell_1, \dots, \ell_n$ are the same. Therefore, setting $\mathbb{N}_0 = \{0, 1, 2, \dots\}$, we have

$$\begin{aligned}\mathbb{P}\{L(t) \le n\} &= \frac{1}{n!} e^{-t^2} t^{-n(n-1)} \sum_{\ell \in \mathbb{N}_0^n} \Delta(\ell)^2 \prod_{j=1}^{n} \frac{t^{2\ell_j}}{(\ell_j!)^2} \\ &= \frac{1}{n!} e^{-t^2} t^{-n(n-1)} \sum_{\ell \in \mathbb{N}_0^n} \det(\ell_j^{i-1})^2_{1\le i,j\le n} \prod_{j=1}^{n} \frac{t^{2\ell_j}}{(\ell_j!)^2}.\end{aligned} \tag{7.17}$$

Applying Andréief's formula Lemma 6.2 with $f_i(m) = g_i(m) = m^{i-1}$ and $d\mu(m) = \frac{t^{2m}}{(m!)^2}$ on the discrete space $X = \mathbb{N}_0 = \{0, 1, 2, \dots\}$, we obtain

$$\mathbb{P}\{L(t) \le n\} = e^{-t^2} t^{-n(n-1)} H_{n-1}, \tag{7.18}$$

where

$$H_{n-1} = \det\left( \sum_{m=0}^{\infty} m^{i+j} \frac{t^{2m}}{(m!)^2} \right)_{0 \le i,j \le n-1} \tag{7.19}$$

is a Hankel determinant with respect to the discrete measure $d\mu(m) = \frac{t^{2m}}{(m!)^2}$ (see (5.6)).

We now show that $t^{-n(n-1)} H_{n-1} = D_{n-1}(\varphi)$. First, suitable row and column operations yield

$$H_{n-1} = \det\left( \sum_{m=0}^{\infty} q_j(m) q_k(m) \frac{t^{2m}}{(m!)^2} \right)_{0 \le j,k \le n-1} \tag{7.20}$$

for any sequence of monic polynomials $q_j = x^j + \cdots$ of degree $j$, $j = 0, 1, 2, \dots$. We make a particular choice of $q_j$,

$$\begin{aligned} &q_0(x) \equiv 1, \\ &q_j(x) \equiv x(x-1)(x-2)\cdots(x-j+1), \qquad j \ge 1, \end{aligned} \tag{7.21}$$

so that

$$q_j(m) a^m = a^j \frac{d^j}{da^j} a^m, \qquad m, j \ge 0. \tag{7.22}$$

We have

$$\sum_{m=0}^{\infty} q_j(m) q_k(m) \frac{t^{2m}}{(m!)^2} = a^j b^k \frac{d^j}{da^j} \frac{d^k}{db^k} \sum_{m=0}^{\infty} \frac{a^m b^m}{(m!)^2} \bigg|_{a=b=t}. \tag{7.23}$$

But a direct computation using the power series of the exponential function shows that

$$\oint_{|z|=1} e^{az^{-1} + bz} \frac{dz}{2\pi i z} = \sum_{m=0}^{\infty} \frac{a^m b^m}{(m!)^2}, \tag{7.24}$$

hence the right-hand side of (7.23) equals

$$t^{j+k} \oint_{|z|=1} z^{-j+k} e^{t(z^{-1}+z)} \frac{dz}{2\pi i z} = t^{j+k} \varphi_{j-k}. \tag{7.25}$$

Thus

$$\begin{aligned} \mathbb{P}\{L(t) \le n\} &= e^{-t^2} t^{-n(n-1)} \det \left( t^{j+k} \varphi_{j-k} \right)_{0 \le j,k \le n-1} \\ &= e^{-t^2} \det(\varphi_{j-k}) = e^{-t^2} D_{n-1}(\varphi) \end{aligned} \tag{7.26}$$

as desired.

## 7.2. Second Proof

We now give a second proof of Gessel's formula. The determinantal formula (3.62) of Frobenius and Young for $F_\lambda$ is

$$F_\lambda = |\lambda|! \det \left( \frac{1}{(\lambda_i - i + j)!} \right)_{1 \le i,j \le n} \tag{7.27}$$

for a partition $\lambda = (\lambda_1, \ldots, \lambda_n)$. Inserting this formula into (7.9) leads to

$$\mathbb{P}\{L(t) \le n\} = e^{-t^2} \sum_{r(\lambda) \le n} t^{2|\lambda|} \left[ \det \left( \frac{1}{(\lambda_i - i + j)!} \right)_{1 \le i,j \le n} \right]^2. \tag{7.28}$$

Setting

$$\ell_j = \lambda_j + n - j, \qquad 1 \le j \le n, \tag{7.29}$$

as in (7.12), (7.28) becomes

$$\begin{aligned} &\mathbb{P}\{L(t) \le n\} \\ &= e^{-t^2} t^{-n(n-1)} \sum_{\ell_1 > \cdots > \ell_n \ge 0} \left[ t^{\sum \ell_j} \cdot \det \left( \frac{1}{(\ell_i - n + j)!} \right)_{1 \le i,j \le n} \right]^2 \\ &= e^{-t^2} \sum_{\ell_1 > \cdots > \ell_n \ge 0} \left[ \det \left( \frac{t^{\ell_i - n + j}}{(\ell_i - n + j)!} \right)_{1 \le i,j \le n} \right]^2 \\ &= e^{-t^2} \sum_{\ell_1 > \cdots > \ell_n \ge 0} \left[ \det \left( \frac{t^{\ell_i - j}}{(\ell_i - j)!} \right)_{0 \le i,j \le n-1} \right]^2. \end{aligned} \tag{7.30}$$

Since the summand in the last sum is symmetric in $\ell_1, \ldots, \ell_n$ and vanishes when two of $\ell_1, \ldots, \ell_n$ are the same, we have, after taking the transpose of the matrix,

$$\mathbb{P}\{L(t) \le n\} = \frac{e^{-t^2}}{n!} \sum_{\ell \in \mathbb{N}_0^n} \left[ \det \left( \frac{t^{\ell_j - i}}{(\ell_j - i)!} \right)_{0 \le i,j \le n-1} \right]^2, \tag{7.31}$$

where $\ell = (\ell_1, \ldots, \ell_n)$.

Using Andréief's formula Lemma 6.2 for the functions $f_i(m) = g_i(m) = \frac{t^{m-i}}{(m-i)!}$, $i = 0, \dots, n$, and counting measure on the discrete space $\mathbb{N}_0$, we obtain

$$\mathbb{P}\{L(t) \le n\} = e^{-t^2} \det \left( \sum_{m=0}^{\infty} \frac{t^{2m-i-j}}{(m-i)!(m-j)!} \right)_{0 \le i,j \le n-1}. \tag{7.32}$$

As $\frac{1}{a!} = 0$ for $a < 0$, we see that

$$\begin{aligned} \sum_{m=0}^{\infty} \frac{t^{2m-i-j}}{(m-i)!(m-j)!} &= \sum_{m=-\infty}^{\infty} \frac{t^{2m-i-j}}{(m-i)!(m-j)!} \\ &= \sum_{m=-\infty}^{\infty} \frac{t^{2m+i-j}}{m!(m+i-j)!} \equiv d_{i-j}. \end{aligned} \tag{7.33}$$

Hence $\mathbb{P}\{L(t) \le n\} = e^{-t^2} \det(d_{i-j})_{0 \le i,j \le n}$. But

$$\begin{aligned} \sum_{k \in \mathbb{Z}} d_k z^k &= \sum_{k \in \mathbb{Z}} \sum_{m \in \mathbb{Z}} \frac{t^{2m+k}}{m!(m+k)!} z^k = \sum_{m \in \mathbb{Z}} \sum_{s \in \mathbb{Z}} \frac{t^{m+s}}{m!s!} z^{s-m} \\ &= e^{tz + tz^{-1}} = \varphi(z). \end{aligned} \tag{7.34}$$

Therefore, $d_k = \varphi_k$, and we obtain $\mathbb{P}\{L(t) \le n\} = e^{-t^2} D_{n-1}(\varphi)$ as desired.

## 7.3. Recurrence Formulae and Differential Equations

The appearance of the Toeplitz determinant in Gessel's formula (7.3) allows us to find various recurrence formulae and differential equations associated to $\mathbb{P}\{L(t) \le n\}$. Four recurrence formulae and differential equations (the Ablowitz-Ladik equation, the Toda lattice, the discrete Painlevé II equation, and the Painlevé III equation) will be obtained in this section. These are all examples of classical integrable systems. The results in this section will not be used in later sections. We have included them in order to illustrate the many deep, unanticipated, and quite remarkable connections between the longest increasing subsequence problem and integrable theory. In particular, and in addition, in Section 7.4 below we show how to use the discrete Painlevé II equation to give an heuristic argument for the convergence of the scaled distribution $L(t)$ to the Tracy-Widom distribution (see (7.150)).

Toeplitz matrices are closely related to orthogonal polynomials on the unit circle as discussed in Section 5.4. Let

$$p_n(z) = p_n(z;t) = \kappa_n z^n + \cdots, \qquad \kappa_n = \kappa_n(t) > 0, \tag{7.35}$$

denote the $n$th orthonormal polynomial with respect to the measure $\varphi\frac{d\theta}{2\pi} = e^{2t\cos\theta}\frac{d\theta}{2\pi}$ on the unit circle. Then, by (5.89),

$$\kappa_n^2 = \frac{D_{n-1}(\varphi)}{D_n(\varphi)}, \qquad n \geq 0. \tag{7.36}$$

Hence, for any $k \geq n$, Gessel's formula implies

$$\mathbb{P}\{L(t) \leq n\} = e^{-t^2}\left[\prod_{j=n}^{k} \kappa_j^2\right] D_k(\varphi). \tag{7.37}$$

Note that the strong Szegő limit theorem (Theorem 5.22) implies that

$$\lim_{k\to\infty} D_k(\varphi) = e^{t^2}. \tag{7.38}$$

This also follows from (7.1) by dominated convergence. Hence, (7.37) implies that

$$\mathbb{P}\{L(t) \leq n\} = \prod_{j=n}^{\infty} (\kappa_j(t))^2. \tag{7.39}$$

Taking the logarithm, one obtains

$$\log \mathbb{P}\{L(t) \leq n\} = \sum_{j=n}^{\infty} \log(\kappa_j(t))^2. \tag{7.40}$$

Recall from Section 5.4 that RHP 5.10 for the above polynomials $p_j(z;t)$ is given as follows:

**Riemann-Hilbert Problem 7.2.** *Find $Y(z)$ satisfying the following conditions:*

- *$Y(z)$ is analytic in $\mathbb{C}\backslash\Sigma$;*
- *$Y_+(z) = Y_-(z)v(z)$, $z \in \Sigma$, where $v(z) = \begin{pmatrix} 1 & z^{-n}e^{t(z+z^{-1})} \\ 0 & 1 \end{pmatrix}$;*
- *$Y(z)\begin{pmatrix} z^{-n} & 0 \\ 0 & z^n \end{pmatrix} \to I$ as $z \to \infty$.*

As noted in Remark 6.19, there is a mantra in the theory of integrable systems that if the jump matrix for an RHP is independent of some associated external parameter, then differentiation with respect to that parameter leads to a differential equation. For example, in Section 6.5 the differential equation that emerges is the Painlevé II equation. By (5.72), $(\kappa_{n-1}(t))^2 = -Y_{21}(0)$, and we see that $(\kappa_j(t))^2$ appears directly in the RHP 7.2. As we will see, RHP 7.2 is equivalent to a number of different RHP's with jump matrices independent of certain external parameters. This then leads to a number of dynamical systems associated to the $(\kappa_j(t))^2$'s, and we may think of (7.40) as expressing $\log \mathbb{P}\{L(t) \leq n\}$ as a sum of its dynamical "degree of freedom". As we will see, an elegant procession of classical integrable systems accompanies the analysis of $\mathbb{P}\{L(t) \le n\}$: the Ablowitz-Ladik lattice, the Toda lattice, the discrete Painlevé II equation, and the Painlevé III equation all show up for the parade!

The above mantra may be viewed as the analogue for RHP's of Noether's well-known theorem for Lagrangian systems: If the Lagrangian is independent of a variable $x_k$, say, then one obtains an associated integral for the system.

**Remark 7.3.** The formula (7.39) can also be used to obtain asymptotic results for $L(t)$. Indeed, in the original proof in [**BDJ99**] of the central limit theorem, Theorem 9.14 below, the authors used (7.39) together with the RHP 5.10 for the $p_j(z;t)$'s and the Deift-Zhou steepest-descent method. In this book, however, we prove the central limit theorem using a Fredholm determinant formula which will be discussed in the next chapter.

Orient $\Sigma = \{z \in \mathbb{C} : |z| = 1\}$ counter-clockwise. From (5.65), the solution $Y = Y(z;n,t)$ of RHP 7.2 is given by

$$Y(z) = \begin{pmatrix} \pi_n(z) & \int_\Sigma \frac{s^{-n}\pi_n(s)\varphi(s)}{s-z}\frac{ds}{2\pi i} \\ -\kappa_{n-1}^2\pi_{n-1}^*(z) & -\kappa_{n-1}^2\int_\Sigma \frac{s^{-n}\pi_{n-1}^*(s)\varphi(s)}{s-z}\frac{ds}{2\pi i} \end{pmatrix}, \tag{7.41}$$

where $\pi_n(z) = \kappa_n^{-1}p_n(z) = z^n + \cdots$ is the monic orthogonal polynomial associated with the weight $\varphi(e^{i\theta})\frac{d\theta}{2\pi}$ and $\pi_{n-1}^*(z) \equiv z^{n-1}\overline{\pi_{n-1}(\bar{z}^{-1})}$ is the reverse polynomial. From Corollary 5.13,

$$Y(0) = \begin{pmatrix} \pi_n(0) & \kappa_n^{-2} \\ -\kappa_{n-1}^2 & \pi_n(0) \end{pmatrix}, \qquad \kappa_{-1} \equiv 0, \tag{7.42}$$

and

$$\pi_n(0)^2 + \left(\frac{\kappa_{n-1}}{\kappa_n}\right)^2 = 1. \tag{7.43}$$

The Verblunsky coefficient

$$\alpha_n \equiv -\overline{\pi_{n+1}(0)}, \qquad n \ge -1, \tag{7.44}$$

is real and satisfies $|\alpha_n| < 1$ for $n \ge 0$ and $\alpha_{-1} = -1$. Recall the Szegő recurrence relation (7.45)

$$z\pi_{n-1}(z) = \pi_n(z) + \bar{\alpha}_{n-1}\pi_{n-1}^*(z). \tag{7.45}$$

**7.3.1. The Ablowitz-Ladik Equation.** Our first differential relation (or more properly, difference relation) follows from the invariance of the jump relation under shifts $n \mapsto n+1$. More precisely, set $W(z;n) \equiv Y(z;n)\left(\begin{smallmatrix} 1 & 0 \\ 0 & z^n \end{smallmatrix}\right)$.

Then $W_+ = W_- \left(\begin{smallmatrix} 1 & e^{t(z+z^{-1})} \\ 0 & 1 \end{smallmatrix}\right)$ for $z \in \Sigma$, and the jump matrix is now clearly independent of $n$. It follows that

$$R \equiv W(z;n+1)W(z;n)^{-1} = Y(z;n+1)\left(\begin{smallmatrix} 1 & 0 \\ 0 & z \end{smallmatrix}\right)Y(z;n)^{-1},$$

is entire. Since

$$Y(z;n)\begin{pmatrix} z^{-n} & 0 \\ 0 & z^n \end{pmatrix} = I + \frac{A(n)}{z} + O(z^{-2}), \qquad z \to \infty, \tag{7.46}$$

for some constant matrix $A(n)$, we see that

$$R(z) = \begin{pmatrix} z & 0 \\ 0 & 1 \end{pmatrix} + A(n+1)\begin{pmatrix} 1 & 0 \\ 0 & 0 \end{pmatrix} - \begin{pmatrix} 1 & 0 \\ 0 & 0 \end{pmatrix} A(n) + O(z^{-1}) \tag{7.47}$$

as $z \to \infty$. By Liouville's theorem, we must have

$$R(z) = \begin{pmatrix} z & 0 \\ 0 & 1 \end{pmatrix} + A(n+1)\begin{pmatrix} 1 & 0 \\ 0 & 0 \end{pmatrix} - \begin{pmatrix} 1 & 0 \\ 0 & 0 \end{pmatrix} A(n), \tag{7.48}$$

or

$$Y(z;n+1)\begin{pmatrix} 1 & 0 \\ 0 & z \end{pmatrix} = \left\{\begin{pmatrix} z & 0 \\ 0 & 1 \end{pmatrix} + A(n+1)\begin{pmatrix} 1 & 0 \\ 0 & 0 \end{pmatrix} - \begin{pmatrix} 1 & 0 \\ 0 & 0 \end{pmatrix} A(n)\right\} Y(z;n). \tag{7.49}$$

Right multiplying by $Y(z;n)^{-1}$ on both sides and setting $z = 0$, we obtain, using (7.42),

$$A(n+1)\begin{pmatrix} 1 & 0 \\ 0 & 0 \end{pmatrix} - \begin{pmatrix} 1 & 0 \\ 0 & 0 \end{pmatrix} A(n) = \begin{pmatrix} \pi_{n+1}(0)\pi_n(0) & -\kappa_n^{-2}\pi_{n+1}(0) \\ -\kappa_n^2\pi_n(0) & 0 \end{pmatrix}. \tag{7.50}$$

Inserting this relation into (7.49), we obtain the basic difference equation for $Y(z;n)$,

$$Y(z;n+1)\begin{pmatrix} 1 & 0 \\ 0 & z \end{pmatrix} = \left\{\begin{pmatrix} z & 0 \\ 0 & 1 \end{pmatrix} + \begin{pmatrix} \alpha_n\alpha_{n-1} & \kappa_n^{-2}\alpha_n \\ \kappa_n^2\alpha_{n-1} & 0 \end{pmatrix}\right\} Y(z;n), \tag{7.51}$$

for $n \geq 0$.

By the result of Killip and Nenciu ([**KN07**], in particular Section 6), the time dependence of the weight $w_t = e^{t(z+z^{-1})} = e^{2t\cos\theta}$ implies that RHP 7.2 gives rise to a solution of the *Ablowitz-Ladik equation* [**AL75, AL76**]

$$i\frac{\partial g_n}{\partial t} = (1 - |g_n|^2)(g_{n-1} + g_{n+1}) \tag{7.52}$$

or, if we set $h_n = g_n e^{2it}$,

$$i\frac{\partial h_n}{\partial t} = (h_{n+1} + h_{n-1} - 2h_n) - |h_n|^2(h_{n+1} + h_{n-1}), \tag{7.53}$$

which is clearly a discretization of the defocusing nonlinear Schrödinger equation, $ih_t = h_{xx} - 2|h|^2 h$.

We derive (7.52) directly from the RHP, together with the associated initial conditions (see (7.76) below).

Set

$$T = T(z;n) \equiv Y(z;n)\begin{pmatrix} e^{t(z+z^{-1})} & 0 \\ 0 & 1 \end{pmatrix}. \tag{7.54}$$

Then $T$ is analytic in $\mathbb{C} \setminus (\Sigma \cup \{0\})$ and satisfies the jump relation

$$T_+ = T_- \begin{pmatrix} 1 & z^{-n} \\ 0 & 1 \end{pmatrix}, \qquad z \in \Sigma. \tag{7.55}$$

As the jump matrix is independent of $t$, we may differentiate with respect to $t$ and conclude as before that

$$S(z) \equiv \dot{T}(z)T(z)^{-1} = \dot{Y}Y^{-1} + Y\begin{pmatrix} z+z^{-1} & 0 \\ 0 & 0 \end{pmatrix}Y^{-1} \tag{7.56}$$

is analytic in $\mathbb{C} \setminus \{0\}$. As $z \to 0$, we have

$$\begin{aligned} S(z) &= z^{-1}Y(0;n)\begin{pmatrix} 1 & 0 \\ 0 & 0 \end{pmatrix}Y(0;n)^{-1} + O(1) \\ &= z^{-1}\begin{pmatrix} \pi_n(0)^2 & -\kappa_n^{-2}\pi_n(0) \\ -\kappa_{n-1}^2\pi_n(0) & \kappa_n^{-2}\kappa_{n-1}^2 \end{pmatrix} + O(1), \end{aligned} \tag{7.57}$$

using (7.42). As $z \to \infty$, we have from (7.46)

$$S(z) = \begin{pmatrix} z & 0 \\ 0 & 0 \end{pmatrix} + \begin{pmatrix} 0 & -A_{12}(n) \\ A_{21}(n) & 0 \end{pmatrix} + O(z^{-1}). \tag{7.58}$$

By Liouville's theorem,

$$S(z) = \begin{pmatrix} z & -A_{12}(n) \\ A_{21}(n) & 0 \end{pmatrix} + z^{-1}\begin{pmatrix} \pi_n(0)^2 & -\kappa_n^{-2}\pi_n(0) \\ -\kappa_{n-1}^2\pi_n(0) & \kappa_n^{-2}\kappa_{n-1}^2 \end{pmatrix}. \tag{7.59}$$

The terms $A_{12}(n)$ and $A_{21}(n)$ are the off-diagonal entries of the identity (7.50):

$$A_{12}(n) = \kappa_n^{-2}\pi_{n+1}(0), \qquad A_{21}(n) = -\kappa_{n-1}^2\pi_{n-1}(0). \tag{7.60}$$

Inserting these relations into (7.59), (7.56), we obtain for $n \geq 0$,

$$\dot{T}(z;n) = \left\{\begin{pmatrix} z & \kappa_n^{-2}\alpha_n \\ \kappa_{n-1}^2\alpha_{n-2} & 0 \end{pmatrix} + z^{-1}\begin{pmatrix} \alpha_{n-1}^2 & \kappa_n^{-2}\alpha_{n-1} \\ \kappa_{n-1}^2\alpha_{n-1} & \kappa_n^{-2}\kappa_{n-1}^2 \end{pmatrix}\right\}T(z;n), \tag{7.61}$$

which is the basic differential equation for $T(z;n,t)$ as a function of $t$. Note that for $n = 0$, $\alpha_{n-2} = \alpha_{-2}$ is not defined, but $\kappa_{n-1}^2 = \kappa_{-1}^2 \equiv 0$, so we can take $\alpha_{-2}$ to be any value. For definiteness, set $\alpha_{-2} \equiv 0$.

Following the canonical Lax-pair methodology, the compatibility of (7.51) and (7.61) gives rise to the Ablowitz-Ladik equation as follows. From (7.51), we have

$$T(z;n+1) = \begin{pmatrix} z + \alpha_n\alpha_{n-1} & \kappa_n^{-2}\alpha_n \\ \kappa_n^2\alpha_{n-1} & 1 \end{pmatrix}T(z;n)\begin{pmatrix} 1 & 0 \\ 0 & z^{-1} \end{pmatrix}. \tag{7.62}$$

Let $\mathcal{F}$ denote the shift operator on sequences $(c(0), c(1), c(2), \dots)$ taking $c(k) \mapsto c(k+1)$. Then we must have

$$\mathcal{F}\frac{d}{dt}T(n) = \frac{d}{dt}\mathcal{F}T(n), \tag{7.63}$$

and so, after some algebra, we must have for $n \geq 0$,

(7.64)

$$\left\{ \begin{pmatrix} z & \kappa_{n+1}^{-2}\alpha_{n+1} \\ \kappa_n^2\alpha_{n-1} & 0 \end{pmatrix} + z^{-1}\begin{pmatrix} \alpha_n^2 & \kappa_{n+1}^{-2}\alpha_n \\ \kappa_n^2\alpha_n & \kappa_{n+1}^{-2}\kappa_n^2 \end{pmatrix} \right\} \begin{pmatrix} z+\alpha_n\alpha_{n-1} & \kappa_n^{-2}\alpha_n \\ \kappa_n^2\alpha_{n-1} & 1 \end{pmatrix}$$
$$= \begin{pmatrix} \frac{d}{dt}(\alpha_n\alpha_{n-1}) & \frac{d}{dt}(\kappa_n^{-2}\alpha_n) \\ \frac{d}{dt}(\kappa_n^2\alpha_{n-1}) & 0 \end{pmatrix}$$
$$+ \begin{pmatrix} z+\alpha_n\alpha_{n-1} & \kappa_n^{-2}\alpha_n \\ \kappa_n^2\alpha_{n-1} & 1 \end{pmatrix} \left\{ \begin{pmatrix} z & \kappa_n^{-2}\alpha_n \\ \kappa_{n-1}^2\alpha_{n-2} & 0 \end{pmatrix} \right.$$
$$\left. + z^{-1}\begin{pmatrix} \alpha_{n-1}^2 & \kappa_n^{-2}\alpha_{n-1} \\ \kappa_{n-1}^2\alpha_{n-1} & \kappa_n^{-2}\kappa_{n-1}^2 \end{pmatrix} \right\}.$$

In particular, we find, using (7.43), that for $n \geq 0$,

$$\frac{d}{dt}(\kappa_n^{-2}\alpha_n) = \kappa_{n+1}^{-2}\alpha_{n+1} - \kappa_n^{-2}\alpha_n^2\alpha_{n-1} - \kappa_n^{-2}\alpha_{n-1} \tag{7.65}$$

and

$$\frac{d}{dt}(\kappa_n^2\alpha_{n-1}) = \kappa_n^2\alpha_n\alpha_{n-1}^2 + \kappa_n^2\alpha_n - \kappa_{n-1}^2\alpha_{n-2}. \tag{7.66}$$

We leave it as an exercise to show that all the other relations contain no further information. Setting $\rho_n \equiv \kappa_n^2$, (7.65) becomes

$$\dot{\alpha}_n - \frac{\dot{\rho}_n}{\rho_n}\alpha_n = \frac{\rho_n}{\rho_{n+1}}\alpha_{n+1} - \alpha_n^2\alpha_{n-1} - \alpha_{n-1}. \tag{7.67}$$

Replacing $n$ by $n+1$ in (7.66), we obtain

$$\dot{\alpha}_n + \frac{\dot{\rho}_{n+1}}{\rho_{n+1}}\alpha_n = \alpha_{n+1}\alpha_n^2 + \alpha_{n+1} - \frac{\rho_n}{\rho_{n+1}}\alpha_{n-1}. \tag{7.68}$$

Adding (7.67) and (7.68) and using (7.43), we obtain

$$2\dot{\alpha}_n - \alpha_n\frac{d}{dt}(\log(1-\alpha_n^2)) = 2(\alpha_{n+1} - \alpha_{n-1}), \tag{7.69}$$

and hence

$$\dot{\alpha}_n = (\alpha_{n+1} - \alpha_{n-1})(1-\alpha_n^2), \qquad n \geq 0 \tag{7.70}$$

$(\alpha_{-1} = -1)$.

Set

$$g_n = i^n\alpha_n, \qquad n \geq -1. \tag{7.71}$$

Using the fact that $\alpha_n$ is real, we find

$$i\dot{g}_n = (1 - |g_n|^2)(g_{n+1} + g_{n-1}), \qquad n \geq 0 \tag{7.72}$$

$(g_{-1} \equiv i)$, which is the Ablowtiz-Ladik equation advertised in (7.52).

**Exercise 7.4.** Use (7.67) and (7.70) to show that

$$\frac{d}{dt} \log \kappa_n(t) = \alpha_{n-1}\alpha_n, \qquad n \geq 0. \tag{7.73}$$

**Remark 7.5.** Equation (7.70) is referred to as the Schur flow (see [**AG94**]).

In order to specify the initial conditions for (7.70) and (7.72), we need to compute $\pi_n(0)$ for the weight $w_0(z) = 1$. In this case we clearly have

$$\pi_n(z) = z^n, \qquad n \geq 0, \tag{7.74}$$

so the initial conditions for (7.70) and (7.72) are given by

$$\alpha_n(t=0) = -\overline{\pi_{n+1}(0)} = -\delta_{n,-1}, \qquad n \geq -1, \tag{7.75}$$

and

$$g_n(t=0) = i\delta_{n,-1}, \qquad n \geq -1, \tag{7.76}$$

respectively. Also, clearly,

$$\kappa_n(t=0) = 1, \qquad n \geq 0. \tag{7.77}$$

Summarizing, we have the following:

**Theorem 7.6.** *Let $\alpha_n(t) = -\overline{\pi_{n+1}(0)}$ as in (7.44). Set $\alpha_{-1}(t) \equiv -1$. Then*

$$\frac{d}{dt}\alpha_n = (\alpha_{n+1} - \alpha_{n-1})(1 - \alpha_n^2), \qquad n = 0, 1, 2, \ldots, \tag{7.78}$$

*and*

$$\alpha_n(0) = -\delta_{n,-1}. \tag{7.79}$$

*Setting $g_n(t) \equiv i^n \alpha_n(t)$, $g_n(t)$ solves the Ablowitz-Ladik equation*

$$i\frac{d}{dt}g_n = (1 - |g_n|^2)(g_{n+1} + g_{n-1}), \qquad n = 0, 1, 2, \ldots \tag{7.80}$$

*with the initial condition $g_n(0) = i\delta_{n,-1}$. Moreover, $\kappa_n(t)$ satisfies*

$$\frac{d}{dt} \log \kappa_n = \alpha_{n-1}\alpha_n, \qquad n = 0, 1, 2, \ldots, \tag{7.81}$$

*and $\kappa_n(0) = 1$ for all $n$.*

**7.3.2. Toda Lattice.** The appearance of the Toda lattice can be anticipated in two different ways. Firstly, let $\rho(x) \geq 0$ be a weight function on the interval $-1 \leq x \leq 1$. Then there is a canonical relation (the *Geronimus relations*) between the polynomials orthogonal with respect to the weight $\rho(x)\frac{dx}{\sqrt{1-x^2}}$ on $[-1, 1]$ and the polynomials orthogonal with respect to the measure $\rho(\cos\theta)\frac{d\theta}{2\pi}$ on the unit circle $\Sigma$; see [**Sze75, Sim05a, Sim05b**]. In particular, for $\rho(\cos\theta) = e^{2t\cos\theta}$, we see that the $p_k(z;t)$'s are associated with polynomials orthogonal with respect to $e^{2tx}\frac{dx}{\sqrt{1-x^2}}$ on $[-1, 1]$. But the time dependence $e^{2tx}$ is precisely the time evolution for the spectral measure of a Jacobi operator (see (5.17))

$$J = \begin{pmatrix} a_0 & b_0 & 0 & \cdots \\ b_0 & a_1 & b_1 & \cdots \\ 0 & b_1 & a_2 & \cdots \\ \vdots & \vdots & \vdots & \ddots \end{pmatrix} \tag{7.82}$$

under the Toda flow (see e.g. [**Dei99b**])

$$\frac{da_m}{dt} = 2(b_m^2 - b_{m-1}^2), \qquad \frac{db_m}{dt} = b_m(a_{m+1} - a_m) \tag{7.83}$$

for $m \geq 0$ where $b_{-1} \equiv 0$. Thus, the Toda lattice is simply the push forward of the Ablowitz-Ladik lattice under the Geronimus relations.

Alternatively, if $Y(z;n)$, $n \geq 0$, is given by (7.41), and we set

(7.84)

$$Q(z) = \begin{cases} \begin{pmatrix} 0 & -1 \\ 1 & 0 \end{pmatrix} Y(z;n) \begin{pmatrix} z^{-n}e^{tz^{-1}} & 0 \\ 0 & z^n e^{-tz^{-1}} \end{pmatrix} \begin{pmatrix} 0 & 1 \\ -1 & 0 \end{pmatrix}, & |z| > 1, \\ \begin{pmatrix} 0 & -1 \\ 1 & 0 \end{pmatrix} Y(z;n) \begin{pmatrix} e^{tz} & 0 \\ 0 & e^{-tz} \end{pmatrix}, & |z| < 1, \end{cases}$$

then $Q(z)$ solves the following RHP.

**Riemann-Hilbert Problem 7.7.** *The matrix $Q(z)$ satisfies the following conditions:*

- *$Q(z)$ is analytic in $\mathbb{C} \setminus \Sigma$;*
- *for $z \in \Sigma$,*

$$Q(z)_+ = Q_-(z) \begin{pmatrix} 0 & -z^{-n}e^{-t(z-z^{-1})} \\ z^n e^{t(z-z^{-1})} & 1 \end{pmatrix}; \tag{7.85}$$

- *$Q(z) \to I$ $z \to \infty$.*

In [**Kam93**], Kamvissis considered the doubly infinite Toda lattice (7.83) with $-\infty < m < \infty$, and showed that, in the soliton free case, the solution of the lattice could be constructed from the solution $U(z)$ of the following RHP on $\Sigma$:

**Riemann-Hilbert Problem 7.8.** *The matrix $U(z)$ is analytic in $\mathbb{C}\setminus\Sigma$, $U(z)\to I$ as $z\to\infty$, and satisfies*

(7.86)
$$U(z)_+ = U_-(z)\begin{pmatrix} 1-|r(z)|^2 & -\overline{r(z)}z^{2m}e^{-t(z-z^{-1})} \\ r(z)z^{-2m}e^{t(z-z^{-1})} & 1 \end{pmatrix}, \qquad z\in\Sigma,$$

*where $r(z)$ is the reflection coefficient associated with the Lax spectral operator for the flow.*

We note that, by the general methods of RHP's (see Sections 5.3 and 5.5), $Q(z)$ and $U(z)$ are the unique solutions of their respective RHP's.

We see that for even $n$, RHP 7.7 is just a special case of the RHP 7.8 with $r(z)\equiv 1$. This gives another explanation for the appearance of the Toda flow.

In the original variables, the Toda lattice is a Hamiltonian system with (formal) Hamiltonian

$$H \equiv \frac{1}{2}\sum_k y_k^2 + \sum_k e^{x_k - x_{k+1}}. \tag{7.87}$$

The sum must be renormalized depending on the context: for example, if $x_k, y_k \to 0$ as $|k|\to\infty$, then we must replace $e^{x_k-x_{k+1}}$ with $e^{x_k-x_{k+1}}-1$ in the second term, etc. Hamilton's equations,

$$\dot{x}_k = H_{y_k} = y_k, \qquad \dot{y}_k = -H_{x_k} = e^{x_{k-1}-x_k} - e^{x_k-x_{k+1}} \tag{7.88}$$

are transformed (see [**Fla74, Man74**]) to (7.83) by the change of variables $a_k = -\frac{1}{2}y_k$, $b_k = \frac{1}{2}e^{(x_k-x_{k+1})/2}$.

In the article [**Kam93**] we find the following prescription. Let $u(z) = (u_1(z;m), u_2(z;m))$ be the row vector solution of the RHP with the same jump condition as in (7.86) such that

$$u(z) = u(z^{-1})\begin{pmatrix} 0 & 1 \\ 1 & 0 \end{pmatrix}, \qquad z\in\mathbb{C}\setminus\Sigma, \tag{7.89}$$

and

$$u(z) \to (u_1(\infty;m), u_2(\infty;m)) \qquad \text{as } z\to\infty, \tag{7.90}$$

where

$$u_1(\infty), u_2(\infty) > 0, \qquad u_1(\infty)u_2(\infty) = 1. \tag{7.91}$$

(Such a solution exists and is unique.) Then

$$x_m = 2\log u_1(\infty, m), \qquad -\infty < m < \infty, \tag{7.92}$$

solves the Toda equation (7.88)

$$\ddot{x}_m = e^{x_{m-1}-x_m} - e^{x_m - x_{m+1}}, \qquad -\infty < m < \infty. \tag{7.93}$$

We follow this prescription, taking into account that our case is singular, $r(z) \equiv 1$. In [**Kam93**], $|r(z)| < 1$, except possibly at $z = \pm 1$.

Let $Q(z)$ be the solution of the normalized RHP 7.7 given by (7.84). As noted above, this solution is unique. Set

$$u(z) = (c_1, c_2)Q(z), \tag{7.94}$$

where the constants $c_1, c_2$ are to be determined. Now, it follows (we leave this as an exercise) from the symmetry property (5.68) that

$$Q(z) = \begin{pmatrix} 0 & 1 \\ 1 & 0 \end{pmatrix} Q(0)^{-1} Q(z^{-1}) \begin{pmatrix} 0 & 1 \\ 1 & 0 \end{pmatrix}, \qquad z \in \mathbb{C} \setminus (\Sigma \cup \{0\}). \tag{7.95}$$

Thus for $u(z)$ to satisfy (7.89), we must have

$$(c_1, c_2) = (c_1, c_2) \begin{pmatrix} 0 & 1 \\ 1 & 0 \end{pmatrix} Q(0)^{-1}; \tag{7.96}$$

i.e., $(c_1, c_2)$ is an eigenvector corresponding to eigenvalue 1 for the matrix

$$\begin{pmatrix} 0 & 1 \\ 1 & 0 \end{pmatrix} Q(0)^{-1} = \begin{pmatrix} 0 & 1 \\ 1 & 0 \end{pmatrix} Y(0)^{-1} \begin{pmatrix} 0 & 1 \\ -1 & 0 \end{pmatrix} = \begin{pmatrix} \alpha_{n-1} & \kappa_{n-1}^2 \\ \kappa_n^{-2} & -\alpha_{n-1} \end{pmatrix} \tag{7.97}$$

by (7.42). As $\det Y(0) = 1$ (alternatively, recall (5.75) and (5.80)), the spectrum of this matrix is clearly $\{-1, 1\}$. Hence, $(c_1, c_2)$ is uniquely given up to a scalar multiplies by

$$(c_1, c_2) = \lambda(\kappa_n^{-2}, 1 - \alpha_{n-1}), \qquad n \geq 0. \tag{7.98}$$

Note that as $-1 < \alpha_{n-1} < 1$ for $n \geq 1$ and $\alpha_{-1} = -1$, we have $1 - \alpha_n > 0$ for all $n \geq -1$. Choosing $\lambda = \frac{\kappa_n}{\sqrt{1-\alpha_{n-1}}}$, we see that

$$u(z) = \Big(\frac{1}{\kappa_n\sqrt{1-\alpha_{n-1}}}, \kappa_n\sqrt{1-\alpha_{n-1}}\Big)Q(z) \tag{7.99}$$

is a row vector solution of RHP 7.7 satisfying (7.89) and (7.90).

**Exercise 7.9.** Show that $u(z)$ with properties (7.89) and (7.90) is indeed unique.

For $m \leq 0$ and $n = -2m$, set

$$x_m = 2\log u_1(\infty, m) = -\log\left(\kappa_{-2m}^2(1 - \alpha_{-2m-1})\right) \tag{7.100}$$

as in (7.92). Using (7.70) and (7.73), we find from (7.99)

$$\dot{x}_m = -\alpha_{-2m-2}(1 + \alpha_{-2m-1}) + \alpha_{-2m}(1 - \alpha_{-2m-1}) \tag{7.101}$$

and, differentiating again,

$$\begin{aligned}\ddot{x}_m =&(1-\alpha_{-2m-1})(1-\alpha_{-2m}^2)(1+\alpha_{-2m+1})\\&-(1+\alpha_{-2m-1})(1-\alpha_{-2m-2}^2)(1-\alpha_{-2m-3}).\end{aligned} \tag{7.102}$$

Using (7.100) and (7.43), we have

$$\begin{aligned}e^{x_m-x_{m+1}} &= \frac{\kappa_{-2m-2}^2(1-\alpha_{-2m-3})}{\kappa_{-2m}^2(1-\alpha_{-2m-1})}\\&=(1+\alpha_{-2m-1})(1-\alpha_{-2m-2}^2)(1-\alpha_{-2m-3}).\end{aligned} \tag{7.103}$$

Thus, as advertised, we obtain the Toda lattice

$$\ddot{x}_m = e^{x_{m-1}-x_m} - e^{x_m-x_{m+1}}, \qquad m \le 0, \tag{7.104}$$

where $e^{x_0-x_1} \equiv 0$ as $1+\alpha_{-1} \equiv 0$.

By (7.75), (7.77), and (7.100),

$$x_m(t=0) = -\delta_{m,0}\log 2, \qquad m \le 0, \tag{7.105}$$

and from (7.101),

$$\dot{x}_m(t=0) = 0, \qquad m \le 0. \tag{7.106}$$

Relations (7.105) and (7.106) provide the initial conditions for (7.104).

**Remark 7.10.** The fact that we obtain the Toda lattice on the semi-axis $m \le 0$ rather than the full line $-\infty < m < \infty$ as in [**Kam93**] is a consequence of the singular nature of $r$, $r(z) \equiv 1$.

**Exercise 7.11.** In the above calculation, we only utilized even values of $n$. Show that odd values of $n$ also lead to a solution of the Toda lattice, but with different initial conditions.

We have obtained the following result.

**Theorem 7.12.** *Set* $x_m = -\log\left(\kappa_{-2m}^2(1-\alpha_{-2m-1})\right)$ *for* $m = 0, -1, -2, \ldots$. *Then* $x_m$ *solves the Toda lattice*

$$\ddot{x}_m = e^{x_{m-1}-x_m} - e^{x_m-x_{m+1}}, \qquad m = 0, -1, -2, \ldots, \tag{7.107}$$

*where* $e^{x_0-x_1} \equiv 0$, *and*

$$x_m(0) = -\delta_{m,0}\log 2, \qquad \dot{x}_m(0) = 0, \qquad m = 0, -1, -2, \ldots. \tag{7.108}$$

**7.3.3. The Discrete Painlevé II Equation and the Painlevé III Equation.** Our third basic differential relation is a differential equation for $Q(z)$ as a function of $z$. As we will see, compatibility of this relation with (7.51) and (7.61) leads to the discrete Painlevé II equation and the Painlevé III equation, respectively.

From (7.42) and (7.84), we have

$$Q(0;n) = \begin{pmatrix} \kappa_{n-1}^2 & \alpha_{n-1} \\ -\alpha_{n-1} & \kappa_n^2 \end{pmatrix}, \qquad n \geq 0. \tag{7.109}$$

As $z \to 0$,

$$Q(z;n) = Q(0;n)\big(I + Q_1^0(n)z + O(z^2)\big) \tag{7.110}$$

for some matrix $Q_1^0(n)$. On the other hand, by (7.46) and (7.84), as $z \to \infty$,

$$Q(z;n) = I + \frac{Q_1^\infty(n)}{z} + O(z^{-2}) \tag{7.111}$$

for some constant matrix $Q_1^\infty(n)$. As $\det Q(z) \equiv 1$, we must have $\operatorname{tr} Q_1^\infty(n) = 0$. Set

$$Q_1^\infty(n) = \begin{pmatrix} -\lambda(n) & \nu(n) \\ \mu(n) & \lambda(n) \end{pmatrix}. \tag{7.112}$$

Taking $z \to 0$ in the symmetry relation (7.95) and collecting the terms of order $O(z)$, we obtain

$$Q_1^0(n) = \begin{pmatrix} 0 & 1 \\ 1 & 0 \end{pmatrix} Q_1^\infty(n) \begin{pmatrix} 0 & 1 \\ 1 & 0 \end{pmatrix} = \begin{pmatrix} \lambda(n) & \mu(n) \\ \nu(n) & -\lambda(n) \end{pmatrix}. \tag{7.113}$$

Set

$$\Psi(z) = \Psi(z;n) \equiv Q(z) \begin{pmatrix} 1 & 0 \\ 0 & z^n e^{t(z-z^{-1})} \end{pmatrix}, \qquad z \in \mathbb{C} \setminus \Sigma. \tag{7.114}$$

Then $\Psi(z)$ is analytic in $\mathbb{C} \setminus (\Sigma \cup \{0\})$ and

$$\Psi_+(z) = \Psi_-(z) \begin{pmatrix} 0 & -1 \\ 1 & 1 \end{pmatrix}, \qquad z \in \Sigma. \tag{7.115}$$

As the jump matrix for $\Psi$ is independent of $z$, we may differentiate with respect to $z$ and conclude, as above, that $P(z;n) \equiv \frac{\partial \Psi(z;n)}{\partial z} \Psi(z;n)^{-1}$ is analytic in $\mathbb{C} \setminus \{0\}$. At $z = 0$, $P(z;n)$ has a double pole. Set

$$J_1 = \begin{pmatrix} 0 & 0 \\ 0 & 1 \end{pmatrix}, \qquad J_0 = \begin{pmatrix} 1 & 0 \\ 0 & 0 \end{pmatrix}. \tag{7.116}$$

We have from (7.114)

$$P(z;n) = \frac{\partial Q}{\partial z} Q^{-1} + (\frac{n}{z} + t + \frac{t}{z^2}) Q J_1 Q^{-1}. \tag{7.117}$$

It follows from Liouville's theorem and the asymptotics (7.110), (7.111) that

$$P(z;n) = tJ_1 + \frac{P_1(n)}{z} + \frac{P_2(n)}{z^2} \tag{7.118}$$

for suitable constant matrices $P_1(n)$ and $P_2(n)$. The equation

$$\frac{\partial \Psi}{\partial z} = P(z;n)\Psi \tag{7.119}$$

is our third differential equation. On the other hand, from (7.111) and (7.114), we find that $H(z;n) \equiv \Psi(z;n+1)\Psi(z;n)^{-1}$ is of the form

$$H(z;n) = zJ_1 + H_0(n) \tag{7.120}$$

for some constant matrix $H_0(n)$. Note that the relation

$$\Psi(z;n+1) = H(z;n)\Psi(z;n) \tag{7.121}$$

is in fact equivalent to the equation (7.51).

Implementing the compatibility of (7.119) and (7.121) as earlier (cf. (7.63)) $\mathcal{F}\frac{\partial}{\partial z}\Psi = \frac{\partial}{\partial z}\mathcal{F}\Psi$, we obtain the relation

$$P(z;n+1)H(z;n) = \frac{\partial H(z;n)}{\partial z} + H(z;n)P(z;n). \tag{7.122}$$

Inserting (7.118) and (7.120) into this relation, we obtain, in particular, the relation

$$J_1 + J_1 P_1(n) - P_1(n+1)J_1 + t[H_0(n), J_1] = 0. \tag{7.123}$$

The matrices $P_1(n)$ and $H_0(n)$ are computed by substituting (7.111) into (7.117), and by substituting (7.110) into (7.120), respectively. We obtain

$$P_1(n) = t[Q_1^{\infty}(n), J_1] + nJ_1 \tag{7.124}$$

and

$$H_0(n) = Q(0;n+1)J_0 Q(0;n)^{-1}. \tag{7.125}$$

Inserting (7.124) and (7.125) into (7.123) and computing the (12) and (21) entries, we obtain the relations

$$\mu(n) = \kappa_n^{-2}\alpha_n, \qquad \nu(n+1) = -\kappa_n^2 \alpha_{n-1}. \tag{7.126}$$

On the other hand, substituting (7.110) into (7.118), we obtain

$$P_1(n) = Q(0;n)\{t[Q_1^0(n), J_1] + nJ_1\}Q(0;n)^{-1}. \tag{7.127}$$

Together with (7.112) and (7.113), the equality of this formula and the right-hand-side of (7.124) implies that

$$t(-\mu(n)\kappa_{n-1}^2 + \nu(n)\kappa_n^{-2}) - n\alpha_{n-1} = 0. \tag{7.128}$$

Substituting (7.126) into (7.128) and using (7.43), $\alpha_{n-1}^2 + \kappa_n^{-2}\kappa_{n-1}^2 = 1$, we obtain finally the discrete Painlevé II equation

$$\frac{n}{t}\alpha_{n-1} + (\alpha_n + \alpha_{n-2})(1 - \alpha_{n-1}^2) = 0; \tag{7.129}$$

see [**Bai03, Bor03, AvM03**].

We may think of (7.129) as a recurrence relation

$$\alpha_n = -\alpha_{n-2} - \frac{n}{t} \cdot \frac{\alpha_{n-1}}{1 - \alpha_{n-1}^2}, \qquad n \geq 1, \tag{7.130}$$

where

$$\alpha_{-1} \equiv -1, \qquad \alpha_0 = -\pi_1(0; t) = \frac{d}{dt} \log \int_{-\pi}^{\pi} e^{t\cos\theta} d\theta. \tag{7.131}$$

Finally, we note that the compatibility of (7.53) and (7.119) leads to the Painlevé III equation,

$$\frac{d^2 f}{dx^2} = \frac{1}{f}\left(\frac{df}{dx}\right)^2 - \frac{1}{x}\frac{df}{dx} + \frac{4}{x}(\theta_0 f^2 + 1 - \theta_\infty) + 4f^3 - \frac{4}{f}, \tag{7.132}$$

where $\theta_0$ and $\theta_\infty$ are constants. Indeed, by direct inspection we see that RHP 7.7 for $Q(z)$ is precisely of the kind that arises in the analysis of PIII in the work of Fokas, Mugan, and Zhou ([**FMZ92**] and also [**BDJ99**] and [**FIKN06**]), in the case that the monodromy data (see [**FMZ92**, Figure 4 and pp. 765–766]) is given by

$$\begin{aligned} \theta_\infty &= -\theta_0 = n, \\ a_0 &= b_0 = a_\infty = b_\infty = 0, \\ E &= \begin{pmatrix} 1 & 1 \\ -1 & 0 \end{pmatrix}, \end{aligned} \tag{7.133}$$

and

$$x = it. \tag{7.134}$$

Set

$$y(x) \equiv -ix(Q_1^\infty(n))_{12} = -ix\nu(n) = ix\kappa_{n-1}^2\alpha_{n-2}. \tag{7.135}$$

Here we have used (7.112) and (7.126). Then by the prescription in [**FMZ92**]

$$f(x) = \frac{2}{\frac{d}{dx}\log(x^{-n}y(x))} = \frac{2}{\frac{d}{dx}\log(x^{-n+1}\kappa_{n-1}^2\alpha_{n-2})} \tag{7.136}$$

solves PIII (7.132) with $\theta_\infty = -\theta_0 = n$. The proof of this result uses the compatibility of (7.53) and (7.119) in a now familiar way and the details are left to the reader; see [**FMZ92**].

It is well known that in the appropriate scaling limit, solutions of PIII go over to solutions of PII. This is the case, in particular, in [**BDJ99**] and accounts for the appearance of PII via the Tracy-Widom distribution in Theorem 9.14.

**Theorem 7.13.** *The sequence* $\{\alpha_n\}$ *solves the discrete Painlevé II equation*

$$(n+1)\alpha_n + t(\alpha_{n-1}+\alpha_{n+1})(1-\alpha_n^2) = 0, \qquad n = 0,1,2,\ldots, \tag{7.137}$$

*with the initial condition*

$$\alpha_{-1} = -1, \qquad \alpha_0 = \frac{d}{dt}\left(\log \int_{-\pi}^{\pi} e^{t\cos\theta}\,d\theta\right). \tag{7.138}$$

*On the other hand,*

$$f(x) \equiv \frac{2}{\frac{d}{dx}\log(x^{-n+1}\kappa_{n-1}^2(-ix)\alpha_{n-2}(-ix))} \tag{7.139}$$

*solves the Painlevé III equation*

$$\frac{d^2f}{dx^2} = \frac{1}{f}\left(\frac{df}{dx}\right)^2 - \frac{1}{x}\frac{df}{dx} + \frac{4}{x}\big(-nf^2+1-n\big) + 4f^3 - \frac{4}{f} \tag{7.140}$$

*with the associated monodromy data given in* (7.133).

**Remark 7.14.** The asymptotic behavior of all the dynamical systems considered in this section with the given initial conditions (for Painlevé III, $f(x)$ is specified by the monodromy data (7.133)) can readily be computed by using the steepest-descent method to analyze RHP 7.2 as in [**BDJ99**].

**Remark 7.15.** Some of the above differential/difference equations were also obtained in the physics literature [**PS90**], [**His96**].

## 7.4. Heuristic Argument for Convergence of the Scaled Distribution for $L(t)$ to the Tracy-Widom Distribution

In the appropriate limit, (7.137) converges to the Painlevé II equation. Indeed, set

$$q_t(x) \equiv (-1)^{[2t+xt^{1/3}]}t^{1/3}\alpha_{[2t+xt^{1/3}-1]}(t), \qquad x\in\mathbb{R}, \tag{7.141}$$

where $[w]$ denotes the integer part of $w$. Then from (7.129),

$$\frac{[2t+xt^{1/3}]}{t}q_t(x) = \big(q_t(x+t^{-1/3}) + q_t(x-t^{-1/3})\big)\big(1-t^{-2/3}q_t^2(x)\big). \tag{7.142}$$

If $q_t(x)$ is a smooth function of $x$, then $q_t(x+t^{-1/3}) = q_t(x) + t^{-1/3}q_t'(x) + \frac{1}{2}t^{-2/3}q_t''(x)+\cdots$, and we obtain from (7.142)

$$q_t''(x) = 2q_t^3(x) + xq_t(x) + O(t^{-1/3}). \tag{7.143}$$

So if $q(x) \equiv \lim_{t\to\infty} q_t(x)$ exists and is smooth, it solves the Painlevé II equation

$$q''(x) = 2q^3(x) + xq(x). \tag{7.144}$$

From (7.40) and (7.43), $Q_n(t) \equiv \log \mathbb{P}\{L(t) \le n\}$ satisfies

$$Q_{n-1}(t) - 2Q_n(t) + Q_{n+1}(t) = \log(1 - (\alpha_{n-1}(t))^2). \tag{7.145}$$

The left-hand side is a second-order discrete difference. For any fixed $x$, consider the scaling limit $n, t \to \infty$ such that $n = [2t + xt^{1/3}]$. If we assume that

$$Q_{[2t+xt^{1/3}]}(t) \to Q(x) \tag{7.146}$$

locally uniformly in $x$ for some smooth function $Q(x)$, then we obtain from (7.145)

$$Q(x - t^{-1/3}) - 2Q(x) + Q(x + t^{-1/3}) \sim \log\left(1 - \frac{(q(x))^2}{t^{2/3}}\right), \tag{7.147}$$

which implies that

$$\frac{Q''(x)}{t^{2/3}} + O(t^{-1}) = -\frac{(q(x))^2}{t^{2/3}} + O(t^{-4/3}). \tag{7.148}$$

Comparing the leading terms, we find that $Q''(x) = -(q(x))^2$. Hence, if we assume that $Q(x), q(x) \to 0$ sufficiently fast as $x \to +\infty$,

$$Q(x) = -\int_x^\infty \int_y^\infty (q(s))^2 ds dy = -\int_x^\infty (s - x)(q(s))^2 ds. \tag{7.149}$$

Therefore, we have arrived at

$$\lim_{t\to\infty} \mathbb{P}\{L(t) \le 2t + xt\} = e^{-\int_x^\infty (s-x)(q(s))^2 ds}. \tag{7.150}$$

The right-hand side is the Tracy-Widom distribution $F(x)$ once we show that $q(x)$ satisfies the boundary condition $q(x) \sim \mathrm{Ai}(x)$, $x \to +\infty$, and hence is the Hastings-McLeod solution to the Painlevé II equation.

The above discussion gives a heuristic argument of the convergence to the Tracy-Widom distribution of scaled $L(t)$. This formal argument can be made rigorous by applying the steepest-descent method to RHP 7.2 as in the original proof of (7.150) in [**BDJ99**]. However, we prove the convergence theorem differently using a Fredholm determinant formula which is discussed in the next chapter.

Chapter 8

# Fredholm Determinant Formula

In Chapter 7, we expressed $\mathbb{P}\{L(t) \leq n\}$ in terms of a Toeplitz determinant of size $n$. We now obtain a different formula for $\mathbb{P}\{L(t) \leq n\}$ in terms of a Fredholm determinant. This formula will be used in Chapter 9 to evaluate rigorously the asymptotics as given in (7.150) of the distribution function for $L(t)$.

We first recall the integral representation for the Bessel function (see, e.g., [**AS64**]),

$$J_n(2t) \equiv \frac{1}{2\pi i}\int_C z^{-n} e^{t(z-z^{-1})} \frac{dz}{z} = \frac{1}{2\pi i}\int_C z^n e^{-t(z-z^{-1})} \frac{dz}{z}, \tag{8.1}$$

where $C$ is a simple closed contour which encloses the origin. Hence, $J_n(2t)$ is the $n$th Fourier coefficient of the function $e^{2it\sin\theta}$:

$$J_n(2t) = \psi_n = (\psi^{-1})_{-n}, \qquad \psi(z) \equiv e^{t(z-z^{-1})} = \sum_{n=-\infty}^{\infty} \psi_n z^n. \tag{8.2}$$

The main result in this chapter is:

**Theorem 8.1.** *For* $n = 0, 1, 2, \ldots$, *let* $\mathcal{K}_n$ *be the operator on*

$$\ell^2(\{n+1, n+2, \ldots\})$$

*defined by the kernel*

$$\mathcal{K}_n(j,k) = \sum_{m=0}^{\infty} J_{j+m}(2t) J_{m+k}(2t), \qquad i, j \geq n+1. \tag{8.3}$$

*Then*

$$\mathbb{P}\{L(t) \leq n\} = \det(1 - \mathcal{K}_n). \tag{8.4}$$

This result was first obtained in [**BOO00**] and [**Joh01a**] independently.

Before we prove the theorem, we discuss some properties of $\mathcal{K}_n$.

**Lemma 8.2.** *The operator $\mathcal{K}_n$ is a self-adjoint, positive, trace class operator, and the eigenvalues of $\mathcal{K}_n$ lie in $[0,1]$.*

**Proof.** Note that $\mathcal{K}_n = A^*A$ where $A : \ell_2(\{n+1, n+2, \dots\}) \to \ell_2(\mathbb{N}_0)$ has kernel $A(m,j) = J_{m+j}(2t)$ for $m \geq 0$ and $j \geq n+1$. Since $J_k(2t)$ is real-valued, we see that $\mathcal{K}_n$ is self-adjoint and positive.

To check that $\mathcal{K}_n$ is trace class, note that the Hilbert-Schmidt norm of $A$ is

$$\sum_{m=0}^{\infty} \sum_{j=n+1}^{\infty} A(m,j)^2 = \sum_{k=0}^{\infty} (k+1) J_{k+n+1}(2t)^2. \tag{8.5}$$

But $J_n(2t)$ decays exponentially as $n \to \infty$ since the $J_n(2t)$'s are the Fourier coefficients of an analytic function (see (8.2)), which implies that the above series converges. Hence $A$ is a Hilbert-Schmidt operator and therefore $\mathcal{K}_n$ is trace class.

Now we verify that the eigenvalues of $\mathcal{K}_n$ are in $[0,1]$. Let $f = \{f_k\}_{k \geq n+1} \in \ell^2(\{n+1, n+2, \dots\})$. Then for $\psi_n = J_n(2t)$,

$$\begin{aligned} 0 \leq (f, \mathcal{K}_n f) &= \sum_{j=n+1}^{\infty} \overline{f}_j \sum_{m=0}^{\infty} \psi_{j+m} \sum_{k=n+1}^{\infty} \psi_{m+k} f_k \\ &= \sum_{m=0}^{\infty} \left| \sum_{j=n+1}^{\infty} f_j \psi_{j+m} \right|^2 \leq \sum_{m=-\infty}^{\infty} \left| \sum_{j=n+1}^{\infty} f_j \psi_{j+m} \right|^2 \end{aligned} \tag{8.6}$$

and

$$\begin{aligned} \sum_{m=-\infty}^{\infty} \left| \sum_{j=n+1}^{\infty} f_j \psi_{j+m} \right|^2 &= \oint \frac{dz}{2\pi i z} \left| \sum_{m=-\infty}^{\infty} z^m \left( \sum_{j=n+1}^{\infty} f_j \psi_{j+m} \right) \right|^2 \\ &= \oint \frac{dz}{2\pi i z} \left| \left( \sum_{j=n+1}^{\infty} f_j z^{-j} \right) \psi(z) \right|^2 \\ &\leq \oint \frac{dz}{2\pi i z} \left| \sum_{j=n+1}^{\infty} f_j z^{-j} \right|^2 = \sum_{j=n+1}^{\infty} |f_j|^2, \end{aligned} \tag{8.7}$$

where $\oint$ denotes the integral over the unit circle, and we have used the unitarity of the Fourier transform $\ell_2 \to L^2([0, 2\pi])$. We have also used the fact that (see (8.2)) $\psi(z) = \sum_{j=-\infty}^{\infty} \psi_j z^j = \sum_{j=-\infty}^{\infty} J_j(2t) z^j = e^{2it\sin\theta}$, $z = e^{i\theta}$, and so $|\psi(z)| = 1$. Thus, $0 \le (f, \mathcal{K}_n f) \le \|f\|^2$, and hence $0 \le \mathcal{K}_n \le 1$, and so the eigenvalues of $\mathcal{K}_n$ lie in $[0, 1]$. □

**Exercise 8.3.** Use the argument in the proof of Lemma 8.2 above to show that in fact the eigenvalues lie in $[0, 1)$, i.e., 1 is not an eigenvalue of $\mathcal{K}_n$. This is also clear from (8.13) below as $D_{n-1}(\varphi) \neq 0$.

The kernel of $\mathcal{K}_n$ can also be written in the following ways.

**Lemma 8.4.** *For $j, k \ge n+1$,*

$$\mathcal{K}_n(j,k) = \frac{-1}{4\pi^2} \int_{C_1} \int_{C_2} \frac{z^{-j} w^{k-1}}{z - w} e^{t(z - z^{-1}) - t(w - w^{-1})} dw dz, \tag{8.8}$$

*where the contour $C_1$ for $z$ is the circle centered at 0 of radius $\rho_1$ and the contour $C_2$ for $w$ is the circle centered at 0 of radius $\rho_2$ such that $\rho_1 > \rho_2$. Both circles are oriented counter-clockwise. We also have*

$$\mathcal{K}_n(j,k) = \frac{tJ_{j-1}(2t)J_k(2t) - tJ_j(2t)J_{k-1}(2t)}{j-k}. \tag{8.9}$$

**Proof.** Inserting (8.1) and (8.2) into (8.3),

$$\begin{aligned} \mathcal{K}_n(j,k) &= \frac{-1}{4\pi^2} \sum_{m=0}^{\infty} \int_{C_1} \int_{C_2} z^{-j-m} w^{m+k} e^{t(z-z^{-1}) - t(w-w^{-1})} \frac{dw}{w} \frac{dz}{z} \\ &= \frac{-1}{4\pi^2} \int_{C_1} \int_{C_2} z^{-j} w^k \frac{1}{1 - z^{-1}w} e^{t(z-z^{-1}) - t(w-w^{-1})} \frac{dw}{w} \frac{dz}{z}. \end{aligned} \tag{8.10}$$

This gives the double integral expression (8.8).

In (8.8), for a fixed $x > 0$, change variables $z \mapsto xz$ and $w \mapsto xw$. Since the integrand is analytic in $\mathbb{C} \setminus \{0\}$, the contours can be kept the same:

$$\mathcal{K}_n(j,k) = \frac{-1}{4\pi^2} \int_{C_1} \int_{C_2} \frac{x^{-j+k} z^{-j} w^{k-1}}{z - w} e^{t(xz - \frac{1}{xz}) - t(xw - \frac{1}{xw})} dw dz. \tag{8.11}$$

Now take the derivative with respect to $x$ and evaluate at $x = 1$. Since the left-hand side vanishes, we find that

$$\begin{aligned} 0 =& (-j+k)\, \mathcal{K}_n(j,k) \\ &+ \frac{-1}{4\pi^2} \int_{C_1} \int_{C_2} t\left(1 - \frac{1}{zw}\right) z^{-j} w^{k-1} e^{t(z - \frac{1}{z}) - t(w - \frac{1}{w})} dw dz \\ =& (-j+k)\, \mathcal{K}_n(j,k) + t\big(J_{j-1}(2t)J_k(2t) - J_j(2t)J_{k-1}(2t)\big). \end{aligned} \tag{8.12}$$

This implies (8.9). □

Formula (8.9) shows that $\mathcal{K}_n$ is an integrable operator (see Section 5.6). Hence we may use the RHP naturally associated with $\mathcal{K}_n$ to find differential/difference equations for $\mathbb{P}\{L(t) \le n\}$ as we did in the case of the Airy operator in Theorem 6.18. This in turn provides alternative proofs of some of the results in Section 7.3. See [**Bor03**] for further discussion in this direction.

We give two proofs of Theorem 8.1 in the next two sections. The first proof derives (8.4) from Gessel's formula. In particular, we will show that

$$D_{n-1}(\varphi) = e^{t^2} \det(1 - \mathcal{K}_n). \tag{8.13}$$

It turns out that (8.13) is a special case of a general identity, now called the Borodin-Okounkov-Geronimo-Case identity, between Toeplitz determinants and Fredholm determinants. We prove this general identity in the next section. The second proof is more direct and does not use Gessel's formula and generalizes easily to derive formulae for other statistics associated with Poissonized Pancherel measure: from such formulae we learn in particular that the Poissonized Plancherel measure gives rise to a determinantal point process.

## 8.1. First Proof: Borodin-Okounkov-Geronimo-Case Identity

In [**BO00**] Borodin and Okounkov obtained a general formula expressing the Toeplitz determinant of a (suitably smooth) symbol $w$ in terms of a Fredholm determinant $(1 - K)$ for some trace class operator $K$; (8.13) above is a particular example of this general result. Quite remarkably, this general formula was discovered earlier by Geronimo and Case [**GC79**] in 1979 in the context of inverse scattering theory for semi-infinite systems; unfortunately the significance of the formula was not appreciated at the time. The formulation of the identity as given in Theorem 8.6 below is due to [**BW00**].

Let $w(z)$ be a smooth, nonvanishing function on the unit circle $\Sigma = \{|z| = 1\}$ with zero winding number. Consider the following scalar Riemann-Hilbert problem $(\Sigma, w)$.

**Riemann-Hilbert Problem 8.5.** *Find $\delta(z)$ satisfying the following conditions:*

- *$\delta(z)$ is analytic in $\mathbb{C} \setminus \Sigma$ and does not vanish in $\{|z| \le 1\}$ and $\{|z| \ge 1\}$,*
- *$\delta_+(z) = \delta_-(z) w(z)$ for $z \in \Sigma$,*
- *$\delta(z) \to 1$ as $z \to \infty$.*

Under appropriate smoothness conditions (see Section 5.5 and below), there is a (unique) solution to the scalar RHP $(\Sigma, w)$. Moreover $\delta(z) \neq 0$ for $z \in \mathbb{C} \setminus \Sigma$ and $\delta_\pm(z) \neq 0$ for $z \in \Sigma$. Set

$$b(z) \equiv \frac{1}{\delta_+(z)\delta_-(z)}, \qquad c(z) \equiv \delta_+(z)\delta_-(z), \qquad z \in \Sigma. \tag{8.14}$$

Let $b(z) = \sum_{k=-\infty}^{\infty} b_k z^k$ and $c(z) = \sum_{k=-\infty}^{\infty} c_k z^k$ be the Fourier series for $b$ and $c$. Define the semi-infinite Hankel matrices

$$H(b) = \begin{pmatrix} b_1 & b_2 & b_3 & \cdots \\ b_2 & b_3 & b_4 & \cdots \\ b_3 & b_4 & b_5 & \cdots \\ \vdots & \vdots & \vdots & \ddots \end{pmatrix} = (b_{j+k+1})_{j,k\geq 0} \tag{8.15}$$

and

$$H(\tilde{c}) = \begin{pmatrix} c_{-1} & c_{-2} & c_{-3} & \cdots \\ c_{-2} & c_{-3} & c_{-4} & \cdots \\ c_{-3} & c_{-4} & c_{-5} & \cdots \\ \vdots & \vdots & \vdots & \ddots \end{pmatrix} = (c_{-j-k-1})_{j,k\geq 0}. \tag{8.16}$$

Both $H(b)$ and $H(\tilde{c})$ are kernels for Hankel operators in $\ell_2(\mathbb{N}_0)$ (see [**Pel80**]): for $u = (u_0, u_1, \dots)^T \in \ell_2(\mathbb{N}_0)$, $(H(b)u)(j) = \sum_{k=0}^{\infty} b_{j+k+1} u_k$, $j \geq 0$, and $(H(\tilde{c})u)(j) = \sum_{k=0}^{\infty} c_{-j-k-1} u_k$, $j \geq 0$.

Since $w(z)$, $z \in \Sigma$, is smooth and does not vanish, and has zero winding number, $\log w(z)$ is also smooth. It does not matter for our purposes which branch we take (see Remark 8.7). Let $\{L_k\}_{k\in\mathbb{Z}}$ be the Fourier coefficients of $\log w$,

$$\log w(z) = \sum_{k=-\infty}^{\infty} L_k z^k. \tag{8.17}$$

We say that $f(z) = \sum f_k z^k \in L^2(\Sigma)$ lies in $H^{1/2}(\Sigma)$ if $\sum_{-\infty}^{\infty} |k||f_k|^2 < \infty$. $H^{1/2}(\Sigma)$ carries the inner product

$$\langle f, g\rangle_{1/2} = \sum_{-\infty}^{\infty} (1+|k|)\overline{f_k} g_k \tag{8.18}$$

and norm

$$\|f\|_{1/2} = \sqrt{\langle f, f\rangle_{1/2}}. \tag{8.19}$$

Let $Q_n$ denote the orthogonal projection of $\ell_2(\mathbb{Z})$ onto $\ell_2(\{n, n+1, \dots\})$. Then the Borodin-Okounkov-Geronimo-Case (BOGC) formula asserts the following:

**Theorem 8.6.** *Suppose that $w(z)$ is a smooth, nonvanishing function on $\Sigma$ with zero winding number. Then for all sufficiently large $n$*

$$D_{n-1}(w) = e^{nL_0} e^{\sum_{k=1}^{\infty} kL_k L_{-k}} \det\left(1 - Q_n H(b) H(\tilde{c}) Q_n\right)_{\ell^2(\mathbb{N}_0)}, \tag{8.20}$$

*where $H(b)$ and $H(\tilde{c})$ are defined in* (8.15) *and* (8.16), *respectively.*

**Remark 8.7.** In (8.17), we may take an arbitrary branch of the logarithm since a different choice of a branch results only in the change $L_0 \mapsto L_0 + 2\pi i m$ for some $m \in \mathbb{Z}$, but $L_0$ appears in the theorem only in $e^{nL_0}$.

**Remark 8.8.** As $w(z)$ is smooth, nonvanishing and without winding, one easily shows (we leave this as an exercise) that $H(b)$ and $H(\tilde{c})$ are Hilbert-Schmidt operators in $\ell_2(\mathbb{N}_0)$. Thus $Q_n H(b) H(\tilde{c}) Q_n$ is trace class in $\ell_2(\mathbb{N}_0)$ for all $n \geq 0$. By a standard result (see Appendix A), it follows that $Q_n H(b) H(\tilde{c}) Q_n \to 0$ in trace norm as $n \to \infty$, and so $\det(1 - Q_n H(b) H(\tilde{c}) Q_n) \to 0$ as $n \to \infty$. Thus, we see that the Szegő strong limit theorem (Theorem 5.22), $e^{-nL_0} D_{n-1}(w) \to e^{\sum_{k=1}^{\infty} kL_k L_{-k}}$, is an immediate consequence of the above formula. We regard the term $\det\left(1 - Q_n H(b) H(\tilde{c}) Q_n\right)_{\ell^2(\mathbb{N}_0)}$ as providing an explicit error term in Szegő's strong limit theorem.

A special case of Theorem 8.6 implies Theorem 8.1.

**Proof of Theorem 8.1.** Take $w(z) = e^{t(z+z^{-1})} = \varphi(z)$. Then

$$\delta(z) = \begin{cases} e^{tz}, & |z| < 1, \\ e^{-tz^{-1}}, & |z| > 1, \end{cases} \tag{8.21}$$

is the (unique) solution to RHP $(\Sigma, w)$ above, and hence $b(z) = e^{-t(z-z^{-1})}$, $c(z) = e^{t(z-z^{-1})}$. From (8.1),

$$b_{-k} = c_k = J_k(2t). \tag{8.22}$$

Hence, for $i, j \geq 0$,

$$\big(H(b)H(\tilde{c})\big)(i,j) = \sum_{k=0}^{\infty} b_{i+k+1} c_{-k-j-1} = \sum_{k=0}^{\infty} J_{-i-k-1}(2t) J_{-k-j-1}(2t). \tag{8.23}$$

It is easy to check from (8.1) that

$$J_k(z) = (-1)^k J_{-k}(z), \tag{8.24}$$

and thus

$$\big(H(b)H(\tilde{c})\big)(i,j) = (-1)^{i+j} \sum_{k=0}^{\infty} J_{i+k+1}(2t) J_{k+j+1}(2t). \tag{8.25}$$

Since $u = (u_0, u_1, u_2, \dots) \mapsto ((-1)^0 u_0, (-1)^1 u_1, (-1)^2 u_2, \dots)$ is a unitary map in $\ell_2(\mathbb{N}_0)$, we see that (note that $Q_n$ is a projection on $\{n, n+1, \dots\}$ and $\mathcal{K}_n$ acts on $\ell^2(\{n+1, n+2, \dots\})$)

$$\det(1 - Q_n H(b) H(\tilde{c}) Q_n) = \det(1 - \mathcal{K}_n). \tag{8.26}$$

On the other hand, $\log w = t(z + z^{-1})$, and hence

$$L_k = \begin{cases} t, & k = 1, -1 \\ 0, & k \neq 1, -1. \end{cases} \tag{8.27}$$

Thus, in this case, (8.20) implies that

$$D_{n-1}(\varphi) = e^{t^2} \det\left(1 - \mathcal{K}_n\right). \tag{8.28}$$

Since $\mathbb{P}\{L(t) \leq n\} = e^{-t^2} D_{n-1}(\varphi)$ from the Gessel's formula (7.3), we obtain (8.4). □

There are several proofs of Theorem 8.6. In addition to [**GC79**] and [**BO00**], proofs under far weaker assumptions than in Theorem 8.6 can be found in [**BW00**], [**Böt01**] and [**Böt02**]. For a proof of the general result (8.20) for a positive, integrable function $w(z)$ just under the additional assumption $\log w \in H^{1/2}$, see Section 6.2 of [**Sim05a**]; in this case (8.20) holds for all $n \geq 0$.

Here we will only verify (8.20) for a restricted class of weights which is sufficient for our present purposes, i.e., the proof of (8.4). Assume in particular that $w$ is smooth and positive. In Section 10.5, however, a more general version of (8.20) allowing for complex valued symbols $w(z)$ is needed in order to prove Gessel's formula for general Schur measures (see below). The proof of (8.20) that we give below replies on the classical connection between Toeplitz determinants and orthogonal polynomials with respect to the positive weight $w(e^{i\theta})\frac{d\theta}{2\pi}$. In Section 10.5 we can of course rely on any of the proofs of Theorem 8.6 mentioned above, but our proof of (8.20) for $w(z) \geq 0$ extends appropriately to the case when $w(z)$ is a complex valued function with zero winding number, using polynomials that are orthogonal with respect to the complex weight $w(e^{i\theta})\frac{d\theta}{2\pi}$ in the manner of [**DIK11**]. We leave the details to the interested reader.

So assume, in particular, that $w(z)$ is smooth and positive. Under these assumptions $D_{n-1}(w) > 0$, and so we anticipate from (8.20) that $1 - Q_n H(b) H(\tilde{c}) Q_n$ is invertible for *all* $n \geq 0$. This is in contrast to the case when $w(z)$ is complex valued and we only expect that $1 - Q_n H(b) H(\tilde{c}) Q_n$ is invertible for $n$ large (and hence (8.20) only holds for $n$ large).

**Lemma 8.9.** *Suppose that $w(z)$ is smooth and positive. Then*

$$1 - Q_n H(b) H(\tilde{c}) Q_n$$

*is invertible for all $n \geq 0$.*

**Proof.** Note first that by (5.120) the solution $\delta(z)$ of the above scalar RHP 8.5 is given by

$$\delta(z) = e^{\int_\Sigma \frac{\log w(s)}{s-z} \frac{ds}{2\pi i}}, \tag{8.29}$$

from which we see, in particular, that

$$\delta(0) = e^{\int_0^{2\pi} \log w(e^{i\theta}) \frac{d\theta}{2\pi}}. \tag{8.30}$$

Also, direct calculation shows that $\delta(z)\overline{\delta(\overline{z}^{-1})} = \delta(0)$ for $z \in \mathbb{C} \setminus \Sigma$ so that $\delta_+(z)\overline{\delta_-(z)} = \delta(0)$ for $z \in \Sigma$. It follows that

$$b(z) = \phi(z)^{-1}\delta(0)^{-1}, \qquad c(z) = \phi(z)\delta(0), \qquad z \in \Sigma, \tag{8.31}$$

where

$$\phi(z) = \delta_+(z)\overline{\delta_+(z)}^{-1}, \qquad z \in \Sigma. \tag{8.32}$$

A simple computation shows that the Fourier coefficients $\phi_k$ and $(\phi^{-1})_k$ are related as follows,

$$\phi_k = \overline{(\phi^{-1})_k}, \qquad k \in \mathbb{Z}. \tag{8.33}$$

Let $\Phi$ denote the following operator acting on sequences $f = \{f_k\}_{k \geq 0} \in \ell_2(\mathbb{N}_0)$,

$$(\Phi f)_k = \sum_{j=0}^{\infty} \phi_{-k-j-1} f_j, \qquad k \geq 0. \tag{8.34}$$

We have $H(\tilde{c}) = \delta(0)\Phi$ and $H(b) = \delta(0)^{-1}\Phi^*$ so that

$$H(b)H(\tilde{c}) = \Phi^*\Phi. \tag{8.35}$$

For any $f \in \ell_2(\mathbb{N}_0)$, we have (cf. Lemma 6.15)

$$(f, Q_n H(b) H(\tilde{c}) Q_n f) = \|\Phi f^{(n)}\|^2 \geq 0, \tag{8.36}$$

where $f^{(n)} = Q_n f$. Using (8.34),

$$\begin{aligned} (f, Q_n H(b) H(\tilde{c}) Q_n f) &= \sum_{k=0}^{\infty} \left| \sum_{j=0}^{\infty} \phi_{-k-j-1} f_j^{(n)} \right|^2 \\ &= \sum_{k=0}^{\infty} \left| \sum_{j=-\infty}^{\infty} \phi_{-k-j-1} f_j^{(n)} \right|^2 \end{aligned} \tag{8.37}$$

as $f_j^{(n)} = 0$ for $j < 0$. Since the last sum over $j$ is the convolution of the Fourier coefficients of $\phi(z)$ and $\widehat{f^{(n)}}(z) = \sum_{j\in\mathbb{Z}} z^j f_j^{(n)}$, we obtain

$$(8.38)\qquad \begin{aligned}(f, Q_nH(b)H(\tilde{c})Q_nf) &= \sum_{k=0}^{\infty}\left|(\phi\widehat{f^{(n)}})_{-k-1}\right|^2\\ &\leq \sum_{k=-\infty}^{\infty}\left|(\phi\widehat{f^{(n)}})_k\right|^2 = \|\phi\widehat{f^{(n)}}\|^2_{L^2(\Sigma)}\\ &= \|\widehat{f^{(n)}}\|^2_{L^2(\Sigma)} = \|f^{(n)}\|_{\ell_2(\mathbb{N}_0)} \leq \|f\|^2_{\ell_2(\mathbb{N}_0)},\end{aligned}$$

where we have used $|\phi(z)| = 1$ for $z \in \Sigma$. From (8.36) and (8.38) we see in particular that $0 \leq Q_nH(b)H(\tilde{c})Q_n \leq 1$, and since $\Phi$ is Hilbert-Schmidt (see Remark 8.8), $Q_nH(b)H(\tilde{c})Q_n$ is compact (in fact, trace class). Hence in order to show that $1 - Q_nH(b)H(\tilde{c})Q_n$ is invertible, it is enough to show that $(1 - Q_nH(b)H(\tilde{c})Q_n)f = 0$, $f \in \ell_2(\mathbb{N}_0)$, implies $f = 0$. But for such $f$, we see from (8.38) that $(\phi\widehat{f^{(n)}})_k = 0$ for $k \geq 0$. Thus $\phi(z)\widehat{f^{(n)}}(z) = \sum_{j=1}^{\infty}\gamma_j z^{-j} \equiv \gamma(z)$ for some $\{\gamma_j\} \in \ell_2(\mathbb{Z})$. Thus

$$(8.39)\qquad \delta_+(z)\widehat{f^{(n)}}(z) = \overline{\delta_+(1/\overline{z})}\gamma(z), \qquad z \in \Sigma.$$

As $\widehat{f^{(n)}} = \sum_{j\in\mathbb{Z}} z^j f_j^{(n)} = \sum_{j=n}^{\infty} z^j f_j$, $n \geq 0$, it follows that the left-hand side of (8.39) has an analytic continuation to $\{|z| < 1\}$, while the right-hand side has a continuation to $\{|z| > 1\}$; furthermore, the right-hand side converges to zero as $z \to \infty$. It follows that $\widehat{f^{(n)}}(z) \equiv 0$ and hence $f = 0$. This proves the invertibility of $1 - Q_nH(b)H(\tilde{c})Q_n$ for all $n \geq 0$. □

The following computation illustrates the connection between the operator $Q_mH(b)H(\tilde{c})Q_m$ and the RHP for orthogonal polynomials.

**Proof of Theorem 8.6 when $w(z)$ is smooth and positive.** We first prove that, for all $m \geq 0$,

$$(8.40)\qquad e^{L_0}\frac{D_{m-1}(w)}{D_m(w)} = \frac{\det\left(1 - Q_mH(b)H(\tilde{c})Q_m\right)_{\ell^2(\mathbb{N}_0)}}{\det\left(1 - Q_{m+1}H(b)H(\tilde{c})Q_{m+1}\right)_{\ell^2(\mathbb{N}_0)}}$$

(recall $D_{-1} \equiv 1$). Let $K_m : \ell^2(\mathbb{N}_0) \to \ell^2(\mathbb{N}_0)$ be the operator defined by the kernel

$$(8.41)\qquad \begin{aligned}K_m(k,\ell) &= \big(H(b)H(\tilde{c})\big)(k+m, \ell+m)\\ &= \sum_{j=0}^{\infty} b_{j+k+m+1}c_{-j-\ell-m-1}, \qquad k,\ell \geq 0.\end{aligned}$$

The Fredholm determinant of $Q_mH(b)H(\tilde{c})Q_m$ equals the Fredholm determinant of $K_m$, and hence $1 - K_m$ is invertible for all $m \geq 0$, and the right-hand side of (8.40) equals

$$\begin{aligned}\frac{\det(1-K_m)}{\det(1-K_{m+1})} &= \det\left(\frac{1}{1-K_{m+1}}(1-K_m)\right) \\ &= \det\left(1-\frac{1}{1-K_{m+1}}(K_m-K_{m+1})\right).\end{aligned} \tag{8.42}$$

From (8.41), for $k,\ell \geq 0$,

$$K_m(k,\ell) - K_{m+1}(k,\ell) = b_{k+m+1}c_{-\ell-m-1}. \tag{8.43}$$

Hence $K_m - K_{m+1}$ is an operator of rank 1:

$$K_m - K_{m+1} = u \otimes v, \tag{8.44}$$

where

$$\begin{aligned} u &= (b_{m+1}, b_{m+2}, b_{m+3}, \dots)^T, \\ v &= (c_{-m-1}, c_{-m-2}, c_{-m-3}, \dots)^T. \end{aligned} \tag{8.45}$$

Therefore (8.42) equals

$$\det\left(1-\frac{1}{1-K_{m+1}}u\otimes v\right) = 1-\sum_{\ell=0}^{\infty}\left(\frac{1}{1-K_{m+1}}u\right)(\ell)v(\ell). \tag{8.46}$$

On the other hand, consider the left-hand side of (8.40). From (5.89) in the theory of orthogonal polynomials

$$\frac{D_{m-1}(w)}{D_m(w)} = \kappa_m^2, \tag{8.47}$$

where $\kappa_m$ is the leading coefficient of the $m$th orthonormal polynomial with respect to the measure $w(e^{i\theta})\frac{d\theta}{2\pi}$ on the unit circle $\Sigma$. Let $Y(z) = Y(z;m+1)$ be the solution to RHP 5.10 with $n$ replaced by $m+1$:

- $Y(z)$ is analytic in $\mathbb{C} \setminus \Sigma$,
- $Y_+(z) = Y_-(z)\begin{pmatrix} 1 & z^{-m-1}w(z) \\ 0 & 1 \end{pmatrix}$ for $z \in \Sigma$,

and

- $Y(z)\begin{pmatrix} z^{-m-1} & 0 \\ 0 & z^{m+1} \end{pmatrix} = I + O(z^{-1})$ as $z \to \infty$.

Then the (21)-entry of $Y$ satisfies (see (5.72))

$$Y_{21}(0;m+1) = -\kappa_m^2. \tag{8.48}$$

Hence, we have

$$\frac{D_{m-1}(w)}{D_m(w)} = -Y_{21}(0;m+1). \tag{8.49}$$

Let $\delta(z)$ be the solution of the above scalar RHP for $w$ and define

$$M(z) = \begin{cases} Y(z)\begin{pmatrix} 0 & -\delta(z) \\ \delta(z)^{-1} & 0 \end{pmatrix}, & |z| < 1, \\ Y(z)\begin{pmatrix} z^{-m-1}\delta(z)^{-1} & 0 \\ 0 & z^{m+1}\delta(z) \end{pmatrix}, & |z| > 1. \end{cases} \tag{8.50}$$

It is easy to check using the RHP $(\Sigma, w)$ for $\delta(z)$ that $M$ solves the *normalized* RHP $(\Sigma, v_M)$, where

$$\begin{aligned} v_M(z) &= \begin{pmatrix} 1 & -z^{m+1}\delta_+(z)\delta_-(z) \\ z^{-m-1}\delta_+(z)^{-1}\delta_-(z)^{-1} & 0 \end{pmatrix} \\ &= \begin{pmatrix} 1 & -z^{m+1}c(z) \\ z^{-m-1}b(z) & 0 \end{pmatrix} \end{aligned} \tag{8.51}$$

for $z \in \Sigma$. From (8.50), (8.49) becomes

$$\frac{D_{m-1}(w)}{D_m(w)} = \delta(0)^{-1} M_{22}(0; m+1). \tag{8.52}$$

By (8.30), $\delta(0) = e^{L_0}$. Thus, (8.40) is proved if we show that

$$M_{22}(0; m+1) = 1 - \sum_{\ell=0}^{\infty} \big((1 - K_{m+1})^{-1} u\big)(\ell) v(\ell). \tag{8.53}$$

The jump matrix $v_M$ can be written as

$$v_M = (I - w^-)^{-1}(I + w^+), \tag{8.54}$$

where

$$w^-(z) = \begin{pmatrix} 0 & 0 \\ z^{-m-1}b(z) & 0 \end{pmatrix}, \qquad w^+(z) = \begin{pmatrix} 0 & -z^{m+1}c(z) \\ 0 & 0 \end{pmatrix}. \tag{8.55}$$

Let $\mu(z) = (\mu_{jk}(z))_{j,k=1}^2 \in I + L^2(\Sigma)$ be the (unique) solution of the equation (see (5.118))

$$\mu - C_+(\mu w^-) - C_-(\mu w^+) = I, \tag{8.56}$$

where C denotes the Cauchy operator on $\Sigma$, and $C_+$ and $C_-$ denote the limits from the inside and the outside, respectively. We will discuss the (unique) solvability of the equation further on in the proof. Assume that equation (8.56) indeed has a solution $\mu$: By the general theory of RHP's (see (5.119)),

$$M(z; m+1) = I + \frac{1}{2\pi i} \int_\Sigma \frac{\mu(s)(w^+(s) + w^-(s))}{s - z} ds. \tag{8.57}$$

Hence,

$$M_{22}(0; m+1) = 1 - \frac{1}{2\pi i} \int_\Sigma \frac{\mu_{21}(s) s^{m+1} c(s)}{s} ds. \tag{8.58}$$

The second row of the equation (8.56) implies that

$$\mu_{21}(z) = \mathrm{C}_+(\mu_{22}b^{(-m)})(z), \qquad \mu_{22}(z) = 1 - \mathrm{C}_-(\mu_{21}c^{(m)})(z), \tag{8.59}$$

where we set

$$b^{(-m)}(s) = s^{-m-1}b(s), \quad c^{(m)}(s) = s^{m+1}c(s).$$

Inserting the second equation into the first equation in (8.59) we find the following equation for $\mu_{21}$:

$$\mu_{21}(z) = \mathrm{C}_+(b^{(-m)})(z) - \mathrm{C}_+(\mathrm{C}_-(\mu_{21}c^{(m)})b^{(-m)})(z). \tag{8.60}$$

We now express this equation in terms of Fourier coefficients. Note that the first equation of (8.59) implies that $\mu_{21}(z)$ has an analytic continuation inside the unit circle. Hence the Fourier series expansion of $\mu_{21}(z)$ is of form

$$\mu_{21}(z) \equiv \sum_{\ell=0}^{\infty} q_k z^k \tag{8.61}$$

for some $\ell^2$ sequence $q_k$, $k = 0, 1, \dots$. Then equation (8.60) implies that

$$\begin{aligned} q_k &= b_{k+m+1} + \sum_{\ell=0}^{\infty}\left\{\sum_{s=-\infty}^{-\ell-m-2} b_{k-s-\ell}c_s\right\}q_\ell \\ &= b_{k+m+1} + \sum_{\ell=0}^{\infty}\left\{\sum_{j=0}^{\infty} b_{j+k+m+2}c_{-j-\ell-m-2}\right\}q_\ell, \qquad k \ge 0. \end{aligned} \tag{8.62}$$

Comparing with (8.41) and (8.45), these equations are equivalent to the following equation: with $q = (q_0, q_2, \dots)^T$,

$$q = u + K_{m+1}q. \tag{8.63}$$

Thus, $q = (1 - K_{m+1})^{-1}u$. Reversing the above argument, the existence of $q$ implies the existence of the solution $(\mu_{21}, \mu_{22}) \in (0,1) + L^2(\Sigma)$ of (8.59). Moreover $(\mu_{21}, \mu_{22})$ is unique. Indeed if $(\tilde\mu_{21}, \tilde\mu_{22}) \in (0,1) + L^2(\Sigma)$ were a second solution, then the above calculations show that $\rho = \tilde\mu_{21} - \mu_{21}$ solves $\rho = K_{m+1}\rho$, and we have $\rho = 0$ by the invertibility of $1 - K_{m+1}$, and then $\tilde\mu_{22} - \mu_{22} = -\mathrm{C}_-(\rho(\cdot)^{m+1}c) = 0$. Thus $(\mu_{21}, \mu_{22}) = (\tilde\mu_{21}, \tilde\mu_{22})$. The existence and uniqueness of $(\mu_{11}, \mu_{12}) \in (1,0) + L^2(\Sigma)$ can be proved similarly. Hence, we see that equation (8.56) is (uniquely) solvable.

From (8.58) and (8.61),

$$M_{22}(0; m+1) = 1 - \sum_{\ell=0}^{\infty} q_\ell c_{-\ell-m-1} = 1 - \sum_{\ell=0}^{\infty}((1 - K_{m+1})^{-1}u)(\ell)v(\ell). \tag{8.64}$$

This is precisely (8.53). Hence, we have obtained (8.40).

Now, successive applications of (8.40) imply that

$$e^{(N-n+1)L_0}\frac{D_{n-1}(w)}{D_N(w)} = \frac{\det\left(1 - Q_n H(b)H(\tilde{c})Q_n\right)_{\ell^2(\mathbb{N}_0)}}{\det\left(1 - Q_{N+1}H(b)H(\tilde{c})Q_{N+1}\right)_{\ell^2(\mathbb{N}_0)}}. \tag{8.65}$$

As $N \to \infty$, $\det\left(1 - Q_{N+1}H(b)H(\tilde{c})Q_{N+1}\right)_{\ell^2(\mathbb{N}_0)} \to 0$ since $Q_{N+1} \to 0$ strongly and $H(b)H(\tilde{c})$ is trace class (see (A.12)). Hence, by taking $N \to \infty$ and using the strong Szegő Theorem 5.22 for the limit of $e^{-(N+1)L_0}D_N(w)$, we obtain (8.20).

Alternatively, one can conclude (8.20) from (8.65) without using the strong Szegő theorem as follows. Considering the identity (8.65) when $n = 0$ and recalling that $D_{-1}(w) \equiv 1$, we see that (8.20) is obtained if we show that

$$\det\left(1 - H(b)H(\tilde{c})\right)_{\ell^2(\mathbb{N}_0)} = e^{-\sum_{k=1}^{\infty} kL_k L_{-k}}. \tag{8.66}$$

This result follows in turn from Widom's elementary but basic identity

$$T(\varphi\psi) = T(\varphi)T(\psi) + H(\varphi)H(\tilde{\psi}) \tag{8.67}$$

(see [**Wid76**]) applied to $\varphi = b$ and $\psi = c = b^{-1}$. Here $T(a)$ denotes the semi-infinite Toeplitz matrix with symbol $a$, $T(a) = (a_{j-k})_{j,k=0}^{\infty}$ (see Proposition 5.25). To see that (8.66) follows from (8.67), note that

$$T(b) = T(\delta_-^{-1}\delta_+^{-1}) = T(\delta_-^{-1})T(\delta_+^{-1}) = T(\delta_-)^{-1}T(\delta_+)^{-1} \tag{8.68}$$

(we leave this as an exercise and cf. Proposition 5.28) and similarly $T(c) = T(\delta_-)T(\delta_+)$. Hence, one obtains from (8.67)

$$1 - H(b)H(\tilde{c}) = T(\delta_-)^{-1}T(\delta_+)^{-1}T(\delta_-)T(\delta_+). \tag{8.69}$$

Now one uses the Helton-Howe formula for bounded operators $A, B$ in a Hilbert space: if the commutator $[A, B] = AB - BA$ is trace class, then

$$\det(e^A e^B e^{-A} e^{-B}) = e^{\operatorname{tr}[A,B]}. \tag{8.70}$$

In our case, from (8.29), $\delta_+ = e^{\mathrm{C}_+(\log w)}$, and so

(8.71)

$$T(\delta_+) = T\Big(\sum_{k=0}^{\infty}\frac{1}{k!}(\mathrm{C}_+(\log w))^k\Big) = \sum_{k=0}^{\infty}\frac{1}{k!}(T(\mathrm{C}_+(\log w)))^k = e^{T(\mathrm{C}_+(\log w))},$$

etc. Now applying the Helton-Howe formula to (8.69), we obtain $\det(1 - H(b)H(\tilde{c}))_{\ell^2(\mathbb{N}_0)} = e^{\operatorname{tr}[T(\mathrm{C}_-(\log w)), T(\mathrm{C}_+(\log w))]}$, which immediately yields (8.66). For more details, see [**Sim05a**, Theorem 6.2.13]. The identity (8.67) is due to Widom [**Wid76**]. □

**Remark 8.10.** Widom's formula (8.66) plays a key role in his proof [**Wid76**] of the Szegő strong limit theorem. Thus both of the arguments given above deriving the BOGC identity from (8.66) are intimately connected to the Szegő strong limit theorem. As noted in Remark 8.8, letting $n \to \infty$ in (8.20), we recover the Szegő theorem. The fact that (8.20) in fact gives an *independent* proof of Szegő's theorem rests, in our proof of (8.20), on the second derivation of the BOGC identity from (8.65), using Widom's formula (8.66).

**Exercise 8.11.** Assume that $b(z)$ is analytic in a neighborhood of $\Sigma$. Using $c(z) = b^{-1}(z)$, show that (cf. (8.8))

$$(Q_n H(b) H(\tilde{c}) Q_n)(j,k) = \frac{-1}{4\pi^2} \int_{C_1} \int_{C_2} \frac{z^{-j-1} w^k}{z-w} \frac{b(z)}{b(w)} dz dw \tag{8.72}$$

for $j, k \geq n$, where the contour $C_1$ for $z$ is the circle centered at 0 of radius $\rho_1$ and the contour $C_2$ for $w$ is the circle centered at 0 of radius $\rho_2$ such that $\rho_1 > \rho_2$. Using the above formula, show that

$$\begin{aligned} &(j-k)(Q_n H(b) H(\tilde{c}) Q_n)(j,k) \\ &= \frac{-1}{4\pi^2} \int_{C_1} \int_{C_2} z^{-j-1} w^k \frac{\frac{zb'(z)}{b(z)} - \frac{wb'(w)}{b(w)}}{z-w} \frac{b(z)}{b(w)} dz dw. \end{aligned} \tag{8.73}$$

This implies that if $\frac{b'(z)}{b(z)}$ is a rational function, $Q_n H(b) H(\tilde{c}) Q_n$ is an integrable operator. This is true in particular for $b(z) = e^{-t(z-z^{-1})}$ as above.

## 8.2. Second Proof

In this section, we give a direct proof of Theorem 8.1 without recourse to the Gessel's formula. As a byproduct, this establishes the Borodin-Okounkov-Geronimo-Case identity (8.20) in the special case $w(z) = e^{t(z+z^{-1})}$. The calculations that follow are modeled on [**Joh03**].

From (2.6),

$$\mathbb{P}(L(t) \leq n) = \mathbb{P}^{P.P}(\lambda_1 \leq n) = \mathbb{E}^{P.P}\left[1_{\lambda_1 \leq n}\right], \tag{8.74}$$

where $P.P.$ means the Poissonized Plancherel measure as before. Theorem 8.1 is a consequence of the following more general result.

**Theorem 8.12.** *Let $g$ be a bounded function on $\mathbb{Z}$ supported on $\{-m_0, -m_0+1, \dots\}$ for some $m_0$. Let $P$ denote the projection in $\ell_2(\mathbb{Z})$ onto the set $\{-m_0, -m_0+1, \dots\}$, and let $M_g$ denote multiplication by $g$. Then*

$$\mathbb{E}^{P.P}\left[\prod_{i=1}^{\infty}(1 + g(\lambda_i - i + 1))\right] = \det(1 + P\mathcal{K}M_g), \tag{8.75}$$

*where the expectation is with respect to the Poissonized Plancherel measure* (2.5) *and $\mathcal{K}$ has the kernel (cf.* (8.3)*)*

$$\mathcal{K}(i,j) = \sum_{\ell=0}^{\infty} J_{i+\ell}(2t)J_{\ell+j}(2t), \qquad i,j \in \mathbb{Z}. \tag{8.76}$$

The reason for the shift $\lambda_i - i + 1$ in the variable of $g$ will be clear in the proof below. Note that for any partition $\lambda = (\lambda_1, \lambda_2, \dots)$, the parts satisfy $\lambda_1 \geq \lambda_2 \geq \cdots$, and hence $\lambda_i - i + 1$ is a strictly decreasing function of $i$.

Clearly, Theorem 8.1 follows from Theorem 8.12 by setting

$$g(m) = \begin{cases} -1, & m \geq n+1, \\ 0, & m \leq n. \end{cases} \tag{8.77}$$

The advantage of this general formula is that by suitably choosing $g$, we obtain, for example, a formula for the joint distribution function of the first $k$ rows of a partition under the Poissonized Plancherel measure, and not just for the first row (see Proposition 8.13 below).

**Proof of Theorem 8.12.** Note that since $g(m) = 0$ for all large negative $m$, the product $\prod_{i=1}^{\infty}(1 + g(\lambda_i - i + 1))$ is finite for all $\lambda$.

We first assume that $g(m) \geq 0$ for all $m$. The general case when $g$ is bounded will be discussed at the end of the proof.

By the monotone convergence theorem (recall that $r(\lambda)$ is the number of rows of $\lambda$)

$$\begin{aligned} &\mathbb{E}^{P.P.}\left[\prod_{i=1}^{\infty}(1 + g(\lambda_i - i + 1))\right] \\ &\quad = \lim_{r\to\infty} \mathbb{E}^{P.P.}\left[1_{r(\lambda)\leq r}\prod_{i=1}^{\infty}(1 + g(\lambda_i - i + 1))\right]. \end{aligned} \tag{8.78}$$

From the definition of the Poissonized Plancherel measure (2.5), this equals

$$e^{-t^2} \lim_{r\to\infty} \sum_{r(\lambda)\leq r} \frac{t^{2|\lambda|}F_\lambda^2}{(|\lambda|!)^2} \prod_{i=1}^{\infty}(1 + g(\lambda_i - i + 1)). \tag{8.79}$$

Note that if $r > m_0$, then for $i > r$ and $\lambda$ satisfying $r(\lambda) \leq r$, $g(\lambda_i - i + 1) = 0$. Hence, the product equals the product for $i = 1$ to $r$:

$$\begin{aligned} &\mathbb{E}^{P.P.}\left[\prod_{i=1}^{\infty}(1 + g(\lambda_i - i + 1))\right] \\ &\quad = e^{-t^2} \lim_{r\to\infty} \sum_{r(\lambda)\leq r} \frac{t^{2|\lambda|}F_\lambda^2}{(|\lambda|!)^2} \prod_{i=1}^{r}(1 + g(\lambda_i - i + 1)). \end{aligned} \tag{8.80}$$

We now use the determinantal formula (3.62) for $F_\lambda$: for a partition $\lambda = (\lambda_1, \dots, \lambda_r)$,

$$F_\lambda = |\lambda|! \det\left(\frac{1}{(\lambda_i - i + j)!}\right)_{i,j=1}^r. \tag{8.81}$$

We insert this formula into (8.80) and change the summation indices to

$$s_i = \lambda_i + r - i. \tag{8.82}$$

Then the sum is over $s_1, \dots, s_r \in \mathbb{N}_0$ such that $s_1 > s_2 > \cdots > s_r \geq 0$. Also, note that with the new indices,

(8.83)
$$\frac{t^{|\lambda|}}{|\lambda|!} F_\lambda = t^{\sum_{i=1}^r (s_i - r + i)} \det\left(\frac{1}{(\lambda_i - i + j)!}\right)_{i,j=1}^r = \det\left(\frac{t^{s_i - r + j}}{(s_i - r + j)!}\right)_{i,j=1}^r.$$

Hence, (8.80) equals

(8.84)
$$e^{-t^2} \lim_{r\to\infty} \sum_{0 \leq s_r < \cdots < s_1} \left[\det\left(\frac{t^{s_i - r + j}}{(s_i - r + j)!}\right)_{i,j=1}^r\right]^2 \prod_{i=1}^r (1 + g(s_i - r + 1))$$
$$= e^{-t^2} \lim_{r\to\infty} \frac{1}{r!} \sum_{(s_1,\dots,s_r)\in\mathbb{N}_0^r} \left[\det\left(\frac{t^{s_i - r + j}}{(s_i - r + j)!}\right)_{i,j=1}^r\right]^2 \prod_{i=1}^r (1 + g(s_i - r + 1)),$$

where the last equality follows by noting that the summand is a symmetric function in $s_1, \dots, s_r$ and vanishes when two of the $s_i$'s are the same. Using Andréief's formula (6.21), this equals

(8.85)
$$e^{-t^2} \lim_{r\to\infty} \det\left[\sum_{s=0}^\infty \frac{t^{s-r+i}}{(s-r+i)!} \frac{t^{s-r+j}}{(s-r+j)!} (1 + g(s - r + 1))\right]_{i,j=1}^r$$
$$= e^{-t^2} \lim_{r\to\infty} \det\left[\sum_{m=-\infty}^\infty \frac{t^{2m+i+j}}{(m+i)!(m+j)!} (1 + g(m))\right]_{i,j=0}^{r-1}.$$

Here in the last equality we used the fact that $g(m) = 0$ for $m < -m_0$ for some $m_0$.

This formula can be expressed as

$$\mathbb{E}^{P.P.}\left[\prod_{i=1}^\infty (1 + g(\lambda_i - i + 1))\right] = e^{-t^2} \lim_{r\to\infty} \det(A_r + B_r M_g C_r), \tag{8.86}$$

where the operators $A_r, B_r, M_g$, and $C_r$ are defined as follows. Let $\mathbb{N}_{(r)} = \{0, 1, \dots, r-1\}$. The operators $B_r : \ell_2(\mathbb{Z}) \to \ell_2(\mathbb{N}_{(r)})$ and $C_r : \ell_2(\mathbb{N}_{(r)}) \to \ell_2(\mathbb{Z})$ are defined by the kernels

(8.87)
$$B_r(i, m) = \frac{t^{i+m}}{(i+m)!}, \qquad C_r(m, j) = \frac{t^{m+j}}{(m+j)!}, \qquad m \in \mathbb{Z}, \quad i, j \in \mathbb{N}_{(r)}.$$

The operator $M_g : \ell_2(\mathbb{Z}) \to \ell_2(\mathbb{Z})$ is multiplication by $g$: $(M_g f)(m) = g(m)f(m)$. Finally, $A_r : \ell_2(\mathbb{N}_r) \to \ell_2(\mathbb{N}_r)$ is defined by $A_r = B_r C_r$. Explicitly,

$$A_r(i,j) = \sum_{m=-\infty}^{\infty} \frac{t^{2m+i+j}}{(m+i)!(m+j)!} = \sum_{m=-\infty}^{\infty} \frac{t^{2m+i-j}}{(m+i-j)!m!}. \tag{8.88}$$

From (7.33) and (7.34) we have

$$A_r(i,j) = \varphi_{i-j}, \tag{8.89}$$

where $\varphi_k$ are the Fourier coefficients of $\varphi(z) = e^{t(z+z^{-1})}$. Thus $A_r = T_{r-1}(\varphi)$, the $r \times r$ Toeplitz matrix of symbol $\varphi$ (see Section 5.4). From the strong Szegő Theorem 5.22 we have

$$\lim_{r\to\infty} \det A_r = e^{t^2}. \tag{8.90}$$

This can also be seen by setting $g = 0$ in (8.86). Thus, from (8.86)

$$\mathbb{E}^{P.P}\left[\prod_{i=1}^{\infty}(1+g(\lambda_i - i + 1))\right] = \lim_{r\to\infty} \det(1 + A_r^{-1} B_r M_g C_r). \tag{8.91}$$

Let $S : \ell_2(\mathbb{Z}) \to \ell_2(\mathbb{N}_0)$ be the operator with kernel

$$S(i,j) = \frac{t^{i+j}}{(i+j)!}, \qquad i \in \mathbb{N}_0, \quad j \in \mathbb{Z}. \tag{8.92}$$

Note that

$$\sum_{i=0}^{\infty} \sum_{j=-m_0}^{\infty} |S(i,j)|^2 < \infty \tag{8.93}$$

for any $m_0 > 0$. Hence, $SP$ is a Hilbert-Schmidt operator where $P$ is the projection onto $\{-m_0, -m_1 + 1, \dots\}$. Therefore, $SM_gPS^*$ is a trace class operator from $\ell_2(\mathbb{N}_0)$ to $\ell_2(\mathbb{N}_0)$. Let $Q_r : \ell_2(\mathbb{N}_0) \to \ell_2(\mathbb{N}_0)$ be the projection onto $\{0, 1, \dots, r-1\}$. With these definitions (8.91) can be written as

$$\mathbb{E}^{P.P}\left[\prod_{i=1}^{\infty}(1+g(\lambda_i - i + 1))\right] = \lim_{r\to\infty} \det(1 + Q_r A_r^{-1} Q_r S M_g P S^* Q_r). \tag{8.94}$$

The operator $Q_r$ converges to the identity operator strongly as $r \to \infty$. On the other hand, since $A_r = T_{r-1}(\varphi)$, Corollary 5.30 and Proposition 5.28 yield that $Q_r A_r^{-1} Q_r$ converges to $A^{-1}$ strongly where

$$A^{-1}(j,k) = (T(\varphi_+^{-1})T(\varphi_-^{-1}))(j,k)$$

with

$$\varphi_+ = e^{tz}, \qquad \varphi_- = e^{tz^{-1}}. \tag{8.95}$$

Therefore, from (A.12),

$$\mathbb{E}^{P.P.}\left[\prod_{i=1}^{\infty}(1+g(\lambda_i-i+1))\right]=\det(1+A^{-1}SM_gPS^*). \tag{8.96}$$

Using the general identity (see (A.39))

$$\det(1-CD)=\det(1-DC) \tag{8.97}$$

when $CD$ and $DC$ are trace class operators, we find

$$\begin{aligned}\mathbb{E}\left[\prod_{i=1}^{\infty}(1+g(\lambda_i-i+1))\right]&=\det(1+PS^*A^{-1}SM_g)\\&=\det(1+PS^*T(\varphi_+^{-1})T(\varphi_-^{-1})SM_g).\end{aligned} \tag{8.98}$$

For $i\in\mathbb{Z}$ and $m\in\mathbb{N}_0$, as $(\varphi_+^{-1})_k=(-t)^k/k!$

$$\begin{aligned}(S^*T(\varphi_+^{-1}))(i,m)&=\sum_{s=0}^{\infty}\frac{t^{i+s}}{(i+s)!}\frac{(-t)^{s-m}}{(s-m)!}\\&=\sum_{s=-\infty}^{\infty}\frac{t^{2s+i+m}(-1)^s}{(s+i+m)!s!}\equiv h(i+m).\end{aligned} \tag{8.99}$$

But

$$\begin{aligned}\sum_{k=-\infty}^{\infty}h(k)z^k&=\sum_{k=-\infty}^{\infty}\sum_{s=-\infty}^{\infty}\frac{t^{2s+k}(-1)^s}{(s+k)!s!}z^k\\&=\sum_{s=-\infty}^{\infty}\sum_{m=-\infty}^{\infty}\frac{(-1)^st^{s+m}z^{m-s}}{m!s!}=e^{t(z-z^{-1})}.\end{aligned} \tag{8.100}$$

Hence (see (8.1))

$$h(k)=\int z^{-k}e^{t(z-z^{-1})}\frac{dz}{2\pi iz}=J_k(2t). \tag{8.101}$$

Similarly,

$$(T(\varphi_-^{-1})S)(m,j)=J_{m+j}(2t). \tag{8.102}$$

Therefore, we find that $S^*T(\varphi_+^{-1})T(\varphi_-^{-1})S(i,j)=\mathcal{K}(i,j)$. Hence, we have proved Theorem 8.12 for bounded nonnegative function $g$ supported on $\{-m_0,-m_0+1,\dots\}$.

Now consider a bounded function $g$ with support $\{-m_0,-m_0+1,\dots\}$ which is not necessarily nonnegative. From the inequality

$$\left|\prod_{i=1}^{\infty}(1+g(\lambda_i-i+1))\right|\le\prod_{i=1}^{\infty}(1+|g(\lambda_i-i+1)|),$$

the theorem with $g$ replaced by $|g|$ shows that the function

$$\lambda \mapsto \prod_{i=1}^{\infty}(1+g(\lambda_i-i+1))$$

is integrable. Hence by the dominated convergence theorem, formula (8.78) still holds and the calculation (8.80) goes through. □

The distribution function for the $k$th part $\lambda_k$ and also the joint distribution function of the first $k$ rows of a partition under the Poissonized Plancherel measure can be obtained from Theorem 8.12. For example, the next result follows by combining Theorem 8.12 and the proof of Proposition 6.7 (we leave this as an exercise).

**Proposition 8.13.** *For $k \in \mathbb{N}$, let*

$$\Lambda_k = \{n=(n_1,\dots,n_k)\in \mathbb{N}_0^k : n_1+\cdots+n_r \le r-1, r=1,2,\dots,k\}. \tag{8.103}$$

*Then for $a_k \le \cdots \le a_1 \le a_0 \equiv \infty$,*

$$\begin{aligned} &\mathbb{P}^{P.P.}\{\lambda_1 \le a_1, \lambda_2-1\le a_2,\dots,\lambda_k-k+1\le a_k\} \\ &= \sum_{n\in\Lambda_k} \frac{1}{n_1!\cdots n_k!}\frac{\partial^{|n|}}{\partial^{n_1}s_1\cdots\partial^{n_k}s_k}\Big|_{s_1=\cdots=s_k=1} \det\Big(1+\big(\sum_{j=1}^{k} s_j 1_{(a_j,a_{j-1}]}\big)\mathcal{K}\Big). \end{aligned} \tag{8.104}$$

Theorem 8.12 also implies that the point process, $\lambda_i-i+1$, $i=1,2,\dots$, forms a determinantal point process (see (11.49) below).

Chapter 9

# Asymptotic Results

It was shown in Chapter 4 (see Propositions 4.8 and 4.9) that

$$\lim_{t\to\infty} \frac{\mathbb{E}[L(t)]}{t} = c, \qquad \lim_{N\to\infty} \frac{\mathbb{E}[\ell_N]}{\sqrt{N}} = c \tag{9.1}$$

for some positive constant $c$. In the same chapter, several estimates of $c$ were given, but a precise value of $c$ was not obtained. In this chapter, we prove that

$$c = 2, \tag{9.2}$$

that all moments of $\frac{L(t)}{t}$ converge as $t \to \infty$, and that $\lim_{t\to\infty} \frac{L(t)}{t} = c$ holds almost surely. In doing so, we also obtain uniform tail estimates for $L(t)$. We then prove the main theorem on $\ell_n$, Theorem 1.2: for $x \in \mathbb{R}$,

$$\lim_{t\to\infty} \mathbb{P}\left\{ \frac{L(t) - 2t}{t^{1/3}} \le x \right\} = F(x), \quad \lim_{t\to\infty} \mathbb{P}\left\{ \frac{\ell_N - 2\sqrt{N}}{N^{1/6}} \le x \right\} = F(x), \tag{9.3}$$

where $F(x)$ is the Tracy-Widom distribution defined in Section 6.5.

Recall that due to the de-Poissonization lemmas in Section 2.2, limit theorems for $L(t)$ with suitable estimates imply corresponding limit theorems for $\ell_N$. Hence we first analyze $L(t)$ asymptotically. For this purpose, we can use either Gessel's formula or the Fredholm determinant discussed in the previous two chapters. As mentioned earlier, in the original work [**BDJ99**] the authors used Gessel's formula, which they analyzed asymptotically by applying the Deift-Zhou nonlinear steepest-descent method to an associated RHP. We do not follow this approach: the interested reader is referred to [**Dei99b**] for more information on the nonlinear steepest-descent method for RHP's. Instead we use the Fredholm determinant formula (8.4). The kernel of the determinant involves integrals with large parameters, whose asymptotics can be obtained by means of the classical steepest-descent method. The limit of the Fredholm determinant then follows from a general convergence theorem for trace class operators. In [**BOO00, Joh01a**], the authors proved these limit theorems by using this method. However, we note that the steepest-descent analysis needed here is rather involved. For the convenience of reader, some basics of the classical steepest-descent method are included in Appendix B.

The Fredholm determinant formula in (8.4) tells us that

$$\mathbb{P}\{L(t) \leq n\} = \det(1 - \mathcal{K}_n), \qquad n = 0, 1, 2, \dots. \tag{9.4}$$

Here the operator $\mathcal{K}_n$ acts on $\ell^2(\{n+1, n+2, \dots\})$ with kernel

$$\mathcal{K}_n(i,j) = \sum_{m=0}^{\infty} J_{i+m}(2t) J_{m+j}(2t), \qquad i, j \geq n+1, \tag{9.5}$$

where $J_k(2t)$ is the Bessel function (8.1)

$$J_k(2t) = \int z^{-k} e^{t(z-z^{-1})} \frac{dz}{2\pi i z} = \int z^k e^{-t(z-z^{-1})} \frac{dz}{2\pi i z}. \tag{9.6}$$

Here, the integral is over a simple closed curve with $z = 0$ in its interior, oriented counter-clockwise.

A few graphs of Bessel functions $J_\nu(x)$ are plotted in Figure 9.1 using Maple. Note that the graph consists of an oscillatory part and a part that is almost equal to zero. Also, the graph appears to move to the right as $\nu$ increases. Hence, if we focus on a neighborhood of a fixed point $x$, say $x = 60$, $J_\nu(x)$ is oscillatory for $\nu = 20$ and 40 and is exponentially small for $\nu = 80$.

The transition from oscillation to exponential decay for $x = 60$ occurs when $\nu$ is about 60. The change of the asymptotic behavior of $J_\nu(x)$ for different regimes of $\nu$ and $x$ is responsible for the proper scaling regimes in the limit theorems of $\mathbb{P}\{L(t) \leq n\}$. We make this precise in the next sections.

## 9.1. Exponential Upper Tail Estimate

The goal of this section is to obtain an exponential upper bound for $\mathbb{P}\{L(t) > \mu t\}$ in the case $\mu > 2$. An exponential upper bound for $\mathbb{P}\{L(t) < \mu t\}$ in the case $\mu < 2$ is obtained in the next section. Combining these bounds, we establish the law of large numbers for $L(t)$ and, as a consequence, for $\ell_N$ in Section 9.3.

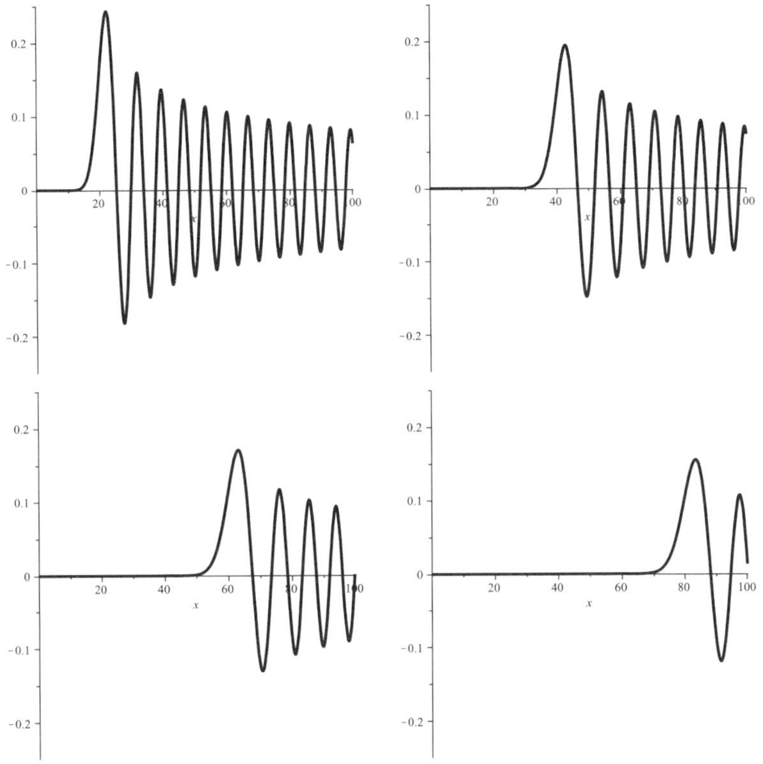

**Figure 9.1.** Graph of the Bessel function $y = J_\nu(x)$ for $\nu = 20, 40, 60, 80$. The function "moves to the right" as $\nu$ increases.

Recall that for trace class operators $A$ and $B$ on a Hilbert space $\mathcal{H}$ (see Appendix A),

$$\big|\det(1+A) - \det(1+B)\big| \le ||A-B||_1 e^{||A||_1+||B||_1+1}. \tag{9.7}$$

Setting $A=0$ and $B=\mathcal{K}_{[\mu t]}$ in (9.7), we find from (9.4) that

$$\begin{aligned}\mathbb{P}\{L(t)>\mu t\} &= 1-\mathbb{P}\{L(t)\le \mu t\} = 1-\det(1-\mathcal{K}_{[\mu t]})\\ &\le ||\,\mathcal{K}_{[\mu t]}\,||_1 e^{||\,\mathcal{K}_{[\mu t]}\,||_1+1}.\end{aligned} \tag{9.8}$$

Since $\mathcal{K}_n$ is a positive trace class operator (See Lemma 8.2), we have

$$\begin{aligned}||\,\mathcal{K}_{[\mu t]}\,||_1 = \operatorname{Tr}\mathcal{K}_{[\mu t]} &= \sum_{i=[\mu t]+1}^{\infty}\sum_{m=0}^{\infty} J_{i+m}(2t)J_{m+i}(2t)\\ &= \sum_{k=[\mu t]+1}^{\infty}(k-[\mu t])J_k(2t)^2.\end{aligned} \tag{9.9}$$

Hence, if we have upper bounds for $J_k(2t)$ as $t\to\infty$ for all $k\ge[\mu t]+1$, then we can find an upper bound for $\mathbb{P}\{L(t)>\mu t\}$.

We have the following estimate. A simple computation shows that $\sqrt{1-s^2} - \log\left(\frac{1+\sqrt{1-s^2}}{s}\right) < 0$ for $0 < s < 1$.

**Lemma 9.1.** *For each* $\nu = 1, 2, \ldots$ *and* $0 < s < 1$, *we have*

$$|J_\nu(\nu s)| \le \frac{\sqrt{\pi} e^{\nu\left\{\sqrt{1-s^2} - \log\left(\frac{1+\sqrt{1-s^2}}{s}\right)\right\}}}{2\sqrt{2\nu}(1-s^2)^{1/4}}. \tag{9.10}$$

**Proof.** Set

$$z_0 \equiv \frac{1+\sqrt{1-s^2}}{s}. \tag{9.11}$$

Let $C_0$ be the circle of radius $z_0$, centered at the origin. From (9.6), the Bessel function can be written as

$$J_\nu(\nu s) = \frac{1}{2\pi i}\int_{C_0} e^{\nu f(z)}\frac{dz}{z}, \qquad f(z) \equiv \frac{s}{2}(z - z^{-1}) - \log z, \tag{9.12}$$

where the log is defined as the principal branch of the logarithm, $\log z = \log|z| + \arg(z)$, $\arg(z) \in (-\pi, \pi]$. For $z = z_0 e^{i\theta} \in C_0$,

$$\operatorname{Re}\left(f(z_0 e^{i\theta}) - f(z_0)\right) = \frac{s}{2}(z_0 - z_0^{-1})(\cos\theta - 1). \tag{9.13}$$

Using the inequality $\cos\theta - 1 \le -\frac{2}{\pi^2}\theta^2$ for $\theta \in [-\pi, \pi]$, and using $s(z_0 - z_0^{-1}) = 2\sqrt{1-s^2}$,

$$\operatorname{Re}(f(z_0 e^{i\theta}) - f(z_0)) \le -\frac{2\sqrt{1-s^2}}{\pi^2}\theta^2 \tag{9.14}$$

for $\theta \in [-\pi, \pi]$. Hence we have

$$\begin{aligned} |J_\nu(\nu s)| &= \left|\frac{e^{\nu f(z_0)}}{2\pi i}\int_{C_0} e^{\nu(f(z)-f(z_0))}\frac{dz}{z}\right| \\ &\le \frac{e^{\nu f(z_0)}}{2\pi}\int_{-\pi}^{\pi} e^{-\frac{2\nu\sqrt{1-s^2}}{\pi^2}\theta^2}\,d\theta \\ &\le \frac{e^{\nu f(z_0)}}{2\pi}\int_{-\infty}^{\infty} e^{-\frac{2\nu\sqrt{1-s^2}}{\pi^2}\theta^2}\,d\theta = \frac{\sqrt{\pi}e^{\nu f(z_0)}}{2\sqrt{2\nu}(1-s^2)^{1/4}}. \end{aligned} \tag{9.15}$$

As $f(z_0) = \sqrt{1-s^2} - \log\left(\frac{1+\sqrt{1-s^2}}{s}\right)$, we obtain (9.10). □

In the above estimate, $z_0$ is a critical point of $f(z)$. Indeed, the critical points of $f$, $f'(z) = 0$, are $z = z_0$ and $z = 1/z_0$. When $s < 1$, the two critical points are distinct, real and positive. Among these two critical points, we use $z_0$. Note that $z_0 > 1$. The curves of constant phase, $\operatorname{Im}(f(z)) = \operatorname{Im}(f(z_0)) = 0$,

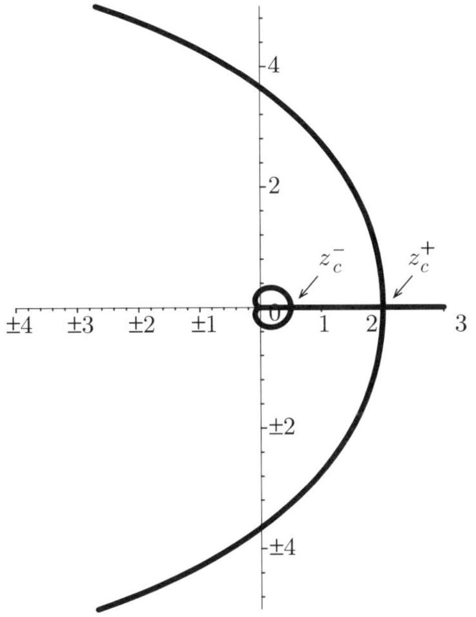


**Figure 9.2.** The curves of constant phase, $\operatorname{Im} f(z) = \operatorname{Im} f(z_0) = \operatorname{Im} f(1/z_0)$, passing the critical points $z_0^{\pm}$ when $s = \frac{4}{5}$

passing through the critical points $z_0$ and $z_0^{-1}$ are given by the parametric equations in polar coordinates

$$(9.16)\quad r = \frac{\theta}{s\sin(\theta)} \pm \frac{\theta}{s\sin(\theta)}\sqrt{1 - \left(\frac{s\sin(\theta)}{\theta}\right)^2}, \qquad z = re^{i\theta},\ \theta \in (-\pi, \pi),$$

together with the straight line $\theta = 0, r > 0$, and are drawn in Figure 9.2 for the case when $s = \frac{4}{5}$. A direct computation shows that $f''(z_0) = \frac{s^2\sqrt{1-s^2}}{(1+\sqrt{1-s^2})^2} > 0$ and $f''(1/z_0) < 0$. From this it is easy to check that the unbounded curve crossing $z_0$ (vertically) corresponds to the curve of steepest descent passing through $z_0$. We call this curve $\Sigma$. On the other hand the curve of steepest descent passing through $1/z_0$ is a part of the positive real line to which the contour for $J_\nu(\nu s)$ in (9.6) cannot be deformed. Accordingly, $\Sigma$ is the appropriate steepest-descent contour for the asymptotic analysis of $J_\nu(\nu s)$ as $\nu \to \infty$. In the above lemma, we used an explicit contour $C_0$ that passes $z_0$ vertically to obtain an explicit upper bound. If we apply the steepest-descent analysis, we obtain (see Appendix B) as $\nu \to \infty$

$$(9.17)\qquad J_\nu(\nu s) \sim \frac{e^{\nu\left\{\sqrt{1-s^2} - \log\left(\frac{1+\sqrt{1-s^2}}{s}\right)\right\}}}{\sqrt{2\pi\nu}(1-s^2)^{1/4}}$$

for each fixed $s \in (0, 1)$. Hence (9.10) matches the leading term up to a multiplicative constant factor.

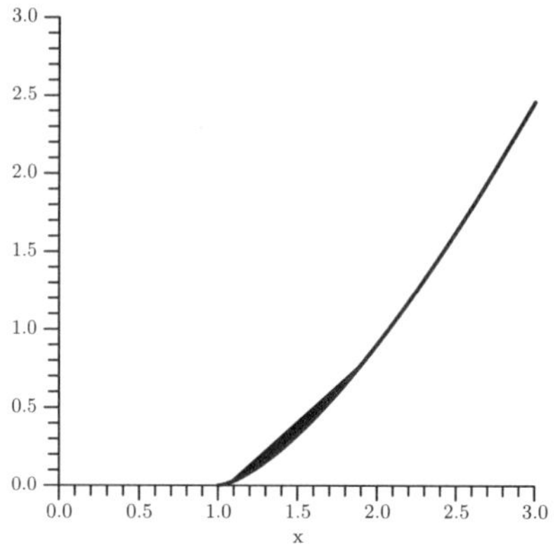


**Figure 9.3.** Graph of $h(x)$ for $x \geq 1$.

Lemma 9.1 implies the following estimate for the upper tail of $L(t)$.

**Proposition 9.2.** *Let $\mu > 2$. For $t \geq \frac{1}{\sqrt{\mu^2-4}}$,*

$$\mathbb{P}\{L(t) > \mu t\} \leq g(\frac{\mu}{2}, t)e^{g(\frac{\mu}{2},t)+1}, \tag{9.18}$$

*where*

$$g(x,t) = \frac{\pi}{16}(\frac{1}{h'(x)} + 2)e^{-4th(x)}, \qquad x > 1, \tag{9.19}$$

*and*

$$\begin{aligned} h(x) &= x\log(x + \sqrt{x^2-1}) - \sqrt{x^2-1}, \\ h'(x) &= \log(x + \sqrt{x^2-1}). \end{aligned} \tag{9.20}$$

Observe that $h(x)$ is increasing and nonnegative for $x \geq 1$. Also, as $x \downarrow 1$, $h(x) = \frac{2\sqrt{2}}{3}(x-1)^{3/2} + O((x-1)^2))$. See Figure 9.3 for the graph of $h(x)$.

**Proof.** For $k \geq [\mu t] + 1$, $s \equiv \frac{2t}{k}$ satisfies $s \leq \frac{2t}{[\mu t]+1} \leq \frac{2t}{\mu t} < 1$. Hence, applying (9.10) with $\nu = k$ and $s = \frac{2t}{k}$, we find from (9.9) that

$$|| \mathcal{K}_{[\mu t]} ||_1 \leq \frac{\pi}{8} \sum_{k=[\mu t]+1}^{\infty} \frac{k - [\mu t]}{\sqrt{k^2 - 4t^2}} e^{2\sqrt{k^2-4t^2} - 2k\log\left(\frac{k+\sqrt{k^2-4t^2}}{2t}\right)}. \tag{9.21}$$

Using $[\mu t] \geq \mu t - 1$, one checks directly that $k - [\mu t] \leq \sqrt{k^2 - 4t^2}$ for $k \geq [\mu t] + 1$ if $t \geq \frac{1}{\sqrt{\mu^2-4}}$. Hence for $t \geq \frac{1}{\sqrt{\mu^2-4}}$,

$$|| \mathcal{K}_{[\mu t]} ||_1 \leq \frac{\pi}{8} \sum_{k=[\mu t]+1}^{\infty} e^{-4th(\frac{k}{2t})}, \tag{9.22}$$

where $h(x)$ is given in (9.20). We first estimate the part $k \geq [\mu t] + 2$ of the sum and then estimate the term $k = [\mu t] + 1$ separately.

For $t \geq \frac{1}{\sqrt{\mu^2-4}}$, since $h(x)$ is increasing in $x \geq 1$, we have

$$
\begin{aligned}
\frac{\pi}{8} \sum_{k=[\mu t]+2}^{\infty} e^{-4th(\frac{k}{2t})} &\leq \frac{\pi}{8} \int_{[\mu t]+1}^{\infty} e^{-4th(\frac{k}{2t})} dk \\
&\leq \frac{\pi}{8} \int_{\mu t}^{\infty} e^{-4th(\frac{k}{2t})} dk = \frac{t\pi}{4} \int_{\frac{\mu}{2}}^{\infty} e^{-4th(x)} dx.
\end{aligned} \tag{9.23}
$$

Integrating by parts,

$$
\begin{aligned}
\int_{\frac{\mu}{2}}^{\infty} e^{-4th(x)} dx &= -\frac{1}{4t} \int_{\frac{\mu}{2}}^{\infty} \frac{1}{h'(x)} \left( \frac{d}{dx} e^{-4th(x)} \right) dx \\
&= \frac{1}{4th'(\frac{\mu}{2})} e^{-4th(\frac{\mu}{2})} - \frac{1}{4t} \int_{\frac{\mu}{2}}^{\infty} e^{-4th(x)} \frac{h''(x)}{h'(x)^2} dx \\
&\leq \frac{1}{4th'(\frac{\mu}{2})} e^{-4th(\frac{\mu}{2})}
\end{aligned} \tag{9.24}
$$

since $h''(x) = \frac{1}{\sqrt{x^2-1}} > 0$. Therefore, for $t \geq \frac{1}{\sqrt{\mu^2-4}}$,

$$
\frac{\pi}{8} \sum_{k=[\mu t]+2}^{\infty} e^{-4th(\frac{k}{2t})} \leq \frac{\pi}{16h'(\frac{\mu}{2})} e^{-4th(\frac{\mu}{2})}. \tag{9.25}
$$

On the other hand, the term in the sum (9.22) corresponding to $k = [\mu t] + 1$ is less than or equal to the same term with $k = \mu t$ since $h(x)$ is an increasing function in $x$. Therefore, together with (9.25), we have

$$
|| \mathcal{K}_{[\mu t]} ||_1 \leq \frac{\pi}{16h'(\frac{\mu}{2})} e^{-4th(\frac{\mu}{2})} + \frac{\pi}{8} e^{-4th(\frac{\mu}{2})}. \tag{9.26}
$$

Thus from (9.8), we obtain (9.18). □

Proposition 9.2 implies that

$$
\limsup_{t\to\infty} \frac{1}{t} \log \mathbb{P}\{L(t) > \mu t\} \leq -2\mu \cosh^{-1}(\frac{\mu}{2}) + 2\sqrt{\mu^2 - 4} \tag{9.27}
$$

for $\mu > 2$. This estimate matches the rate function of the large deviation principle for $L(t)$:

**Proposition 9.3** (Large deviation principle for $L(t)$: upper tail). *For* $\mu > 2$,

$$
\lim_{t\to\infty} \frac{1}{t} \log \mathbb{P}\{L(t) > \mu t\} = -2\mu \cosh^{-1}(\frac{\mu}{2}) + 2\sqrt{\mu^2 - 4}. \tag{9.28}
$$

We also have:

**Proposition 9.4** (Large deviation principle for $\ell_N$: upper tail). *For* $\mu > 2$,

$$(9.29)\qquad \lim_{N\to\infty} \frac{1}{\sqrt{N}} \log \mathbb{P}\{\ell_N > \mu\sqrt{N}\} = -2\mu\cosh^{-1}(\frac{\mu}{2}) + 2\sqrt{\mu^2-4}.$$

The bound

$$(9.30)\qquad \limsup_{N\to\infty} \frac{1}{\sqrt{N}} \log \mathbb{P}\left\{\ell_N \ge \mu\sqrt{N}\right\} \le -2\mu\cosh^{-1}(\frac{\mu}{2}) + 2\sqrt{\mu^2-4}$$

was obtained in [**Kim96**]. The bound

$$(9.31)\qquad \liminf_{N\to\infty} \frac{1}{\sqrt{N}} \log \mathbb{P}\left\{\ell_N \ge \mu\sqrt{N}\right\} \ge -2\mu\cosh^{-1}(\frac{\mu}{2}) + 2\sqrt{\mu^2-4}$$

was obtained in [**DZ99**]. Independently, at the same time, (9.28) was proven by [**Sep98**]. We do not, however, discuss the proofs of these large deviation principles in this text; see [**Sep98**] or [**DZ99**] for details. We note that the upper large deviation principle for $\ell_N$ can be obtained from equation (9.27) by a de-Poissonization argument as in Section 2.2.

## 9.2. Exponential Lower Tail Estimate

The goal of this section is to obtain an exponential lower bound for $\mathbb{P}\{L(t) < \mu t\}$ in the case $0 < \mu < 2$. The corresponding result to Proposition 9.2 is Proposition 9.9 below.

Our calculation follows the method in [**Wid02**]. The following observation is useful (see (A.37) in Appendix A). If all the eigenvalues $\{\lambda_i\}$ of a trace class operator $A$ are real and satisfy $\lambda_i \le 1$, then

$$(9.32)\qquad \det(1-A) \le e^{-\operatorname{Tr} A}.$$

By Lemma 8.2, this estimate applies to $\mathcal{K}_n$, and we find

$$(9.33)\qquad \mathbb{P}\{L(t) \le \mu t\} = \det(1-\mathcal{K}_{[\mu t]}) \le e^{-\operatorname{Tr}\mathcal{K}_{[\mu t]}}.$$

In order to obtain a lower bound of $\operatorname{Tr}\mathcal{K}_{[\mu t]}$, we could use (9.9) together with the asymptotics of the Bessel function as in the previous section on the upper tail. Now we would need asymptotics for $J_k(2t)$ when $\frac{2t}{k} > 1$. In this case, for $0 < \beta < \frac{1}{2}\pi$, the steepest-descent method yields (see [**AS64**] 9.3.3) (cf. Figure 9.1)

$$(9.34)\qquad J_\nu(\nu\sec\beta) = \sqrt{\frac{2}{\pi\nu\tan\beta}}\left\{\cos(\nu\tan\beta - \nu\beta - \frac{1}{4}\pi) + O(\nu^{-1})\right\},$$

where the error term is uniform for $\beta$ in compact subsets of $(0, \frac{1}{2}\pi)$. However, note that due to the oscillatory behavior of the cosine term, the term $O(\nu^{-1})$ is not always smaller than the leading term $\cos(\nu\tan\beta - \nu\beta - \frac{1}{4}\pi)$. Therefore, in computing the sum (9.9), we would need to take special care to ensure that the sum of the error terms is smaller than the sum of the leading terms. Even though this could certainly be done, we take a different approach to estimate $\operatorname{Tr}\mathcal{K}_{[\mu t]}$ which avoids this technicality. Nevertheless, the lower tail estimate remains significantly more complicated than the upper tail estimate.

We use the following double integral formula of $\mathcal{K}_n(j,k)$ in (8.8):

$$\begin{aligned}\operatorname{Tr}\mathcal{K}_n &= \sum_{j=n+1}^{\infty} \mathcal{K}_n(j,j) \\ &= \frac{-1}{4\pi^2}\int_{C_1} dz \int_{C_2} \frac{z^{-n}w^n}{(z-w)^2} e^{t(z-z^{-1})-t(w-w^{-1})} dw,\end{aligned} \tag{9.35}$$

where $C_j$, $j=1,2$, is a circle of radius $\rho_j$ centered at the origin such that $\rho_1 > \rho_2 > 0$. We apply the steepest-descent method directly to this double integral.

**Lemma 9.5.** *Uniformly for $\mu$ in a compact subset of $[0,2)$, we have*

$$\operatorname{Tr}\mathcal{K}_{[\mu t]} = \frac{2t}{\pi}\left\{\sqrt{1-\left(\frac{[\mu t]}{2t}\right)^2} - \frac{[\mu t]}{2t}\arccos\left(\frac{[\mu t]}{2t}\right)\right\} + O(1) \tag{9.36}$$

*as $t \to \infty$.*

When we apply the steepest-descent method to the double integral (9.35), it turns out that the steepest-descent curves for the $z$-variable and the $w$-variable intersect at two points. Thus, on the steepest-descent curves, the integrand of (9.35) does not converge absolutely due to the term $\frac{1}{(z-w)^2}$. However, we can still proceed by interpreting the integral as the principal value. In doing so, we end up with the following principal value integral that captures the main contribution from the intersection point:

$$\lim_{\epsilon\downarrow 0} \frac{-1}{4\pi^2}\int_{[-s_1,s_1]\setminus[-\epsilon,\epsilon]} \left(\int_{[-s_2,s_2]} \frac{e^{-\frac{1}{2}(\zeta_1^2+\zeta_2^2)}}{(\zeta_2+i\zeta_1)^2} d\zeta_2\right) d\zeta_1. \tag{9.37}$$

We have the following uniform estimate:

**Lemma 9.6.** *Fix $s_* > 0$. The integral (9.37) is uniformly bounded for $s_1, s_2 \in [s_*, \infty)$.*

**Proof.** Set $\Gamma_1^\epsilon := [-s_1,s_1]\setminus[-\epsilon,\epsilon]$ and $\Gamma_2 := [-s_2,s_2]$. We can assume that $\epsilon < s_1$. Write (9.37) as

$$\begin{aligned}&\frac{-1}{4\pi^2}\int_{\Gamma_1^{\epsilon,0}}\int_{\Gamma_2^0} \frac{e^{-\frac{1}{2}(\zeta_1^2+\zeta_2^2)}-1}{(\zeta_2+i\zeta_1)^2} d\zeta_2 d\zeta_1 + \frac{-1}{4\pi^2}\int_{\Gamma_1^{\epsilon,0}}\int_{\Gamma_2^0} \frac{1}{(\zeta_2+i\zeta_1)^2} d\zeta_2 d\zeta_1 \\ &+ \frac{-1}{4\pi^2}\int\int_{(\Gamma_1^\epsilon\times\Gamma_2)\setminus(\Gamma_1^{\epsilon,0}\times\Gamma_2^0)} \frac{e^{-\frac{1}{2}(\zeta_1^2+\zeta_2^2)}}{(\zeta_2+i\zeta_1)^2} d\zeta_2 d\zeta_1,\end{aligned} \tag{9.38}$$

where $\Gamma_1^{\epsilon,0} = \Gamma_1^\epsilon \cap \{|\zeta_1| \le 1\}$ and $\Gamma_2^0 = \Gamma_2 \cap \{|\zeta_2| \le 1\}$. The integrand of the first integral is bounded as $\zeta_1, \zeta_2 \to 0$. Hence, the first integral is uniformly bounded as $\epsilon \downarrow 0$ for $s_1, s_2 \ge s_*$. The integrand of the third integral is also bounded, and hence due to the Gaussian decay, the third integral is also bounded uniformly as $\epsilon \downarrow 0$ for $s_1, s_2 \ge s_*$. On the other hand, the second integral can be evaluated explicitly and the evaluation of the integral shows that for $s_1, s_2 \ge s_*$,

$$\lim_{\epsilon \downarrow 0} \frac{-1}{4\pi^2} \int_{\Gamma_1^{\epsilon,0}} \int_{\Gamma_2^0} \frac{1}{(\zeta_2 + i\zeta_1)^2} d\zeta_2 d\zeta_1 \tag{9.39}$$

exists uniformly and is uniformly bounded. Hence we obtain the lemma. □

**Exercise 9.7.** The boundedness of (9.39) relies on the fact that we have $(\xi_1 + i\xi_2)^2$ in the denominator, not $|\xi_1 + i\xi_2|^2$. Show that (9.39) does not converge absolutely. Also show that (9.39) equals $\frac{1}{4\pi}$ when $s_1 = s_2$ for arbitrary $s_1 > 0$.

**Exercise 9.8.** We can also evaluate the integral (9.37) explicitly in the case when the contours are the entire real lines. Show that

$$\lim_{\epsilon \downarrow 0} \int_{\mathbb{R}\setminus(-\epsilon,\epsilon)} \int_{-\infty}^{\infty} \frac{e^{-\frac{1}{2}(\zeta_1^2+\zeta_2^2)}}{(\zeta_2 + i\zeta_1)^2} d\zeta_2 d\zeta_1 = -\pi, \tag{9.40}$$

using

$$\frac{1}{2\pi i} \int_{-\infty}^{\infty} \frac{e^{-\frac{1}{2}(x^2+y^2)}}{y + ix} dy = \begin{cases} \frac{1}{\sqrt{2\pi}} \int_{-\infty}^{x} e^{-\frac{1}{2}s^2} ds, & x < 0, \\ \frac{1}{\sqrt{2\pi}} \int_{\infty}^{x} e^{-\frac{1}{2}s^2} ds, & x > 0. \end{cases} \tag{9.41}$$

We now turn to the proof of Lemma 9.5.

**Proof of Lemma 9.5.** We fix $\mu_* \in (0, 2)$ and prove Lemma 9.5 uniformly for $\mu \in [0, \mu_*]$. We have

$$\operatorname{Tr} \mathcal{K}_{[\mu t]} = \frac{-1}{4\pi^2} \int_{C_1} \int_{C_2} \frac{e^{t(f(z)-f(w))}}{(z-w)^2} dw dz, \tag{9.42}$$

where

$$f(z) \equiv z - z^{-1} - 2a \log z, \qquad a \equiv \frac{[\mu t]}{2t}. \tag{9.43}$$

Here log again denotes the principal branch of the logarithm. Note that since $a \le \frac{\mu}{2}$, $a$ lies in a compact subset $[0, a*] \subset [0, 1)$ for some $a_* \in (0, 1)$.

The critical points of $f(z)$ are $z = e^{\pm i\theta_c}$, where

$$e^{\pm i\theta_c} \equiv a \pm i\sqrt{1 - a^2}, \qquad \theta_c \in (0, \frac{\pi}{2}]. \tag{9.44}$$

Note that since $a \in [0, a_*]$, $\theta_c$ lies in $[(\theta_c)_*, \frac{\pi}{2}]$ for some $(\theta_c)_* \in (0, \frac{\pi}{2})$. At the critical values, we have

$$(9.45) \quad f(e^{\pm i\theta_c}) = \pm 2i(\sin\theta_c - \theta_c\cos\theta_c), \qquad f''(e^{\pm i\theta_c}) = 2\sin\theta_c e^{\mp i(2\theta_c - \frac{\pi}{2})}.$$

Observe that $|f''(e^{\pm i\theta_c})|$ is uniformly bounded away from 0.

Consider the curves of constant phase through the critical points, given by the equations $\operatorname{Im} f(z) = \operatorname{Im} f(e^{i\theta_c})$ and $\operatorname{Im} f(z) = \operatorname{Im} f(e^{-i\theta_c})$. Note that due to the symmetry $\operatorname{Im} f(z) = -\operatorname{Im} f(\bar{z})$, the curve $\operatorname{Im} f(z) = \operatorname{Im} f(e^{-i\theta_c})$ is the reflection of the curve $\operatorname{Im} f(z) = \operatorname{Im} f(e^{i\theta_c})$ about the real axis, and it is easy to see that the curve $\operatorname{Im} f(z) = \operatorname{Im} f(e^{i\theta_c})$ lies in $\mathbb{C}_+$. We focus on the curve of constant phase through $e^{i\theta_c}$. The equation $\operatorname{Im} f(z) = \operatorname{Im} f(e^{i\theta_c})$ has a simple form in terms of polar coordinate $z = re^{i\theta}$:

$$(9.46) \qquad (r + r^{-1})\sin\theta - 2\theta\cos\theta_c = 2\sin\theta_c - 2\theta_c\cos\theta_c.$$

We can solve this equation explicitly and find the parametrization of the curve as $r = r(\theta)$. There are two solutions, one for the curve of steepest descent and the other for the curve of steepest ascent. Using (9.45), we find that the curve of steepest descent through $e^{i\theta_c}$ is given by

(9.47)
$$r(\theta) = \begin{cases} \frac{\sin\theta_c + (\theta-\theta_c)\cos\theta_c}{\sin\theta} - \sqrt{\left(\frac{\sin\theta_c+(\theta-\theta_c)\cos\theta_c}{\sin\theta}\right)^2 - 1}, & 0 < \theta \le \theta_c, \\ \frac{\sin\theta_c + (\theta-\theta_c)\cos\theta_c}{\sin\theta} + \sqrt{\left(\frac{\sin\theta_c+(\theta-\theta_c)\cos\theta_c}{\sin\theta}\right)^2 - 1}, & \theta_c \le \theta < \pi, \end{cases}$$

which is smooth for $0 < \theta < \pi$. Note that $r(\theta) \to 0$ as $\theta \downarrow 0$ and $r(\theta) \to \infty$ as $\theta \uparrow \pi$. The curve of steepest ascent of $f(z)$ (which is the curve of steepest descent of $-f(z)$) through $e^{i\theta_c}$ is given by $\frac{1}{r(\theta)}e^{i\theta_c}$. See Figure 9.4 for the graph when $a = 1/\sqrt{2}$. The situation is similar for any $a$ as in (9.43).

Let $\Sigma_1' = \Sigma_1'(a)$ and $\Sigma_2' = \Sigma_2'(a)$ be the curves analogous to those in Figure 9.4 for general $a \in [0, a*]$. We would like to deform the contours $C_1$ and $C_2$ in (9.42) to $\Sigma_1'$ and $\Sigma_2'$, respectively. However, it is more convenient if we modify $\Sigma_1'$ and $\Sigma_2'$ near the origin and infinity to $\Sigma_1$ and $\Sigma_2$ as in Figure 9.5. Then $\Sigma_j, j = 1, 2$ do not intersect the origin, the essential singularity of $f$, and are of finite length. As noted in Appendix B, such modifications of the curves of steepest descent away from the critical points do not impact the computation of the leading order behavior of the integral. The exact modification is done as follows. We have $\operatorname{Re} f(z) \le 1 - \frac{\operatorname{Re}(z)}{|z|^2} - 2\log|z| \to -\infty$ uniformly in $a \in [0, a_*]$ as $z \to 0$ in a symmetric sector of angle less than $\pi$ in the right half-plane. Also $\operatorname{Re} f(z) \le (1 - |z|^{-2})\operatorname{Re}(z) \to -\infty$ uniformly in $a \in [0, a_*]$ as $z \to \infty$ in a symmetric sector of angle less than $\pi$ in the left half-plane. Since $\theta_c \in [(\theta_c)_*, \pi/2]$ for all $a \in [0, a_*]$, this implies that there exist $\theta_1 \in (0, (\theta_c)_*)$ and $\theta_2 \in (\pi/2, \pi)$ such that $\operatorname{Re} f(z) \le -1$ on the vertical line segment joining the two points

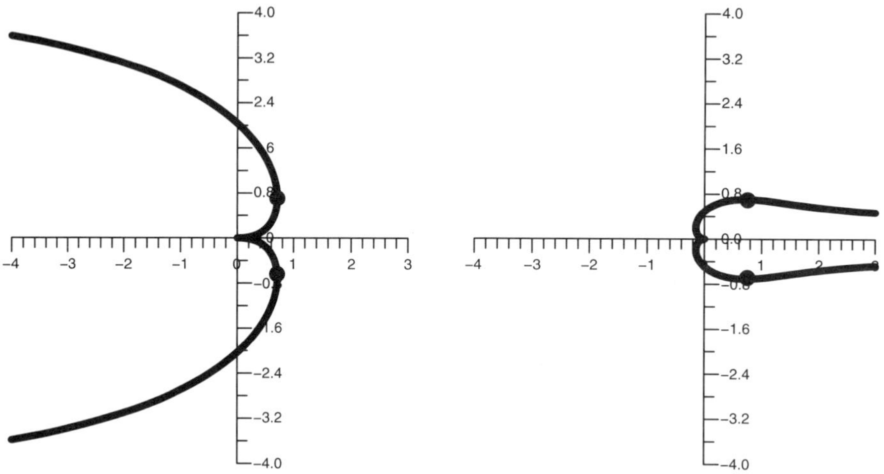


**Figure 9.4.** Let $a = 1/\sqrt{2}$. Left: curve $\Sigma_1'$ of steepest descent for $f(z)$ passing $e^{\pm i\theta_c}$. Right: curve $\Sigma_2'$ of steepest descent for $-f(z)$ passing $e^{\pm i\theta_c}$.

$\Sigma_1' \cap \{z : \arg(z) = \pm\theta_1\}$ and also on the vertical line segment joining the two points $\Sigma_1' \cap \{z : \arg(z) = \pm\theta_2\}$. We define $\Sigma_1$ as the union of the part of $\Sigma_1'$ in the sectors $\theta_1 \le |\arg(z)| \le \theta_2$ and these two vertical line segments; see the left graph in Figure 9.5. We define $\Sigma_2$ be the image of $\Sigma_1$ under the transformation $z \mapsto \frac{1}{z}$. See the right graph in Figure 9.5. The modified contours, $\Sigma_1$ and $\Sigma_2$ are bounded, and are away from the origin, uniformly for $a \in [0, a_*]$. We have $\operatorname{Re} f(z) \le -1$ on $\Sigma_1 \setminus \Sigma_1'$ and $\operatorname{Re}(-f(z)) \le -1$ on

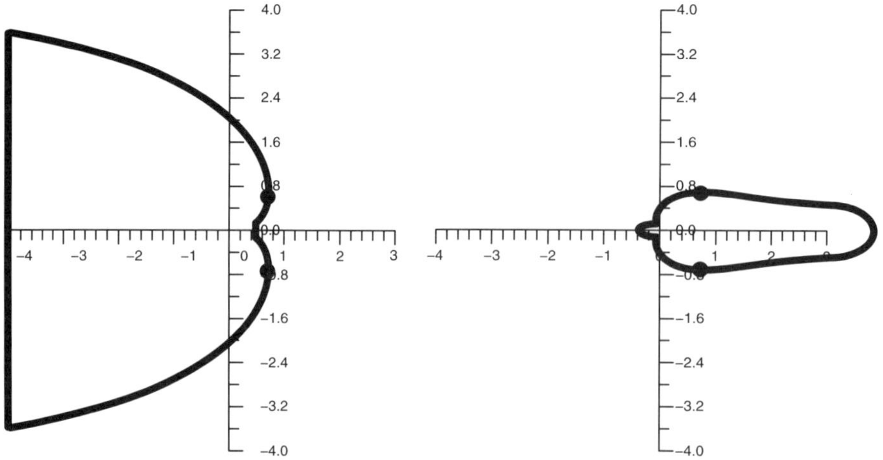


**Figure 9.5.** Left: curve $\Sigma_1$. Right: $\Sigma_2$

$\Sigma_2 \setminus \Sigma_2'$. Also the critical points $e^{\pm i\theta_c}$ are uniformly away from the modified parts of the contours.

We now deform the contours $C_1$ and $C_2$ in (9.42) to $\Sigma_1$ and $\Sigma_2$, respectively. Then, since $\Sigma_1$ and $\Sigma_2$ intersect and the integral contains a factor $\frac{1}{(z-w)^2}$, the new integral does not converge absolutely. Therefore, care is needed in the deformation utilizing the principal value integral as follows. First shrink the contour $C_2$ so that it lies inside $\Sigma_1$. Then deform the contour $C_1$ to $\Sigma_1$. The integral is now over the contour $\Sigma_1 \times C_2$. For a small $\epsilon > 0$, consider two small disks around $e^{\pm i\theta_c}$ of radius $\epsilon$. Let $\Sigma_1^\epsilon$ be the part of $\Sigma_1$ outside these disks. Since the integral is absolutely convergent on $\Sigma_1 \times C_2$, we have

$$\operatorname{Tr} \mathcal{K}_n = \lim_{\epsilon \downarrow 0} \frac{-1}{4\pi^2} \int_{\Sigma_1^\epsilon} \int_{C_2} \frac{e^{t(f(z)-f(w))}}{(z-w)^2} dw dz. \tag{9.48}$$

Now proceed to deform $C_2$ to $\Sigma_2$. Let $\Sigma_1^{\#,\epsilon}$ be the part of $\Sigma_1^\epsilon$ lying in the sector $|\theta| \le \theta_c$. There is a contribution from the pole at $w = z$ for $z \in \Sigma_1^{\#,\epsilon}$ (but no contribution for $z \in \Sigma_1^\epsilon \setminus \Sigma_1^{\#,\epsilon}$), and we obtain

$$\begin{aligned} \operatorname{Tr} \mathcal{K}_n &= \lim_{\epsilon \downarrow 0} \left\{ \frac{-1}{4\pi^2} \int_{\Sigma_1^\epsilon} \int_{\Sigma_2} \frac{e^{t(f(z)-f(w))}}{(z-w)^2} dw dz + \frac{t}{2\pi i} \int_{\Sigma_1^{\#,\epsilon}} f'(z) dz \right\} \\ &= \lim_{\epsilon \downarrow 0} \left\{ \frac{-1}{4\pi^2} \int_{\Sigma_1^\epsilon} \int_{\Sigma_2} \frac{e^{t(f(z)-f(w))}}{(z-w)^2} dw dz \right\} \\ &\quad + \frac{t}{2\pi i} \{ f(e^{i\theta_c}) - f(e^{-i\theta_c}) \}. \end{aligned} \tag{9.49}$$

From (9.45),

$$\begin{aligned} \frac{t}{2\pi i} \{ f(e^{i\theta_c}) - f(e^{-i\theta_c}) \} &= \frac{2t}{\pi} (\sin\theta_c - \theta_c \cos\theta_c) \\ &= \frac{2t}{\pi} \left\{ \sqrt{1 - \left( \frac{[\mu t]}{2t} \right)^2} - \frac{[\mu t]}{2t} \arccos\left( \frac{[\mu t]}{2t} \right) \right\}, \end{aligned} \tag{9.50}$$

which is the first term in (9.36) and is of order $O(t)$.

We now prove that the remaining double integral in (9.49) is $O(1)$. Let $\Sigma_2^+$ and $\Sigma_2^-$ be the parts of $\Sigma_2$ in the upper half-plane and in the lower half-plane, respectively. Let $\Sigma_1^{+,\epsilon}$ and $\Sigma_1^{-,\epsilon}$ be the parts of $\Sigma_1^\epsilon$ in the upper half-plane and in the lower half-plane, respectively. Since the lengths of $\Sigma_1^{+,\epsilon}$ and $\Sigma_2^-$ are uniformly bounded and the curves are uniformly separated, $\frac{1}{d} \ge \operatorname{dist}(\Sigma_1^{+,\epsilon}, \Sigma_2^-) \ge d$ for some $d > 0$ uniformly for all $\epsilon > 0$, and

$$\begin{aligned} \operatorname{Re} f(z) &\le \operatorname{Re} f(e^{\pm i\theta_c}) = 0, \qquad z \in \Sigma_1, \\ \operatorname{Re}(-f(w)) &\le \operatorname{Re}(-f(e^{\pm i\theta_c})) = 0, \qquad w \in \Sigma_2, \end{aligned} \tag{9.51}$$

we have

$$\lim_{\epsilon \downarrow 0} \frac{-1}{4\pi^2} \int_{\Sigma_1^{+,\epsilon}} \int_{\Sigma_2^-} \frac{e^{t(f(z)-f(w))}}{(z-w)^2} dwdz = O(1). \tag{9.52}$$

The same estimate applies to $\Sigma_1^{-,\epsilon} \times \Sigma_2^+$ too from the symmetry $f(z) = \overline{f(\bar{z})}$.

We now estimate the integral over $\Sigma_1^{+,\epsilon} \times \Sigma_2^+$. The integral over $\Sigma_1^{-,\epsilon} \times \Sigma_2^-$ has the same estimate due to symmetry. We split each of $\Sigma_1^{+,\epsilon}$ and $\Sigma_2^+$ into two parts. Let $D \equiv D(e^{i\theta_c}, t^{-\kappa})$ be the disk centered at $e^{i\theta_c}$ and of radius $t^{-\kappa}$, where

$$1/3 < \kappa < 1/2 \tag{9.53}$$

is fixed and $t$ is large enough so that $D \subset \mathbb{C}_+$. We can assume that $\epsilon$ is small enough so that $\epsilon < t^{-\kappa}$. (Note that we first take $\epsilon \downarrow 0$ and then $t \to \infty$.) Set $\Sigma_{1,\text{in}}^{+,\epsilon} \equiv \Sigma_1^{+,\epsilon} \cap D$, $\Sigma_{1,\text{out}}^{+,\epsilon} \equiv \Sigma_1^{+,\epsilon} \cap D^c$, and set $\Sigma_{2,\text{in}}^{+} \equiv \Sigma_2^+ \cap D$ and $\Sigma_{2,\text{out}}^{+} \equiv \Sigma_2^+ \cap D^c$. Of the two points in the intersection of $\Sigma_1^{+,\epsilon}$ and $\partial D$, let $z_*$ be the point at which $\operatorname{Re} f(z)$ is larger. Then $\operatorname{Re}(f(z)) \leq \operatorname{Re}(f(z_*))$ for $z \in \Sigma_{1,\text{out}}^{+,\epsilon}$. (Recall that $\operatorname{Re}(f(z)) \leq -1$ for $z \in \Sigma_1 \setminus \Sigma_1'$ and $\operatorname{Re}(f(z_*)) \approx \operatorname{Re}(f(z_c)) = 0$.)

Since $\operatorname{dist}(\Sigma_{1,\text{out}}^{+,\epsilon}, \Sigma_2^+) \geq ct^{-\kappa}$ for some constant $c$ independent of $a \in [0, a_*]$, we find, using (9.51) for $\Sigma_2^+$, that

$$\left| \frac{-1}{4\pi^2} \int_{\Sigma_{1,\text{out}}^{+,\epsilon}} \int_{\Sigma_2^+} \frac{e^{t(f(z)-f(w))}}{(z-w)^2} dwdz \right| \leq \frac{|\Sigma_1^+||\Sigma_2^+|}{4\pi^2 c^2} t^{2\kappa} e^{t \operatorname{Re}(f(z_*))}. \tag{9.54}$$

Note that the bound is independent of $\epsilon$ as long as $\epsilon < t^{-\kappa}$. We use the Taylor series

$$f(z) - f(e^{i\theta_c}) = \frac{1}{2} f''(e^{i\theta_c})(z - e^{i\theta_c})^2 + g(z - e^{i\theta_c}), \tag{9.55}$$

for some function $g(s) = O(|s|^3)$ uniformly in $a \in [0, a_*]$. Since $z_*$ is on the curve of steepest descent, we have $\operatorname{Im}(f(z_*) - f(e^{i\theta_c})) = 0$ and $\operatorname{Re}(f(z_*) - f(e^{i\theta_c})) < 0$, and hence we find that

$$\begin{aligned} \operatorname{Re}(f(z_*) - f(e^{i\theta_c})) &= f(z_*) - f(e^{i\theta_c}) \\ &\leq -\left|\frac{1}{2} f''(e^{i\theta_c})(z_* - e^{i\theta_c})^2\right| + |g(z_* - e^{i\theta_c})| \\ &= -\sin\theta_c |z_* - e^{i\theta_c}|^2 + O(|z_* - e^{i\theta_c}|^3). \end{aligned} \tag{9.56}$$

Here we have used the fact that if $a, b \in \mathbb{C}$ and $a + b < 0$, then $a + b \leq -|a| + |b|$. As $|z_* - e^{i\theta_c}| = t^{-\kappa}$ and $\operatorname{Re}(f(e^{i\theta_c})) = 0$, this implies that

$$\operatorname{Re}(f(z_*)) \leq -\sin\theta_c t^{-2\kappa} + O(t^{-3\kappa}). \tag{9.57}$$

Therefore we find that
(9.58)
$$\left| \lim_{\epsilon \downarrow 0} \frac{-1}{4\pi^2} \int_{\Sigma_{1,\mathrm{out}}^{+,\epsilon}} \int_{\Sigma_2^+} \frac{e^{t(f(z)-f(w))}}{(z-w)^2} dw dz \right| \le \frac{|\Sigma_1^+||\Sigma_2^+|}{4\pi^2 c^2} t^{2\kappa} e^{-t^{1-2\kappa}\sin\theta_c + O(t^{1-3\kappa})},$$
which tends to 0 as $t \to \infty$ uniformly for $\mu$ in a compact subset of $[0,2)$. It is easy to check that we have the same estimate for the integral on the contour $\Sigma_1^{+,\epsilon} \times \Sigma_{2,\mathrm{out}}^+$.

Now consider $\Sigma_{1,\mathrm{in}}^{+,\epsilon} \times \Sigma_{2,\mathrm{in}}^+$. We first compute the double integral where $f$ is replaced by its quadratic Taylor approximation,

$$\lim_{\epsilon \downarrow 0} \frac{-1}{4\pi^2} \int_{\Sigma_{1,\mathrm{in}}^{+,\epsilon}} \int_{\Sigma_{2,\mathrm{in}}^+} \frac{e^{\frac{t}{2} f''(e^{i\theta_c})\{(z-e^{i\theta_c})^2 - (w-e^{i\theta_c})^2\}}}{(z-w)^2} dw dz. \tag{9.59}$$

By using the changes of variables $z \mapsto \zeta_1$ and $w \mapsto \zeta_2$ defined by

$$\sqrt{tf''(e^{i\theta_c})}(z-e^{i\theta_c}) = \sqrt{2t\sin\theta_c} e^{-i(\theta_c - \frac{\pi}{4})}(z-e^{i\theta_c}) = i\zeta_1$$

and

$$\sqrt{tf''(e^{i\theta_c})}(w-e^{i\theta_c}) = \sqrt{2t\sin\theta_c} e^{-i(\theta_c - \frac{\pi}{4})}(w-e^{i\theta_c}) = -\zeta_2,$$

(9.59) equals

$$\lim_{\epsilon \downarrow 0} \frac{-1}{4\pi^2} \int_{\Gamma_1^\epsilon} \int_{\Gamma_2} \frac{e^{-\frac{1}{2}(\zeta_1^2+\zeta_2^2)}}{(\zeta_2 + i\zeta_1)^2} d\zeta_2 d\zeta_1, \tag{9.60}$$

where $\Gamma_1^\epsilon$ and $\Gamma_2$ are close to line segments on the real line since $\Sigma_{1,\mathrm{in}}^{+,\epsilon}$ and $\Sigma_{2,\mathrm{in}}^+$ are parts of the steepest descent/ascent curves of $f(z)$. The contours $\Gamma_1^\epsilon$ and $\Gamma_2$ may be parametrized as

$$\begin{aligned} \zeta_1 &= \zeta_1(s_1) = \sqrt{2t\sin\theta_c}(s_1 + O(s_1^2)), \\ \zeta_2 &= \zeta_2(s_2) = \sqrt{2t\sin\theta_c}(s_2 + O(s_2^2)), \end{aligned} \tag{9.61}$$

where $\epsilon \le |s_1| \le c_1 t^{-\kappa}$ and $|s_2| \le t^{-\kappa}$ for some constants $c_1, c_2 > 0$, independent of $\mu$ in a compact subset of $[0,2)$. By an elementary estimate, the difference between the integrals in (9.60) and (9.37) is $O(1)$ as $\epsilon \downarrow 0$. Lemma 9.6 then implies that (9.59) is $O(1)$ uniformly in $a \in [0, a_*]$ and as $t \to \infty$.

It remains to consider the error resulting from approximating $f$ by its quadratic Taylor expansion. From (9.55),

$$\begin{aligned} &f(z) - f(w) - \frac{1}{2} f''(e^{i\theta_c})\{(z-e^{i\theta_c})^2 - (w-e^{i\theta_c})^2\} \\ &= O(|z-e^{i\theta_c}|)^3 + O(|w-e^{i\theta_c}|^3). \end{aligned} \tag{9.62}$$

Since the contours $\Sigma_1$ and $\Sigma_2$ intersect perpendicularly at $e^{i\theta_c}$, and $f$ depends smoothly on $\mu$, $\frac{z-e^{i\theta_c}}{z-w} = O(1)$ and $\frac{w-e^{i\theta_c}}{z-w} = O(1)$ for $z \in \Sigma^+_{1,\text{in}}$ and $w \in \Sigma^+_{2,\text{in}}$. Hence (9.62) is of order

(9.63)
$$O(|z-w|^2|z-e^{i\theta_c}|) + O(|z-w|^2|w-e^{i\theta_c}|) = O(t^{-\kappa}|z-w|^2) = O(t^{-3\kappa}),$$

for $z \in \Sigma^{+,\epsilon}_{1,\text{in}}$ and $w \in \Sigma^+_{2,\text{in}}$ As $1-3\kappa < 0$,

(9.64)
$$\begin{aligned}
&\lim_{\epsilon\downarrow 0} \frac{-1}{4\pi^2} \int_{\Sigma^{+,\epsilon}_{1,\text{in}}} \int_{\Sigma^+_{2,\text{in}}} \frac{e^{t(f(z)-f(w))} - e^{\frac{t}{2}f''(e^{i\theta_c})\{(z-e^{i\theta_c})^2-(w-e^{i\theta_c})^2\}}}{(z-w)^2} dwdz \\
&= \lim_{\epsilon\downarrow 0} \frac{-1}{4\pi^2} \int_{\Sigma^{+,\epsilon}_{1,\text{in}}} \int_{\Sigma^+_{2,\text{in}}} \frac{O(t^{1-\kappa}|z-w|^2)e^{t(f(z)-f(w))}}{(z-w)^2} dwdz \\
&= O(t^{1-3\kappa}e^{O(t^{1-3\kappa})})
\end{aligned}$$

using (9.51) and the fact that $\Sigma^+_1$ and $\Sigma^+_2$ are of length $O(t^{-\kappa})$. Since $\kappa > 1/3$, the error is bounded.

Combining (9.49), (9.52), (9.58), and (9.64), we obtain (9.36). □

Estimates (9.33) and (9.36) imply the following uniform estimate for the lower tail.

**Proposition 9.9.** *Let $\mu_0 \in [0, 2)$. There exist $t_0 > 0$ and $C > 0$ such that*

$$\mathbb{P}\{L(t) \le \mu t\} \le Ce^{-tg(\frac{\mu}{2})} \tag{9.65}$$

*for $t \ge t_0$ and $\mu \in [0, \mu_0]$, where*

$$g(u) = \frac{2}{\pi}\big(\sqrt{1-u^2} - u\arccos(u)\big). \tag{9.66}$$

Note that $g(u) = \frac{4\sqrt{2}}{3\pi}(1-u)^{3/2}(1+O(1-u))$ as $u \uparrow 1$. See Figure 9.6 for the graph of $g$.

**Proof.** Estimates (9.33) and (9.36) imply that $\mathbb{P}\{L(t) \le \mu t\} \le Ce^{-tg(\frac{[\mu t]}{2t})}$. The function $g(u)$ is positive and decreasing for $u \in [0, 1)$, and $g(1) = 0$. Since $[\mu t] \le \mu t$, we obtain (9.65). □

Unlike the upper tail estimate, (9.65) is far from sharp. Indeed the following large deviation result due to Seppäläinen [**Sep98**] shows that the sharp decay exponent is $t^2$, not $t$.

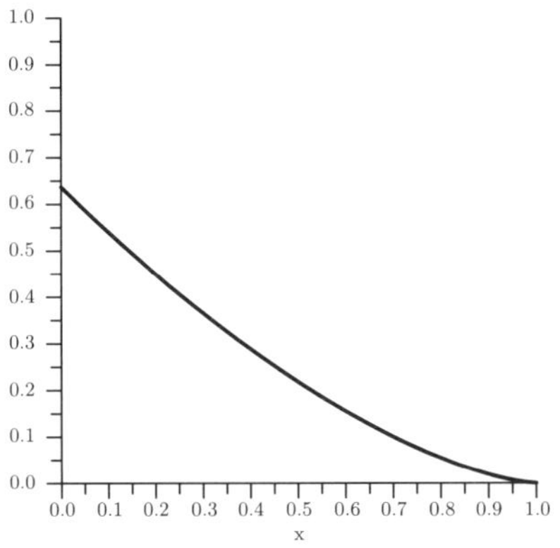


**Figure 9.6.** Function $g(u)$.

**Proposition 9.10** (Large deviation principle for $L(t)$: lower tail). *For* $0 < \mu < 2$,

$$\lim_{t\to\infty} \frac{1}{t^2} \log \mathbb{P}\{L(t) \le \mu t\} = -1 + 2\mu - \frac{3}{4}\mu^2 - \frac{\mu^2}{2}\log\frac{2}{\mu} \equiv U(\mu). \tag{9.67}$$

The reason that the exponent in (9.65) is not sharp is due to the fact that the estimate (9.33) is far from optimal. This is because the eigenvalues $\lambda_j(t)$ of $\mathcal{K}_{[\mu t]}$ converge to 1 as $t \to \infty$ for each $j$: in this case, $\prod_{j=1}^{\infty}(1-\lambda_j(t))$ is much smaller than $e^{-\sum_{j=1}^{\infty}\lambda_j(t)}$, and this accounts for the discrepancy between (9.67) and (9.65).

The rate function for the large deviation principle for $\ell_N$ is different from that for $L(t)$.

**Proposition 9.11** (Large deviation principle for $\ell_N$: lower tail). *For* $0 < \mu < 2$,

$$\lim_{t\to\infty} \frac{1}{N} \mathbb{P}\{\ell_N \le \mu\sqrt{N}\} = \frac{1}{2} - \frac{\mu^2}{8} - \log\frac{2}{\mu} + \Big(1+\frac{\mu^2}{4}\Big)\log\Big(\frac{2\mu^2}{4+\mu^2}\Big) =: -V(\mu). \tag{9.68}$$

The fact that the left-hand side is bounded above by $-V(\mu)$ was first obtained in [**LS77**]. The complimentary bound from the below was proven by [**DZ99**]. Using this result for $\ell_N$, [**Sep98**] derived Proposition 9.10. Indeed the functions $U(\mu)$ and $V(\mu)$ are related as

$$U(\mu) = \inf_{0\le x\le 1}\Big\{xV\Big(\frac{\mu}{x}\Big) + x\log x - x + 1\Big\}. \tag{9.69}$$

To see how to obtain the relation between (9.68) and (9.67), see [**Sep98**].

## 9.3. Convergence of $L(t)/t$ and $\ell_N/\sqrt{N}$

Propositions 9.2 and 9.9 immediately imply weak convergence,

$$\lim_{t\to\infty} \mathbb{P}\left\{\left|\frac{L(t)}{t}-2\right| \geq \epsilon\right\} = 0, \tag{9.70}$$

for arbitrary $\epsilon > 0$. This implies the weak convergence of $\ell_N$ from the de-Poissonization lemmas in Section 2.2 by the following argument. From Lemma 2.7, for fixed $n \geq 0$, $a_N = \mathbb{P}\{\ell_N \leq n\}$ is a decreasing sequence in $N$. For this sequence $\{a_N\}_{N=0}^{\infty}$, $\phi(t)$ in equation (2.7) is given by $\phi(t) = \mathbb{P}\{L(t) \leq n\}$. From Lemma 2.6,

$$\mathbb{P}\{\ell_N \leq n\} \leq 5\,\mathbb{P}\{L((N-\sqrt{N})^{1/2}) \leq n\}, \qquad n \geq 0. \tag{9.71}$$

Taking $n = [(2-\epsilon)\sqrt{N}]+1$,

$$\begin{aligned} \mathbb{P}\{\ell_N \leq (2-\epsilon)\sqrt{N}\} &\leq \mathbb{P}\{\ell_N \leq n\} \\ &\leq 5\,\mathbb{P}\{L((N-\sqrt{N})^{1/2}) \leq [(2-\epsilon)\sqrt{N}]+1\} \to 0 \end{aligned} \tag{9.72}$$

as $N \to \infty$ from (9.70). Similarly, from Lemma 2.6 and (9.70),

$$\begin{aligned} \mathbb{P}\{\ell_N > (2+\epsilon)\sqrt{N}\} &\leq \mathbb{P}\{\ell_N > [(2+\epsilon)\sqrt{N}]\} \\ &\leq 5\,\mathbb{P}\{L((N+\sqrt{N})^{1/2}) > [(2+\epsilon)\sqrt{N}]\} \to 0. \end{aligned} \tag{9.73}$$

Hence we obtain

$$\lim_{N\to\infty} \mathbb{P}\left\{\left|\frac{\ell_N}{\sqrt{N}}-2\right| \geq \epsilon\right\} = 0. \tag{9.74}$$

As we now show, the tail estimates for $L(t)$ in the last two sections also imply the convergence of all moments for $L(t)$ and $\ell_N$ and almost sure convergence for $L(t)$ (see Remark 2.1).

**Theorem 9.12.** *For any $m \geq 1, 2, 3, \dots$,*

$$\lim_{t\to\infty} \mathbb{E}\left[\left|\frac{L(t)}{t}-2\right|^m\right] = 0, \qquad \lim_{N\to\infty} \mathbb{E}\left[\left|\frac{\ell_N}{\sqrt{N}}-2\right|^m\right] = 0. \tag{9.75}$$

*In addition, $\frac{L(t)}{t} \to 2$ almost surely.*

**Proof.** Set $F_t(x) = \mathbb{P}\{\frac{L(t)}{t} - 2 \leq x\}$. Then for $m = 1, 2, \dots$, integrating by parts,

$$\begin{aligned} \mathbb{E}\left[\left|\frac{L(t)}{t}-2\right|^m\right] &= \int_{-\infty}^{\infty} |x|^m dF_t(x) \\ &= \int_{(-\infty,0)} m|x|^{m-1}F_t(x)dx + \int_{[0,\infty)} mx^{m-1}(1-F_t(x))dx. \end{aligned} \tag{9.76}$$

Note that the boundary terms vanish since $x^m(1-F_t(x)) \to 0$ as $x \to +\infty$ from (9.19) and $|x|^m F_t(x) = 0$ for $x < -2$. We prove that the integrals converge to 0 using the dominated convergence theorem. From (9.70), the integrands in (9.76) converge to 0 for each fixed $x \neq 0$ as $t \to \infty$. In Proposition 9.2, for $x \geq 2$, $h'(1+\frac{x}{2}) \geq \log(3)$, so $g(1+\frac{x}{2}, t) \leq e^{-4th(1+\frac{x}{2})}$. Hence (9.18) implies that

$$(9.77)\quad 1 - F_t(x) = \mathbb{P}\{L(t) > (2+x)t\} \leq e^2 e^{-4h(1+\frac{x}{2})}, \qquad x \geq 2, \quad t \geq 1.$$

For negative $x$, since $L(t) \geq 0$, we have the trivial bound

$$F_t(x) = \mathbb{P}\{L(t) \leq (2+x)t\} \leq 1_{[-2,\infty)}(x).$$

Therefore the integrands in (9.76) are uniformly bounded by integrable functions for all $t \geq 1$, and hence the dominated convergence theorem implies that (9.76) converges to 0.

For the convergence of moments of $\ell_N$, set $G_N(t) = \mathbb{P}\{\frac{\ell_N}{\sqrt{N}} - 2 \leq x\}$. By (9.73), for $x > 0$

$$(9.78)\quad \begin{aligned} 1 - G_N(x) &= \mathbb{P}\{\ell_N > (2+x)\sqrt{N}\} \\ &\leq 5\,\mathbb{P}\left\{L((N+\sqrt{N})^{1/2} > [(2+x)\sqrt{N}]\right\}. \end{aligned}$$

For all large enough $N$, $\frac{\sqrt{N}}{(N+\sqrt{N})^{1/2}} \geq \frac{3}{4}$. Also for $x \geq 2$, $\frac{3}{4}(2+x) \geq 2 + \frac{x}{2}$. Hence, from (9.77),

(9.79)

$$\begin{aligned} &1 - G_N(x) \\ &\leq 5\,\mathbb{P}\left\{L((N+\sqrt{N})^{1/2}) > (2+x)\frac{\sqrt{N}}{(N+\sqrt{N})^{1/2}}(N+\sqrt{N})^{1/2} - 1\right\} \\ &\leq 5\,\mathbb{P}\left\{L((N+\sqrt{N})^{1/2}) > (2+\frac{x}{2} - \frac{1}{(N+\sqrt{N})^{1/2}})(N+\sqrt{N})^{1/2}\right\} \\ &\leq 5\,\mathbb{P}\left\{L((N+\sqrt{N})^{1/2}) > (2+\frac{x}{3})(N+\sqrt{N})^{1/2}\right\} \\ &\leq 5e^2 e^{-4h(1+\frac{x}{6})} \end{aligned}$$

for all large enough $N$ and for $x \geq 6$. The function $G_N(x)$ is also bounded by $1_{[-2,\infty)}(x)$. As in equation (9.76), the dominated convergence theorem together with (9.74) again implies that $\mathbb{E}[|\frac{\ell_N}{\sqrt{N}} - 2|^m] \to 0$ as $N \to \infty$.

It remains to prove the almost sure convergence for $L(t)$. From Propositions 9.2 and 9.9, we have for any $\epsilon > 0$,

$$(9.80)\qquad \mathbb{P}\left\{\left|\frac{L(t)}{t} - 2\right| > \epsilon\right\} \leq ce^{-c't}, \qquad t > 0$$

for suitable constants $c = c(\epsilon) > 0$ and $c' = c'(\epsilon) > 0$. This implies that

$$(9.81)\qquad \int_0^\infty \mathbb{P}\left\{\left|\frac{L(t)}{t} - 2\right| > \epsilon\right\} dt < \infty.$$

Set $f_t(\omega) \equiv \left|\frac{L(t,\omega)}{t} - 2\right|$, $\mathcal{A}_\epsilon = \{\omega : \int \chi_{\{f_t(\omega)>\epsilon\}} dt < \infty\}$, and $\mathcal{A} = \bigcap_n \mathcal{A}_{\frac{1}{n}}$. Fubini's theorem implies that $\mathbb{P}\{\mathcal{A}\} = 1$.

In order to prove the almost sure convergence of $\frac{L(t)}{t}$, we observe the following. Let $T > 0$, $\delta > 0$, and $\epsilon > 0$, and assume that $t' > t > T$, $t' - t < \delta$, and $\frac{L(t,\omega)}{t} - 2 > \epsilon$. Then by the monotonicity of $L(t,\omega)$,

$$\begin{aligned}\frac{L(t',\omega)}{t'} - 2 &\geq \frac{L(t',\omega)}{t}\left(1 - \frac{\delta}{T}\right) - 2 \geq \left(\frac{L(t',\omega)}{t} - 2\right)\left(1 - \frac{\delta}{T}\right) - \frac{2\delta}{T} \\ &\geq \left(\frac{L(t,\omega)}{t} - 2\right)\left(1 - \frac{\delta}{T}\right) - \frac{2\delta}{T} \geq \left(1 - \frac{\delta}{T}\right)\epsilon - \frac{2\delta}{T}.\end{aligned} \tag{9.82}$$

Given $\epsilon > 0$, for any fixed $\delta > 0$, one can choose $T > 0$ sufficiently large so that the right-hand side of equation (9.82) is greater than $\frac{\epsilon}{2}$.

Suppose there exists an $\omega \in \mathcal{A}$, an $\epsilon > 0$, and a sequence of $t_j \to \infty$ such that $\frac{L(t_j,\omega)}{t_j} - 2 > \epsilon$. Without loss of generality, assume $t_{j+1} - t_j > 1$. Given this $\epsilon$, fix $0 < \delta < 1$ and choose $T$ as above. Observe that by the estimate (9.82) and the remark following it,

$$\int \chi_{\{f_t(\omega)>\epsilon/2\}} dt \geq \sum_j \int_{t_j}^{t_j+\delta} dt = \infty, \tag{9.83}$$

contradicting the assumption that $\omega \in \mathcal{A}$. A similar argument can be made for the case $\frac{L(t,\omega)}{t} - 2 < -\epsilon$. □

**Remark 9.13.** From Lemma 2.6 and Propositions 9.2 and 9.9, we obtain the following analogue of (9.80) for $\ell_N$. For any $\epsilon > 0$,

$$\mathbb{P}\left\{\left|\frac{\ell_N}{\sqrt{N}} - 2\right| > \epsilon\right\} \leq Ce^{-C'\sqrt{N}} \tag{9.84}$$

for some constants $C = C(\epsilon) > 0$ and $C' = C'(\epsilon) > 0$.

When $m = 1$ in (9.75), we have, in particular,

$$\lim_{t\to\infty} \frac{\mathbb{E}[L(t)]}{t} = 2, \qquad \lim_{N\to\infty} \frac{\mathbb{E}[\ell_N]}{\sqrt{N}} = 2, \tag{9.85}$$

and thus obtain (9.1), (9.2).

## 9.4. Central Limit Theorem

We now consider the limit of the scaled random variables $\frac{L(t)-2t}{t^\beta}$ and $\frac{\ell_N - 2\sqrt{N}}{(\sqrt{N})^\beta}$ for an appropriate $\beta > 0$. As discussed in the Introduction, it turns out that when $\beta = 1/3$, the scaled random variable converges to a nontrivial random variable. The goal of this section is to prove the following limit theorems.

Let

$$F(x) = \det(1 - \mathbb{A}_x) \tag{9.86}$$

be the Tracy-Widom distribution (6.97). Recall from (6.87) that the kernel of $\mathbb{A}_x$ can be written as

$$\mathbb{A}(u, v) = \int_0^\infty \mathrm{Ai}(u+s)\,\mathrm{Ai}(v+s)ds, \tag{9.87}$$

where Ai is the Airy function.

**Theorem 9.14.** *For any $x \in \mathbb{R}$,*

$$\lim_{t\to\infty} \mathbb{P}\left\{\frac{L(t) - 2t}{t^{1/3}} \le x\right\} = F(x). \tag{9.88}$$

*Moreover, convergence is uniform for $x$ in compact sets.*

The analogous result for $\ell_N$ is:

**Theorem 9.15** (Central Limit Theorem). *For any $x \in \mathbb{R}$,*

$$\lim_{N\to\infty} \mathbb{P}\left\{\frac{\ell_N - 2\sqrt{N}}{N^{1/6}} \le x\right\} = F(x), \tag{9.89}$$

*where the convergence is uniform for $x$ in compact sets.*

The above results were first obtained in [**BDJ99**] by analyzing Gessel's formula (7.3) using the Deift-Zhou steepest-descent method for Riemann-Hilbert problems. Here we prove the above theorems using Fredholm determinants, as was done in [**BOO00, Joh01a**]. This is a considerable simplification as only the classical steepest-descent method for integrals is needed.

We use the following fact (see item (iv) after (A.33) in Appendix A): for Hilbert-Schimit operators $A, A_n : \mathcal{H}_2 \to \mathcal{H}_1$ and $B, B_n : \mathcal{H}_1 \to \mathcal{H}_2$, $n = 1, 2, \ldots$. If $A_n \to A$ and $B_n \to B$ in the Hilbert-Schmidt norm as $n \to \infty$, then

$$\det(1 - A_n B_n) \to \det(1 - AB). \tag{9.90}$$

From (9.4) and (9.5), $\mathbb{P}\{L(t) \le n\} = \det(1 - \mathcal{K}_n)$, where

$$\mathcal{K}_n(i, j) = \sum_{m=0}^{\infty} J_{i+m}(2t) J_{m+j}(2t), \qquad \text{for } i, j \ge n+1. \tag{9.91}$$

Note that $\mathcal{K}_n$ is a product of two (Hilbert-Schmidt) operators whose kernels are $J_{i+m}(2t)$, $i \ge n+1, m \ge 0$, and $J_{m+j}(2t)$, $m \ge 0, j \ge n+1$. These operators act on discrete spaces. Our first step is to replace $\mathcal{K}_n$ by an operator acting on $L^2((x, \infty))$.

Let $\lceil z \rceil$ denote the smallest integer greater than or equal to $z$. Hence $z \le \lceil z \rceil < z + 1$. Let $U : \ell_2(n+1, n+2, \ldots) \to L^2((n, \infty))$ be the isometry taking $e_j \to \chi_j$, $j \geq n+1$, where $\chi_j(s)$ is the characteristic function of the interval $(j-1, j]$. Let $\tilde{\mathcal{K}}_n$ be the operator on $L^2((n,\infty))$ with kernel

$$\begin{aligned}\tilde{\mathcal{K}}_n(z,w) &= \mathcal{K}_n(\lceil z\rceil, \lceil w\rceil) \\ &= \sum_{m=0}^{\infty} J_{\lceil z\rceil+m}(2t)J_{m+\lceil w\rceil}(2t) \\ &= \int_0^\infty J_{\lceil z\rceil+\lceil s\rceil-1}(2t)J_{\lceil s\rceil+\lceil w\rceil-1}(2t)ds, \qquad z,w>n.\end{aligned} \tag{9.92}$$

As $\tilde{\mathcal{K}}_n$ is again a product of Hilbert-Schmidt operators, it is a trace class on $L^2(n,\infty)$. A direct computation shows that $U\mathcal{K}_n = \tilde{\mathcal{K}}_n U$, and hence $\det(1-\mathcal{K}_n) = \det(1-\tilde{\mathcal{K}}_n)$ as $\tilde{\mathcal{K}}_n = 0$ on $(\mathrm{Ran}U)^\perp$. Thus,

$$\mathbb{P}\{L(t)\leq n\} = \det(1-\tilde{\mathcal{K}}_n) \tag{9.93}$$

and hence

$$\mathbb{P}\left\{\frac{L(t)-2t}{t^{1/3}}\leq x\right\} = \det(1-\tilde{\mathcal{K}}_{[2t+xt^{1/3}]}), \tag{9.94}$$

where $\tilde{\mathcal{K}}_{[2t+xt^{1/3}]}$ acts on the space $L^2(([2t+xt^{1/3}],\infty))$. In view of (9.87) and (9.90), the limit theorem (9.88) can be obtained if we show that $J_\nu(2t)$ converges to the Airy function in an appropriate scaling regime.

**9.4.1. Airy Asymptotics of Bessel Functions.** The appearance of the Airy function in the scaling asymptotics of the modified Bessel function can be anticipated by looking at the graphs in Figure 9.1. Near the point $x=\nu$, the graphs in Figure 9.1 look quite similar if we scale both $x$ and $y$ coordinates appropriately. Figure 9.1 should be compared, after changing $x \to -x$, with the graph of the Airy function given in Figure 9.7.

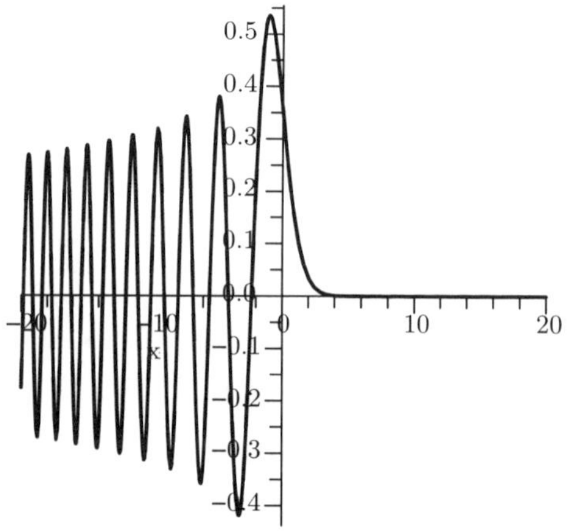


**Figure 9.7.** The Airy function $\mathrm{Ai}(x)$.

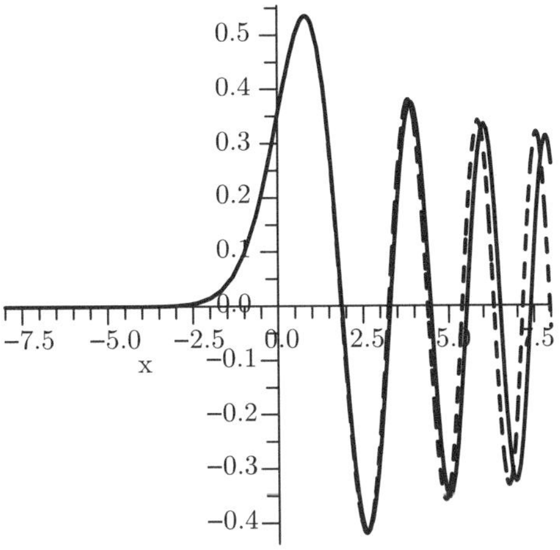


**Figure 9.8.** The solid curve is the graph of the scaled Bessel function $2^{-1/3}\nu^{1/3}J_\nu(\nu + \nu^{1/3}w)$ when $\nu = 500$. The dashed curve is the graph of the scaled Airy function $\mathrm{Ai}(-2^{1/3}w)$.

The precise scaling turns out to be $x = \nu + w\nu^{1/3}$, and we have the following asymptotics.

**Proposition 9.16.** *For $w \in \mathbb{C}$,*

$$\lim_{\nu\to\infty} 2^{-1/3}\nu^{1/3}J_\nu(\nu + w\nu^{1/3}) = \mathrm{Ai}(-2^{1/3}w). \tag{9.95}$$

*The convergence is uniform for $w$ in compact sets.*

See Figure 9.8 for a graphical comparison of the scaled Bessel function and the Airy function.

Before we proceed to prove Proposition 9.16, we recall the following properties of the Airy function. These can be found, for example, in [**AS64**]. The Airy function (see (1.12)) has an integral representation,

$$\mathrm{Ai}(x) = \frac{1}{2\pi}\int_C e^{i(xs+\frac{1}{3}s^3)}ds, \tag{9.96}$$

where $C$ is a contour from $\infty e^{i\theta_1}$ to $\infty e^{i\theta_2}$ for some $\theta_1 \in (2\pi/3, \pi)$, $\theta_2 \in (0, \pi/3)$. The contour is chosen so that $\mathrm{Re}(is^3) < 0$ as $|s| \to \infty$ which makes the integral convergent. The asymptotics of $\mathrm{Ai}(x)$ are as follows: as $x \to +\infty$,

$$\mathrm{Ai}(x) = \frac{e^{-\frac{2}{3}x^{3/2}}}{2\sqrt{\pi}x^{1/4}}\left(1 + O\left(x^{-3/2}\right)\right), \tag{9.97}$$

and as $x \to -\infty$,

$$\text{(9.98)}\quad \begin{aligned} \operatorname{Ai}(x) = & \frac{1}{\sqrt{(\pi)}}(-x)^{-1/4}\left\{\sin\left(\frac{2}{3}(-x)^{3/2}+\frac{\pi}{4}\right)\left(1+O\left(x^{-3}\right)\right)\right. \\ & \left. -\cos\left(\frac{2}{3}(-x)^{-3/2}+\frac{\pi}{4}\right)\left(\frac{5}{48}(-x)^{-3/2}+O\left(x^{-9/2}\right)\right)\right\}. \end{aligned}$$

The behavior of the Airy function changes from (super-)exponential decay to oscillatory as $x$ runs from $+\infty$ to $-\infty$.

Proposition 9.16 can be found, for example, in [**AS64**, 9.3.4]. For the convenience of the reader, we will give a proof using the steepest-descent method as described in Appendix B. We first sketch the argument and illustrate how the exponent $\nu^{1/3}$, which is responsible for $\beta = 1/3$ (see the discussion at the beginning of Section 9.4), is chosen; a rigorous argument will follow the sketch. From (9.6), for an integer $\nu$,

$$\text{(9.99)}\qquad J_\nu(\nu + w\nu^\beta) = \frac{1}{2\pi i}\oint_{\{|z|=1\}} e^{\nu f(z) + \frac{w\nu^\beta}{2}(z - z^{-1})}\frac{dz}{z},$$

where (cf. (9.43)) and, as before, log denotes the principal branch)

$$\text{(9.100)}\qquad f(z) = \frac{1}{2}(z - z^{-1}) - \log z.$$

If $\nu$ is not an integer, an additional term (see [**AS64**, 9.1.22]) must be added to (9.99). However, it turns out that this term is of order $\nu^{-1}$ as $\nu \to \infty$, uniformly for $w$ in compact sets, and hence does not contribute to the limit in (9.95). The details for general $\nu$ are left to the reader. The only critical

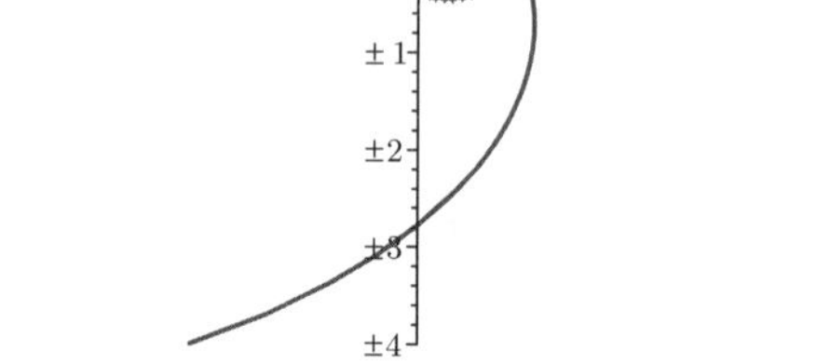


**Figure 9.9.** Curves of constant phase passing the critical point $z = 1$. The solid curves are curves of steepest descent and and the dashed curves are curves of steepest ascent.

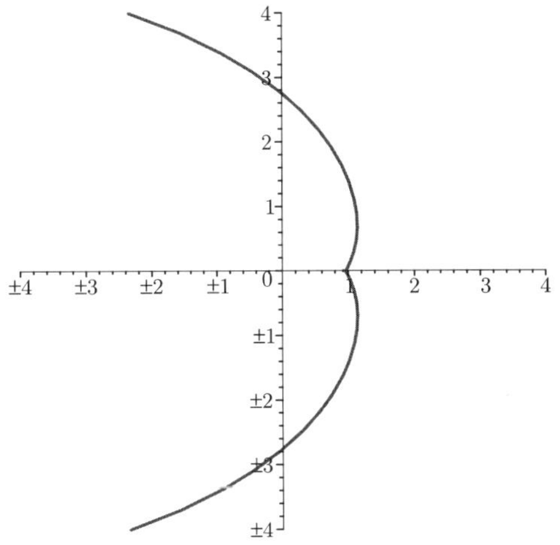


**Figure 9.10.** Curves $\Gamma$ of steepest descent passing $z = 1$

point of $f$ is $z = 1$, and it is a double root:

$$f'(1) = f''(1) = 0, \quad f^{(3)}(1) = 1, \quad f(1) = 0. \tag{9.101}$$

The curve of constant phase passing through the critical point (the curve $\operatorname{Im} f(z) = \operatorname{Im} f(1) = 0$) is given in polar coordinates by

$$(r + r^{-1}) \sin\theta - 2\theta = 0. \tag{9.102}$$

See Figure 9.9. Let $\Gamma$ be the part of the curve corresponding to steepest descent: $\Gamma$ is given by the following solution of (9.102) (see Figure 9.10):

$$r(\theta) = \frac{\theta}{\sin\theta} + \sqrt{\left(\frac{\theta}{\sin\theta}\right)^2 - 1}. \tag{9.103}$$

We expect that that the main contribution to the integral comes from a neighborhood of $z = 1$. Using the Taylor approximation for $f$ near $z = 1$,

$$\begin{aligned} f(z) &= \frac{1}{6}(z-1)^3 + O(|z-1|^4), \\ z - z^{-1} &= 2(z-1) + O(|z-1|^2), \\ \frac{1}{z} &= 1 + O(|z-1|), \end{aligned} \tag{9.104}$$

we anticipate that as $\nu \to \infty$

(9.105)

$$J_\nu(\nu + w\nu^\beta) \sim \frac{1}{2\pi i}\int_{\Gamma_0} e^{\frac{1}{6}\nu(z-1)^3 + w\nu^\beta(z-1) + O(\nu|z-1|^4) + O(\nu^\beta|z-1|^2)}\big(1 + O(|z-1|)\big)dz.$$

where $\Gamma_0$ is the part of $\Gamma$ which lies inside a small neighborhood of $z = 1$. Ignoring the $O$-terms and substituting $2^{-1/3}\nu^{1/3}(z-1) = s$, we expect that

$$J_\nu(\nu + w\nu^\beta) \sim \frac{2^{1/3}\nu^{-1/3}}{2\pi i} \int e^{\frac{1}{3}s^3 + 2^{1/3} w \nu^{\beta - \frac{1}{3}} s} ds. \tag{9.106}$$

When $\beta = 1/3$, we see that the second term of the exponent is of order $O(1)$; hence, it yields a nontrivial limit. With this choice $\beta = 1/3$, setting $s = -iu$, extending the contour to infinity, and then reversing the orientation,

$$J_\nu(\nu + w\nu^\beta) \sim \frac{2^{1/3}\nu^{-1/3}}{2\pi} \int e^{i(\frac{1}{3}s^3 - 2^{1/3} ws)} ds = 2^{1/3}\nu^{-1/3}\,\mathrm{Ai}(-2^{1/3}w), \tag{9.107}$$

where the contour is from $\infty e^{\frac{5}{6}\pi i}$ to $\infty e^{\frac{1}{6}\pi i}$.

Now we make this argument rigorous.

**Proof of Proposition 9.16.** Let $f(z) = \frac{1}{2}(z - z^{-1}) - \log z$ be as in (9.100), and let $\Gamma$ be the curve of steepest decent of $f(z)$ passing through the critical point $z = 1$, given as above by equation (9.103) (see Figure 9.10). $\Gamma$ passes through $z = 1$ at the angles $\pi/3$ and $-\pi/3$. We have for $\nu = 1, 2, 3, \ldots,$

$$\nu^{1/3} J_\nu(\nu + w\nu^{1/3}) = \frac{\nu^{1/3}}{2\pi i} \int_\Gamma e^{\nu f(z) + \frac{w\nu^{1/3}}{2}(z - z^{-1})} \frac{dz}{z}. \tag{9.108}$$

As before, we modify $\Gamma$ to $\Gamma'$ of finite length as follows. Since $\mathrm{Re}\, f(x+iy) = \frac{x}{2}\left(1 - \frac{1}{x^2+y^2}\right) - \frac{1}{2}\log(x^2+y^2)$, it is easy to check that for $x \le -2$ and for all $y$, $\mathrm{Re}\, f(x+iy)$ is an increasing function of $x$ and $\mathrm{Re}\, f(x+iy) \le -\frac{4}{3}$. Define $\Gamma'$ to be the union of the part of $\Gamma$ satisfying $\mathrm{Re}\, z \ge -2$ and the segment of $\mathrm{Re}\, z = -2$ which closes the contour (see Figure 9.11). Orient

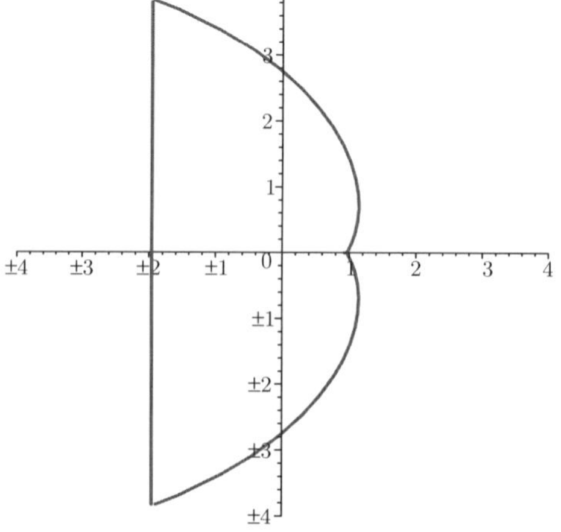


**Figure 9.11.** The curve $\Gamma'$

$\Gamma'$ counter-clockwise. The integral (9.108) is unchanged if we deform the contour to this new one. Note that as $\Gamma$ is a curve of steepest descent, and as $\operatorname{Re} f(x+iy) \le -\frac{4}{3} < 0$ for $x \le -2$,

$$\operatorname{Re}(f(z)) \le \operatorname{Re}(f(1)) = 0, \qquad z \in \Gamma'. \tag{9.109}$$

Note also that $\Gamma'$ is determined by the function (9.100) and does not depend on any parameters (such as $\nu$ and $w$).

Since $\Gamma'$ intersects $z = 1$ at angles $\pi/3$ and $-\pi/3$, there is a constant $\epsilon_1 \in (0,1)$ such that the part of $\Gamma'$ inside the (closed) disk $D(1,\epsilon_1)$ of radius $\epsilon_1$ centered at the critical point $z = 1$, lies in the sectors

$$\begin{aligned} &\frac{3\pi}{12} < \arg(z-1) < \frac{5\pi}{12}, && z \in \Gamma' \cap D(1,\epsilon_1) \cap \mathbb{C}_+, \\ &\frac{-5\pi}{12} < \arg(z-1) < \frac{-3\pi}{12}, && z \in \Gamma' \cap D(1,\epsilon_1) \cap \mathbb{C}_-. \end{aligned} \tag{9.110}$$

From the Taylor series for $f(z)$, we have

$$f(z) = \frac{1}{6}(z-1)^3 + O((z-1)^4). \tag{9.111}$$

Since $f^{(4)}(1) = -6$, there is $0 < \epsilon_2 \le \min\{\epsilon_1, \frac{1}{4}\}$ such that the $O$-term satisfies that

$$|O((z-1)^4)| \le \frac{1}{3}|z-1|^4, \qquad |z-1| \le \epsilon_2. \tag{9.112}$$

Recall that $\operatorname{Im} f(z) = \operatorname{Im} f(1) = 0$ for $z \in \Gamma$, and $\operatorname{Re} f(z) \le \operatorname{Re} f(1) = 0$. Hence for $z \in \Gamma' \cap D(1,\epsilon_2)$, $f(z)$ is real and negative. Note again that if $a, b \in \mathbb{C}$ satisfies that $a + b$ is real and $a + b \le 0$, then $a + b \le -|a| + |b|$. Therefore, for $z \in \Gamma' \cap D(1,\epsilon_2)$, we have

$$\begin{aligned} \operatorname{Re} f(z) = f(z) &= \frac{1}{6}(z-1)^3 + O((z-1)^4) \\ &\le -\frac{1}{6}|z-1|^3 + |O((z-1)^4)| \\ &\le -\frac{1}{6}|z-1|^3 + \frac{1}{3}|z-1|^4 \le -\frac{1}{12}|z-1|^3. \end{aligned} \tag{9.113}$$

Let $C_1$ be the part of $\Gamma'$ that lies outside the disk $D(1,\epsilon_2)$. Let $z_*$ denote the point on $\Gamma' \cap \mathbb{C}_+$ that intersects the circle $\partial D(1,\epsilon_2)$ (see Figure 9.12). Hence, since $\Gamma$ is a curve of steepest descent, we find

$$\operatorname{Re}(f(z)) \le \operatorname{Re}(f(z_*)) = Re(f(\overline{z^*})) \le -\frac{1}{12}|z_* - 1|^3 = -\frac{1}{12}\epsilon_2^3, \tag{9.114}$$

for $z \in C_1 \cap \Gamma$. As $\operatorname{Re} f(x+iy) \le -\frac{4}{3}$ for $x \le -2$, the inequality (9.114) is trivially true for $z \in C_1 \setminus \Gamma$. Since $C_1$ is bounded and away from 0, we

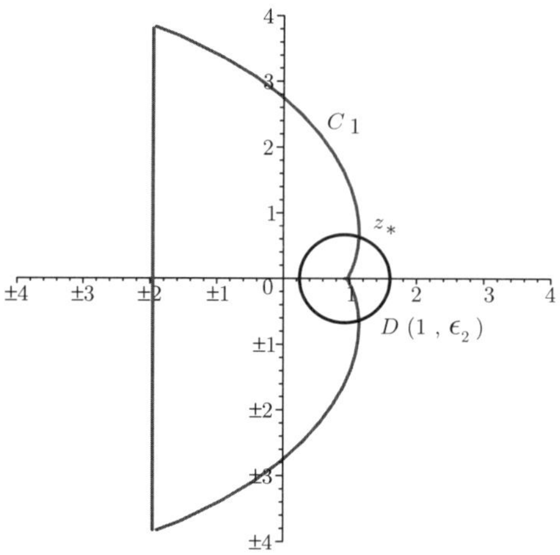


**Figure 9.12.** Contour $C_1$ and point $z_*$

have $|\frac{1}{2}(z - z^{-1})| \le c$ for some constant $c > 0$. Therefore, for some constant $c' > 0$,

$$\left| \frac{\nu^{1/3}}{2\pi i} \int_{C_1} e^{\nu f(z) + \frac{w\nu^{1/3}}{2}(z - z^{-1})} \frac{dz}{z} \right| \le \nu^{1/3} c' e^{-\frac{1}{12}\nu\epsilon_2^3 + c|w|\nu^{1/3}} \to 0 \tag{9.115}$$

as $\nu \to \infty$, uniformly for $w$ in a compact set.

The part of the integral (9.108) over $\Gamma' \setminus C_1$ is unchanged if we deform the contour $\Gamma' \setminus C_1$ to be the union of the straight line from 1 to $z_*$ and the straight line from $\overline{z_*}$ to 1. Call this new contour $C_2$. Fix a number $\kappa$ such that

$$1/4 < \kappa < 1/3. \tag{9.116}$$

We can assume that $\nu$ is large enough so that $\nu^{-\kappa} < \epsilon_2$. Let $C_2 = C_2^1 \cup C_2^{2,+} \cup C_2^{2,-}$ where $C_2^1 = C_2 \cap D(1, \nu^{-\kappa})$, $C_2^{2,+} = C_2 \cap D(1, \nu^{-\kappa})^c \cap \mathbb{C}_+$, and $C_2^{2,-} = \overline{C_2^{2,+}}$; see Figure 9.13. We consider the integrals over these three parts separately. Note that (9.113) holds for all $z \in C_2$.

For $z \in C_2^{2,+}$ and $|z| > 1$ there is a constant $c > 0$ such that $\mathrm{Re}(\frac{1}{2}(z - z^{-1})) \le c|z - 1|$. Hence, using (9.113),

$$\begin{aligned} &\left| \frac{\nu^{1/3}}{2\pi i} \int_{C_2^{2,+}} e^{\nu f(z) + \frac{w\nu^{1/3}}{2}(z - z^{-1})} \frac{dz}{z} \right| \\ &\le \frac{\nu^{1/3}}{2\pi} \int_{C_2^{2,+}} e^{-\frac{1}{12}\nu|z-1|^3 + c|w|\nu^{1/3}|z-1|} |dz| \\ &= \frac{1}{2\pi} \int_{\nu^{1/3-\kappa}}^{\nu^{1/3}\epsilon_2} e^{-\frac{1}{12}r^3 + c|w|r} dr \to 0 \end{aligned} \tag{9.117}$$

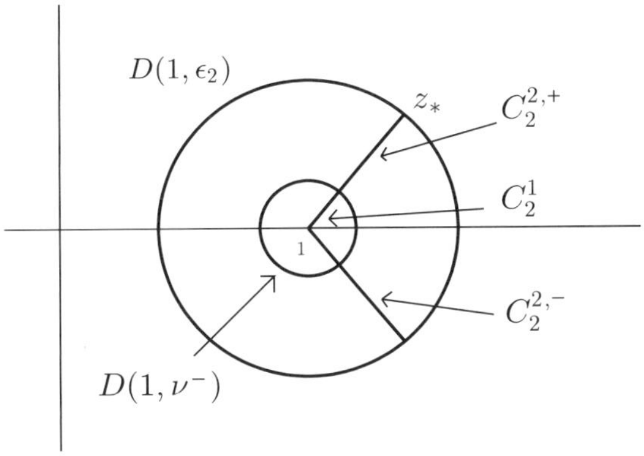


**Figure 9.13.** Contour $C_2 = C_2^1 \cup C_2^{2,+} \cup C_2^{2,-}$

as $\nu \to \infty$, uniformly for $w$ in a compact set. We have the same estimate for the integral over $C_2^{2,-}$.

The integral over $C_2^1$ is the part that gives the Airy asymptotics. Using (9.111) and (9.103), we see that when $\nu$ is large enough,

(9.118)
$$\begin{aligned}
&\frac{\nu^{1/3}}{2\pi i}\int_{C_2^1} e^{\nu f(z)+\frac{w\nu^{1/3}}{2}(z-z^{-1})}\frac{dz}{z} \\
&= \frac{\nu^{1/3}}{2\pi i}\int_{C_2^1} e^{\frac{\nu}{6}(z-1)^3+w\nu^{1/3}(z-1)+O(\nu|z-1|^4)+O(|w|\nu^{1/3}|z-1|^2)}\big(1+O(|z-1|)\big)dz \\
&= \frac{\nu^{1/3}}{2\pi i}\int_{C_2^1} e^{\frac{\nu}{6}(z-1)^3+w\nu^{1/3}(z-1)}\big(1+O(\nu^{-\kappa})+O(\nu^{1-4\kappa})+O(\nu^{1/3-2\kappa})\big)dz
\end{aligned}$$

since $|z-1|\le \nu^{-\kappa}$ for $z\in C_2^1$ and hence all $O$-terms go to zero as $\nu\to\infty$. Using the change of variables $2^{-1/3}\nu^{1/3}(z-1) = -is$,

$$\frac{\nu^{1/3}}{2\pi i}\int_{C_2^1} e^{\frac{\nu}{6}(z-1)^3+w\nu^{1/3}(z-1)}dz = \frac{-2^{1/3}}{2\pi}\int_{C'} e^{i(\frac{1}{3}s^3-2^{1/3}ws)}ds, \tag{9.119}$$

where $C'$ is the image of the contour $C_2^1$ under the change of variables. Note that $C'$ is a contour of finite length, $\nu^{1/3-\kappa}$, which goes to infinity as $\nu\to\infty$. Also due to (9.110), $C'$ lies in the sector $\frac{1}{12}\pi < \arg(s) < \frac{3}{12}\pi$ for $\mathrm{Re}(s)>0$, and $\frac{9}{12}\pi < \arg(s) < \frac{11}{12}\pi$ for $\mathrm{Re}(s)<0$, in which $\mathrm{Re}(is^3)<0$. Hence uniformly for $w$ in a compact set,

$$\begin{aligned}
\lim_{\nu\to\infty}\frac{-2^{1/3}}{2\pi}\int_{C'} e^{i(\frac{1}{3}s^3-2^{1/3}ws)}ds &= \frac{-2^{1/3}}{2\pi}\int_{C''} e^{i(\frac{1}{3}s^3-2^{1/3}ws)}ds \\
&= 2^{1/3}\,\mathrm{Ai}(-2^{1/3}w),
\end{aligned} \tag{9.120}$$

where $C''$ is the contour which is obtained by extending $C'$ to infinity and is oriented from the right to the left. Hence combined with (9.118), we obtain

$$\lim_{\nu\to\infty}\frac{\nu^{1/3}}{2\pi i}\int_{C_2^1} e^{\nu f(z)+\frac{w\nu^{1/3}}{2}(z-z^{-1})}\frac{dz}{z} = 2^{1/3}\,\mathrm{Ai}(-2^{1/3}w) \tag{9.121}$$

uniformly for $w$ in compact sets. Here we use the fact that $\left|e^{i\left(\frac{1}{3}s^3-2^{\frac{1}{3}}ws\right)}\right|$ is integrable on $C'$ and $O(\nu^{-k}), O(\nu^{1-4\kappa}), O(\nu^{\frac{1}{3}-2\kappa}) \to 0$ uniformly on $C'$ as $\nu\to\infty$. The estimates (9.115), (9.117), and (9.121) now imply (9.95). □

The above proof does not give an explicit bound, but as we will see, it is sufficient for our purposes. A more detailed analysis yields (see [**AS64**], 9.3.4)

$$2^{-1/3}\nu^{1/3}J_\nu(\nu+w\nu^{1/3}) = \mathrm{Ai}(-2^{1/3}w)+O(\nu^{-2/3}), \tag{9.122}$$

where the error is uniform for $w$ in compact sets.

We also need a uniform estimate for negative $w$ in an interval which grows with $\nu$.

**Lemma 9.17.** *For* $-\nu^{2/3} < w < 0$,

$$|2^{-1/3}\nu^{1/3}J_\nu(\nu+\nu^{1/3}w)| \le \frac{\sqrt{\pi}e^{-\frac{1}{3}(-w)^{3/2}}}{2^{11/6}(-w)^{1/4}}. \tag{9.123}$$

**Proof.** We use Lemma 9.1 when $s = 1+\nu^{-2/3}w$. Note that $s\in(0,1)$ due to the condition on $w$. It is straightforward to check that

$$\sqrt{1-s^2}-\log\left(\frac{1+\sqrt{1-s^2}}{s}\right) \le -\frac{1}{3}(1-s^2)^{3/2} \tag{9.124}$$

for $0 < s \le 1$. As $1-s^2 \ge 1-s$ for $0\le s\le 1$, we have from (9.10),

$$|J_\nu(\nu s)| \le \frac{\sqrt{\pi}e^{-\frac{1}{3}\nu(1-s)^{3/2}}}{2\sqrt{2\nu}(1-s)^{1/4}}, \qquad 0<s<1. \tag{9.125}$$

Setting $s = 1+\nu^{-2/3}w$ we obtain (9.123). □

Proposition 9.16 implies the following lemma.

**Lemma 9.18.** *For* $w$ *in a compact set,* $|w|\le w_0<\infty$, *and* $\nu>0$,

$$\left|\nu^{\frac{1}{3}}J_\nu(\nu+\nu^{\frac{1}{3}}w)\right| \le c < \infty. \tag{9.126}$$

Combining Lemmas 9.17 and 9.18, we obtain Corollary 9.19.

**Corollary 9.19.** *Let $w_0$ be a fixed positive number. There exists a constant $c > 0$ such that*

$$\left|\nu^{1/3} J_\nu(\nu + \nu^{1/3} w)\right| \le c \frac{e^{-\frac{1}{3}|w|^{3/2}}}{(1+|w|)^{1/4}} \tag{9.127}$$

*for $-\nu^{2/3} < w \le w_0 < \infty$ and $\nu \ge 0$.*

**9.4.2. Central Limit Theorem.** We now prove Theorems 9.14 and 9.15.

**Proof of Theorem 9.14.** We scale the operator in (9.94) so that the new operator acts on $L^2((x,\infty))$ which does not depend on $t$. More precisely, set $\delta' = \delta'_{x,t} \equiv 2t + xt^{1/3} - [2t + xt^{1/3}] \in [0,1)$. Define $\mathbf{K}_t$ to be the operator on $L^2((x,\infty))$ with kernel

$$\mathbf{K}_t(u,v) = t^{1/3} \tilde{\mathcal{K}}_{[2t+xt^{1/3}]}(2t + ut^{1/3} - \delta', 2t + vt^{1/3} - \delta'). \tag{9.128}$$

Then the map $f(z) \to (Vf)(u) = t^{\frac{1}{6}} f(2t + ut^{\frac{1}{3}} - \delta')$ is a unitary map from $L^2([2t + xt^{\frac{1}{3}}], \infty)$ onto $L^2(x,\infty)$ and $V\tilde{\mathcal{K}}_{[2t+xt^{\frac{1}{3}}]} = \mathbf{K}_t V$. Hence,

$$\det(1 - \tilde{\mathcal{K}}_{[2t+xt^{1/3}]}) = \det(1 - \mathbf{K}_t). \tag{9.129}$$

From (9.92), the explicit form of the kernel $\mathbf{K}_t$ is, for $u, v > x$,

$$\begin{aligned} &\mathbf{K}_t(u,v) \\ &= t^{1/3} \int_0^\infty J_{\lceil 2t+ut^{1/3}-\delta'\rceil + \lceil s\rceil - 1}(2t) J_{\lceil 2t+vt^{1/3}-\delta'\rceil + \lceil s\rceil - 1}(2t) ds \\ &= t^{2/3} \int_0^\infty J_{\lceil 2t+ut^{1/3}-\delta'\rceil + \lceil st^{1/3}\rceil - 1}(2t) J_{\lceil 2t+vt^{1/3}-\delta'\rceil + \lceil st^{1/3}\rceil - 1}(2t) ds. \end{aligned} \tag{9.130}$$

From (9.93)

$$\mathbb{P}\left\{ \frac{L(t) - 2t}{t^{1/3}} \le x \right\} = \det(1 - \mathbf{K}_t). \tag{9.131}$$

The kernel of the operator $\mathbf{K}_t$ on $L^2((x,\infty))$ can be written as

$$\mathbf{K}_t(u,v) = \int_0^\infty \mathbf{H}_t(u,s)\, \mathbf{H}_t(v,s) ds, \tag{9.132}$$

where

$$\begin{aligned} \mathbf{H}_t(u,s) &= t^{1/3} J_{\lceil 2t+ut^{1/3}-\delta'\rceil + \lceil st^{1/3}\rceil - 1}(2t) \\ &= t^{1/3} J_{2t+(u+s)t^{1/3}+\delta}(2t). \end{aligned} \tag{9.133}$$

Here $\delta = \delta(u,s,x,t) \in [-2,1]$. Hence $\mathbf{K}_t = \mathbf{H}_t \mathbf{H}'_t$, where $\mathbf{H}'_t$ denotes the adjoint operator and $\mathbf{H}_t : L^2((0,\infty)) \to L^2((x,\infty))$ is the Hilbert-Schmidt operator defined by the kernel $\mathbf{H}_t(u,s)$. We need to show that $\mathbf{H}_t$ converges in the Hilbert-Schmidt norm as $t \to \infty$.

Now

$$\mathbf{H}_t(u,s) = t^{\frac{1}{3}} J_{2t+(u+s)t^{1/3}+\delta}(2t) = \left(\frac{t}{\nu}\right)^{1/3} \nu^{1/3} J_\nu(\nu + \nu^{1/3} w), \tag{9.134}$$

where $\nu$ and $w$ are defined through the relations

$$\nu = 2t + (u+s)t^{1/3} + \delta \tag{9.135}$$

and

$$\nu + w\nu^{1/3} = 2t. \tag{9.136}$$

We have $u \geq x$ where $x \in \mathbb{R}$ is fixed, and $s \geq 0$. Thus, from (9.136) and (9.135), for $t$ sufficiently large

$$-w = \frac{\nu - 2t}{\nu^{1/3}} = \frac{(u+s)t^{1/3} + \delta}{(2t + (u+s)t^{1/3} + \delta)^{1/3}} \geq \frac{xt^{1/3} + \delta}{(2t + xt^{1/3} + \delta)^{1/3}}. \tag{9.137}$$

Hence, for $t$ sufficiently large, $w \leq 1 + |x|$ for all $u \geq x$ and $s \geq 0$. Also, as $t > 0$, we have $\nu + w\nu^{1/3} > 0$ from (9.136), and hence

$$-\nu^{2/3} < w \leq 1 + |x|. \tag{9.138}$$

For fixed $u$ and $s$, we see from (9.135) and (9.136) that as $t \to \infty$, $w$ remains in a compact set and $w \to -2^{-1/3}(u+s)$. Hence, by Proposition 9.16 and Vitali's theorem, we see that

$$\mathbf{H}_t(u,s) \to \mathrm{Ai}(u+s) \tag{9.139}$$

for each fixed $u \geq x$ and $s \geq 0$. On the other hand, as $-\nu^{2/3} < w \leq 1 + |x|$ and as $\nu \geq 2t + xt^{\frac{1}{3}} + \delta \to \infty$, we see from Corollary 9.19 and (9.134) that

$$|\,\mathbf{H}_t(u,s)| \leq c\frac{e^{-\frac{1}{3}|w|^{3/2}}}{(1+|w|)^{1/4}} \leq ce^{-\frac{1}{3}|w|^{3/2}} \tag{9.140}$$

for sufficiently large $t$, where $w$ is given by (9.137).

Now suppose $u + s \geq 2$. Then a simple calculation shows that $-w \geq 2^{-1/3}(u+s)^{2/3}$ for $t$ sufficiently large. Thus, for $u+s \geq 2$ we see from (9.140) that

$$|\,\mathbf{H}_t(u,s)| \leq ce^{-\frac{1}{3\sqrt{2}}(u+s)}. \tag{9.141}$$

If $x \geq 2$, then $u + s \geq x \geq 2$. On the other hand, if $x < 2$, then for $u+s \in [x, 2]$, $w$ is bounded, and hence from (9.140) we see that $|\,\mathbf{H}_t(u,s)| \leq c$ for $t$ sufficiently large. Combining these two estimates, we find that there exists a constant $C > 0$ such that

$$|\,\mathbf{H}_t(u,s)| \leq Ce^{-\frac{1}{3\sqrt{2}}(u+s)} \tag{9.142}$$

for all $u \geq x, s \geq 0$; hence, $\mathbf{H}_t(u,s)$ is bounded uniformly as $t \to \infty$ by an integrable function on $(x,\infty) \times (0,\infty)$. The function $|\operatorname{Ai}(u+s)|^2$ is also an integrable function, so by the dominated convergence theorem, we have

$$\int_x^\infty \int_0^\infty |\mathbf{H}_t(u,s) - \operatorname{Ai}(u+s)|^2 ds du \to 0. \tag{9.143}$$

Hence $\mathbf{H}_t$ converges to the operator with kernel $\operatorname{Ai}(u+s)$ from $L^2(0,\infty)$ to $L^2(x,\infty)$ in the Hilbert-Schmidt norm. Therefore, $\mathbf{K}_x$ converges to the Airy operator in trace norm. This proves Theorem 9.14. □

Theorem 9.15 follows from Theorem 9.14 by the de-Poissonization lemma.

**Proof of Theorem 9.15.** From Lemma 2.7, for any $n \geq 0$, $a_N = \mathbb{P}\{\ell_N \leq n\}$ is a decreasing sequence in $N$. Fix $s > 1$, and let $\mu_N = N + 2\sqrt{sN \log N}$ and $\nu_N = N - 2\sqrt{sN \log N}$ be as in (2.18). From Corollary 2.5, we find

$$\begin{aligned} \mathbb{P}\{L(\sqrt{\mu_N}) \leq n\} - \frac{1}{N^s} &\leq \mathbb{P}\{\ell_N \leq n\} \\ &\leq \mathbb{P}\{L(\sqrt{\nu_N}) \leq n\} + \frac{1}{N^s}. \end{aligned} \tag{9.144}$$

As $N \to \infty$,

$$\begin{aligned} &\mathbb{P}\{L(\sqrt{\nu_N}) \leq 2\sqrt{N} + xN^{1/6}\} \\ &= \mathbb{P}\left\{ \frac{L(\sqrt{\nu_N}) - 2\sqrt{\nu_N}}{\nu_N^{1/6}} \leq \frac{2\sqrt{N} + xN^{1/6} - 2\sqrt{\nu_N}}{\nu_N^{1/6}} \right\} \to F(x) \end{aligned} \tag{9.145}$$

as $\frac{2\sqrt{N}+xN^{1/6}-2\sqrt{\nu_N}}{\nu_N^{1/6}} \to x$. Hence the upper inequality in (9.144) implies that

$$\limsup_{N\to\infty} \mathbb{P}\left\{ \frac{\ell_N - 2\sqrt{N}}{N^{1/6}} \leq x \right\} \leq F(x). \tag{9.146}$$

Similarly we obtain

$$F(x) \leq \liminf_{N\to\infty} \mathbb{P}\left\{ \frac{\ell_N - 2\sqrt{N}}{N^{1/6}} \leq x \right\}. \tag{9.147}$$

□

## 9.5. Uniform Tail Estimates and Convergence of Moments

In Sections 9.1 and 9.2 we considered the tail estimates of $L(t) - \mu t$ for $\mu \neq 2$. The tail estimates for the scaled random variable $\frac{L(t)-2t}{t^{1/3}}$ can also be obtained by following the same steps as in Sections 9.1 and 9.2. However, this requires considerably more detailed asymptotic analysis which we omit. We only state the results but discuss two consequences of the estimates. The first is the convergence of moments, in this section, and the second is the order of the transversal fluctuations, in the next section.

We have the following upper and lower tail estimates.

**Proposition 9.20.** *Let $t \geq 1$. There exist constants $C, c > 0$ such that*

$$\mathbb{P}\left\{\frac{L(t)-2t}{t^{1/3}} > x\right\} \leq Ce^{-cx^{3/2}} \tag{9.148}$$

*for $2 \leq x \leq 4t^{2/3}$.*

**Proposition 9.21.** *For a given $M > 0$, there are $\delta > 0$, $t_0 > 0$, $c_1 > 0$, and $c_2 \in \mathbb{R}$ such that for $t \geq t_0$ and for $x$ satisfying*

$$M \leq x \leq \delta t^{2/3}, \tag{9.149}$$

*we have*

$$\mathbb{P}\left\{\frac{L(t)-2t}{t^{1/3}} \leq -x\right\} \leq e^{-c_1|x|^{3/2}-c_2}. \tag{9.150}$$

As in Section 9.2, this lower tail bound is far from sharp. The bound can be improved to $e^{-c_1|x|^3-c_2}$. See [**Wid02**] for a proof of Proposition 9.21 following the method of Section 9.2, and see [**BDJ99**] for the proof of the improved bound by using Gessel's formula and the Riemann-Hilbert method. In addition, moderate deviation results analogous to Propositions 9.3 and 9.10 can be found in [**LM01**] and [**LMR02**].

The estimates (9.148), (9.150), (9.65), and the de-Poissonizaztion Lemmas 2.5 or 2.6 imply the following result for the convergence of moments (cf. proof of Theorem 9.12).

**Theorem 9.22** (Convergence of moments [**BDJ99**]). *Setting $F(x) = \det(1-\mathbb{A}_x)$, for any $m = 1, 2, 3, \ldots$, we have*

$$\lim_{t\to\infty} \mathbb{E}\left[\left(\frac{L(t)-2t}{t^{1/3}}\right)^m\right] = \int_{-\infty}^{\infty} x^m dF(x) \tag{9.151}$$

*and*

$$\lim_{N\to\infty} \mathbb{E}\left[\left(\frac{\ell_N-2\sqrt{N}}{N^{1/6}}\right)^m\right] = \int_{-\infty}^{\infty} x^m dF(x). \tag{9.152}$$

## 9.6. Transversal Fluctuations

In this section, we study the so-called transversal fluctuations of longest up/right paths in the Hammersley process. Consider a rate 1 Poisson process in $\mathbb{R}^2$, and denote the length of the longest up/right path from $(0,0)$ to $(t,t)$ by $L(t)$; more generally, given two points, $x_1, x_2 \in \mathbb{R}^2_+$, with $x_1$ below and to the left of $x_2$, and a realization $\omega$ of the Poisson process, we denote by $d(x_1, x_2; \omega)$ the length of the longest up/right path from $x_1$ to $x_2$ in the rectangle with opposite corners $x_1, x_2$ and sides parallel to the axes, corresponding to the realization $\omega$.

We have shown in Theorem 9.14 that *longitudinal* fluctuations—the fluctuations of $L(t)$—are of order $O(t^\chi)$ where $\chi = 1/3$. We now examine the geometry of typical maximal paths. In particular, we analyze the the magnitude of the fluctuations of maximal paths from the diagonal line $y = x$. The *transversal* fluctuation turns out to be of order $O(t^\xi)$ where $\xi = 2/3$.

The following argument and proof are based on the paper [**Joh00b**]; Johansson's argument in [**Joh00b**] is in turn an adaptation of the method of [**NP95, LNP96**] (see also [**Cha13**] and [**AD14**]).

We begin with a heuristic argument for the scaling of the transversal fluctuations. We know from Theorem 9.12 that asymptotically the expected length of a maximal path in a rectangle of area $A$ is $2\sqrt{A}$. Fix $0 < a < 1$ and let $\delta > 0$ be small. Consider the event in which a maximal path from $(0,0)$ to $(t,t)$ passes the point $(t(a+\delta), t(a-\delta))$, i.e., a small deviation from the diagonal point $(ta, ta)$. Then the length of the maximal path is approximately

$$2t\sqrt{a^2-\delta^2} + 2t\sqrt{(1-a)^2-\delta^2}. \tag{9.153}$$

Now the magnitude of the difference between $2t$, the expected length of $L(t)$, and (9.153) must be of the order of the longitudinal fluctuation, $t^\chi$, $\chi = \frac{1}{3}$. Hence

$$2t - \left(2t\sqrt{a^2-\delta^2} + 2t\sqrt{(1-a)^2-\delta^2}\right) = O(t^\chi), \tag{9.154}$$

and we find that $\delta^2 = O(a(1-a)t^{\chi-1})$. So, for a fixed $a \in (0,1)$, the transversal fluctuations are of order $t\delta = O(t^{\frac{1+\chi}{2}})$, hence we arrive at the relation for the transversal fluctuation exponent $\xi$

$$2\xi = \chi + 1. \tag{9.155}$$

From $\chi = 1/3$, it follows that $\xi = 2/3$.

We now define precisely the *transversal fluctuation exponent* and state the theorem that we will eventually prove. For $\gamma > 0$ and $t > 0$, define the cylinder

$$C(\gamma, t) = \{(x,y) \in \mathbb{R}^2 : 0 \le x + y \le 2t, -\sqrt{2}t^\gamma \le -x + y \le \sqrt{2}t^\gamma\} \tag{9.156}$$

of width $2t^\gamma$ and height $\sqrt{2}t$. Let $0 = (0,0)$ and $e_t = (t,t)$. These two points are the midpoints on the sides of length $2t^\gamma$ of the cylinder. Define $A_t^\gamma$ to be the event that all maximal paths from $0$ to $e_t = (t,t)$ are contained in the cylinder $C(\gamma, t)$.

**Theorem 9.23.** *Define the transversal fluctuation exponent for Hammersley's processes* $\xi$ *by*

$$\xi \equiv \inf\{\gamma > 0 : \lim_{t\to\infty} \mathbb{P}\{A_t^\gamma\} = 1\}. \tag{9.157}$$

*We have*

$$\xi = \frac{2}{3}. \tag{9.158}$$

We prove the theorem basically by making the above heuristic argument rigorous. In this argument, the uniform tail estimates obtained in the previous section play a key role. We repeatedly use estimates for $d(z_1, z_2)$, the longest path in a rectangle with opposite corners at $z_1$ and $z_2$ as above. Note that since the relative positions of points in a rectangle are invariant under horizontal and vertical scaling, the results in the previous section are not restricted to points in a square and can be applied to points in a rectangle. Hence Proposition 9.20, together with Proposition 9.2, implies that there exist constants $C, c > 0$ such that

$$\mathbb{P}\left\{\frac{d(z_1, z_2) - 2\sqrt{\mu(z_1, z_2)}}{\mu(z_1, z_2)^{1/6}} > x\right\} \le Ce^{-c\min\{x^{3/2},\, 2x\mu(z_1,z_2)^{\frac{1}{6}}\}} \tag{9.159}$$

for $x \ge 2$ and $\mu(z_1, z_2) > 0$. Here $\mu(z_1, z_2)$ denotes the area of the rectangle with corners at $z_1$ and $z_2$.

**Remark 9.24.** Note that in applying Proposition 9.2 to prove (9.159) in the case $x^{3/2} > 2x\mu^{\frac{1}{6}}$, i.e., $x > 4\mu^{\frac{1}{3}}$, where $\mu \equiv \mu(z_1, z_2)$, we need to verify the conditions $2 + x\mu^{-\frac{1}{3}} >$ constant $> 2$ and $\mu^{\frac{1}{2}} > ((2 + (x\mu^{-\frac{1}{3}})^2) - 4)^{-1/2}$, i.e., $4x\mu^{\frac{2}{3}} + x^2\mu^{\frac{1}{3}} > 1$. As $x\mu^{-\frac{1}{3}} > 4$, the first condition is clearly fulfilled, and if $x\mu^{\frac{1}{6}} > 1$, the second condition is also fulfilled, and we obtain the desired bound on the right-hand side of (9.159) from (9.18), (9.19). On the other hand, if $4\mu^{\frac{1}{3}} < x \le \mu^{-\frac{1}{6}}$, then $e^{-c\min\{x^{3/2},\, 2x\mu^{\frac{1}{6}}\}} = e^{-c2x\mu^{\frac{1}{6}}} > e^{-c}$, and hence (9.159) is trivially satisfied for $C > e^c$.

Similarly, Proposition 9.21, together with Proposition 9.9, implies that there exist constants $\delta > 0$ and $C, c > 0$ such that

$$\mathbb{P}\left\{\frac{d(z_1, z_2) - 2\sqrt{\mu(z_1, z_2)}}{\mu(z_1, z_2)^{1/6}} < -x\right\} \le Ce^{-c\min\{x^{3/2},\delta x\mu(z_1,z_2)^{\frac{1}{6}}\}} \tag{9.160}$$

for $x \ge 2$ and $\mu(z_1, z_2) > 0$.

The proof is divided into two parts: First we show that $\xi \le \frac{2}{3}$, and then we show that $\xi \ge \frac{2}{3}$. For the remainder of this section, $\Omega$ denotes the probability space of Poisson configurations $\omega$.

**Proof of $\xi \le \frac{2}{3}$.** The bound $\xi \le \frac{2}{3}$ is obtained if we show that for any $\gamma \in (\frac{2}{3}, 1)$,

$$\lim_{t\to\infty} \mathbb{P}\{\Omega \setminus A_t^\gamma\} = 0. \tag{9.161}$$

Fix $\frac{2}{3} < \gamma < 1$. If $\omega \in \Omega \setminus A_t^\gamma$, then there is a maximal path $\pi_0(\omega)$ not contained in $C(\gamma, t)$. Let $\pi_0^*(\omega)$ be the piecewise linear interpolation of $\pi_0(\omega)$. Then $\pi_0^*(\omega)$ intersects the upper side $AB$ or the low side $CD$ of the cylinder $C(\gamma, t)$ in the square $[0, t] \times [0, t]$ (see Figure 9.14).

Suppose that $\pi_0^*(\omega)$ intersects $AB$. We describe the points of intersection as follows. Divide the part of $AB$ that lies in $[0, t] \times [0, t]$ into $K$ small segments. Let

$$z_j \equiv \left(\frac{js}{K}, \frac{js}{K} + \sqrt{2}t^\gamma\right), \qquad j = 0, 1, \ldots, K, \tag{9.162}$$

where $s \equiv t - \sqrt{2}t^\gamma$, be points on $AB \cap ([0, t] \times [0, t])$. Set

$$K \equiv [2\sqrt{2}t^{1+\gamma}] + 1. \tag{9.163}$$

Let $D_j$, $j = 1, \ldots, K$, denote the parallelogram with corners $z_j$, $z_{j-1}$, $z_j - (0, 2\sqrt{2}t^\gamma)$, and $z_{j-1} - (0, 2\sqrt{2}t^\gamma)$ (note that the two latter points lie on the line through $CD$). For $\omega \in \Omega \setminus A_t^\gamma$, set

$$z(\omega) = z(\pi_0^*(\omega)) = z_{j-1}, \qquad 1 \le j \le K, \tag{9.164}$$

where $z_{j-1}z_j$ is the left-open line segment in which the path $\pi_0^*(\omega)$ intersects the upper side for the first time. If $\pi_0^*(\omega)$ intersects the upper side at $z_0$ (a zero probability event), we set $z(\omega) = z_0$. If $\pi_0^*(\omega)$ does not intersect the upper side, let $z'_{j-1}z'_j$, $z'_j = (\frac{js}{K} + \sqrt{2}t^\gamma, \frac{js}{K})$, $j = 0, 1, \ldots, K$, be the corresponding line segment in the lower side in which $\pi_0^*(\omega)$ intersects the

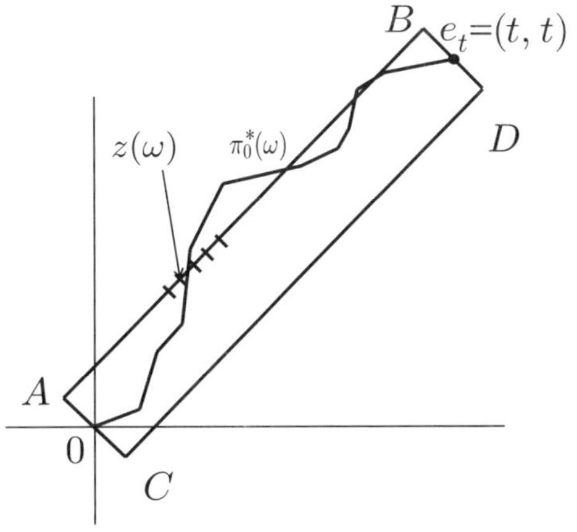


**Figure 9.14.** $C(\gamma, t)$ and $\pi_0^*$

lower side for the *last* time. In this case we set $z(\omega) = z(\pi_0^*(\omega)) = z'_{j-1}$, $1 \le j \le K$, and again $z(\omega) = z'_0$ if $\pi_0^*(\omega)$ intersects the lower side at $z'_0$.

From the definition of $d(x, y; \omega)$,

$$d(0, e_t; \omega) \le d(0, z(\omega); \omega) + d(z(\omega), e_t; \omega) + \max_{1\le j\le K} \#(D_j; \omega), \tag{9.165}$$

where $\#(Q)$ denotes the number of Poisson points in a region $Q$. The next lemma provides a useful estimate on the random variable $\max_{1\le j\le K} \#(D_j)$.

**Lemma 9.25.** *Set*

$$\Lambda_t \equiv \{w \in \Omega : \max_{1\le j\le K} \#(D_j; \omega) \le 3\log t\}. \tag{9.166}$$

*Then for an arbitrary $\epsilon > 0$, there is $t_0 > 0$ such that for $t \ge t_0$,*

$$\mathbb{P}\{\Lambda_t\} \ge 1 - \epsilon. \tag{9.167}$$

**Proof.** The area of $D_j$ is at most $\frac{2\sqrt{2}t^\gamma s}{K}$. Since the random variables $\#(D_j)$, $j = 1, \dots, K$, are independent,

$$\begin{aligned} \mathbb{P}\{\Omega \setminus \Lambda_t\} &= \mathbb{P}\left\{\max_{1\le j\le K} \#(D_j) > 3\log t\right\} \\ &\le K\,\mathbb{P}\{X > 3\log t\} \le K\frac{\mathbb{E}[e^X]}{t^3}, \end{aligned} \tag{9.168}$$

where $X$ is a Poisson random variable with parameter $\lambda \equiv \frac{2\sqrt{2}t^\gamma s}{K} \le 1$ (see (9.163)). Hence $\mathbb{E}e^X = e^{\lambda(e-1)} \le e^2$, and we conclude that

$$\mathbb{P}(\Omega \setminus \Lambda_t) \le K\frac{e^2}{t^3} \to 0 \tag{9.169}$$

as $t \to \infty$. This concludes the proof of the lemma. □

Let $\Delta \equiv \{z_0, \dots, z_K\}$ and let $\Delta' \equiv \{z'_0, \dots, z'_K\}$. Recall that for $x_1, x_2 \in \mathbb{R}^2$, $\mu(x_1, x_2)$ denotes the area of the rectangle with opposite corners at $x_1$ and $x_2$.

**Lemma 9.26.** *For $\frac{2}{3} < \gamma < 1$, fix $\delta$ such that $\frac{1}{3} < \delta < 2\gamma - 1$. For each $z \in \Delta \cup \Delta'$, set*

$$\begin{aligned} E_z \equiv \{\omega \in \Omega :\ & d(0, z; \omega) \le 2\sqrt{\mu(0, z)} + \mu(0, z)^{\delta/2} + t^\delta \\ & \textit{and} \quad d(z, e_t; \omega) \le 2\sqrt{\mu(z, e_t)} + \mu(z, e_t)^{\delta/2} + t^\delta\}. \end{aligned} \tag{9.170}$$

*Then for an arbitrary $\epsilon > 0$, there is $t_0$ such that for $t \ge t_0$,*

$$\mathbb{P}\left\{\bigcup_{z\in\Delta\cup\Delta'} (\Omega \setminus E_z)\right\} \le \epsilon. \tag{9.171}$$

**Proof.** Note that the scaling properties $d(x, y)$ and $d(x', y')$ have the same distribution as long as the rectangle with corners at $x$ and $y$ and the rectangle with corners at $x'$ and $y'$ have the same area. From this, we can use the upper tail estimate (9.159) to find that $\mathbb{P}(\Omega \setminus E_z)$ is bounded by a term that decays more quickly than any polynomial in $K$. Since there are only $2K$ points in $\Delta \cup \Delta'$, we obtain the result. □

Set

$$B_t^\gamma \equiv \Lambda_t \cap \left( \bigcap_{z \in \Delta \cup \Delta'} E_z \right). \tag{9.172}$$

Fix $\epsilon > 0$. From Lemmas 9.25 and 9.26, there is a $t_0$ such that $\mathbb{P}\{(B_t^\gamma)^c\} \le 2\epsilon$ for $t \ge t_0$. This implies that for $t \ge t_0$

$$\mathbb{P}\{\Omega \setminus A_t^\gamma\} \le \mathbb{P}\{(\Omega \setminus A_t^\gamma) \cap B_t^\gamma\} + 2\epsilon. \tag{9.173}$$

We now estimate $\mathbb{P}\{(\Omega \setminus A_t^\gamma) \cap B_t^\gamma\}$. For $\omega \in (\Omega \setminus A_t^\gamma) \cap B_t^\gamma$, from (9.165), (9.166), and (9.170), we obtain

$$\begin{aligned} d(0, e_t; \omega) \le & 2\sqrt{\mu(0, z(\omega); \omega)} + 2\sqrt{\mu(z(\omega), e_t; \omega)} \\ & + \mu(0, z(\omega); \omega)^{\delta/2} + \mu(z(\omega), e_t; \omega)^{\delta/2} + 2t^\delta + 3\log t. \end{aligned} \tag{9.174}$$

Since $\mu(0, z) \le t^2$ and $\mu(z, e_t) \le t^2$ for all $z \in \Delta \cup \Delta'$, we find that

$$d(0, e_t; \omega) \le 2\sqrt{\mu(0, z(\omega); \omega)} + 2\sqrt{\mu(z(\omega), e_t; \omega)} + 4t^\delta + 3\log t. \tag{9.175}$$

We now use the following lemma.

**Lemma 9.27.** *There is $t_0 > 0$ such that for any $z \in \Delta \cup \Delta'$,*

$$\sqrt{\mu(0, z)} + \sqrt{\mu(z, e_t)} - \sqrt{\mu(0, e_t)} \le -t^{2\gamma - 1} \tag{9.176}$$

*for $t \ge t_0$.*

**Proof.** Fix $j = 1, \ldots, K$. Set $x = \frac{js}{Kt}$ and $y = \sqrt{2}t^{\gamma-1}$. There is $t_0 > 0$ such that $y < 1$ and $0 \le x \le 1 - y$ for $t \ge t_0$. Since $\mu(0, z_j) = t^2 x(x + y)$, $\mu(z_j, e_t) = t^2(1 - x)(1 - x - y)$ and $\mu(0, e_t) = t^2$,

$$\sqrt{\mu(0, z_j)} + \sqrt{\mu(z_j, e_t)} - \sqrt{\mu(0, e_t)} = t f(x, y), \tag{9.177}$$

where $f(x, y) = \sqrt{x(x + y)} + \sqrt{(1 - x)(1 - x - y)} - 1$. For each $y \in [0, 1]$, $f(x, y)$ has its maximum $\sqrt{1 - y^2} - 1$ at $x = \frac{1-y}{2}$. Since $\sqrt{1 - y^2} - 1 \le -\frac{1}{2}y^2$, we obtain the lemma for $z \in \Delta$. The same estimate holds for $z \in \Delta'$ by symmetry. □

Thus we have for $\omega \in (\Omega \setminus A_t^\gamma) \cap B_t^\gamma$,

$$d(0, e_t; \omega) \le 2\sqrt{\mu(0, e_t)} - 2t^{2\gamma - 1} + 4t^\delta + 3\log t \tag{9.178}$$

and hence
(9.179)
$$\mathbb{P}\{(\Omega \setminus A_t^\gamma) \cap B_t^\gamma\} \le \mathbb{P}\left\{d(0,e_t) - 2\sqrt{\mu(0,e_t)} \le -2t^{2\gamma-1} + 4t^\delta + 3\log t\right\}.$$

The lower tail estimate (9.160) implies that this probability tends to 0 as $t \to \infty$ since $0 < \delta < 2\gamma - 1$. Combined with (9.173), we find that there is $t_0 > 0$ such that

$$\mathbb{P}\{\Omega \setminus A_t^\gamma\} \le 3\epsilon \tag{9.180}$$

for all $t \ge t_0$. Therefore, we obtain (9.161) and hence $\xi \le \frac{2}{3}$. □

**Proof of $\xi \ge \frac{2}{3}$.** We have shown that that $\xi \le \frac{2}{3}$. Fix $\gamma$ such that $\xi < \gamma < 1$. If we show that necessarily $\gamma \ge \frac{2}{3}$, then $\xi \ge \frac{2}{3}$ is proven.

Let

$$\eta \equiv \frac{1}{3}F(0)(1 - F(0)) > 0, \tag{9.181}$$

where $F(x)$ is the Tracy-Widom distribution. From the definition of $\xi$, there is $t_0 > 0$ such that for all $t > t_0$

$$\mathbb{P}\{A_t^\gamma\} \ge 1 - \frac{1}{10}\eta. \tag{9.182}$$

If $\omega \in A_t^\gamma$, all maximal paths from 0 to $e_t$ are contained in $C_1 \equiv C(\gamma, t)$. Hence we have for all $t > t_0$

$$\mathbb{P}\left\{d^{C_1}(0,e_t) = d(0,e_t)\right\} \ge 1 - \frac{1}{10}\eta. \tag{9.183}$$

Here, $d^R(z,z')$ denotes length of the maximal path in the region $R$ from $z$ to $z'$. Since $d(0,e_t) \ge d^{C_1}(0,e_t)$ always holds, the previous inequality implies that for all $t > t_0$

$$\mathbb{P}\left\{d^{C_1}(0,e_t) \ge d(0,e_t)\right\} \ge 1 - \frac{1}{10}\eta. \tag{9.184}$$

We now estimate $d^{C_1}(0,e_t) - d(0,e_t)$ in a different way (see (9.201) below). Set

$$m_t \equiv 3t^\gamma\left(-\frac{1}{\sqrt{2}}, \frac{1}{\sqrt{2}}\right), \qquad C_2 \equiv C_1 + m_t. \tag{9.185}$$

Note that $C_1$ and $C_2$ do not overlap (see Figure 9.15). We write $d^{C_1}(0,e_t) - d(0,e_t) = (d^{C_1}(0,e_t) - d^{C_2}(m_t, m_t + e_t)) + (d^{C_2}(m_t, m_t+e_t) - d(0,e_t))$.

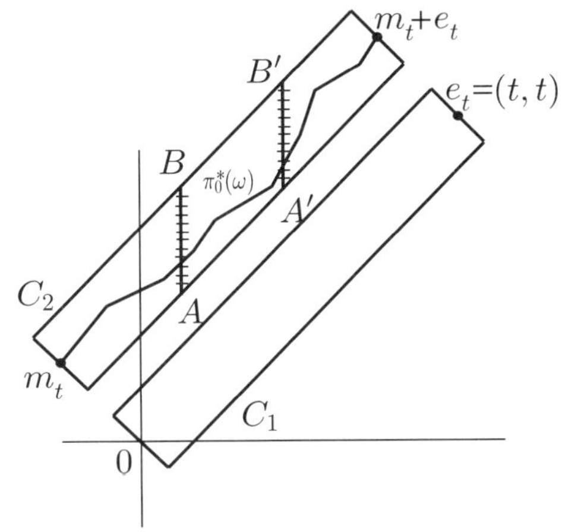


**Figure 9.15.** $C_1$ and $C_2$

**Lemma 9.28.** *Fix $\kappa_1 \in (0, \frac{1}{3})$. There is $t_0 > 0$ such that for $t \geq t_0$,*

$$\mathbb{P}\left\{d^{C_1}(0, e_t) - d^{C_2}(m_t, m_t + e_t) \leq -t^{\kappa_1}\right\} \geq \eta. \tag{9.186}$$

**Proof.** Since $C_1$ and $C_2$ do not overlap, $d_1 \equiv d^{C_1}(0, e_t)$ and $d_2 \equiv d^{C_2}(m_t, m_t + e_t)$ are independent and identically distributed random variables. Hence

$$\begin{aligned} &\mathbb{P}\{d_1 - d_2 \leq -t^{\kappa_1}\} \\ &\quad \geq \mathbb{P}\{d_1 - 2t \leq 0 \text{ and } d_2 - 2t \geq t^{\kappa_1}\} \\ &\quad = \mathbb{P}\{d_1 - 2t \leq 0\}\,\mathbb{P}\{d_2 - 2t \geq t^{\kappa_1}\} \\ &\quad \geq \mathbb{P}\{\{d_1 - 2t \leq 0\} \cap A_t^\gamma\}\,\mathbb{P}\{\{d_1 - 2t \geq t^{\kappa_1}\} \cap A_t^\gamma\}. \end{aligned} \tag{9.187}$$

Since $d^{C_1}(0, e_t) = d(0, e_t)$ for $\omega \in A_t^\gamma$, we have from the inclusion/exclusion principle that

$$\mathbb{P}\{\{d_1 - 2t \leq 0\} \cap A_t^\gamma\} \geq \mathbb{P}\{d(0, e_t) - 2t \leq 0\} + \mathbb{P}\{A_t^\gamma\} - 1 \tag{9.188}$$

and

$$\mathbb{P}\{\{d_1 - 2t \geq t^{\kappa_1}\} \cap A_t^\gamma\} \geq \mathbb{P}\{d(0, e_t) - 2t \geq t^{\kappa_1}\} + \mathbb{P}\{A_t^\gamma\} - 1. \tag{9.189}$$

As $t \to \infty$, $\mathbb{P}\{d(0, e_t) - 2t \leq 0\} \to F(0)$ and $\mathbb{P}\{d(0, e_t) - 2t \geq t^{\kappa_1}\} \to 1 - F(0)$ since $\kappa_1 < \frac{1}{3}$. Hence, there is $t_0 > 0$ such that for all $t \geq t_0$, (9.188) is larger than $(F(0) - \frac{1}{10}\eta) + (1 - \frac{1}{10}\eta) - 1$ and (9.189) is larger than $(1 - F(0) - \frac{1}{10}\eta) + (1 - \frac{1}{10}\eta) - 1$. Therefore, we obtain for $t > t_0$,

$$\mathbb{P}\{d_1 - d_2 \leq -t^{\kappa_1}\} > \big(F(0) - \frac{1}{5}\eta)\big)\big(1 - F(0) - \frac{1}{5}\eta\big) \geq \eta. \tag{9.190}$$

□

We now consider $d^{C_2}(m_t, m_t + e_t) - d(0, e_t)$. Fix $b$ such that $\gamma < b < 1$. We will assume that $t$ is large enough so that $t^b - 4t^\gamma > 0$. Set

$$A = \frac{1}{\sqrt{2}}(t^b, t^b + 4t^\gamma), \quad B = \frac{1}{\sqrt{2}}(t^b, t^b + 8t^\gamma). \tag{9.191}$$

These are two points on the boundary of $C_2$ with the same $x$-coordinate (see Figure 9.15). Since $b > \gamma$, the point $A$ is "not near the bottom point $m_t$". A maximal path $\pi_0^*(\omega)$ from $m_t$ to $m_t + e_t$ in $C_2$ crosses the line segment $AB$. We will study the intersection point. Divide the line segment $AB$ into $\tilde{K}$ intervals as follows. Let

$$z_j \equiv A + (0, \frac{2\sqrt{2}t^\gamma j}{\tilde{K}}), \qquad j = 0, 1, \ldots, \tilde{K}. \tag{9.192}$$

Hence $z_0 = A$ and $z_{\tilde{K}} = B$. Let $F_j$ be the parallelogram obtained by the intersection of $C_2$ and the horizontal lines passing $z_j$ and $z_{j-1}$. (Note that these horizontal lines cross the boundary of $C_2$ above the corner point $(-2\sqrt{2}t^\gamma, 2\sqrt{2}t^\gamma)$.) Given $\omega \in \Omega$, the maximal path $\pi_0^*(\omega)$ crosses one of the line segments $z_{j-1}z_j$. Define

$$z(\omega) = z(\pi_0^*(\omega)) = z_j. \tag{9.193}$$

(If the path crosses at some $z_j$, then set $z(\pi_0^*(\omega)) = z_j$.) Let the points $A'$, $B'$ and $z_j'$ be the translation of the points $A$, $B$ and $z_j$ by the vector $(\sqrt{2}t - 6t^\gamma - 2t^b)(\frac{1}{\sqrt{2}}, \frac{1}{\sqrt{2}})$. The parallelogram $F_j'$ is defined in a similar way.

For $\omega \in \Omega$, similarly define $z'(\omega) = z_j'$, where the maximal path in $C_2$ crosses the segment $z_{j-1}'z_j'$. Then

$$\begin{aligned} d^{C_2}(m_t, m_t + e_t; \omega) \leq d^{C_2}(m_t, z(\omega); \omega) + d^{C_2}(z(\omega), z'(\omega); \omega) \\ + d^{C_2}(z'(\omega), m_t + e_t; \omega) + D_t(\omega), \end{aligned} \tag{9.194}$$

where

$$D_t(\omega) = \max_{1 \leq j \leq \tilde{K}} \#(F_j; \omega) + \max_{1 \leq j \leq \tilde{K}} \#(F_j'; \omega). \tag{9.195}$$

For $\omega \in \Omega$,

$$d(0, e_t) \geq d(0, z(\omega)) + d(z(\omega), z'(\omega)) + d(z'(\omega), e_t). \tag{9.196}$$

Subtracting (9.196) from (9.194) and using that $d^{C_2}(a, b) \leq d(a, b)$ for any points $a$ and $b$, we have

$$\begin{aligned} d^{C_2}(m_t, m_t + e_t) - d(0, e_t) \leq D_t(\omega) + d^{C_2}(m_t, z(\omega)) - d(0, z(\omega)) \\ + d^{C_2}(z'(\omega), m_t + e_t) - d(z'(\omega), e_t). \end{aligned} \tag{9.197}$$

Using $d^{C_2}(a,b) \le d(a,b)$ for any $a, b$ one more time, we find

$$(9.198)\quad \begin{aligned} d^{C_2}(m_t, m_t + e_t) - d(0, e_t) \le D_t(\omega) + d(m_t, z(\omega)) - d(0, z(\omega)) \\ + d(z'(\omega), m_t + e_t) - d(z'(\omega), e_t). \end{aligned}$$

We have the following estimates.

**Lemma 9.29.** *Let $\tilde{K} = [8t^{2\gamma}] + 1$. Fix $\kappa_2 \in (\frac{1}{3}, 1)$. There is $t_0 > 0$ such that for $t \ge t_0$,*

(9.199)

$$\mathbb{P}\left\{|d(0, z(\omega); \omega) - 2\sqrt{\mu(0, z(\omega))}| \le t^{b\kappa_2}\right\} \ge 1 - \frac{1}{10}\eta,$$

$$\mathbb{P}\left\{|d(m_t, z(\omega); \omega) - 2\sqrt{\mu(m_t, z(\omega))}| \le t^{b\kappa_2}\right\} \ge 1 - \frac{1}{10}\eta,$$

$$\mathbb{P}\left\{|d(z'(\omega), e_t; \omega) - 2\sqrt{\mu(z'(\omega)), e_t)}| \le t^{b\kappa_2}\right\} \ge 1 - \frac{1}{10}\eta,$$

$$\mathbb{P}\left\{|d(z'(\omega), m_t + e_t; \omega) - 2\sqrt{\mu(z'(\omega), m_t + e_t)}| \le t^{b\kappa_2}\right\} \ge 1 - \frac{1}{10}\eta.$$

**Proof.** We have

$$(9.200)\quad \begin{aligned} &\mathbb{P}\left\{|d(0, z(\omega); \omega) - 2\sqrt{\mu(0, z(\omega))}| \le t^{b\kappa_2}\right\} \\ &\ge \mathbb{P}\Big\{\bigcap_{j=1}^{\tilde{K}} \{|d(0, z_j) - 2\sqrt{\mu(0, z_j)}| \le t^{b\kappa_2}\}\Big\} \\ &\ge 1 - \sum_{j=1}^{\tilde{K}} \mathbb{P}\left\{|d(0, z_j) - 2\sqrt{\mu(0, z_j)}| > t^{b\kappa_2}\right\}. \end{aligned}$$

Note that $\frac{1}{2}t^b(t^b + 4t^\gamma) \le \mu(0, z_j) \le \frac{1}{2}t^b(t^b + 8t^\gamma)$. Since $\kappa_2 > \frac{1}{3}$, the exponential tail estimates (9.159) and (9.160) imply that the sum tends to 0 as $t \to \infty$ since $\tilde{K}$ grows as a polynomial in $t$.

Since $\frac{1}{2}(t^b + 3t^\gamma)(t^b + t^\gamma) \le \mu(m_t, z_j) \le \frac{1}{2}(t^b + 3t^\gamma)(t^b + 5t^\gamma)$, we have a similar estimate for $d(m_t, z(\omega); \omega) - 2\sqrt{\mu(m_t, z(\omega))}$. The proofs of the remaining two estimates are analogous. □

Hence, using the inequality $\mathbb{P}\{A \cap B\} \ge \mathbb{P}\{A\} + \mathbb{P}\{B\} - 1$ for any events $A, B$ repeatedly, we find that

(9.201)

$$\begin{aligned} d^{C_2}(m_t, m_t + e_t) - d(0, e_t) \le D_t(\omega) + 2\sqrt{\mu(m_t, z(\omega))} - 2\sqrt{\mu(0, z(\omega))} \\ + 2\sqrt{\mu(z'(\omega), m_t + e_t)} - 2\sqrt{\mu(z'(\omega)), e_t)} + 4t^{b\kappa_2} \end{aligned}$$

holds with probability at least $1 - \frac{4}{10}\eta$. Set $\Delta \equiv \{z_1, \dots, z_{\tilde{K}}\}$ and $\Delta' \equiv \{z'_1, \dots, z'_{\tilde{K}}\}$. The following lemma is simple calculus.

**Lemma 9.30.** *There is $t_0 > 0$ such that for $t \geq t_0$,*

$$\begin{aligned} &\sqrt{\mu(m_t, z)} - \sqrt{\mu(0, z)} \leq 10t^{2\gamma - b}, \qquad z \in \Delta, \\ &\sqrt{\mu(z', m_t + e_t)} - \sqrt{\mu(z', e_t)} \leq 10t^{2\gamma - b}, \qquad z' \in \Delta'. \end{aligned} \tag{9.202}$$

**Proof.** Let $z = z_j \in \Delta$. Since

$$\mu(0, z_j) = \frac{1}{2} t^b (t^b + 4t^\gamma (1 + \frac{j}{\tilde{K}})) \tag{9.203}$$

and

$$\mu(m_t, z_j) = \frac{1}{2}(t^b + 3t^\gamma)(t^b + 4t^\gamma(1 + \frac{j}{\tilde{K}}) - 3t^\gamma), \tag{9.204}$$

we have

$$\begin{aligned} \sqrt{\mu(m_t, z_j)} - \sqrt{\mu(0, z_j)} &= \frac{\mu(m_t, z_j) - \mu(0, z_j)}{\sqrt{\mu(m_t, z_j)} + \sqrt{\mu(0, z_j)}} \\ &\leq \frac{3t^{2\gamma}(1 + \frac{4j}{\tilde{K}})}{2\sqrt{2}t^b}, \end{aligned} \tag{9.205}$$

which is less than $10t^{2\gamma - b}$ since $j \leq \tilde{K}$. The other inequality is similar. □

Therefore, we find that

$$d^{C_2}(m_t, m_t + e_t) - d(0, e_t) \leq D_t(\omega) + 40t^{2\gamma - b} + 4t^{b\kappa_2} \tag{9.206}$$

holds with probability at least $1 - \frac{4}{10}\eta$. Finally, we estimate a probability involving $D_t$.

**Lemma 9.31.** *Let $\tilde{K} = [8t^{2\gamma}] + 1$. Then there is $t_1 > 0$ such that for $t \geq t_1$,*

$$\mathbb{P}\{D_t \leq 5 \log t\} \geq 1 - \frac{1}{10}\eta. \tag{9.207}$$

**Proof.** The parallelogram $F_j$ has base of length $2\sqrt{2}t^\gamma$ and height of length $2\sqrt{2}t^\gamma / \tilde{K}$. Hence, the area of $F_j$ is $\lambda \equiv 8t^{2\gamma}/\tilde{K} \leq 1$. Thus,

$$\begin{aligned} &\mathbb{P}\{D_t > 5 \log t\} \\ &\leq \mathbb{P}\left\{\max_j \#(F_j) > \frac{5}{2}\log t\right\} + \mathbb{P}\left\{\max_j \#(F_j') > \frac{5}{2}\log t\right\} \\ &\leq 2\tilde{K}\, \mathbb{P}\{X > \frac{5}{2}\log t\} \leq 2\tilde{K}\frac{\mathbb{E}[e^X]}{t^{5/2}}, \end{aligned} \tag{9.208}$$

where $X$ is a Poisson random variable with parameter $\lambda$. As in the estimate (9.168) of the proof of Lemma 9.25, we conclude that

$$\mathbb{P}\{D_t > 5 \log t\} \leq 2\tilde{K}\frac{e^2}{t^{5/2}} = 2e^2 \frac{[8t^{2\gamma}] + 1}{t^{5/2}} \to 0 \tag{9.209}$$

as $t \to \infty$ since $\gamma < 1$. This concludes the proof of the lemma. □

Hence, we find that for $t$ sufficiently large,

$$d^{C_2}(m_t, m_t + e_t) - d(0, e_t) \leq 5 \log t + 40t^{2\gamma - b} + 4t^{b\kappa_2} \tag{9.210}$$

with probability at least $1 - \frac{5}{10}\eta$.

Combining this estimate with Lemma 9.28, we find that

$$\begin{aligned}
&\mathbb{P}\left\{d^{C_1}(0, e_t) - d(0, e_t) \leq -t^{\kappa_1} + 4t^{b\kappa_2} + 40t^{2\gamma-b} + 5\log t\right\} \\
&\geq \mathbb{P}\left\{d^{C_1}(0, e_t) - d^{C_2}(m_t, m_t + e_t) \leq -t^{\kappa_1} \text{ and} \right. \\
&\qquad \left. d^{C_2}(m_t, m_t + e_t) - d(0, e_t) \leq 5\log t + 40t^{2\gamma-b} + 4t^{b\kappa_2}\right\} \\
&\geq \mathbb{P}\left\{d^{C_1}(0, e_t) - d^{C_2}(0, e_t) \leq -t^{\kappa_1}\right\} \\
&\quad + \mathbb{P}\{d^{C_2}(m_t, m_t + e_t) - d(0, e_t) \leq 5\log t + 40t^{2\gamma-b} \mid 4t^{b\kappa_2}\} - 1 \\
&\geq \eta + (1 - \frac{5}{10}\eta) - 1 = \frac{1}{2}\eta.
\end{aligned} \tag{9.211}$$

Since (9.184) implies that

$$\mathbb{P}\left\{d^{C_1}(0, e_t) - d(0, e_t) < 0\right\} \leq \frac{1}{10}\eta, \tag{9.212}$$

this gives a contradiction unless $-t^{\kappa_1} + 4t^{b\kappa_2} + 40t^{2\gamma-b} + 5\log t \geq 0$ for all large $t$. This necessarily implies that

$$\kappa_1 \leq \max\{b\kappa_2, 2\gamma - b\}. \tag{9.213}$$

Since $\kappa_1 \in (0, \frac{1}{3})$ and $\kappa_2 \in (\frac{1}{3}, 1)$ are arbitrary, letting $\kappa_1 \uparrow \frac{1}{3}$ and $\kappa_2 \downarrow \frac{1}{3}$, we obtain $\frac{1}{3} \leq \max\{\frac{1}{3}b, 2\gamma - b\}$. Since $b < 1$, this implies that $\frac{1}{3} \leq 2\gamma - b$. Letting $b \uparrow 1$, we obtain $\gamma \geq \frac{2}{3}$, as desired. This completes the proof of Theorem 9.23. □

**Remark 9.32.** If the longest up/right path is unique with probability 1, then it is interesting to consider the location of the intersection of the path and the line segment $L_t := \{(x, t - x) : x \in [0, 1]\}$. Theorem 9.23 implies that this intersection is near the midpoint of $L_t$ with fluctuation of over $t^{2/3}$. The limiting distribution of the location is not yet determined. However, see [**MFQR13**] for the distribution for a related model.

# Chapter 10

# Schur Measure and Directed Last Passage Percolation

Plancherel measure on the set of partitions has a natural generalization known as *Schur measure*. Gessel's formula and the Fredholm determinant formula for Plancherel measure, discussed in Chapter 3, extend naturally to Schur measure. Schur measure arises, in particular, in the so-called *directed last passage percolation* model in two dimensions, which is an analogue of the longest up/right path problem in $\mathbb{R}^2_+$ for Plancherel measure. In this chapter, we develop the theory of Schur measure and present some applications.

## 10.1. Schur Functions

Here we develop those parts of the theory of symmetric functions that are needed in the text below. In particular, Schur functions are introduced and play a distinguished role. Most of the following material is taken from [**Sag01**, Chapter 4].

**10.1.1. Simple Symmetric Functions.** Let $x = \{x_1, x_2, x_3, \dots\}$ be a finite or infinite set of variables and consider the *formal power series ring*

$$(10.1)\quad \mathbb{C}\big[[x]\big] = \big\{\sum a_{i_1,\dots,i_\ell;\lambda_1,\dots,\lambda_\ell} x_{i_1}^{\lambda_1} \cdots x_{i_\ell}^{\lambda_\ell} : \lambda_j \geq 0, a_{i_1,\dots,i_\ell;\lambda_1,\dots,\lambda_\ell} \in \mathbb{C}\big\},$$

which is indeed a ring under the usual operations of addition and multiplication. The *monomial* $x_{i_1}^{\lambda_1} \cdots x_{i_\ell}^{\lambda_\ell}$ is said to have *degree* $n$ if $\sum_{i=1}^{\ell} \lambda_i = n$. We say $f(x) \in \mathbb{C}\big[[x]\big]$ is *homogeneous of degree* $n$ if every monomial in $f(x)$ has degree $n$.

For every $n$, there is a natural action of $\pi \in S_n$ on $f(x) \in \mathbb{C}[[x]]$, viz.,

$$\pi f(x_1, x_2, \dots) = f(x_{\pi(1)}, x_{\pi(2)}, \dots) \tag{10.2}$$

where $\pi$ is extended naturally by setting $\pi(i) \equiv i$ for $i > n$. A function $f(x) \in \mathbb{C}[[x]]$ is *symmetric* if $\pi f = f$ for all $\pi \in S_n$, $n = 2, 3, \dots$. The simplest symmetric functions are obtained by symmetrizing a monomial.

**Definition 10.1.** Let $\lambda = (\lambda_1, \lambda_2, \dots, \lambda_\ell)$ be a partition. The *monomial symmetric function* corresponding to $\lambda$ is

$$m_\lambda = m_\lambda(x) = \sum x_{i_1}^{\lambda_1} \cdots x_{i_\ell}^{\lambda_\ell}, \qquad i_r \neq i_s \text{ if } r \neq s, \tag{10.3}$$

where the sum is taken over *all distinct* monomials having exponents $\lambda_1, \dots, \lambda_\ell$. Clearly $m_\lambda$ is indeed symmetric and is homogeneous of degree $n$ when $\lambda \vdash n$. Note that if the number of variables $x = (x_1, \dots, x_n)$ is finite, then $m_\lambda(x)$ is a polynomial.

For example, for $\lambda = (2, 1)$

$$m_{(2,1)}(x) = x_1^2 x_2 + x_2^2 x_1 + x_1^2 x_3 + x_3^2 x_1 + x_2^2 x_3 + x_3^2 x_2 + \cdots \tag{10.4}$$

and $m_{(2,1)}$ is homogeneous of degree 3. For $\lambda = (2, 2)$,

$$m_{(2,2)}(x) = x_1^2 x_2^2 + x_1^2 x_3^2 + x_2^2 x_3^2 + \cdots . \tag{10.5}$$

**Definition 10.2.** The *ring of symmetric functions* is given by

$$\Lambda = \Lambda(x) = \mathbb{C} m_\lambda, \tag{10.6}$$

the vector space spanned by all the $m_\lambda$.

**Exercise 10.3.** Show that $\Lambda = \{f(x) = \sum_{|\lambda| \leq c} a_{i,\lambda} x_{i_1}^{\lambda_1} \cdots x_{i_\ell}^{\lambda_\ell}$, for some $c = c(f) < \infty : \pi f = f$ for all $\pi \in S_n$, $n = 0, 1, 2, \dots\}$.

**Remark 10.4.** $\Lambda$ is indeed a ring as it is closed under products. *Caveat*: There exist elements in $\mathbb{C}[[x]]$ that are invariant under $S_n$, but are not in $\Lambda$. For example

$$\prod_{i=1}^{\infty} (1 + x_i) = 1 + x_1 + x_2 + \cdots + x_1 x_2 + x_1 x_3 + \cdots \tag{10.7}$$

cannot be a *finite* linear combination of $m_\lambda$'s.

We have the decomposition

$$\Lambda = \bigoplus_{n \geq 0} \Lambda^n, \tag{10.8}$$

where $\Lambda^n$ is the space spanned by all $m_\lambda$ of degree $n$. In fact, this is a *grading* of $\Lambda$ for if $f \in \Lambda^n$, $g \in \Lambda^m$, then $fg \in \Lambda^{m+n}$. As the $m_\lambda$'s are independent, we have the following result:

**Proposition 10.5.** *The space* $\Lambda^n$ *has basis*

$$\{m_\lambda \colon \lambda \vdash n\} \tag{10.9}$$

*and has dimension* $\#(n)$*, the number of partitions of* $n$*.*

**Exercise 10.6.** Prove Euler's famous theorem

$$\sum_{n\geq 0} \#(n)x^n = \prod_{i=1}^{\infty} \frac{1}{1-x^i} = (1+x+x^2+\cdots)(1+x^2+(x^2)^2+\cdots)\cdots. \tag{10.10}$$

Thus $\prod_{i=1}^{\infty} \frac{1}{1-x^i}$ is a *generating function* for $\#(n)$.

Several other bases for $\Lambda^n$ are of interest.

**Definition 10.7.** (a) The *nth power sum symmetric function* is $m_\lambda$ for $\lambda = (n)$, i.e.,

$$p_n(x) = m_{(n)} = \sum_{i\geq 1} x_i^n. \tag{10.11}$$

(b) The *nth elementary symmetric function* is $m_\lambda$ for $\lambda = (1, 1, \ldots, 1)$;

$$e_n(x) = m_{(1,1,\ldots,1)} = \sum_{i_1<i_2<\cdots<i_n} x_{i_1}x_{i_2}\cdots x_{i_n}. \tag{10.12}$$

(c) The *nth complete homogeneous symmetric function*

$$\begin{aligned} h_n(x) &= \sum_{\lambda\vdash n} m_\lambda = \sum_{\lambda\vdash n} \sum_{\text{distinct monomials}} x_{i_1}^{\lambda_1}\cdots x_{i_\ell}^{\lambda_\ell} \\ &= \sum_{i_1\leq i_2\leq\cdots\leq i_n} x_{i_1}x_{i_2}\cdots x_{i_n}. \end{aligned} \tag{10.13}$$

**Exercise 10.8.** Prove that indeed in (10.13)

$$\sum_{\lambda\vdash n} \sum_{\text{distinct monomials}} x_{i_1}^{\lambda_1}\cdots x_{i_\ell}^{\lambda_\ell} = \sum_{i_1\leq i_2\leq\cdots\leq i_n} x_{i_1}x_{i_2}\cdots x_{i_n}. \tag{10.14}$$

For example, for $n = 3$

$$\begin{aligned} p_3(x) &= x_1^3 + x_2^3 + x_3^3 + \cdots, \\ e_3(x) &= x_1x_2x_3 + x_1x_2x_4 + x_1x_3x_4 + x_2x_3x_4 + \cdots, \\ h_3(x) &= x_1^3 + x_2^3 + x_3^3 + \cdots + x_1^2x_2 + x_1^2x_3 + \cdots \\ &\quad + x_1x_2x_3 + x_1x_2x_4 + \cdots. \end{aligned} \tag{10.15}$$

All three of the symmetric functions in Definition 10.7 have explicit and useful generating functions.

**Exercise 10.9.** Prove the following generating function formulae (see [**Sag01**]):

(i) $\sum_{n\geq 1} p_n(x)\frac{t^n}{n} = \log\left(\prod_{i\geq 1}\frac{1}{1-x_it}\right)$.

(ii) $\sum_{n\geq 0} e_n(x)t^n = \prod_{i\geq 1}(1+x_it)$.

(iii) $\sum_{n\geq 0} h_n(x)t^n = \prod_{i\geq 1}\frac{1}{1-x_it}$.

**Definition 10.10.** For a partition $\lambda = (\lambda_1, \ldots, \lambda_\ell)$, let

$$f_\lambda(x) \equiv f_{\lambda_1}(x)f_{\lambda_2}(x)\cdots f_{\lambda_\ell}(x), \tag{10.16}$$

where $f = p$, $e$, or $h$.

For example, for $\lambda = (2,1)$

$$p_{(2,1)} = p_2p_1 = (x_1^2 + x_2^2 + \cdots)(x_1 + x_2 + \cdots). \tag{10.17}$$

The following basic result is proved in [**Sag01**].

**Theorem 10.11.** *The following form bases for* $\Lambda^n$*:*

*(i)* $\{p_\lambda \colon \lambda \vdash n\}$,

*(ii)* $\{e_\lambda \colon \lambda \vdash n\}$,

*(iii)* $\{h_\lambda \colon \lambda \vdash n\}$.

**Proof.** We leave this as an exercise. □

**10.1.2. Schur functions.** There is a fifth basis for $\Lambda^n$ that plays a very important role in this text. This basis is constructed from the *Schur functions.*

A *composition of* $n$ is an ordered sequence of nonnegative integers

$$\mu = (\mu_1, \ldots, \mu_m) \tag{10.18}$$

such that $\sum_i \mu_i = n$.

Thus $(1,3,2)$ and $(3,2,1)$ are both compositions of 6 but only the latter is a partition. Some entries $\mu_i$ may be zero: $(0,2,0,1,3)$ is also a composition of 6. If $\lambda = (\lambda_1, \ldots, \lambda_\ell) \vdash n$ is a partition, a *generalized Young tableau of shape* $\lambda$ is an array $T$ obtained by placing positive integers into the boxes of the Young diagram of shape $\lambda$, repetitions allowed. The *type* of $T$, denoted by type$(T)$, is the composition $\mu = (\mu_1, \mu_2, \ldots, \mu_m)$, where $\mu_i$ equals the number of $i$'s in $T$. Clearly,

$$\sum_1^m \mu_i = \sum_1^\ell \lambda_i = n. \tag{10.19}$$

For example, for $\lambda = (3,2) \vdash 5$,

$$T = \begin{array}{|c|c|c|}\hline 2 & 1 & 1 \\ \hline 3 & 2 \\ \cline{1-2}\end{array} \tag{10.20}$$

is a generalized Young tableau with $\mathrm{type}(T) = (2,2,1)$.

Given any composition $\mu = (\mu_1, \dots, \mu_m)$, there is a corresponding monomial in $\mathbb{C}[[x]]$

$$x^{\mu} = x_1^{\mu_1} \cdots x_m^{\mu_m}. \tag{10.21}$$

Now consider any generalized Young tableau $T$ of shape $\lambda$: it gives rise to a monomial

$$x^T \equiv x^{\mathrm{type}(T)} = \prod_{(i,j)\in\lambda} x_{T_{i,j}}, \tag{10.22}$$

where $T_{i,j}$ is the entry in the box $(i,j)$ of $T$. For example, $T = \begin{array}{|c|c|c|}\hline 4 & 1 & 4 \\ \hline 1 & 3 \\ \cline{1-2}\end{array}$ gives rise to the monomial

$$x^T = x_1^2 x_3 x_4^2. \tag{10.23}$$

**Definition 10.12.** A tableau $T$ is *semi-standard* (SSYT) if the entries are *weakly* increasing along the rows and *strictly* increasing down the columns.

For example, $\begin{array}{|c|c|c|}\hline 1 & 1 & 3 \\ \hline 2 & 4 \\ \cline{1-2}\end{array}$ is a semi-standard tableau.

**Definition 10.13** (Combinatorial definition of the Schur function). Given a partition $\lambda$ the associated *Schur function* is

$$s_\lambda(x) = \sum_T x^T, \tag{10.24}$$

where the sum is over all the semi-standard tableaux $T$ of shape $\lambda$.

Clearly, $s_\lambda$ is homogeneous of degree $|\lambda|$.

By way of illustration, if $\lambda = (2,1)$, then some of the possible tableaux are

(10.25)

$$T: \begin{array}{|c|c|}\hline 1 & 1 \\ \hline 2 \\ \cline{1-1}\end{array} \quad \begin{array}{|c|c|}\hline 1 & 2 \\ \hline 2 \\ \cline{1-1}\end{array} \quad \begin{array}{|c|c|}\hline 1 & 1 \\ \hline 3 \\ \cline{1-1}\end{array} \quad \begin{array}{|c|c|}\hline 1 & 3 \\ \hline 3 \\ \cline{1-1}\end{array} \quad \begin{array}{|c|c|}\hline 1 & 2 \\ \hline 3 \\ \cline{1-1}\end{array} \quad \begin{array}{|c|c|}\hline 1 & 3 \\ \hline 2 \\ \cline{1-1}\end{array} \quad \begin{array}{|c|c|}\hline 1 & 2 \\ \hline 4 \\ \cline{1-1}\end{array} \quad \begin{array}{|c|c|}\hline 1 & 4 \\ \hline 2 \\ \cline{1-1}\end{array} \quad \cdots$$

and so

$$s_{(2,1)}(x) = x_1^2x_2 + x_1x_2^2 + x_1^2x_3 + x_1x_3^2 + \cdots + 2x_1x_2x_3 + 2x_1x_2x_4 + \cdots. \tag{10.26}$$

In the case that $x = (x_1, \dots, x_\ell, 0, 0, \dots)$, $x^T = 0$ for any SSYT $T$ of shape $\lambda$ if $r(\lambda) > \ell$ (where $r(\lambda)$ is the number of rows of $\lambda$) since the first column of $T$ contains an entry which is larger than $\ell$. In this case, $s_\lambda(x) = 0$ unless $r(\lambda) \le \ell$. Also note that in this case $s_\lambda(x)$ is a polynomial.

Clearly (why?)

$$s_{(n)}(x) = h_n(x) \tag{10.27}$$

and

$$s_{(1,\dots,1)}(x) = e_n(x). \tag{10.28}$$

We also note that if $\lambda \vdash n$ is arbitrary, then

$$s_\lambda(x) = \cdots + F_\lambda x_1 x_2 \cdots x_n + \cdots , \tag{10.29}$$

where $F_\lambda = \#\{\text{SYT's with shape } \lambda\}$.

We first verify that the Schur functions are indeed symmetric.

**Proposition 10.14.** *The function $s_\lambda(x)$ is symmetric.*

**Proof (due to Knuth).** It suffices to show that

$$(i, i+1)s_\lambda(x) = s_\lambda(x) \tag{10.30}$$

for each adjacent transposition. To this end, we introduce an involution on the semi-standard $\lambda$ tableaux,

$$T \to T'. \tag{10.31}$$

We construct the map $T \to T'$ as follows. Given $T$, each column contains either an $i$, $i+1$ pair, exactly one of $i$ or $i+1$, or neither. Regard the pairs as fixed and all other occurrences of $i$ or $i+1$ as free. In each row switch the number of free $i$'s and $(i+1)$'s, i.e., if the row consists of $k$ free $i$'s and $\ell$ free $(i+1)$'s, then replace them by $\ell$ free $i$'s followed by $k$ free $(i+1)$'s. To illustrate, if $i = 2$ and

$$T = \begin{array}{|c|c|c|c|c|c|c|c|c|c|} \hline 1 & 1 & 1 & 1 & 2 & 2 & 2 & 2 & 2 & 3 \\ \hline 2 & 2 & 3 & 3 & 3 & 3 \\ \cline{1-6} 3 \\ \cline{1-1} \end{array} , \tag{10.32}$$

then there are three fixed occurrences of $(2,3)$, and three free 2's and one free 3 in row 1, and one free 2 and two free 3's in row 2. Thus

$$T' = \begin{array}{|c|c|c|c|c|c|c|c|c|c|} \hline 1 & 1 & 1 & 1 & 2 & 2 & 2 & 3 & 3 & 3 \\ \hline 2 & 2 & 2 & 3 & 3 & 3 \\ \cline{1-6} 3 \\ \cline{1-1} \end{array} , \tag{10.33}$$

By the definition of "free", $T'$ will always (check this) be semi-standard and $(i, i+1)(x^T + x^{T'}) = x^T + x^{T'}$. Indeed in the above example

$$x^T + x^{T'} = x_1^4 x_2^7 x_3^6 + x_1^4 x_2^6 x_3^7 \tag{10.34}$$

and clearly, $(2,3)(x^T + x^{T'}) = x^T + x^{T'}$. Note that $T = T'$ if and only if $\#(\text{free } i) = \#(\text{free } i+1)$ in each row, and if $T' \neq T$, $\#\{i\text{'s in } T'\} = \#\{(i+1)\text{'s in } T\}$ and $\#\{(i+1)\text{'s in } T'\} = \#\{i\text{'s in } T\}$. Finally, $T \to T'$ is clearly an involution, and so we are done. $\square$

Next we prove:

**Theorem 10.15.** *The set $\{s_\lambda \colon \lambda \vdash n\}$ is a basis for $\Lambda^n$.*

We will need the following concept:

**Definition 10.16.** Suppose $\lambda = (\lambda_1, \lambda_2, \ldots, \lambda_\ell)$ and $\mu = (\mu_1, \ldots, \mu_m)$ are partitions of $n$. Then $\lambda$ *dominates* $\mu$, written $\lambda \trianglerighteq \mu$, if

$$\lambda_1 + \cdots + \lambda_i \geq \mu_1 + \cdots + \mu_i \tag{10.35}$$

for all $i \geq 1$. If $i > \ell$ (resp., $i > m$), then we take $\lambda_i$ (resp., $\mu_i$) to be zero.

Note that $\trianglerighteq$ is a partial ordering on $\{\mu \vdash n\}$. Intuitively, "short and fat" $\trianglerighteq$ "tall and skinny".

**Proof of Theorem 10.15.** Suppose $\lambda = (\lambda_1, \ldots, \lambda_\ell) \vdash n$. As $s_\lambda$ is symmetric by Theorem 10.14, it follows from Exercise 10.3 that $s_\lambda \in \Lambda^n$ and hence has the form

$$s_\lambda = \sum_{\mu \vdash n} k_{\lambda,\mu} m_\mu \tag{10.36}$$

for suitable coefficients $k_{\lambda,\mu}$. Suppose $k_{\lambda,\mu} \neq 0$ for some $\mu \vdash n$. Then from the definition of $s_\lambda$, $\lambda$ must have an associated SSYT $T$ of type $\mu$. Now as $T$ is column strict, all occurrences of the numbers $1, \ldots, i$, if any, must be in the first $i$ rows of $T$. Hence for any $i$

$$\begin{aligned} \mu_1 + \cdots + \mu_i &= \# \text{ of occurrences of } 1, \ldots, i \\ &\leq \# \text{ boxes in first } i \text{ rows} \\ &= \lambda_1 + \cdots + \lambda_i. \end{aligned} \tag{10.37}$$

Thus $\mu \trianglelefteq \lambda$. Hence

$$s_\lambda = \sum_{\mu \trianglelefteq \lambda} k_{\lambda,\mu} m_\lambda. \tag{10.38}$$

We show that $k_{\lambda,\lambda} = 1$. Consider the SSYT $T = (T_{ij})$ defined by $T_{ij} = i$ for $1 \leq i \leq \ell$ and $1 \leq j \leq \lambda_i$, which looks like

$$T = \begin{array}{|c|c|c|c|c|} \hline 1 & 1 & 1 & \cdot\cdot & 1 \\ \hline 2 & 2 & 2 \\ \cline{1-3} 3 & 3 & 3 \\ \cline{1-3} \cdot\cdot & \cdot\cdot \\ \cline{1-2} \ell & \ell \\ \cline{1-2} \end{array} .$$

Then $T$ is a SSYT of type $\lambda = (\lambda_1, \lambda_2, \ldots, \lambda_\ell)$ and hence $x^T = x_1^{\lambda_1} \cdots x_\ell^{\lambda_\ell}$. As $s_\lambda$ contains such a monomial, the same must be true for the right-hand side of (10.38). If $\mu \trianglelefteq \lambda$, $\mu \neq \lambda$, then $m_\mu$ does not contain such a monomial. It follows that $k_{\lambda,\lambda} = 1$.

We now show that $\{s_\kappa\colon \kappa \vdash n\}$ spans $\Lambda^n$. Suppose that for some $\lambda \vdash n$, $m_\lambda$ is not in $\mathrm{sp_n} \equiv \mathrm{span}\{s_\kappa\colon \kappa \vdash \mathrm{n}\}$. Now suppose there exists $\lambda' \trianglelefteq \lambda$, $\lambda' \neq \lambda$, such that $m_{\lambda'}$ is not in $\mathrm{sp_n}$. Then suppose there exists $\lambda'' \trianglelefteq \lambda' \trianglelefteq \lambda$, $\lambda'' \neq \lambda'$ such that $m_{\lambda''}$ is not in $\mathrm{sp_n}$, etc. Eventually, we must obtain some $\lambda^* \trianglelefteq \lambda$ such that $m_{\lambda^*}$ is not in $\mathrm{sp_n}$ but $m_\mu \in \mathrm{sp_n}$ for all $\mu \trianglelefteq \lambda^*$, $\mu \neq \lambda^*$. Now by (10.38)

$$s_{\lambda^*} = \sum_{\mu \trianglelefteq \lambda^*} k_{\lambda^*,\mu} m_\mu = m_{\lambda^*} + \sum_{\mu \trianglelefteq \lambda^*, \mu \neq \lambda^*} k_{\lambda^*,\mu} m_\mu. \tag{10.39}$$

Since both $s_{\lambda^*}$ and the sum on the right-hand side is in $\mathrm{sp_n}$, this implies that $m_{\lambda^*} \in \mathrm{sp_n}$. This is a contradiction. Hence $\mathrm{sp_n} = \Lambda^n$.

Finally,

$$\dim \Lambda^n = \dim \mathrm{sp_n} = \#\{\text{independent vectors } s_\lambda\} \leq \#\{\mu\colon \mu \vdash n\} = \dim \Lambda^n.$$

It follows that the $s_\lambda$'s are independent, and hence $\{s_\lambda\colon \lambda \vdash n\}$ is a basis for $\Lambda^n$. □

**10.1.3. Jacobi-Trudi Identities.** We now give determinantal formulae, called the Jacobi-Trudi identities, for $s_\lambda$ in terms of the elementary and complete symmetric functions $e_n$ and $h_n$, respectively. These formulae are analogous to the determinantal formula (3.62) for $F_\lambda$. Indeed, (3.62) can be derived as a corollary of the Jacobi-Trudi identities; see Subsection 10.1.4 below.

For a partition $\lambda$, denote the conjugate by $\lambda'$: $\lambda' = (\lambda'_1, \dots, \lambda'_m)$ where $\lambda'_i$ is the length of the $i$th *column* of $\lambda$.

**Theorem 10.17** (Jacobi-Trudi). *Let $\lambda = (\lambda_1, \lambda_2, \dots, \lambda_\ell)$ be a partition. Then*

$$s_\lambda = \det(h_{\lambda_i - i + j})_{1 \leq i,j \leq \ell} \tag{10.40}$$

*and*

$$s_{\lambda'} = \det(e_{\lambda_i - i + j})_{1 \leq i,j \leq \ell}, \tag{10.41}$$

*where $\lambda'$ is the conjugate of $\lambda$.*

**Remark 10.18.** As noted in (10.27) and (10.28), $s_{(n)} = h_n$ and $s_{(1,1,\dots,1)} = e_n$. The special case of (10.40) when $\lambda = (n)$ is the trivial identity $h_n = h_n$, and the special case of (10.41) when $\lambda' = (1, \dots, 1)$ is the trivial identity $e_n = e_n$.

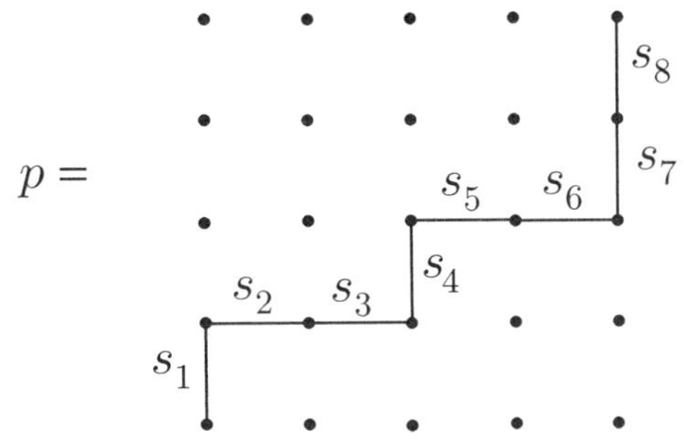


**Figure 10.1.** Path $p = s_1 s_2 \cdots s_8$

**Proof of Theorem 10.17.** ([**Sag01**] following Lindström [**Lin73**]) As we will see, the proof will have many similarities with our first proof of formula (3.62) for $F_\lambda$. The key idea is that one can view both tableaux and determinants as lattice paths: again only nonintersecting paths make a contribution. Consider the plane $\mathbb{Z} \times \mathbb{Z}$ of integer lattice points. We consider (possibly infinite) paths in this plane

$$p = s_1, s_2, s_3, \ldots, \tag{10.42}$$

where each step $s_i$ is of unit length northward (N) or eastward (E); for example see Figure 10.1. Label the *eastward* steps of $p$ using one of two labelings. The *e-labeling* assigns to each eastward $s_i$ the label

$$L(s_i) = i, \tag{10.43}$$

The *h-labeling* gives each eastward $s_i$ the label

$$\tilde{L}(s_i) = \#\{\text{northward } s_j \text{ preceding } s_i\} + 1. \tag{10.44}$$

Intuitively, in the $h$-labeling, all the eastward steps on the line through the origin of $p$ are labeled 1, all those on the line one unit above are labeled 2, and so on; see Figure 10.2 for the labelings for the path $p$ in Figure 10.1.

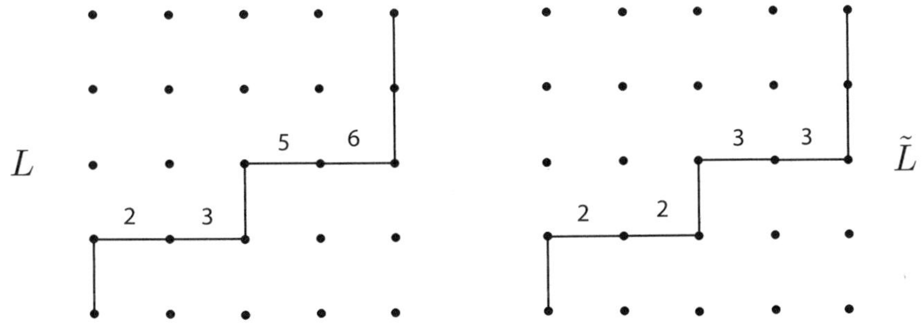


**Figure 10.2.** $e$-labeling and $h$-labeling

It is convenient to extend $\mathbb{Z} \times \mathbb{Z}$ by the addition of some points at $\infty$. Specifically, for each $x \in \mathbb{Z}$, add a point $(x, \infty)$ above each point on the vertical line with coordinate $x$. Clearly, a path can reach $(x, \infty)$ only by ending with an infinite number of consecutive northward steps along this line.

If $p$ starts at a vertex $u$ and ends at a vertex $v$ (which may be a point at $\infty$), then we write $u \xrightarrow{p} v$.

There are two weightings of paths corresponding to the two labelings. If $p$ has only a finite # of eastward steps, define

$$x^p \equiv \prod x_{L(s_i)}, \qquad \tilde{x}^p \equiv \prod x_{\tilde{L}(s_i)}, \tag{10.45}$$

where each product is taken over the eastward $s_i$ in $p$. For the example $p$ in Figure 10.1, $x^p = x_2 x_3 x_5 x_6$ and $\tilde{x}^p = x_2^2 x_3^2$. Note that $x^p$ is always square free, whereas $\tilde{x}^p$ can be any monomial. So we have

$$e_n(x) = \sum_p x^p, \qquad h_n(x) = \sum_p \tilde{x}^p, \tag{10.46}$$

where both sums are over all paths $(a, b) \xrightarrow{p} (a + n, \infty)$ for any fixed initial vertex $(a, b)$.

Just as all paths between one pair of points $\{(a, b), (a + n, \infty)\}$ corresponds to an elementary or complete symmetric function, all $\ell$-tuples of paths between $\ell$ pairs of points will be used to evaluate the $\ell$-fold products contained in the Jacobi-Trudi determinants. Let $u_1, u_2, \ldots, u_\ell$ and $v_1, v_2, \ldots, v_\ell$ be fixed sets of initial and final vertices. Fix $\pi \in S_\ell$, and consider a family of paths $\mathcal{P} = (p_1, \ldots, p_\ell)$, where, for each $i$

$$u_i \xrightarrow{p_i} v_{\pi(i)}. \tag{10.47}$$

Give weight to $\mathcal{P}$ by setting

$$x^{\mathcal{P}} = \prod_i x^{p_i} \tag{10.48}$$

with a similar definition for $\tilde{x}^{\mathcal{P}}$. Also define the sign of $\mathcal{P}$ to be

$$(-1)^{\mathcal{P}} = \operatorname{sgn} \pi. \tag{10.49}$$

For the example of the family of paths in Figure 10.3, $\pi = 2314$, $x^{\mathcal{P}} = (x_3)(x_3)(x_2 x_3 x_4 x_5)(x_4) = x_2 x_3^3 x_4^2 x_5$, $\tilde{x}^{\mathcal{P}} = (x_3)(x_3)(x_2^4) x_4 = x_2^4 x_3^2 x_4$, and $(-1)^{\mathcal{P}} = 1$.

We now concentrate on proving

$$s_\lambda = \det(h_{\lambda_i - i + j}). \tag{10.50}$$

The proof of (10.41) for $s_{\lambda'}$ is similar and is left as an exercise (see Exercise 10.20 below).

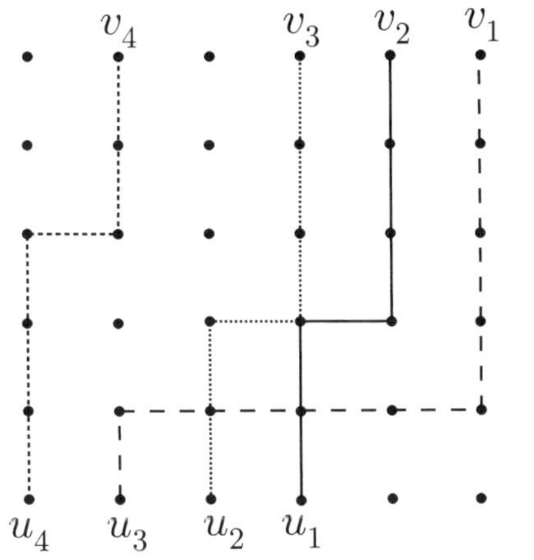


**Figure 10.3.** Family of paths $\mathcal{P}$

Given $\lambda = (\lambda_1, \ldots, \lambda_\ell)$, pick initial points

(10.51) $$u_i = (1 - i, 0), \qquad i = 1, \ldots, \ell,$$

and final points

(10.52) $$v_i = (\lambda_i - i + 1, \infty), \qquad i = 1, \ldots, \ell.$$

In Figure 10.3, $\lambda = (2, 2, 2, 1)$, and

$$v_1 = (2, \infty), \quad v_2 = (1, \infty), \quad v_3 = (0, \infty), \quad v_4 = (-2, \infty)$$

and

$$u_1 = (0, 0), \quad u_2 = (-1, 0), \quad u_3 = (-2, 0), \quad u_4 = (-3, 0).$$

From the choice of vertices, since $(\lambda_i - i + 1) - (-j + 1) = \lambda_i + j - i$, (10.46) implies that

(10.53) $$h_{\lambda_i - i + j} = \sum_{u_j \xrightarrow{p} v_i} \tilde{x}^p.$$

(Of course, if $\lambda_i - i + j < 0$, then both sides of (10.53) are zero; consider, e.g., $\lambda_4 - 4 + 1 = -2$ in Figure 10.3.) Thus

(10.54) $$\begin{aligned} &\det(h_{\lambda_i - i + j})_{1 \le i, j \le \ell} \\ &= \sum_{\sigma \in S_\ell} \operatorname{sgn} \sigma \cdot h_{\lambda_{\sigma(1)} - \sigma(1) + 1} h_{\lambda_{\sigma(2)} - \sigma(2) + 2} \cdots h_{\lambda_{\sigma(\ell)} - \sigma(\ell) + \ell} \\ &= \sum_{\sigma \in S_\ell} \operatorname{sgn} \sigma \cdot \Bigg( \sum_{u_1 \xrightarrow{p_1} v_{\sigma(1)}} \tilde{x}^{p_1} \Bigg) \cdots \Bigg( \sum_{u_\ell \xrightarrow{p_\ell} v_{\sigma(\ell)}} \tilde{x}^{p_\ell} \Bigg). \end{aligned}$$

Therefore,

(10.55) $$\det(h_{\lambda_i - i+j})_{1\le i,j\le \ell} = \sum_{\mathcal{P}} (-1)^{\mathcal{P}} \tilde{x}^{\mathcal{P}},$$

where the sum is over all the families of paths $\mathcal{P} = (p_1, \ldots, p_\ell)$ such that $u_i \xrightarrow{p_i} v_{\sigma(i)}$ for some $\sigma \in S_\ell$.

Next we show that all the terms on the right-hand side of equation (10.55) cancel in pairs except for those corresponding to $\ell$-tuples of nonintersecting paths. To do this, we need a weight preserving involution

(10.56) $$\tau : \mathcal{P} \to \mathcal{P}'$$

such that if $\mathcal{P}$ is nonintersecting, then $\mathcal{P}' = \mathcal{P}$, and if $\mathcal{P}$ is intersecting, then $\mathcal{P}' \neq \mathcal{P}$ and $(-1)^{\mathcal{P}} = -(-1)^{\mathcal{P}'}$.

**Definition 10.19.** Given a family of paths $\mathcal{P}$, define $\tau\mathcal{P} = \mathcal{P}'$ as follows:

(i) If $p_i \cap p_j = \emptyset$ for all $i, j$, define $\mathcal{P}' = \mathcal{P}$.

(ii) Otherwise, find the smallest index $i$ such that the path $p_i$ intersects some other path. Let $v_0$ be the first (SW-most) intersection on $p_i$ and let $p_j$ be the other path through $v_0$. (If there is more than one, choose $p_j$ such that $j$ is minimal.) Now define $\mathcal{P}' = \mathcal{P}$ with $p_i$, $p_j$ replaced by $p_i'$, $p_j'$ where $p_i' = u_i \xrightarrow{p_i} v_0 \xrightarrow{p_j} v_{\pi(j)}$ and $p_j' = u_j \xrightarrow{p_j} v_0 \xrightarrow{p_i} v_{\pi(i)}$.

See Figure 10.4 for an example.

Clearly, the map is invertible with itself as inverse. Nonintersecting paths are fixed by definition. For the intersecting $\ell$-tuples, the sign clearly changes in passing from $\mathcal{P} \to \mathcal{P}'$. Weight is preserved due to the fact that all paths start on the same horizontal axis. Thus the set of $\tilde{L}$-labels is unchanged

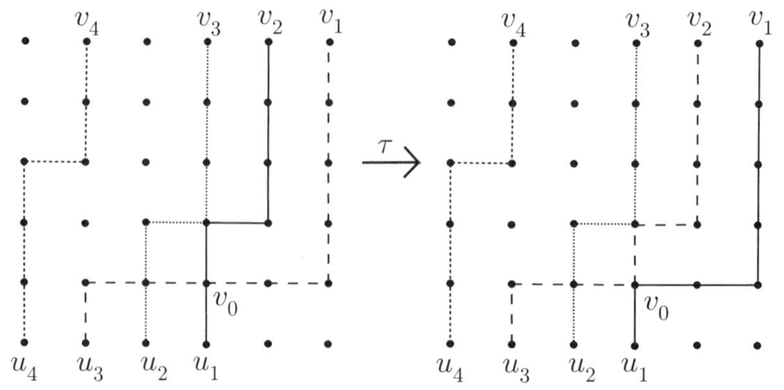


**Figure 10.4.** Mapping $\tau : \mathcal{P} \to \mathcal{P}'$

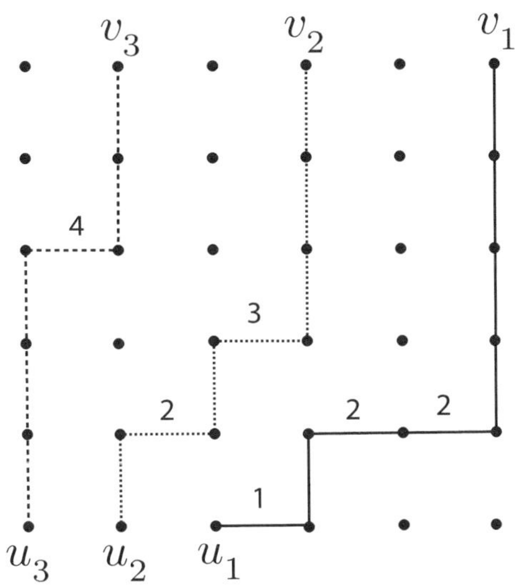


**Figure 10.5.** Nonintersecting paths corresponding to the SSYT (10.58)

by $\tau$. Therefore, (10.55) holds where the sum is now restricted to the family of nonintersecting paths.

All that remains to show is that

$$s_\lambda(x) = \sum_{\mathcal{P}} (-1)^{\mathcal{P}} \tilde{x}^{\mathcal{P}}, \tag{10.57}$$

where the sum is over all $\ell$-tuples of nonintersecting paths. But by our choice of initial and final vertices, $\mathcal{P} = (p_1, \ldots, p_\ell)$ is nonintersecting only if it corresponds to the identity permutation. Thus $(-1)^{\mathcal{P}} = +1$ and $u_i \xrightarrow{p_i} v_i$ for all $i$.

There is a simple bijection between such $\mathcal{P}$'s and SSYT's. Given $\mathcal{P}$, merely use the $h$-labels of the $i$th path, listed in increasing order, for the $i$th row of the tableaux. For example, the nonintersecting paths in Figure 10.5 are associated to the SSYT

$$T = \begin{array}{|c|c|c|}\hline 1 & 2 & 2 \\ \hline 2 & 3 \\ \cline{1-2} 4 \\ \cline{1-1}\end{array}\ . \tag{10.58}$$

In view of (10.51) and (10.52), $T$ has shape $\lambda$. By definition, the entries in the rows of $T$ increase weakly. Finally, the entries in the columns must increase strictly because the nonintersecting condition and the choice of the $u_i$'s force the $j$th eastward step of $p_{i+1}$ to be higher than the corresponding step on $p_i$. More precisely, after $j$ eastward moves $p_i$ has the form as in Figure 10.6. Now $T_{i+1,j}$ corresponds to the $j$th eastward move for $p_{i+1}$.

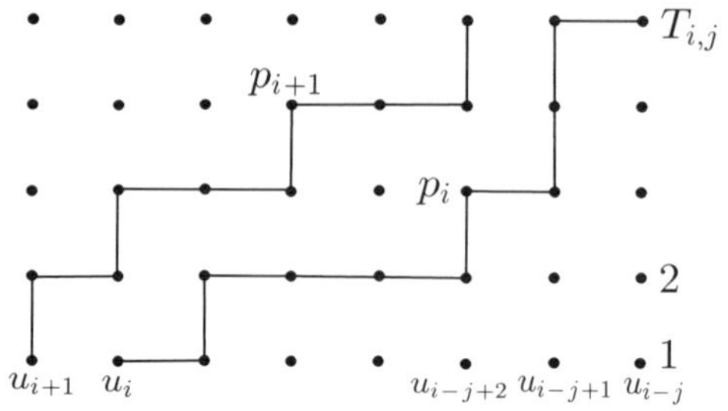


**Figure 10.6.** $j$ east moves of $p_i$

During this move, the $x$-coordinate of $p_{i+1}$ moves from $u_{i-j+2}$ to $u_{i-j+1}$, and it is clear from Figure 10.6 that unless $T_{i+1,j} > T_{i,j}$, the path $p_{i+1}$, which starts at $u_{i+1}$, must touch the path $p_i$, which is a contradiction, and hence the columns in $T$ must increase strictly. Thus we have a well defined map $\varphi\colon \mathcal{P} \to T$ with the property that $T$ is a $\lambda$-SSYT and

$$\tilde{x}^{\mathcal{P}} = x^T. \tag{10.59}$$

Furthermore the map is one-to-one as the entries $(T_{i1}, T_{i2}, \ldots)$ for each row $i$ clearly determine $p_i$ in $\mathcal{P}$ uniquely.

Conversely, given any $\lambda$-SSYT $T$ we can clearly use $(T_{i1}, T_{i2}, \ldots)$ to define paths $p_i\colon u_i \to v_i$, $i = 1, \ldots, \ell$. Now $p_{i+1}$ and $p_i$ cannot touch, for if they did after step $j$, say, then clearly $T_{i+1,j} \le T_{i,j}$, which is a contradiction (see Figure 10.7). Thus $p_i$ and $p_{i+1}$ cannot touch and, as $u_{i+1}$ is to the left of $u_i$, $p_{i+1}$ remains strictly to the left of $p_i$. Thus $p_1$ lies to the right of $p_2$, $p_2$ to the right of $p_3$, etc., and $\mathcal{P} = (p_1, \ldots, p_\ell)$ consists of $\ell$ nonintersecting paths, $u_i \xrightarrow{p_i} v_i$, and clearly $\mathcal{P} \to T$ under the map $\varphi$.

So, $\varphi$ is bijective, and hence by (10.55), (10.59), Definition 10.12 of $s_\lambda$, and the preceding comments,

$$\det(h_{\lambda_i - i + j})_{1 \le i,j \le \ell} = \sum_{\mathcal{P}} (-1)^{\mathcal{P}} \tilde{x}^{\mathcal{P}} = \sum_{T \text{ is a } \lambda\text{-SSYT}} x^T = s_\lambda(x). \tag{10.60}$$

This proves (10.40). □

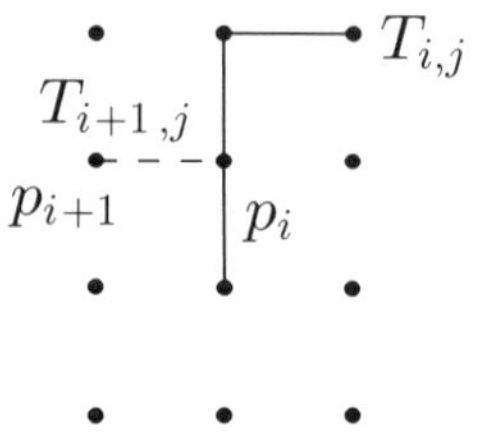


**Figure 10.7.** $p_i$ and $p_{i+1}$

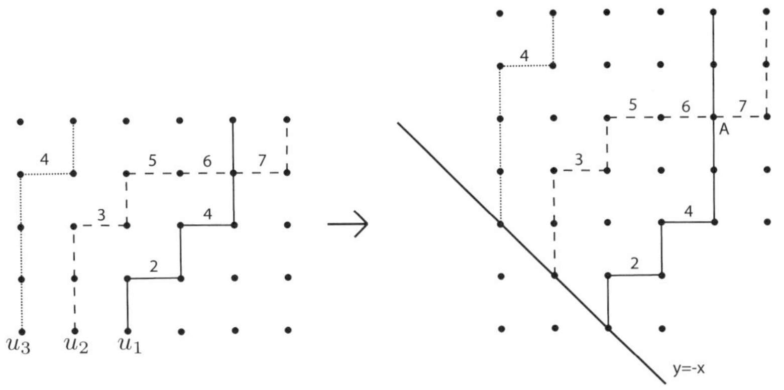


**Figure 10.8.** $e$-labeling of paths

**Exercise 10.20.** Prove (10.41). As in our first proof of the formula for $F_\lambda$ (see Theorem 3.31) it is useful to shift the paths appropriately. Instead of $e$-labeling as above, shift the paths up using the line $y = -x$; see Figure 10.8 for an example. Labeling in this way, we ensure that when two paths meet at A, both paths have taken the same number of steps (in this example, six steps) to get there. One then maps $\mathcal{P} \to \mathcal{P}'$ as before. Clearly now $\mathcal{P}'$ has the same $e$-weight but opposite signature to $\mathcal{P}$, etc.

**10.1.4. $F_\lambda$ Redux: Third Proof.** We use the Jacobi-Trudi identity (10.40) to give another proof of the determinantal formula (3.62) for $F_\lambda$, the number of SYT's of shape $\lambda$. This gives a third proof of Theorem 3.28.

Let $\lambda \vdash n$. Recall from (10.29) that the coefficient of $x_1x_2\cdots x_n$ in $s_\lambda(x)$ is $F_\lambda$. We now simply evaluate the coefficient of $x_1x_2\cdots x_n$ in the right-hand side of formula (10.40). Here we follow [**Sta99**, Section 7.8]. We first define the algebra homomorphism

$$\mathrm{ex} : \Lambda \to \mathbb{C}[[t]], \tag{10.61}$$

called the *exponential specialization*, for a variable $t$ as follows. For a symmetric function $f(x_1, x_2, \ldots) \in \Lambda$, $\mathrm{ex}(f)$ is the function in $t$ obtained from $f$ by the (formal) substitution

$$\begin{aligned} p_1(x) &= x_1 + x_2 + \cdots = t \\ p_j(x) &= x_1^j + x_2^j + \cdots = 0, \qquad j \geq 2. \end{aligned} \tag{10.62}$$

As $\{p_\lambda\}$ forms a basis for $\Lambda$, this clearly defines a homomorphism. For example, for $f = h_2 = x_1^2 + x_2^2 + \cdots + x_1x_2 + x_1x_3 + \cdots = \frac{1}{2}(p_1^2 + p_2)$, we have $\mathrm{ex}(h_2) = \frac{1}{2}t^2$. For $f = e_2 = x_1x_2 + x_1x_3 + x_2x_3 + \cdots = \frac{1}{2}(p_1^2 - p_2)$, we have $\mathrm{ex}(e_2) = \frac{1}{2}t^2$. It is easy to check that ex is indeed a homomorphism.

**Lemma 10.21.** *For $f \in \Lambda$, let $c_n(f)$ be the coefficient of $x_1 x_2 \cdots x_n$ in $f$. Then*

$$\mathrm{ex}(f) = \sum_{n=0}^{\infty} c_n(f) \frac{t^n}{n!}. \tag{10.63}$$

**Proof.** Since both ex and the map $f \mapsto c_n(f)$ are linear, it is enough to check (10.63) for a basis of $\Lambda$. In particular, the power sum symmetric functions $p_\mu = p_{\mu_1} \cdots p_{\mu_\ell} = (x_1^{\mu_1} + x_2^{\mu_1} + \cdots) \cdots (x_1^{\mu_\ell} + x_2^{\mu_\ell} + \cdots)$. By expanding the product, we see that if $\mu \neq (1, 1, \dots, 1)$, then $c_n(p_\mu) = 0$ for all $n$, and if $\mu = (1, 1, \dots, 1)$, then $c_n(p_\mu) = n!\delta_{n,|\mu|}$. On other other hand, from (10.62),

$$\mathrm{ex}(p_\mu) = \begin{cases} 0, & \text{unless } \mu = (1, 1, \dots, 1), \\ t^{|\mu|}, & \text{if } \mu = (1, 1, \dots, 1). \end{cases} \tag{10.64}$$

The relation (10.63) follows immediately. □

**Third Proof of Theorem 3.28.** From (10.29), $c_k(s_\lambda) = F_\lambda \delta_{k,|\lambda|}$. Hence Lemma 10.21 implies that

$$\mathrm{ex}(s_\lambda) = F_\lambda \frac{t^{|\lambda|}}{|\lambda|!}. \tag{10.65}$$

Now we apply ex to the right-hand side of (10.40). But $c_n(h_k) = \delta_{nk}$ from the definition of the complete symmetric functions, and hence

$$\mathrm{ex}(h_k) = \frac{t^k}{k!} \tag{10.66}$$

from (10.63). Since ex is a homomorphism, this implies that

$$\begin{aligned} \mathrm{ex}\left( \det \left(h_{\lambda_i - i + j}\right)_{1 \le i,j \le \ell} \right) &= \det \left( \mathrm{ex}(h_{\lambda_i - i + j})\right)_{1 \le i,j \le \ell} \\ &= \det \left( \frac{t^{\lambda_i - i + j}}{(\lambda_i - i + j)!)} \right)_{1 \le i,j \le \ell}. \end{aligned} \tag{10.67}$$

Together with (10.65), this proves Theorem 3.28. □

**10.1.5. Classical Definition of the Schur Function.** Our definition (10.24) of the Schur function

$$s_\lambda(x) = \sum_T x^T \tag{10.68}$$

is combinatorial. Schur, however, did not originally use this definition. What Schur used is the following definition, which is actually due to Jacobi: First we restrict to $\ell$ variables, $x = (x_1, \dots, x_\ell)$. Said differently, we consider functions in the polynomial ring $\mathbb{C}[x_1, \ldots, x_\ell]$. We say that a function $f$ in $\mathbb{C}[x_1, \ldots, x_\ell]$ is skew-symmetric if

$$\pi f = (\operatorname{sgn} \pi) f \tag{10.69}$$

for all $\pi \in \mathcal{S}_\ell$, where $\pi f(x_1, \ldots, x_\ell) \equiv f(x_{\pi(1)}, \ldots, x_{\pi(\ell)})$. If $\mu = (\mu_1, \mu_2, \ldots, \mu_\ell)$ is a composition of $N$ (i.e., an ordered set of nonnegative integers with $\sum_1^\ell \mu_i = N$), we associate to $\mu$ the monomial

$$x^\mu = x_1^{\mu_1} x_2^{\mu_2} \cdots x_\ell^{\mu_\ell}. \tag{10.70}$$

Define the *alternant*

$$a_\mu(x_1, \ldots, x_\ell) = \sum_{\pi \in \mathcal{S}_\ell} (\operatorname{sgn} \pi) \pi x^\mu. \tag{10.71}$$

It is easy to verify that $a_\mu$ is skew-symmetric. Also clearly

$$a_\mu = \det(x_i^{\mu_j})_{1 \le i,j \le \ell}. \tag{10.72}$$

For example,

$$\begin{aligned} a_{(4,2,1)} &= x_1^4 x_2^2 x_3 - x_1^4 x_3^2 x_2 - x_2^4 x_1^2 x_3 + x_2^4 x_3^2 x_1 + x_3^4 x_1^2 x_2 - x_3^4 x_2^2 x_1 \\ &= \begin{vmatrix} x_1^4 & x_1^2 & x_1 \\ x_2^4 & x_2^2 & x_2 \\ x_3^4 & x_3^2 & x_3 \end{vmatrix}. \end{aligned} \tag{10.73}$$

In particular for the partition $\delta = (\ell - 1, \ell - 2, \ldots, 0)$,

$$a_\delta = \det(x_i^{\ell - j})_{1 \le i,j \le \ell} = \prod_{1 \le i < j \le \ell} (x_i - x_j), \tag{10.74}$$

the Vandermonde determinant (we have it as an exercise for the reader to check this). Note from (10.71) (or (10.72)) that if $\mu$ has two equal parts, i.e., $\mu_i = \mu_j$ for some $i \ne j$, then $a_\mu = 0$. Thus we restrict our attention to $a_\mu$'s for which the $\mu$'s have distinct parts. Note that a partition $\tilde{\lambda}$ has distinct parts if and only if it can be written (uniquely) as a vector sum $\lambda + \delta$ where $\lambda$ is again a partition.

Suppose $\lambda + \delta$ is a partition with distinct parts. Note that if $x_i = x_j$ for some $i \ne j$, $a_{\lambda+\delta}(x_1, \ldots, x_\ell) = 0$. It follows then that $a_{\lambda+\delta}(x)$ is divisible by all the factors in (10.74) and $\frac{a_{\lambda+\delta}}{a_\delta}$ is a (symmetric) polynomial in $x_1, \ldots, x_\ell$. The remarkable fact is that this is actually the symmetric function $s_\lambda$.

To prove this statement, recall the generating functions for the elementary symmetric functions and the complete symmetric functions in Exercise 10.9:

$$E(t) \equiv \sum_{n \ge 0} e_n(x) t^n = \prod_{i \ge 1} (1 + x_i t), \tag{10.75}$$

$$H(t) \equiv \sum_{n\geq 0} h_n(x)t^n = \prod_{i\geq 1} \frac{1}{1-x_i t}. \tag{10.76}$$

Thus

$$H(t)E(-t) \equiv 1 \tag{10.77}$$

as a formal power series. For $1 \leq j \leq \ell$, let $e_\ell^{(j)}$ denote the elementary symmetric function in the variables $x_1, \ldots, x_{j-1}, x_{j+1}, \ldots, x_\ell$ ($x_j$ omitted). Then

**Lemma 10.22.** *Let $\mu = (\mu_1, \ldots, \mu_\ell)$ be any composition. Consider the $\ell \times \ell$ matrices*
(10.78)

$$A_\mu = (x_j^{\mu_i})_{1\leq i,j\leq \ell}, \quad H_\mu = (h_{\mu_i-\ell+j})_{1\leq i,j\leq \ell}, \quad E = ((-1)^{\ell-i} e_{\ell-i}^{(j)})_{1\leq i,j\leq \ell}.$$

*Then*

$$A_\mu = H_\mu E. \tag{10.79}$$

**Proof.** Set $E^{(j)}(t) \equiv \sum_{n=0}^{\infty} e_n^{(j)} t^n = \sum_{n=0}^{\ell-1} e_n^{(j)} t^n$. From (10.75), $E^{(j)}(t) = \prod_{i\neq j}(1+x_i t) = \frac{E(t)}{1+x_j t}$. Then by (10.77) above, $H(t)E^{(j)}(-t) = \frac{1}{1-x_j t}$. Extracting the coefficient of $t^{\mu_i}$ on both sides, we obtain

$$\sum_{k=1}^{\ell} h_{\mu_i-\ell+k}(-1)^{\ell-k} e_{\ell-k}^{(j)} = x_j^{\mu_i}, \tag{10.80}$$

which is the desired result. □

**Theorem 10.23** (Classical definition of the Schur function). *Let $\lambda$ be a partition of length at most $\ell$. Then*

$$s_\lambda(x_1, \ldots, x_\ell) = \frac{a_{\lambda+\delta}}{a_\delta} = \frac{\det(x_i^{\lambda_j+\ell-j})_{1\leq i,j\leq \ell}}{\det(x_i^{\ell-j})_{1\leq i,j\leq \ell}}. \tag{10.81}$$

**Proof.** From the lemma above,

$$a_\mu = \det A_\mu = \det H_\mu \cdot \det E. \tag{10.82}$$

Now for $\mu = \delta$, $H_\mu = (h_{j-i})$ is upper triangular with 1's on the diagonal. Thus $a_\delta = \det E$.

Now for $\mu = \lambda + \delta$ in (10.82)

$$\frac{a_{\lambda+\delta}}{a_\delta} = \det H_{\lambda+\delta} = \det(h_{\lambda_i-i+j}), \tag{10.83}$$

which equals $s_\lambda$ by the Jacobi-Trudi identity (10.40). □

**Exercise 10.24.** Verify (10.81) for $\lambda = (2,1)$, $\ell = \ell(\lambda) = 2$.

The next identity, equation (10.84), is a generalization of the identity $\sum_\lambda F_\lambda^2 = N!$ obtained in (3.17) (see Exercise 10.27 below). There are many proofs of equation (10.84). We will give two of them (one immediately below Proposition 10.25 and another at the end of Subsection 10.2.2).

**Proposition 10.25** (Cauchy's identity). *We have the formal power series identity*

$$\sum_\lambda s_\lambda(x)s_\lambda(y) = \prod_{i,j} \frac{1}{1-x_iy_j}. \tag{10.84}$$

*Here the summation is over all partitions $\lambda$.*

**Proof.** Let us first consider the case when $x = (x_1, \dots, x_\ell, 0, \dots)$ and $y = (y_1, \dots, y_\ell, 0, \dots)$. Then $s_\lambda(x) = 0 = s_\lambda(y)$ if $r(\lambda) > \ell$. Here $r(\lambda)$ is the length of $\lambda$. Thus, the sum can be restricted to the partitions satisfying $r(\lambda) \le \ell$. From the classical definition (10.81) of the Schur functions,

$$\begin{aligned} V_\ell(x)V_\ell(y)\sum_\lambda s_\lambda(x)s_\lambda(y) &= \sum_{\lambda_1\ge\cdots\ge\lambda_\ell\ge 0} \det(x_i^{\lambda_j+\ell-j})\det(y_i^{\lambda_j+\ell-j}) \\ &= \sum_{m_1>\cdots>m_\ell\ge 0} \det(x_i^{m_j})\det(y_i^{m_j}), \end{aligned} \tag{10.85}$$

where $V_\ell(x) = \det(x_i^{\ell-j})_{i,j=1}^\ell$ and we have substituted $m_j = \lambda_j + \ell - j$. Since the last summand is symmetric in $m_j$'s and vanishes when two of $m_j$'s are the same, the above equals the sum over $m_1, \dots, m_\ell = 0, 1, \dots$ divided by $\ell!$ Thus using Andréief's formula (6.21) with discrete measure $\sum_{m=0}^\infty \delta_m$, the above sum equals

$$\det\left(\sum_{m=0}^\infty x_i^m y_j^m\right)_{i,j=1}^\ell = \det\left(\frac{1}{1-x_iy_j}\right)_{i,j=1}^\ell \tag{10.86}$$

if $x_i$, $y_j$'s are small enough. It is an exercise (see Exercise 10.26 below) to check that the last determinant equals

$$V_\ell(x)V_\ell(y)\prod_{i,j=1}^\ell \frac{1}{1-x_iy_j}. \tag{10.87}$$

Hence we obtain (10.84) when $x$ and $y$ have only $\ell$ nonzero components and $x_i, y_j$'s are small enough. Observe from equation (10.24) that if $x_i, y_i > 0$, then $s_\lambda(x), s_\lambda(y)$ are positive. The general case follows from the monotone convergence theorem when $x_i, y_j$ are nonnegative and small enough, and the product $\prod_{i,j}(1-x_iy_j)$ is convergent. Hence the Cauchy identity is obtained as a formal power series. □

**Exercise 10.26.** Prove the formula of the determinant of the Cauchy matrix

$$\det\left(\frac{1}{a_i - b_j}\right)_{i,j=1}^{\ell} = \prod_{1\le i<j\le \ell}(a_i - a_j)(b_j - b_i)\prod_{i,j=1}^{\ell}\frac{1}{a_i - b_j}. \tag{10.88}$$

The equality of the last determinant in (10.86) and (10.87) follows from this by setting $a_i = 1/x_i$ and $b_i = y_i$.

**Exercise 10.27.** Use equation (10.65) to show that the exponential specialization of the $y$-variable in the Cauchy identity implies that

$$\sum_{\lambda} s_\lambda(x) F_\lambda \frac{t^{|\lambda|}}{|\lambda|!} = \prod_i e^{tx_i}. \tag{10.89}$$

One more application of the exponential specialization of the $x$-variable implies the identity $\sum_{\lambda\vdash n} F_\lambda^2 = n!$.

**Remark 10.28.** The proof of Proposition 10.25 also directly applies to a function $f(z) = \sum_{m=0}^{\infty} c_m z^m$ which is analytic at $z = 0$ in addition to the function $\frac{1}{1-z}$ used in the proof. We find that for small enough $x = (x_1, \dots, x_\ell)$ and $y = (y_1, \dots, y_\ell)$,

$$\sum_{\lambda} s_\lambda(x) s_\lambda(y) \prod_{j=1}^{n} c_{\lambda_j+\ell-j} = \frac{\det\left(f(x_i y_j)\right)_{i,j=1}^{\ell}}{V_\ell(x)V_\ell(y)}. \tag{10.90}$$

**Exercise 10.29.** Evaluate equation (10.90) for $f(z) = \frac{1}{1-z^2}$.

**10.1.6. Schur Measure.** For two sequences $\{x_i\}$, $\{y_i\}$ such that $x_i, y_i \in [0,1)$ and $\prod_{i,j}(1 - x_i y_j) \in (0,\infty)$, we can use the Cauchy identity to define a probability measure on the set of partitions. From the proof of Proposition 10.25, we see that if $x = (x_1, x_2, \dots), y = (y_1, y_2, \dots)$ and $0 \le x_i, y_i < 1$ with $0 < \prod_{i,j}(1 - x_i y_j)^{-1} < \infty$, then $s_\lambda(x)$ and $s_\lambda(y)$ exist as finite positive functions and equation (10.24) holds as an equality of convergent expressions.

**Definition 10.30** (Schur measure)**.** Let $0 \le x_i, y_i < 1$ and let $0 < \prod_{i,j}(1 - x_i y_j) < \infty$. *Schur measure* (with parameters $\{x_i\}, \{y_i\}$) is the probability measure on the set of partitions $\lambda$ (of arbitrary length) given by

$$\mathbb{P}^{\text{Schur}}(\lambda) = \prod_{i,j}(1 - x_i y_j) \cdot s_\lambda(x) s_\lambda(y). \tag{10.91}$$

Schur measure was introduced in [**Oko01**] (see Subsection 10.3.1 below). The Poissonized Plancherel measure can be realized as a special case of Schur measure. See Section 10.3 for this and other special cases.

The length of the first part $\lambda_1$ of a partition $\lambda$ under the Poissonized Plancherel measure has a natural interpretation as the length of the longest up/right path in a two-dimensional Poisson process. (See equation (2.6).) Likewise $\lambda_1$ under Schur measure also has a natural geometric interpretation. We describe this interpretation in the next section.

## 10.2. RSK and Directed Last Passage Percolation

**10.2.1. Robinson-Schensted-Knuth (RSK) Algorithm.** Recall that the Robinson-Schensted (RS) algorithm (Section 3.1) provides a bijection $\varphi$ from $\mathcal{S}_N$ to pairs $(P, Q)$ of standard Young tableaux with the same shape $\lambda \vdash N$. Quite remarkably, Knuth realized that the RS algorithm can be extended to all so-called *generalized permutations*. Generalized permutations are 2-line arrays of positive integers

$$\pi = \begin{matrix} i_1 & i_2 & \cdots & i_n \\ j_1 & j_2 & \cdots & j_n \end{matrix} \tag{10.92}$$

whose columns are lexicographically ordered, with the top entry taking precedence, i.e., $i_1 \leq i_2 \leq \cdots \leq i_n$ and if $i_k = i_{k+1}$ for some $k$, then $j_k \leq j_{k+1}$. Note that if $i_\ell = \ell$ and the $j_\ell$s are distinct elements of $\{1, \ldots, n\}$, then $\pi \in S_n$. We say that $\{j_{q_1}, j_{q_2}, \ldots, j_{q_k}\}$ is a *weakly increasing subsequence* of $\pi$ of length $k$ if $q_1 < q_2 < \cdots < q_k$ and $j_{q_1} \leq j_{q_2} \leq \cdots \leq j_{q_k}$. An example of a generalized permutation is

$$\pi = \begin{matrix} 1 & 1 & 1 & 2 & 2 & 2 & 3 \\ 2 & 2 & 4 & 1 & 2 & 2 & 3 \end{matrix} , \tag{10.93}$$

and 22223 is the longest weakly increasing subsequence for $\pi$ and has length 5. The *type* of the upper row, type(upper row), is the sequence $(\mu_1, \mu_2, \ldots)$, where $\mu_i$ is the number of $i$'s in the upper row. The type of the lower row, type(lower row), is defined in a similar way. For the example in (10.93), type(upper row)=$(3, 3, 1, 0, 0, \ldots)$ and type(lower row)=$(1, 4, 1, 1, 0, 0, \ldots)$. If the generalized permutation $\pi$ has type(upper row)=$\alpha$ and type(lower row)=$\beta$, then we say that $\pi$ is of type $(\alpha, \beta)$. Note that necessarily $\sum_i \alpha_i = \sum_j \beta_j$.

Recall from Subsection 10.1.2 that the type, type$(T)$, of a semi-standard Young tableau $T$ is defined as the composition $\mu = (\mu_1, \ldots, \mu_m)$, where $\mu_i$ equals the number of $i$'s in $T$.

The Robinson-Schensted-Knuth (RSK) correspondence claims the following and was proved by Knuth in 1970.

**Theorem 10.31** (RSK). *Let*

$$\alpha = (\alpha_1, \ldots, \alpha_m, 0, 0, \ldots) \quad \textit{and} \quad \beta = (\beta_1, \ldots, \beta_n, 0, 0, \ldots),$$

*where $\alpha_i, \beta_j$ are nonnegative integers. Assume that $\sum_i \alpha_i = \sum_j \beta_j$. Then there is a bijection*

$$\Phi : \pi \mapsto (T, U) \tag{10.94}$$

*between generalized permutations $\pi$ of type $(\alpha, \beta)$ and pairs $(T, U)$ of semi-standard tableaux with the same shape such that* $\operatorname{type}(U) = \alpha$ *and* $\operatorname{type}(T) = \beta$. *Moreover, the length of the first row of* $\operatorname{sh}(T)$ *equals the length of the longest weakly increasing subsequence of $\pi$.*

The proof of this result is very similar to the proof of the RS result and is left as an exercise. We illustrate how $\Phi$ acts on the generalized permutation

$$\pi = \begin{matrix} 1 & 1 & 1 & 2 & 2 & 2 & 3 \\ 2 & 2 & 4 & 1 & 2 & 2 & 3 \end{matrix}$$

above. As before $T$, $U$ are constructed in steps which are the natural analogues of the steps in the RS algorithm:

$$\begin{array}{|c|}\hline 2\\\hline\end{array} \quad \begin{array}{|c|c|}\hline 2&2\\\hline\end{array} \quad \begin{array}{|c|c|c|}\hline 2&2&4\\\hline\end{array} \quad \begin{array}{|c|c|c|}\hline 1&2&4\\\hline 2\\\cline{1-1}\end{array} \quad \begin{array}{|c|c|c|}\hline 1&2&2\\\hline 2&4\\\cline{1-2}\end{array} \quad \begin{array}{|c|c|c|c|}\hline 1&2&2&2\\\hline 2&4\\\cline{1-2}\end{array} \quad \begin{array}{|c|c|c|c|c|}\hline 1&2&2&2&3\\\hline 2&4\\\cline{1-2}\end{array} = T$$

$$\begin{array}{|c|}\hline 1\\\hline\end{array} \quad \begin{array}{|c|c|}\hline 1&1\\\hline\end{array} \quad \begin{array}{|c|c|c|}\hline 1&1&1\\\hline\end{array} \quad \begin{array}{|c|c|c|}\hline 1&1&1\\\hline 2\\\cline{1-1}\end{array} \quad \begin{array}{|c|c|c|}\hline 1&1&1\\\hline 2&2\\\cline{1-2}\end{array} \quad \begin{array}{|c|c|c|c|}\hline 1&1&1&2\\\hline 2&2\\\cline{1-2}\end{array} \quad \begin{array}{|c|c|c|c|c|}\hline 1&1&1&2&3\\\hline 2&2\\\cline{1-2}\end{array} = U.$$

Observe that $\operatorname{type}(U) = (3, 3, 1, 0, 0, \dots) = \operatorname{type}(\text{upper row})$ and $\operatorname{type}(T) = (1, 4, 1, 1, 0, 0, \dots) = \operatorname{type}(\text{lower row})$. In addition, the first row of $T$ has five boxes while the length of the longest weakly increasing subsequence of $\pi$ is also 5.

**Remark 10.32.** If we restrict $\Phi$ to $\pi$'s of type $\alpha = (1, \dots, 1, 0, 0, \dots)$ and $\beta = (1, \dots, 1, 0, 0, \dots)$, i.e.,

$$\pi = \begin{matrix} 1 & 2 & \dots & N \\ \pi_1 & \pi_2 & \dots & \pi_N \end{matrix}, \tag{10.95}$$

where $i \mapsto \pi_i$ is a permutation of $\{1, \dots, N\}$, then $\Phi$ becomes the RS map $\phi$, with range $= \{(P, Q) \colon P$ and $Q$ are SYT's, $\operatorname{sh} P = \operatorname{sh} Q = \lambda \vdash N\}$. If we restrict $\Phi$ to $\pi$'s of type $\alpha = (1, \dots, 1, 0, 0, \dots)$ and $\beta = (\beta_1, \dots, \beta_m, 0, 0, \dots)$, where $\sum_i \beta_i = N$, i.e., $\pi$'s are of the form

$$\pi = \begin{matrix} 1 & 2 & \dots & N \\ \pi_1 & \pi_2 & \dots & \pi_N \end{matrix}, \tag{10.96}$$

where $1 \le \pi_i \le m$ and $\pi_j = k$ for $\beta_k$ values of $i$, then $\operatorname{Ran} \Phi = \{(P, Q) \colon \operatorname{sh} P = \operatorname{sh} Q = \lambda \vdash N, P$ is SSYT, $Q$ is SYT, $\operatorname{type}(P) = \beta\}$. Taking a union over such $\beta$ with $m$ fixed, we then have a bijection from all $\pi$'s of the form

$$\pi = \begin{matrix} 1 & 2 & \dots & N \\ \pi_1 & \pi_2 & \dots & \pi_N \end{matrix}, \tag{10.97}$$

where $1 \leq \pi_i \leq m$ onto Ran $\Phi$, the set of pairs $(P, Q)$ such that $\operatorname{sh} P = \operatorname{sh} Q = \lambda \vdash N$, $P$ is SSYT, $Q$ is SYT, and $P$ has entries from $\{1, \ldots, m\}$. Note that such $\pi$'s are words with $N$ letters obtained from an alphabet of size $m$. This latter result is due to Schensted and was the motivation for Knuth's more general correspondence.

**10.2.2. Geometric Interpretation of RSK.** A way to visualize a generalized permutation is to consider a nonnegative integer valued matrix. To a generalized permutation $\pi$ we associate a matrix $A$ where the entry $A_{ij}$ is given by the number of occurrences of the column $\substack{i \\ j}$ in $\pi$. Note that $A$ is $M \times N$ where $M$ (resp. $N$) is the largest number in the upper (resp. lower) row of $\pi$. For example, for the generalized permutation

$$\pi = \begin{matrix} 1 & 1 & 1 & 2 & 2 & 2 & 3 \\ 2 & 2 & 4 & 1 & 2 & 2 & 3 \end{matrix} \tag{10.98}$$

in (10.93), the associated matrix is

$$A = \begin{pmatrix} 0 & 2 & 0 & 1 \\ 1 & 2 & 0 & 0 \\ 0 & 0 & 1 & 0 \end{pmatrix}. \tag{10.99}$$

The $(1, 2)$-entry of $A$ is 2 since the column $\substack{1 \\ 2}$ occurs twice in (10.98), etc.

Given a nonnegative integer valued matrix $A$, let $\operatorname{row}(A) = (r_1, r_2, \ldots)$ and $\operatorname{col}(A) = (c_1, c_2, \ldots)$ be the sequences of row-sums and column-sums defined, respectively, by

$$r_i = \sum_j A_{ij}, \qquad c_j = \sum_i A_{ij}. \tag{10.100}$$

It is easy to see that

$$\operatorname{row}(A) = \operatorname{type}(\text{upper row of } \pi), \quad \operatorname{col}(A) = \operatorname{type}(\text{lower row of } \pi). \tag{10.101}$$

For the example above, $\operatorname{row}(A) = (3, 3, 1)$ and $\operatorname{col}(A) = (1, 4, 1, 1)$, which are equal to type(upper row) and type(lower row), respectively. In addition it is simple to see that the set of generalized permutations of types $\alpha$ and $\beta$ is in one-to-one correspondence with the set of nonnegative integer valued matrices $A$ such that $\operatorname{row}(A) = \alpha$ and $\operatorname{col}(A) = \beta$.

We now discuss the meaning of the weakly longest increasing subsequence of $\pi$ in $A$. If we circle the entries of $A$ corresponding to the weakly increasing subsequence $\begin{matrix} 1 & 1 & 2 & 2 & 3 \\ 2 & 2 & 2 & 2 & 3 \end{matrix}$ of (10.98), we have

$$A = \begin{pmatrix} 0 & \textcircled{2} & 0 & 1 \\ 1 & \textcircled{2} & 0 & 0 \\ 0 & 0 & \textcircled{1} & 0 \end{pmatrix}. \tag{10.102}$$

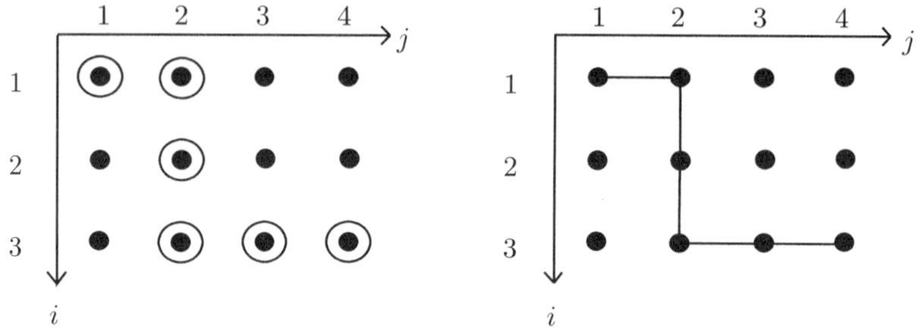


**Figure 10.9.** Directed path

The sum of the circled entries equals the length of the above weakly increasing subsequence. In general, we have the following. A *directed path* $p = \{(i_k, j_k)\}_{k=1}^{n}$ is a collection of points (sites) of $\mathbb{N}^2$ such that $(i_{k+1}, j_{k+1}) - (i_k, j_k) = (1,0)$ or $(0,1)$. For example,

$$p = \{(1,1),(1,2),(2,2),(3,2),(3,3),(3,4)\}$$

is the directed path shown in the left figure of Figure 10.9. Here we arrange the coordinates so that the directed path is moving down/right. To aid the visualization, it is sometimes convenient to connect the vertices by directed line segments; see the right figure in Figure 10.9. Let $\Pi_{M,N}$ be the collection of all directed paths $p$ from the point $(1,1)$ to the point $(M,N)$. It is a simple exercise to verify that the cardinality of $\Pi_{M,N}$ is $\binom{M+N-2}{M-1}$. It is easy to see that for a given $p = \{(i_k, j_k)\}_{k=1}^{M+N-1} \in \Pi_{M,N}$, if we circle the nonzero entries of $\{A_{i_k,j_k}\}$, they give rise to a weakly increasing subsequence of $\pi$, and we see that the length of the weakly increasing subsequence equals $\sum_{(i,j)\in p} A_{ij}$. On the other hand, every weakly increasing subsequence is contained in a weakly increasing subsequence that arises from a directed path. Hence we find that the length of the longest weakly increasing subsequence equals

$$\max_{p\in\Pi_{M,N}} \Big( \sum_{(i,j)\in p} A_{ij} \Big). \tag{10.103}$$

From the RSK correspondence, we obtain the following:

**Corollary 10.33.** *Given two finite sequences* $r = (r_1, \dots, r_M)$ *and* $c = (c_1, \dots, c_N)$ *of nonnegative integers such that* $\sum_i r_i = \sum_j c_j$, *there is a bijection* $\Psi$ *taking the nonnegative integer-valued matrices* $A = (A_{ij})_{i=1,\dots,M}^{j=1,\dots,N}$ *such that* $\operatorname{row}(A) = r$ *and* $\operatorname{col}(A) = c$ *onto the set of pairs* $(T, U)$ *of SSYT's with the same shape such that* $\operatorname{type}(U) = r$ *and* $\operatorname{type}(T) = c$. *Moreover, if we set* $\lambda = \operatorname{sh}(T)$, *then*

$$\lambda_1 = \max_{p\in\Pi_{M,N}} \Big( \sum_{(i,j)\in p} A_{ij} \Big). \tag{10.104}$$

We finish this subsection by giving another proof of the Cauchy identity in Proposition 10.25 using Corollary 10.33.

**Another Proof of Proposition 10.25.** For $x_i y_j$ sufficiently small, the right-hand side of (10.84) is

$$\begin{aligned}\prod_{i,j}\Big(\sum_{A_{ij}=0}^{\infty}(x_iy_j)^{A_{ij}}\Big) &= \sum_{A_{11},A_{12}\cdots\geq 0}^{\infty}\prod_{i,j}(x_iy_j)^{A_{ij}}\\ &= \sum_{A_{11},A_{12}\cdots\geq 0}^{\infty}\Big(\prod_i x_i^{\sum_j A_{ij}}\Big)\Big(\prod_j y_j^{\sum_i A_{ij}}\Big).\end{aligned} \tag{10.105}$$

Here the sum is over all $A_{ij}=0,1,2,\dots$ for $i,j\geq 1$. After expanding out the product, the coefficient of $x^ry^c = x_1^{r_1}\cdots x_n^{r_n}y_1^{c_1}\cdots y_m^{c_m}$ is the number of $A_{ij}$'s such that $\sum_j A_{ij}=r_i$ and $\sum_i A_{ij}=c_j$, i.e., the number of nonnegative integer matrices $A=(A_{ij})$ such that $\mathrm{row}(A)=r$ and $\mathrm{col}(A)=c$.

On the other hand, the left-hand side of (10.84) is

$$\sum_{\lambda}\Big(\sum_{\substack{U:SSYT\\ \mathrm{sh}(U)=\lambda}}x^U\Big)\Big(\sum_{\substack{T:SSYT\\ \mathrm{sh}(T)=\lambda}}y^T\Big) = \sum_{(T,U):\mathrm{sh}(T)=\mathrm{sh}(U)}x^Uy^T. \tag{10.106}$$

The coefficient of $x^ry^c$ is the number of pairs $(T,U)$ of SSYT's of the same shape such that $\mathrm{type}(U)=r$ and $\mathrm{type}(T)=c$. From Corollary 10.33, this equals the number of nonnegative integer matrices $A=(A_{ij})$ such that $\mathrm{row}(A)=r$ and $\mathrm{col}(A)=c$, which is same as the coefficient of (10.105). Hence the Cauchy identity is proved. □

### 10.2.3. Directed Last Passage Percolation.

If the matrix $A$ is random, the corresponding $\lambda=\lambda(T)$ in Corollary 10.33 becomes a random variable. There is a special ensemble of such random matrices which gives rise to Schur measure on $\lambda$. Before discussing this ensemble, we first consider *general directed last passage percolation models* and *last passage times*. Let

$$\Pi_{M,N} := \{p=\{(i_k,j_k)\}_{k=1}^{M+N-1} : (i_{k+1},j_{k+1})-(i_k,j_k)=(1,0) \text{ or } (0,1)\} \tag{10.107}$$

be the set of the directed paths from $(1,1)$ to $(M,N)$ as before.

**Definition 10.34** (Last passage time). Given a collection of random variables $a(i,j)$, $i=1,\dots,M$, $j=1,\dots,N$, the random variable

$$L_{M,N} \equiv \max_{p\in\Pi_{M,N}}\Big(\sum_{(i,j)\in p}a(i,j)\Big) \tag{10.108}$$

is called the (directed) *last passage time* from $(1,1)$ to $(M,N)$.

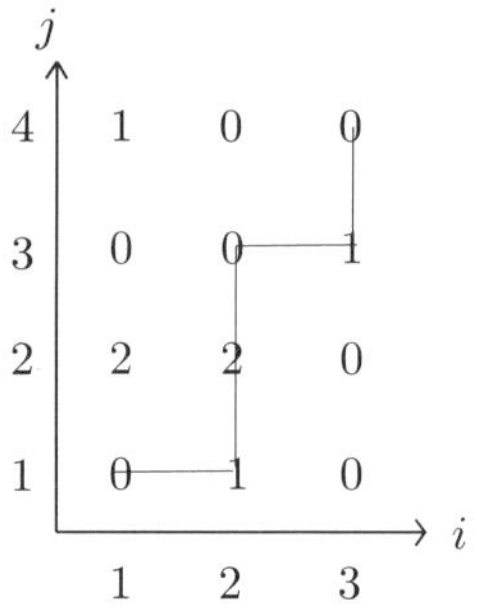


**Figure 10.10.** Direct last passage percolation

The above probabilistic model with random variable $L_{M,N}$ is called a (two-dimensional) directed last passage percolation model and is an important nonequilibrium model in statistical physics. Such models are related to randomly growing surfaces and interacting particle systems. See Section 10.7 where these connections are discussed in greater detail.

**Exercise 10.35.** Prove the recursion formula

$$L_{m,n} = \max\{L_{m-1,n}, L_{m,n-1}\} + a(m,n), \tag{10.109}$$

where $L_{m,n} = 0$ by definition for $m \leq 0$ or $n \leq 0$.

In the previous subsection we considered $p \in \Pi_{M,N}$ as a down/right path but in directed passage percolation models, it is more customary to regard $p$ as a directed up/right patht by rotating the matrix $A = (a(i,j))$ counter-clockwise by 90 degrees. See Figure 10.10.

If we interpret $a(i,j)$ as the time it takes for a traveller to pass the site $(i,j)$, the sum $\sum_{(i,j)\in p} a(i,j)$ can be thought of as the time it takes for the traveller to travel from $(1,1)$ to $(M,N)$ using the path $p$. Then $L_{M,N}$ is the largest time it takes to travel from $(1,1)$ to $(M,N)$ using an *admissible* path, hence the name *last passage percolation.* We often call $a(i,j)$ the weights of the directed last passage percolation model.

It is a remarkable fact that the generalized Schur measure arises from directed last passage percolation models with an appropriate choice of *independent* weights $a(i,j)$. More precisely, let $\Psi : A \mapsto (T(A), U(A))$ denote the bijective mapping in Corollary 10.33, let $\lambda(A)$ denote the shape of $T = T(A)$, and let $\nu$ denote a measure on the matrices $\{A\}$ with independent entries induced by the distributions on the matrix entries given by equation (10.110) below. We now show that such ensembles of matrices $\{A\}$ indeed push forward to Schur measure on the space of partitions. Moreover, all Schur measures are obtained in this way.

Let $x_1, \dots, x_M$ and $y_1, \dots, y_N$ be numbers in $[0,1)$, and let $\{a(i,j)\}$ be independent random variables with probability distribution

$$\mathbb{P}(a(i,j) = k) = (1 - x_i y_j)(x_i y_j)^k, \qquad k = 0, 1, 2, \dots, \tag{10.110}$$

i.e., $a(i,j)$ is a geometric random variable with parameter $x_i y_j$. For a given realization of the random variables $\{a(i,j)\}$, $A = (a(i,j))_{i=1,\dots,M}^{j=1,\dots,N}$ is an $M \times N$ nonnegative integer-valued matrix and $\lambda(A)$ is the induced partition valued random variable described above. We now compute the induced probability measure under the map $A \mapsto \lambda(A)$. Using independence, for a fixed nonnegative integer-valued matrix $B = (b_{ij})$,

$$\begin{aligned} \nu(B) &= \prod_{i,j} \mathbb{P}(a(i,j) = b_{ij}) = \prod_{i,j} (1 - x_i y_j)(x_i y_j)^{b_{ij}} \\ &= \prod_{i,j} (1 - x_i y_j) \cdot \prod_i x_i^{\sum_j b_{ij}} \cdot \prod_j y_j^{\sum_i b_{ij}} \\ &= \prod_{i,j} (1 - x_i y_j) \cdot \prod_i x_i^{r_i(B)} \cdot \prod_j y_j^{c_j(B)} \\ &= \prod_{i,j} (1 - x_i y_j) \cdot x^{U(B)} y^{T(B)} \end{aligned} \tag{10.111}$$

from (10.100) and Corollary 10.33. For a partition $\mu$, the induced probability that $\lambda = \mu$ is

$$\mathbb{P}(\mu) = \sum_{B:\lambda(B)=\mu} \nu(B) = \prod_{i,j} (1 - x_i y_j) \cdot \sum_{B:\lambda(B)=\mu} x^{U(B)} y^{T(B)}. \tag{10.112}$$

By taking the union over all $r$ and $c$ such that $\sum_i r_i = \sum_j c_j$ in Corollary 10.33, the set of nonnegative integer-valued matrices $B$ of size $M \times N$ is in bijection with the pair of SSYT's $(U, T)$ such that $\mathrm{sh}(U) = \mathrm{sh}(T)$ and $U_{ij} \le M$ and $T_{ij} \le N$. Hence we have

$$\mathbb{P}(\mu) = \prod_{i,j} (1 - x_i y_j) \cdot \sum_{\substack{U:\mathrm{sh}(U)=\mu \\ T:\mathrm{sh}(T)=\mu}} x^U y^T. \tag{10.113}$$

Here since $x = (x_1, \dots, x_M)$, $x^U = 0$, if $U_{ij} > M$ for some $i, j$. Similarly $y^T = 0$ if $T_{ij} > N$ for some $i, j$. Thus the sum is over $U$ and $T$ such that $U_{ij} \le M$, $T_{ij} \le N$ and $\mathrm{sh}(U) = \mu = \mathrm{sh}(T)$. From the combinatorial definition (10.24) of the Schur function, the sum above equals $s_\mu(x) s_\mu(y)$. Hence we find that $\mathbb{P}(\mu)$ equals the Schur measure of $\mu$. Therefore, in particular, we obtain:

**Theorem 10.36.** *Let $L_{N,M}$ be the last passage time of the directed passage percolation model where $a(i,j)$ are independent and distributed as*

$$\mathbb{P}(a(i,j) = k) = (1 - x_i y_j)(x_i y_j)^k, \qquad k = 0, 1, 2, \dots. \tag{10.114}$$

*Then*

$$\mathbb{P}(L_{M,N} \le n) = \mathbb{P}^{\text{Schur}}(\lambda_1 \le n) \tag{10.115}$$

*where* $\mathbb{P}^{\text{Schur}}$ *denotes Schur measure with the parameters* $\{x_1, \dots, x_M\}$ *and* $\{y_1, \dots, y_N\}$.

## 10.3. Special Cases of Directed Last Passage Percolation

In this section, we consider a few special cases of the directed last passage percolation model in Theorem 10.36. The Hammersley process gives rise to the Poissonized Plancherel measure on partitions which is the scaling limit of equations (10.114) and (10.115) with the special choice $x_i = y_j = \frac{t}{N}$, $1 \le i, j \le N$, as $N \to \infty$. (See Subsection 10.3.2 below.) Another special case is related to the Wishart ensemble in random matrix theory which arises from sample covariance matrices in statisitcs. (See Subsection 10.3.3 below.)

**10.3.1. Directed Last Passage Percolation Model with iid Geometric Weights.** In Theorem 10.36, the weights $a(i,j)$ are independent but are not identically distributed. If we take them to be independent and identically distributed (iid) geometric weights, we arrive at the model studied in detail by Johansson in [**Joh00a**]. Indeed, Schur measure was introduced by Okounkov [**Oko01**] as a generalization of the measure on the partitions corresponding to the directed last passage percolation model with the iid geometric weights of Johansson.

Fix $q \in (0,1)$. Suppose $M \ge N$. Let

$$x_1 = \dots = x_M = \sqrt{q}, \qquad y_1 = \dots = y_N = \sqrt{q}. \tag{10.116}$$

Then

$$\mathbb{P}(a(i,j) = k) = (1-q)q^k, \qquad k = 0, 1, 2, \dots. \tag{10.117}$$

This is the distribution in [**Joh00a**].

In the rest of this subsection, we evaluate Schur measure explicitly in this special case. We will see that the measure is closely related to the density function of the Gaussian unitary ensemble (GUE). (See Section 6.1.)

From the combinatorial definition (10.24) of the Schur function,

$$s_\lambda(\underbrace{\sqrt{q}, \dots, \sqrt{q}}_{M}) = \sum_{T \in SSYT_{\lambda,M}} \sqrt{q}^{\alpha_1(T)} \cdots \sqrt{q}^{\alpha_M(T)} = q^{|\lambda|/2} d_\lambda(M), \tag{10.118}$$

where $\text{type}(T) = (\alpha_1(T), \alpha_2(T), \dots)$ and $SSYT_{\lambda,M}$ is the set of SSYT's of shape $\lambda$ whose elements are taken from $\{1, 2, \dots, M\}$, and

$$d_\lambda(M) := \# SSYT_{\lambda,M}. \tag{10.119}$$

Hence, by (10.91)

$$\mathbb{P}^{\text{Schur}}(\lambda) = (1-q)^{MN} q^{|\lambda|} d_\lambda(M) d_\lambda(N). \tag{10.120}$$

Note that the measure is supported only on the partitions such that $r(\lambda) \leq \min\{M, N\} = N$.

We have the following formula for $d_\lambda(M)$.

**Lemma 10.37.** *For a partition* $\lambda = (\lambda_1, \lambda_2, \ldots, \lambda_M)$,

$$d_\lambda(M) = \prod_{1 \leq i < j \leq M} \left( \frac{\lambda_i - \lambda_j + j - i}{j - i} \right). \tag{10.121}$$

**Remark 10.38.** By definition, $d_\lambda(M) = 0$ if the length of $\lambda$ is greater than $M$. On the other hand, if $\lambda = (\lambda_1, \ldots, \lambda_\ell)$ for $\ell \leq M$, (10.121) still applies by setting $\lambda_{\ell+1} = \cdots = \lambda_M = 0$.

**Proof.** Consider the Schur polynomial $s_\lambda(z)$ in $M$ variables $z_1, \ldots, z_M$ associated with the partition $\lambda$. From the two definitions (10.24) and (10.81) of Schur polynomials, we have

$$s_\lambda(z) = \sum_T z^T = \frac{\det(z_j^{\lambda_i + M - i})_{1 \leq i,j \leq M}}{\det(z_j^{M-i})_{1 \leq i,j \leq M}}, \tag{10.122}$$

where the sum is over all SSYT's with elements in $\{1, \ldots, M\}$ and $z^T = z_1^{m_1} \cdots z_M^{m_M}$ with $m_j$ the # of times $j$ occurs in $T$. The first equality of equation (10.122) implies that when $z_1 = \cdots = z_M = 1$, $s_\lambda(1, \ldots, 1) = \sum_T 1 = d_\lambda(M)$. By the properties of Vandermonde determinants, the second equality of equation (10.122) implies that (we leave this as an exercise)

$$s_\lambda(1, c, \ldots, c^{M-1}) = c^r \prod_{1 \leq i < j \leq M} \frac{c^{\lambda_i - \lambda_j + j - i} - 1}{c^{j-i} - 1}, \tag{10.123}$$

where $r = \sum_{i=1}^M (i-1)\lambda_i$. By taking the limit $c \to 1$, we obtain

$$s_\lambda(1, 1, \ldots, 1) = \prod_{1 \leq i < j \leq M} \left( \frac{\lambda_i - \lambda_j + j - i}{j - i} \right). \tag{10.124}$$

This proves (10.121). □

Recall that Schur measure, $\mathbb{P}^{\text{Schur}}(\lambda)$, is supported only on the partitions such that $r(\lambda) \leq \min\{M, N\} = N$. For such a partition,

$\lambda = (\lambda_1, \dots, \lambda_N, \underbrace{0, \dots, 0}_{M-N})$, and hence from (10.121),

$$
\begin{aligned}
d_\lambda(M) &= d_\lambda(N) \prod_{i=1}^{N} \prod_{j=N+1}^{M} \left( \frac{\lambda_i + j - i}{j - i} \right) \\
&= d_\lambda(N)[(M-N)!]^N \prod_{i=1}^{N} \binom{\lambda_i - i + M}{M - N} \frac{(N-i)!}{(M-i)!}.
\end{aligned}
\tag{10.125}
$$

Thus for $\lambda$ with $r(\lambda) \leq N$, (10.120) becomes

$$
\begin{aligned}
\mathbb{P}^{\text{Schur}}(\lambda) &= \frac{(1-q)^{MN}[(M-N)!]^N}{\prod_{i=1}^{N}(M-i)!(N-i)!} \\
&\quad \times \prod_{1 \leq i < j \leq N} (\lambda_i - \lambda_j + j - i)^2 \prod_{i=1}^{N} q^{\lambda_i} \binom{\lambda_i - i + M}{M - N}.
\end{aligned}
\tag{10.126}
$$

This is a probability distribution function on a set of $N$ ordered *particles* $\lambda_1 \geq \cdots \geq \lambda_N \geq 0$.

If we make the change of variables $\xi_i \equiv \lambda_i - i + N$, (10.126) becomes the probability distribution

$$
\mathbb{P}(\xi_1, \dots, \xi_N) = \frac{1}{Z_N} \prod_{1 \leq i < j \leq N} (\xi_i - \xi_j)^2 \prod_{i=1}^{N} w(\xi_i)
\tag{10.127}
$$

on the set of $N$ distinct particles $\xi_1 > \xi_2 > \cdots > \xi_N \geq 0$, $\xi_i \in \mathbb{Z}$, where

$$
w(s) := q^s \binom{s + M - N}{M - N}, \qquad s \in \mathbb{N}_0 = \{0, 1, 2, \dots\},
\tag{10.128}
$$

and

$$
Z_n = \frac{q^{N(N-1)/2} \prod_{i=1}^{N} [(M-i)!(N-i)!]}{(1-q)^{MN}[(M-N)!]^N}.
\tag{10.129}
$$

This probability measure has same structural features as GUE (see Section 6.1):

$$
\mathbb{P}(\xi_1, \dots, \xi_N) \propto \Delta(\xi_1, \dots, \xi_N)^2 \prod_{j=1}^{N} f(\xi_j),
\tag{10.130}
$$

for some function $f$. It follows that the Gaudin-Mehta orthogonal polynomial technique discussed in Section 6.2, which made it possible to compute explicitly the statistics of the eigenvalues in GUE, can also be applied to the density $\mathbb{P}(\xi_1, \dots, \xi_N)$, with the change that the particles are now restricted to lie on the integers with (discrete) weights $w(s)$, $s \in \mathbb{N}_0$. This leads, in particular, to explicit formulae for the correlation functions, the distribution function of the largest particle $\xi_1$, and the gap probabilities in terms of the orthogonal polynomials with respect to the discrete weights (10.128). Thus (see (6.38)), for $n \geq N-1$,

$$\mathbb{P}(L_{M,N} \leq n - N + 1) = \mathbb{P}(\xi_1 \leq n) = \det(1 - K)_{\ell^2\{n+1,n+2,\dots\}}, \tag{10.131}$$

where

$$K(u,v) = \frac{\gamma_{N-1}}{\gamma_N} \frac{p_N(u)p_{N-1}(v) - p_{N-1}(u)p_N(v)}{u - v} \sqrt{w(u)w(v)} \tag{10.132}$$

for $u, v \in 0, 1, 2, \dots$. Here $p_k(u)$ are orthonormal polynomials with respect to the weight (10.128) whose leading coefficients are denoted by $\gamma_k$. These polynomials are proportional to the Meixner polynomials [**KS, NSU91**], which are classical discrete orthogonal polynomials and the above particle probability space is called the *Meixner ensemble*. From the algebraic point of view, the fact that the Meixner particles live on the discrete set $\mathbb{N}_0$, as opposed to GUE where the particles live on the continuum $\mathbb{R}$, plays no special algebraic role. However the asymptotic analysis of the discrete system as the number of particles goes to infinity, which we will not present here, involves additional, complicating features beyond the continuum case of GUE (see [**Joh00a, BKMM07**] for a general discussion). These additional features will also be apparent in the asymptotic analysis of the Aztec diamond in Chapter 12.

### 10.3.2. Poissonized Plancherel Measure as a Limit of Schur Measure.

We now show that the Poissonized Plancherel measure $\mathbb{P}^{P.P.}$ (see (2.5)) arises as a special limit of Schur measure with parameters as in (10.116) and with $q$ chosen appropriately. Fix $t > 0$. Take $M = N$ and choose the the parameters of Schur measure as

$$x_1 = \cdots = x_N = \frac{t}{N}, \qquad y_1 = \cdots = y_N = \frac{t}{N}. \tag{10.133}$$

Then from (10.120),

$$\mathbb{P}^{\text{Schur}}(\lambda) = \left(1 - \frac{t^2}{N^2}\right)^{N^2} \left(\frac{t}{N}\right)^{2|\lambda|} d_\lambda(N)^2. \tag{10.134}$$

Recall that $F_\lambda$ is the number of SYT's of shape $\lambda$. (See Section 3.1.)

**Lemma 10.39.** *For a partition* $\lambda$,

$$\lim_{N\to\infty} \frac{d_\lambda(N)}{N^{|\lambda|}} = \frac{F_\lambda}{|\lambda|!}. \tag{10.135}$$

**Proof.** Fix a partition $\lambda = (\lambda_1, \dots, \lambda_r)$. As $N \to \infty$, we may assume that $N \geq r$. Writing $\lambda = (\lambda_1, \dots, \lambda_r, \underbrace{0, \dots, 0}_{N-r})$, we obtain from Lemma 10.37 that (see (10.125))

$$d_\lambda(N) = \frac{\prod_{1\le i<j\le r}(\lambda_i - \lambda_j + j - i)}{\prod_{i=1}^r (\lambda_i + r - i)!} \prod_{i=1}^r \frac{(\lambda_i + N - i)!}{(N-i)!}. \tag{10.136}$$

Since $\frac{(n+x)!}{n!} = n^x(1+O(n^{-1}))$ as $n \to \infty$ for a fixed $x$, we obtian

$$\lim_{N\to\infty} \frac{d_\lambda(N)}{N^{|\lambda|}} = \frac{\prod_{1\le i<j\le r}(\lambda_i - \lambda_j + j - i)}{\prod_{i=1}^r (\lambda_i + r - i)!} \tag{10.137}$$

as $N \to \infty$. The right-hand side is $\frac{F_\lambda}{|\lambda|!}$ from the formula of $F_\lambda$ in Corollary 3.31. □

Hence we obtain

$$\lim_{N\to\infty} \mathbb{P}^{\text{Schur}}(\lambda) = e^{-t^2}\left(\frac{t^{|\lambda|}F_\lambda}{|\lambda|!}\right)^2 = \mathbb{P}^{P.P.}(\lambda) \tag{10.138}$$

for the parameters of Schur measure as in (10.133).

**Remark 10.40.** The Poissonized Plancherel measure is also the exponential specialization (see Subsection 10.1.4) of Schur measure. Recall that the exponential specialization ex is a homomorphism on the symmetric functions. Hence

$$\text{ex}\left(\mathbb{P}^{\text{Schur}}(\lambda)\right) = \text{ex}\left(\prod_{i,j}(1 - x_i y_j)\right) \text{ex}\left(s_\lambda(x)\right) \text{ex}\left(s_\lambda(x)\right). \tag{10.139}$$

From (10.65), $\text{ex}(s_\lambda) = F_\lambda \frac{t^{|\lambda|}}{|\lambda|!}$. On the other hand,

$$\begin{aligned} \text{ex}\left(\prod_{i,j}(1 - x_i y_j)\right) &= \text{ex}\left(e^{-\sum_{i,j}\sum_k \frac{x_i^k y_j^k}{k}}\right) \\ &= \text{ex}\left(e^{-\sum_k \frac{1}{k}(\sum_i x_i^k)(\sum_j y_j^k)}\right). \end{aligned} \tag{10.140}$$

From (10.62), $\text{ex}(\sum_i x_i^k)) = t$ if $k = 1$ and $= 0$ if $k \ge 2$. Hence (10.140) equals $e^{-t^2}$. Therefore,

$$\text{ex}(\mathbb{P}^{\text{Schur}}(\lambda)) = e^{-t^2} F_\lambda^2 \frac{t^{2|\lambda|}}{(|\lambda|!)^2} = \mathbb{P}^{P.P.}(\lambda). \tag{10.141}$$

To see how the parameters (10.133) are related to the exponential specialization, observe that

$$\begin{aligned} &x_1 + x_2 + \cdots = y_1 + y_2 + \cdots = t, \\ &x_1^k + x_2^k + \cdots = y_1^k + y_2^k + \cdots = \frac{t^k}{N^{k-1}} \to 0, \qquad k \ge 2, \end{aligned} \tag{10.142}$$

as $N \to \infty$, which is consistent with (10.62).

**Remark 10.41.** The limit (10.138) can also be understood from the directed last passage percolation models. The random variables $a(i,j)$, $i,j = 1,\dots,N$, of the directed last passage percolation model corresponding to Schur measure are geometric random variables with parameter $q = \frac{t^2}{N^2}$. The model is defined on the lattice $\{1,\dots,N\}^2$. As $N \to \infty$, the probability that all of $a(i,j)$ take values 0 or 1 becomes $((1-q)+(1-q)q)^{N^2} = (1-\frac{t^4}{N^4})^{N^2} \to 1$ as $N \to \infty$. Hence the probability that $n$ of $a(i,j)$ are nonzero becomes $\binom{N^2}{n}(1-q)^{N^2}q^n \to e^{-t^2}\frac{t^{2n}}{n!}$. This is the probability distribution of a Poisson random variable with parameter $t^2$. From this it is not difficult to see that $\mathbb{P}^{\text{Schur}}(L_{N,N} \le x) \to \mathbb{P}^{P.P.}(L(t) \le x)$ as $N \to \infty$.

To summarize, for iid geometric weights, we have found a Fredholm determinant formula for $\mathbb{P}^{\text{Schur}}(L_{M,N} \le x)$ in (10.131). The kernel $K$ involves the Meixner polynomials. There are explicit integral representations for the Meixner polynomials [**KS, NSU91**]. Using such representations, if we take the (Poisson) limit with $M = N$ as in Remark 10.41, we can derive the Fredholm determinant formula for $\mathbb{P}^{P.P.}(L(t) \le x)$ which we obtained in Chapter 8. See [**Joh01a**] for details.

**10.3.3. Directed Last Passage Percolation Model with Exponential Weights and the Wishart Ensemble.** In the previous section, we took the limit $x_i \to 0$, $y_i \to 0$. Another special case of Schur measure is obtained if we take $x_i \to 1$ and $y_i \to 1$.

If $X$ is a geometric random variable with parameter $q = 1 - \frac{\alpha}{L}$, then

$$\mathbb{P}\big(\frac{X}{L} > y\big) = \sum_{k>Ly}^{\infty} \frac{\alpha}{L}\big(1-\frac{\alpha}{L}\big)^k \to \int_y^{\infty} \alpha e^{-\alpha s} ds \tag{10.143}$$

as $L \to \infty$. Hence $\frac{X}{L}$ converges in distribution to an exponential random variable of parameter $\alpha$.

Let $a_1, a_2, \dots, a_M$ and $b_1, b_2, \dots, b_N$ be two sequences of positive numbers. Let the parameters for Schur measure be

$$x_i = 1 - \frac{a_i}{L}, \quad y_j = 1 - \frac{b_j}{L}, \qquad i = 1,\dots,M, \quad j = 1,\dots,N, \tag{10.144}$$

where $L > 0$. As $L \to \infty$, the random variable $a(i,j)$ converges to an exponential random variable of parameter $a_i + b_j$. Consider the directed last passage percolation model where $a(i,j)$ are independent and are distributed as an exponential random variable of parameter $a_i + b_j$. Then it also follows that the last passage time from $(1,1)$ to $(M,N)$ is the limit in distribution of the scaled last passage time $\frac{L_{N,M}}{L}$ of the directed last passage model with geometric random variables with parameters given by (10.144).

We now evaluate the limit of Schur measure with parameters as in (10.144). For simplicity we set $M = N$. The case when $M > N$ can be obtained by setting some of the $b_j$'s equal to $\infty$. We will discuss this more explicitly at the end of this section.

From (10.144), $\det(x_i^{N-j}) = L^{-N(N-1)/2}\prod_{1\le i<j\le N}(a_j - a_i)$. Let $\eta_1 > \cdots > \eta_N \ge 0$ be real numbers and consider the partition

$$\lambda = ([L\eta_1], \dots, [L\eta_N]). \tag{10.145}$$

Then $\det(x_i^{\lambda_j+N-j}) = \det((1-\frac{a_i}{L})^{[L\eta_j]+N-j})_{i,j=1}^N \to \det(e^{-a_i\eta_j})$ as $L \to \infty$ uniformly for $\eta_1, \dots, \eta_N$ in compact sets. Note that $N$ is fixed. Hence from the classical definition of Schur functions (10.81),

$$\lim_{L\to\infty} L^{-N(N-1)/2} s_\lambda(x_1, \dots, x_N) = \frac{\det(e^{-a_i\eta_j})_{1\le i,j\le N}}{\prod_{1\le i<j\le N}(a_j - a_i)} \tag{10.146}$$

uniformly for $\eta_j$'s in compact sets. By (10.91), this implies that

$$\begin{aligned} &\lim_{L\to\infty} L^N\, \mathbb{P}^{\text{Schur}}([L\eta_1], \dots, [L\eta_N]) \\ &= \frac{\prod_{i,j=1}^N (a_i + b_j)}{\prod_{i<j}(a_j - a_i)\prod_{i<j}(b_j - b_i)} \det(e^{-a_i\eta_j})\det(e^{-b_i\eta_j}) \end{aligned} \tag{10.147}$$

uniformly for $\eta_j$'s in compact sets. Note that the right-hand side is symmetric in the $\eta_i$'s. Together with equation (10.115), the above calculation implies the following:

**Proposition 10.42.** *Let $a_1, \dots, a_N$ and $b_1, \dots, b_N$ be positive real numbers. Consider the directed last passage percolation model where the weights $a(i,j)$ are independent and distributed as exponential random variables with parameter $a_i + b_j$ (hence, with expected value equal to $\frac{1}{a_i+b_j}$) for $i, j = 1, \dots, N$. Let $L(N,N)$ be the last passage time from $(1,1)$ to $(N,N)$. Then the probability distribution function of $L(N,N)$ is given by the integral*

$$\mathbb{P}(L(N,N) \le s) = \frac{1}{Z_N}\int_{[0,s]^N} \det(e^{-a_i\eta_j})\det(e^{-b_i\eta_j})\,d\eta_1\cdots d\eta_N, \tag{10.148}$$

*where*

$$Z_N = N!\frac{V_N(a)V_N(b)}{\prod_{i,j=1}^N (a_i + b_j)}. \tag{10.149}$$

*Again, $V_N(a)$ denotes the Vandermonde determinant*

$$V_N(a) := \prod_{1\le i<j\le N}(a_i - a_j) = \det(a_i^{N-j})_{i,j=1}^N. \tag{10.150}$$

**Remark 10.43.** The distribution function $\mathbb{P}(L(M,N) \leq s)$ when $M > N$ can be obtained by replacing $N$ by $M$ in the above formula and setting $b_{N+1} = \cdots = b_M = \infty$:

$$\mathbb{P}(L(M,N) \leq s) = \lim_{b_{N+1}\to\infty} \cdots \lim_{b_M\to\infty} \mathbb{P}(L(M,M) \leq s). \tag{10.151}$$

We leave it as an exercise (see Exercise 10.44) to show that

$$\mathbb{P}(L(M,N) \leq s) = \int_{[0,s]^N} p_{M,N}(\eta_1,\ldots,\eta_N) d\eta_1 \cdots d\eta_N, \tag{10.152}$$

where

$$p_{M,N}(\eta_1,\ldots,\eta_N) = \frac{1}{Z_{M,N}} \det(f_{ij}(a,\eta))_{i,j=1}^M \det(e^{-b_i\eta_j})_{i,j=1}^N \tag{10.153}$$

with

$$f_{ij}(a,\eta) = \begin{cases} e^{-a_i\eta_j}, & j = 1,\ldots,N, \\ (-a_i)^{M-j}, & j = N+1,\ldots,M, \end{cases} \tag{10.154}$$

for $i = 1,\ldots,M$ and

$$Z_{M,N} = N! \frac{V_M(a)V_N(b)}{\prod_{i=1}^M \prod_{j=1}^N (a_i + b_j)}. \tag{10.155}$$

**Exercise 10.44.** Show that

$$\lim_{\eta_{N+1},\ldots,\eta_M\to 0} \frac{\det(e^{-a_i\eta_j})_{i,j=1}^M}{\prod_{N+1\leq i<j<M}(\eta_i - \eta_j)} = \frac{\det(f_{ij}(a,\eta))_{i,j=1}^M}{\prod_{j=0}^{M-N-1} j!}, \tag{10.156}$$

where $f_{ij}$ is given in equation (10.154) above.

The density function $p_{M,N}$ in (10.152) also arises in random matrix theory. Namely, let $X$ be an $M \times N$ matrix whose entries $X_{ij}$ are independent complex normal random variables with mean 0 and variance $\frac{1}{a_i+b_j}$. Borodin and Péché [**BP08**] showed that joint density of the nonzero eigenvalues $\eta_1,\ldots,\eta_N$ of the matrix $XX^*$ is precisely given by (10.153). Thus we have:

**Proposition 10.45.** *Let $X$ be $M \times N$ matrix with independent entries given by*

$$X_{ij} \sim \mathbb{C}\mathcal{N}\left(0, \frac{1}{a_i + b_j}\right). \tag{10.157}$$

*Then $L(M,N)$ is distributed as the largest eigenvalue of $XX^*$.*

The above random matrix $XX^*$ is a generalization of a sample covariance matrix in statistics. Let $Y$ be an $M \times N$ matrix whose column vectors are independent and distributed as a complex normal vector with covariance $\Sigma$ (which is $M \times M$ Hermitian and positive definite). A statistical interpretation of $Y$ is that the columns represent the $N$ samples of an observation consisting of $M$ components. Each observation is a complex Gaussian random vector of covariance $\Sigma$. In order to estimate $\Sigma$ from observation, we define the *sample covariance* as $S := \frac{1}{N}YY^*$ and the random matrices $S$ are said to form the (complex) *Wishart ensemble* (with covariance $\Sigma$). Now note that $Y = \Sigma^{1/2}Z$ for a matrix $Z$ whose entries are iid complex standard normal. By the spectral theorem, $\Sigma = U\Lambda U^*$, where $\Lambda$ is diagonal and $U$ is unitary, and we see that the eigenvalues of $YY^* = \Sigma^{1/2}ZZ^*\Sigma^{1/2}$ are same as those of $\Lambda^{1/2}U^*ZZ^*U\Lambda^{1/2}$. Since $Z$ is an iid Gaussian matrix, so is $U^*Z$, and we find that the eigenvalues of the sample covariance matrix $S$ have the same distribution as the eigenvalues of $\frac{1}{N}XX^*$, where the entries of $X = \Lambda^{1/2}Z$ are iid, and $\mathbb{C}\mathcal{N}(0, \sigma_i^{1/2})$, where $\sigma_1, \dots, \sigma_M$ are the eigenvalues of $\Sigma$. Hence the above proposition implies that the largest eigenvalue of a complex Wishart matrix with covariance $\Sigma$ has the same distribution as $L(M, N)$ when $a_i = \sigma_i^{-1/2}$ and $b_j = 0$.

A further special case is $\Sigma = I$ (i.e., $a_i = 1$, $b_j = 0$). In this case, the random matrix ensemble of $ZZ^*$ is called the *Laguerre unitary ensemble.* This name is due to the fact that the eigenvalue density function below (see (10.160)) is of a form similar to that of GUE but the associated orthogonal polynomials are now Laguerre polynomials.

**Exercise 10.46.** Using Exericse 10.47 below, show that when the parameters are specialized, the density function in (10.153) takes the following forms.

(i) When $b_j = 0$ for all $j$, then

$$p_{M,N}(\eta) = \frac{\prod_{i=1}^{M} a_i^N}{\prod_{j=0}^{N} j! V_M(a)} \cdot \det(f_{ij}(a, \eta))_{i,j=1}^{M} \cdot V_N(\eta), \tag{10.158}$$

where $f_{ij}$ is defined in equation (10.154).

(ii) When $a_i = 0$ for all $i$, then

(10.159)

$$p_{M,N}(\eta) = \frac{\prod_{i=1}^{N} b_i^M}{N! \prod_{j=M-N}^{M-1} j! \cdot V_N(b)} \cdot V_N(\eta) \cdot \det(e^{-b_i \eta_j})_{i,j=1}^{N} \cdot \prod_{i=1}^{N} \eta_i^{M-N}.$$

(iii) When $a_i = 1$ for all $i$ and $b_j = 0$ for all $j$,

$$p_{M,N}(\eta) = \frac{1}{\prod_{j=M-N}^{M-1} j! \cdot \prod_{j=0}^{N} j!} \cdot \left|V_N(\eta)\right|^2 \prod_{i=1}^{N} \eta_i^{M-N} e^{-\eta_i}. \tag{10.160}$$

**Exercise 10.47.** Show that for given distinct $y_1, \dots, y_n$,

$$\frac{\det(e^{x_i y_j})_{i,j=1}^{n}}{V_n(x)} \to \frac{V_n(y)}{0!1! \cdots (N-1)!} \tag{10.161}$$

as $x_1, \ldots, x_n \to 0$. *Hint*: Use Remark 10.28 with $f(z) = e^z$,

$$\frac{\det(e^{x_i y_j})_{i,j=1}^n}{V_n(x)V_n(y)} = \sum_\lambda s_\lambda(x)s_\lambda(y)\prod_{j=1}^n \frac{1}{(\lambda_j + n - j)!} \tag{10.162}$$

whose leading term is attained when $\lambda = \emptyset$.

**10.3.4. Reduction to the Gaussian Unitary Ensemble.** We leave it as an exercise to check that when $M \to \infty$ and $N$ is fixed, case (iii) of Exercise 10.46 reduces to the GUE (Gaussian unitary ensemble).

**Exercise 10.48.** Recall the density function $p_{M,N}(\eta)$ in (10.160).

(i) Set

$$\eta_i = M + \xi_i\sqrt{M}. \tag{10.163}$$

Show that as $M \to \infty$ with $N$ fixed, $p_{M,N}(\eta)$ in (10.160) satisfies

$$(\sqrt{M})^N p_{M,N}(\eta) \to p_N^{GUE}(\xi) \equiv \frac{1}{(2\pi)^{N/2}\prod_{j=0}^N j!}|V_N(\xi)|^2 \prod_{j=1}^N e^{-\frac{1}{2}\xi_i^2} \tag{10.164}$$

for fixed $\xi_i$'s. Note that $p_N(\xi)$ is the eigenvalue density of the $N \times N$ GUE matrix.

(ii) Consider the directed last passage percolation model with iid exponential weights with parameter 1. Let $L(M,N)$ be the last passage time from $(1,1)$ to $(M,N)$. Show that as $M \to \infty$ with $N$ fixed,

$$\mathbb{P}\left(\frac{L(M,N) - M}{\sqrt{M}} \le s\right) \to \int_{(-\infty,s]^N} p_N^{GUE}(\xi)d^N\xi. \tag{10.165}$$

The right-hand side is the cumulative distribution function of the largest eigenvalue of an $N \times N$ GUE matrix.

(iii) Prove a result similar to (ii) for the directed last passage percolation with iid geometric weights.

**Exercise 10.49.** Show that if we repeat Exercise 10.48 for the density $p_{M,N}(\eta)$ in (10.159), but with

$$b_i = 1 + \frac{B_i}{\sqrt{M}} \tag{10.166}$$

in addition to (10.163), $(\sqrt{M})^N p_{M,N}(\eta)$ converges to

$$\frac{\prod_{j=1}^N e^{-\frac{1}{2}B_j^2}}{(2\pi)^{N/2}N!V_N(B)} \cdot V_N(\xi)\det(e^{-B_i\xi_j})_{i,j=1}^N \prod_{j=1}^N e^{-\frac{1}{2}\xi_i^2}. \tag{10.167}$$

This is the density function of the GUE with an external source, i.e., this is the eigenvalue density function of the matrix $H + \mathcal{B}$, where $H$ is a GUE matrix and $\mathcal{B}$ is a fixed Hermitian matrix with eigenvalues $B_1, \ldots, B_N$. (See, for example, [**ABDF11**, Chapter 19].)

## 10.4. Gessel's Formula for Schur Measure

In this and the next section we evaluate the distribution function of the largest part $\lambda_1$ of a partition under Schur measure. We will see that the Toeplitz determinant formula and the Fredholm determinant formula for the Poissonized Plancherel measure (see Chapters 7 and 8) extend to the case of generalized Schur measure.

The (original) *Gessel identity* is the following [**Ges90**]. We will see that Theorem 7.1 is a special case.

**Theorem 10.50** (Gessel's formula: the general case). *Let $x = (x_1, \ldots, x_M)$ and $y = (y_1, \ldots, y_M)$ be two sequences of complex numbers, some of which may be zero. Let $\psi$ be the function*

$$\psi(z) = \prod_{\ell \geq 1} \left(1 + \frac{x_\ell}{z}\right)(1 + y_\ell z), \tag{10.168}$$

*and let $D_{n-1}(\psi)$ denote the $n \times n$ Toeplitz determinant with symbol $\psi$. Then*

$$\sum_{\lambda_1 \leq n} s_\lambda(x) s_\lambda(y), = D_{n-1}(\psi), \tag{10.169}$$

*where the sum on the left-hand side is taken over all partitions $\lambda = (\lambda_1, \lambda_2, \ldots)$ such that $\lambda_1 \leq n$.*

Thus, for general parameters $x$ and $y$, which defines a Schur measure, we obtain

$$\mathbb{P}^{\text{Schur}}(\lambda_1 \leq n) = \prod_{i,j}(1 - x_i y_j) \cdot D_{n-1}(\psi). \tag{10.170}$$

**Remark 10.51.** The above formula is the analogue of Cauchy's identity (10.84) when the summation over partitions is restricted to $\lambda_1 \leq n$. The Cauchy identity is equivalent to the Szegő strong limit theorem in the current context. Let $x_i, y_j$ be real with $|x_i|, |y_j| < 1$. Formula (10.169) together with the Cauchy identity tells us that the right-hand side of (10.170) is a bona fide distribution function, and hence it converges to 1 as $n \to \infty$. This implies that $D_{n-1}(x) \to \prod_{i,j}(1 - x_i y_j)^{-1}$ as $n \to \infty$, which, one may check, is precisely Szegő's strong limit theorem for general symbols of the type (10.168). Conversely, if we assume Szegő's strong limit theorem, then letting $n \to \infty$ in (10.169), we see that

$$\sum_{\lambda} s_\lambda(x) s_\lambda(y) = \lim_{n\to\infty} \sum_{\lambda_1 \leq n} s_\lambda(x) s_\lambda(y) = \prod_{i,j} \frac{1}{1 - x_i y_j}, \tag{10.171}$$

which is just Cauchy's identity.

**Proof.** Recall the Jacobi-Trudi identity (10.41) for Schur functions. For a partition $\lambda = (\lambda_1, \ldots, \lambda_r)$ with $\lambda_1 \leq n$,

$$s_\lambda(x) = \det \left(e_{\lambda'_i - i + j}(x)\right)_{1 \leq i,j \leq n}, \tag{10.172}$$

where $\lambda'$ denotes the transpose of $\lambda$. Here the $e_k$'s denote the elementary symmetric functions, which satisfy the relation (see Exercise 10.9 (ii))

$$\sum_{m=-\infty}^{\infty} e_m(x) z^m = \prod_{\ell=1}^{M} (1 + x_\ell z), \qquad x = (x_1, \ldots, x_M), \tag{10.173}$$

where $e_m = 0$ for $m < 0$ by convention.

Using (10.172),

$$\sum_{\lambda_1 \leq n} s_\lambda(x) s_\lambda(y) = \sum_{\lambda_1 \leq n} \det \left(e_{\lambda'_i - i + j}(x)\right)_{i,j=1}^{n} \det \left(e_{\lambda'_i - i + j}(y)\right)_{i,j=1}^{n}. \tag{10.174}$$

As $\lambda_1 \leq n$, the number of parts of $\lambda'$ is $\leq n$. Hence we may denote $\lambda' = (\lambda'_1, \ldots, \lambda'_n)$ where $\lambda'_1 \geq \cdots \geq \lambda'_n \geq 0$. Setting

$$m_i \equiv \lambda'_i + n - i, \qquad 1 \leq i \leq n, \tag{10.175}$$

the sum over all $\lambda$ such that $\lambda_1 \leq n$ is equivalent to the sum over

$$m_1 > m_2 > \cdots > m_n \geq 0. \tag{10.176}$$

Hence (10.173) equals

$$\sum_{m_1 > m_2 > \cdots > m_n \geq 0} \det \left(e_{m_i + j - n}(x)\right)_{i,j=1}^{n} \det \left(e_{m_i + j - n}(y)\right)_{i,j=1}^{n}. \tag{10.177}$$

This equals

$$\sum_{m_1 > m_2 > \cdots > m_n \geq 0} \det \left(e_{m_{i+1} - j}(x)\right)_{i,j=0}^{n-1} \det \left(e_{m_{i+1} - j}(y)\right)_{i,j=0}^{n-1}. \tag{10.178}$$

The summand is a symmetric function in the $m_i$'s and also vanishes if $m_r = m_s$ for some $r \neq s$. Hence the above sum equals

$$\frac{1}{n!} \sum_{m_1, \ldots, m_n \geq 0} \det \left(e_{m_{i+1} - j}(x)\right)_{i,j=0}^{n-1} \det \left(e_{m_{i+1} - j}(y)\right)_{i,j=0}^{n-1}. \tag{10.179}$$

Applying the Andréief formula (6.21) with the discrete measure $d\mu(x) = \sum_{m=0}^{\infty} \delta_m(x)$, we obtain

$$\sum_{\lambda_1 \leq n} s_\lambda(x) s_\lambda(y) = \det \left( \sum_{m=0}^{\infty} e_{m-i}(x) e_{m-j}(y) \right)_{i,j=0}^{n-1}. \tag{10.180}$$

It remains to show that the last determinant is $D_{n-1}(\psi)$. Since $e_k = 0$ for $k < 0$, the $(i, j)$ entry of the determinant equals

$$\sum_{m=-\infty}^{\infty} e_{m-i}(x)e_{m-j}(y) = \sum_{m=-\infty}^{\infty} e_{m-(i-j)}(x)e_m(y), \quad i, j \geq 0. \tag{10.181}$$

Denoting the above sum by $d_{i-j}$ and using (10.173), we find that

$$\begin{aligned}\sum_{k=-\infty}^{\infty} d_k z^k &= \sum_m \big[\sum_k e_{m-k}(x)z^{-m+k}\big] e_m(y)z^m \\ &= \prod_{\ell=1}^{M} \left(1 + \frac{x_\ell}{z}\right)(1 + y_\ell z). = \psi(z).\end{aligned} \tag{10.182}$$

Note that all sums are finite, which means that all the operations in (10.182) are justified. Therefore the $(i, j)$ entry of the determinant in (10.180) is $\hat{\psi}_{i-j}$, and the Gessel identity is proved. □

We now show that Gessel's formula (7.3) for Poissonized Plancherel measure follows as a special case of Theorem 10.50. Set $x_1 = \cdots = x_M = \frac{t}{M}$ and $y_1 = \cdots = y_M = \frac{t}{M}$ as in (10.133), and let $M \to \infty$ in (10.168). Then

$$\psi(z) \to e^{t(z+z^{-1})} = \varphi(z), \tag{10.183}$$

and hence $D_{n-1}(\psi) \to D_{n-1}(\varphi)$. On the other hand,

$$\sum_{\lambda_1 \leq n} s_\lambda(x)s_\lambda(y) = \sum_{\lambda_1 \leq n} \frac{t^{2|\lambda|} d_\lambda(M)^2}{M^{2|\lambda|}} \tag{10.184}$$

by (10.118). The summand converges pointwise to $t^{2|\lambda|}\frac{F_\lambda^2}{(|\lambda|!)^2}$ by (10.135). It remains to show that we can take the limit under the sum. To prove this, we use the dominated convergence theorem. The above sum equals

$$\sum_{r=0}^{\infty} Q_r(M), \qquad Q_r(M) := \sum_{\lambda_1 \leq n, r(\lambda)=r} \frac{t^{2|\lambda|} d_\lambda(M)^2}{M^{2|\lambda|}}. \tag{10.185}$$

Then $Q_r(M) = 0$ for $r > M$. Suppose $r \leq M$. Then, for a partition $\lambda$ such that $r(\lambda) = r$, we may write $\lambda = (\lambda_1, \ldots, \lambda_r, \underbrace{0, \ldots, 0}_{M-r})$. For this case, from (10.136) and Corollary 3.31,

$$d_\lambda(M) = \frac{F_\lambda}{|\lambda|!} \prod_{i=1}^{r} \frac{(\lambda_i + M - i)!}{(M-i)!} \leq \frac{F_\lambda}{|\lambda|!} \prod_{i=1}^{r} \frac{(\lambda_i + M)!}{M!}. \tag{10.186}$$

Since

$$\frac{(s+M)!}{M^s M!} \leq \frac{(s+M)^s}{M^s} = (1 + \frac{s}{M})^s \leq e^{\frac{s^2}{M}} \tag{10.187}$$

for all integers $M, s > 0$, we find that

$$\frac{d_\lambda(M)}{M^{|\lambda|}} \le \frac{F_\lambda}{|\lambda|!} \prod_{i=1}^{r} e^{\frac{\lambda_i^2}{M}} \le \frac{F_\lambda}{|\lambda|!} e^{2n^2} \tag{10.188}$$

for $\lambda$ such that $r(\lambda) \le r$ and $\lambda_1 \le n$. Thus, for all $r \ge 0$, $Q_r(M)$ can be bounded above by

$$\left( \sum_{\lambda_1 \le n, r(\lambda) = r} \frac{t^{2|\lambda|} F_\lambda^2}{(|\lambda|!)^2} \right) e^{2n^2}. \tag{10.189}$$

Equation (2.6), relating Poissonized Plancharel measure (see Definition 2.2) to Hammersley's process, leads to the inequality

$$\sum_{r=0}^{\infty} \sum_{\lambda_1 \le n, r(\lambda) = r} t^{2|\lambda|} \frac{F_\lambda^2}{(|\lambda|!)^2} = e^{t^2} \cdot \mathbb{P}(L(t) \le n) \le e^{t^2} < \infty, \tag{10.190}$$

hence the dominated convergence theorem applies and we obtain

$$\lim_{M \to \infty} \sum_{\lambda_1 \le n} \frac{t^{2|\lambda|} d_\lambda(M)^2}{M^{2|\lambda|}} = \sum_{\lambda_1 \le n} \frac{t^{2|\lambda|}}{(|\lambda|!)^2} F_\lambda^2. \tag{10.191}$$

Therefore,

$$\sum_{\lambda_1 \le n} \frac{t^{2|\lambda|}}{(|\lambda|!)^2} F_\lambda^2 = D_{n-1}(\varphi), \tag{10.192}$$

which is (7.3).

**Remark 10.52.** We can also obtain (10.192) by using exponential specialization. Set $y = x$ in (10.169). It was shown in (10.65) that $\mathrm{ex}(s_\lambda) = F_\lambda \frac{t^{|\lambda|}}{|\lambda|!}$. Using the fact that ex is an algebraic homomorphism,

$$\mathrm{ex} \left( \sum_{\lambda_1 \le n} s_\lambda(x) s_\lambda(x) \right) = \sum_{\lambda_1 \le n} \frac{t^{2|\lambda|} F_\lambda^2}{(|\lambda|!)^2}. \tag{10.193}$$

On the other hand, since each of the Fourier coefficients of $\psi(z)$ is a symmetric function in $x$, we have

$$\mathrm{ex}(D_{n-1}(\psi)) = D_{n-1}(\mathrm{ex}(\psi)), \tag{10.194}$$

where

$$\mathrm{ex}(\psi) = \mathrm{ex} \left( \prod_\ell (1 + x_\ell z^{-1}) \right) \mathrm{ex} \left( \prod_\ell (1 + x_\ell z) \right). \tag{10.195}$$

But by Exercise 10.9 (ii), the generating function of the elementary symmetric function is given by

$$\prod_\ell (1 + x_\ell z) = \sum_n e_n(x) z^n. \tag{10.196}$$

From (10.63) and the definition of $e_n(x)$, we have $\mathrm{ex}(e_n) = \frac{t^n}{n!}$. Hence

$$\mathrm{ex}(\psi(z)) = e^{t(z+z^{-1})} = \varphi(z). \tag{10.197}$$

This implies that

$$\sum_{\lambda_1 \le n} F_\lambda^2 \frac{t^{2|\lambda|}}{(|\lambda|!)^2} = D_{n-1}(\varphi). \tag{10.198}$$

## 10.5. Fredholm Determinant Formula

In the proof of the Gessel formula, we used the Jacobi-Trudy identity which relates the Schur functions and the elementary symmetric functions. If we use the other Jacobi-Trudy identity which relates the Schur functions and the complete symmetric functions, one obtains a Fredholm determinant formula as in (10.199) below. This is the analogue of the result obtained in Chapter 8 for the Poissonized Plancherel measure, and the proof given below is an extension of the one given in Section 8.2. Another way to prove this Fredholm determinant formula is to use the Borodin-Okounkov-Geronimo-Case identity together with Gessel's formula. We leave this calculation as an exercise. The following theorem was first obtained by Okounkov [**Oko01**] using a different method.

**Theorem 10.53.** *Let $g$ be a bounded function on $\mathbb{Z}$ supported on $\{-m_0, -m_0+1, \dots\}$ for some $m_0$. Let $P$ denote the projection of $\ell_2(\mathbb{Z})$ onto $\ell_2(\{-m_0, -m_0+1, \dots\})$, and let $\mathcal{M}_g$ be the multiplication by $g$. Let $g(m)$ be a bounded function on $\mathbb{Z}$ which vanishes for all large negative $m$'s. Under Schur measure with parameters $x = (x_1, \dots, x_N)$ and $y = (y_1, \dots, y_N)$ satisfying $0 \le x_i, y_j < 1$ for all $i, j$, we have*

$$\mathbb{E}^{\mathrm{Schur}}\left[\prod_{i=1}^{\infty}(1 + g(\lambda_i - i + 1))\right] = \det(1 + PK\mathcal{M}_g), \tag{10.199}$$

*where $K$ has the kernel*

$$K(i,j) \equiv \sum_{m=0}^{\infty} \Psi_{i+m}(\Psi^{-1})_{-m-j}, \qquad i, j \in \mathbb{Z}, \tag{10.200}$$

*and*

$$\Psi(z) \equiv \prod_{i \ge 1} \frac{1 - x_i z^{-1}}{1 - y_i z}. \tag{10.201}$$

*Here, $\{f_k\}$ denotes the Fourier coefficients of a function $f$ on $|z| = 1$.*

Taking $g(m) = -1$ for $m \ge n+1$ and $g(m) = 0$ for $m \le n$, we find that

$$\mathbb{P}^{\mathrm{Schur}}(\lambda_1 \le n) = \det(1 - K_n), \tag{10.202}$$

where $K_n$ is the operator on $\ell^2\{n+1, n+2, \dots\}$ whose kernel is $K_n(i,j) = K(i,j)$ for $i, j \geq n+1$.

**Proof.** We proceed as in the proof of Theorem 8.12. We first assume that $g(m) \geq 0$ for all $m$. The argument leading to (8.80) in the proof of Theorem 8.12 can be applied without any change, and we obtain

$$\begin{aligned} &\mathbb{E}^{\text{Schur}}\left[\prod_{i=1}^{\infty}(1+g(\lambda_i - i + 1))\right] \\ &= \prod_{i,j}(1 - x_i y_j) \cdot \lim_{r\to\infty} \sum_{\lambda: r(\lambda) \leq r} s_\lambda(x) s_\lambda(y) \prod_{i=1}^{r}(1 + g(\lambda_i - i + 1)). \end{aligned} \tag{10.203}$$

From the Jacobi-Trudi identity (10.40), for $\lambda$ such that $r(\lambda) \leq r$,

$$s_\lambda = \det(h_{\lambda_i - i + j})_{i,j=1}^{r}, \tag{10.204}$$

where $h_\ell$ is the complete symmetric function. Recall that $h_\ell = 0$ for $\ell < 0$. Inserting this formula into equation (10.203), changing the variables $s_j = \lambda_j - j + 1$, and using Andréief's formula, we find, as in (8.85), that (10.203) equals

$$\prod_{i,j}(1 - x_i y_j) \cdot \lim_{r\to\infty} \det\left[\sum_{m=-\infty}^{\infty} h_{m+i}(x) h_{m+j}(y)(1 + g(m))\right]_{i,j=0}^{r-1}. \tag{10.205}$$

The last determinant can be written as

$$\det(A_r + B_r \mathcal{M}_g C_r), \tag{10.206}$$

where the operators are defined as follows. We use the notation $\mathbb{N}_r = \{0, 1, \dots, r-1\}$. The operators $B_r : \ell_2(\mathbb{Z}) \to \ell^2(\mathbb{N}_r)$ and $C_r : \ell_2(\mathbb{N}_r) \to \ell_2(\mathbb{Z})$ are defined by the kernels, for $m \in \mathbb{Z}$ and $i, j \in \mathbb{N}_r$,

$$B_r(i, m) = h_{m+i}(x), \qquad C_r(m, j) = h_{m+j}(y). \tag{10.207}$$

On the other hand, $A_r \equiv B_r C_r : \ell_2(\mathbb{N}_r) \to \ell_2(\mathbb{N}_r)$ whose kernel is, for $i, j \in \mathbb{N}_r$,

$$A_r(i,j) = \sum_{m=-\infty}^{\infty} h_{m+i}(x) h_{m+j}(y) = \sum_{m=-\infty}^{\infty} h_{m+i-j}(x) h_m(y). \tag{10.208}$$

As $A_r(i,j)$ is a function of $i - j$, $A_r$ is a Toeplitz matrix. The symbol of this Toepliz matrix is

$$f(z) \equiv \sum_{k=-\infty}^{\infty} \left( \sum_{m=-\infty}^{\infty} h_{m+k}(x) h_m(y) \right) z^k. \tag{10.209}$$

Using the formula of the generating function of $h_m$ in Exercise 10.9 (iii), we find that

$$f(z) = \prod_{i\geq 1} \frac{1}{(1-x_i z)(1-y_i z^{-1})}. \tag{10.210}$$

We have $A_r = T_{r-1}(f)$ and $\det(A_r) = D_{r-1}(f)$.

Note that the symbol $f$ is complex valued for $|z| = 1$ but has zero winding number since $0 \leq x_i, y_i < 1$. Hence the strong Szegő limit theorem in Theorem 5.22 (see Remark 5.23) applies, and we find that $\det(A_r) \to \prod_{i,j}(1-x_i y_j)^{-1}$ as $r \to \infty$. Hence (10.205) equals

$$\lim_{r\to\infty} \det(1 + A_r^{-1} B_r M_g C_r). \tag{10.211}$$

Theorem 5.29 implies that for each $i, j \geq 0$,

$$A_r^{-1}(i,j) \to (T(f_+^{-1})T(f_-^{-1}))(i,j) \qquad \text{as } r \to \infty, \tag{10.212}$$

where

$$f_+(z) = \prod_{i\geq 1} \frac{1}{1-x_i z}, \qquad f_-(z) = \prod_{i\geq 1} \frac{1}{1-y_i z^{-1}}. \tag{10.213}$$

The same argument as in (8.96) now implies that (10.211) equals

$$\det(1 + PCT(f_+^{-1})T(f_-^{-1})BM_g), \tag{10.214}$$

where

$$B(j,i) \equiv h_{i+j}(x), \quad C(i,j) \equiv h_{i+j}(y), \quad i \in \mathbb{Z}, \quad j = 0, 1, \ldots, \tag{10.215}$$

and again $P$ is the projection in $\ell^2(\mathbb{Z})$ onto $\ell^2(\{-m_0, -m_0+1, \ldots\})$. Now for $i \in \mathbb{Z}$ and $m = 0, 1, \ldots,$

$$\begin{aligned} (CT(f_+^{-1}))(i,m) &= \sum_{k=0}^{\infty} h_{i+k}(y)(f_+^{-1})_{k-m} \\ &= \sum_{\ell=-m}^{\infty} h_{i+m+\ell}(y)(f_+^{-1})_\ell. \end{aligned} \tag{10.216}$$

Using the fact that $(f_+^{-1})_\ell = 0$ for $\ell < 0$, this can be written as

$$(CT(f_+^{-1}))(i,m) = \sum_{\ell=-\infty}^{\infty} h_{i+m+\ell}(y)(f_+^{-1})_\ell \equiv c_{i+m}. \tag{10.217}$$

Using the generating function formula for the complete symmetric functions and using $f_+^{-1}(z) = \sum_\ell (f_+^{-1})_\ell z^\ell$, we find that

$$(10.218)\qquad \begin{aligned} \sum_{k=-\infty}^{\infty} c_k z^k &= \sum_{\ell=-\infty}^{\infty} \left[ \sum_{k=-\infty}^{\infty} h_{k+\ell}(y) z^{\ell+k} \right] (f_+^{-1})_\ell z^{-\ell} \\ &= \left( \prod_{i\geq 1} \frac{1}{1-y_i z} \right) f_+^{-1}(z^{-1}) = \Psi(z). \end{aligned}$$

Therefore, we obtain

$$(10.219)\qquad (CT(f_+^{-1}))(i,m) = \Psi_{i+m}, \qquad i \in \mathbb{Z}, \quad m = 0,1,\dots .$$

Similarly, we obtain

$$(10.220)\qquad (T(f_-^{-1})B)(m,j) = (\Psi^{-1})_{-m-j}, \qquad m = 0,1,\dots, \quad j \in \mathbb{Z},$$

and the theorem is proved for nonnegative $g$.

When $g$ is not necessarily positive, the proof follows from the argument in the last paragraph of Proof of Theorem 8.12. □

**Exercise 10.54.** Consider the directed last passage percolation problem with exponential weights with parameter $a_i + b_j$ at site $(i,j)$. Let $L(M,N)$ be the last passage time from $(1,1)$ to $(M,N)$. Using (10.202) and taking the limit $L \to \infty$ with the choice of parameters (10.144), show that

$$(10.221)\qquad \mathbb{P}(L(M,N) \leq x) = \det(1 - K_x), \qquad x > 0,$$

where $K_x$ acts on $L^2((x,\infty))$ and is given by the kernel

$$(10.222)\qquad K_x(u,v) = \int_0^\infty f(u+s)g(s+v)ds, \qquad u,v > 0.$$

Here, for an arbitrary fixed $\epsilon \in (0, \pi/2)$,

$$(10.223)\qquad f(s) \equiv \frac{1}{2\pi} \int_{\infty\cdot e^{i(\pi-\epsilon)}}^{\infty\cdot e^{i\epsilon}} e^{is\xi} \psi(\xi) d\xi, \qquad s > 0,$$

and

$$(10.224)\qquad g(s) \equiv \frac{1}{2\pi} \int_{\infty\cdot e^{i(\pi+\epsilon)}}^{\infty\cdot e^{-i\epsilon}} e^{-is\xi} \frac{1}{\psi(\xi)} d\xi, \qquad s > 0,$$

where

$$(10.225)\qquad \psi(\xi) \equiv \frac{\prod_{j=1}^M (a_j + i\xi)}{\prod_{j=1}^N (b_j - i\xi)}.$$

## 10.6. Asymptotics of Directed Last Passage Percolation

Using Gessel's formula or the Fredholm determinant formula, we can study the asymptotics of the first row $\lambda_1$ of a Young diagram under Schur measure. The result of course depends on the choice of the parameters for Schur measure. In this section, we discuss some asymptotic results for directed last passage percolation models with iid geometric weights introduced in Subsection 10.3.1. This corresponds to the case when the parameters of Schur measure are $x_i = y_j = \sqrt{q}$, $1 \le i \le M$, $1 \le j \le N$, and $x_i = y_j = 0$ otherwise. Asymptotic results for other choices of parameters can be found, for example, in [**BR00**], [**OR03**], [**OR07**], [**Joh08**] and [**BP08**].

Let $a(i,j)$ be iid geometric random variables with parameter $q$, $0 < q < 1$, i.e., $\mathbb{P}\{a(i,j) = k\} = (1-q)q^k$, $k = 0, 1, \dots$. Let $L_{M,N}$ denote the last passage time from $(1,1)$ to $(M,N)$ defined as in (10.108):

$$L_{M,N} = \max_{p \in \Pi_{M,N}} \left\{ \sum_{(i,j)\in p} a(i,j) \right\}, \tag{10.226}$$

where $\Pi_{M,N}$ is the set of up/right paths from $(1,1)$ to $(M,N)$. Combining Theorem 10.36 and (10.202), we have $\mathbb{P}(L_{M,N} \le n) = \det(1 - K_n)$, where $K_n$ is the operator on $\ell_2(\{n+1, n+2, \dots\})$ with kernel (10.200) where

$$\Psi(z) = (1 - \sqrt{q}z^{-1})^M (1 - \sqrt{q}z)^{-N}. \tag{10.227}$$

Another Fredholm determinant formula for $\mathbb{P}(L_{M,N} \le n)$ was already given in (10.131) in terms of Meixner polynomials. Using either of these formulae, one can apply the steepest-descent analysis to the associated operator as in Section 9.4. This implies that (cf. [**Joh00a**]) when $M/N \to \gamma \in (0,\infty)$,

$$\lim_{M,N\to\infty} \mathbb{P}\left( \frac{L(M,N) - \mu(M,N)}{\sigma(M,N)} \le x \right) = \det(1 - \mathbb{A}_x) = F(x), \tag{10.228}$$

where $F(x)$ is the Tracy-Widom distribution function and

$$\begin{aligned} \mu(M,N) &= \frac{2\sqrt{MNq} + (M+N)q}{1-q}, \\ \sigma(M,N) &= \frac{q^{1/6}(\sqrt{M} + \sqrt{Nq})^{2/3}(\sqrt{N} + \sqrt{Mq})^{2/3}}{(1-q)(MN)^{1/6}}. \end{aligned} \tag{10.229}$$

Johansson also proved that the same result holds for the directed last passage percolation model with the iid exponential weights introduced in Subsection 10.3.3 with

$$\mu(M,N) = (\sqrt{M} + \sqrt{N})^2, \qquad \sigma(M,N) = \frac{(\sqrt{M} + \sqrt{N})^{4/3}}{(MN)^{1/6}} \tag{10.230}$$

if the exponential weight has mean 1.

It is expected that the limit in (10.228), with appropriate choice of $\mu$ and $\sigma$, should hold for a wide class of iid weights. However, there must be some restriction on the weights. For example, consider the case when the weights are Bernoulli with parameter $p$. If $p$ is close to 1, there are many sites with value 1; hence it is reasonable to expect that there exists a path from $(1,1)$ to $(M,N)$ consisting of sites almost all of which have value 1. In this case, the fluctuation of $L_{M,N}$ is exponentially small and (10.228) does not hold. The conjecture is that if the weight does not take its largest value with high probability, then, with a suitable moment condition, the limit in (10.228) should still hold. However when $M$ and $N$ are of the same order, this limit theorem has been proven only for the geometric and exponential weights so far. The identification of $\mu$ and $\sigma$ in (10.228) for a given weight is also a significant challenge.

When $M, N \to \infty$ but $N = o(M)$, there are some general results. Glynn and Whitt [**GW91**] proved that for random variables with finite second moment, denoting the mean of the weight by $\mu$,

$$\lim_{M,N\to\infty} \frac{L(M,N) - \mu M}{\sqrt{MN}} = \alpha, \qquad N = o(M), \tag{10.231}$$

almost surely for some constant $\alpha$. It was later proven by Seppäläinen [**Sep97**] (see also [**Mar04**]) that

$$\alpha = 2\sigma, \tag{10.232}$$

where $\sigma$ is the standard deviation of the weight.

The limit theorem for general random variables is known also for "thinner" rectangles. In particular, suppose that the weight has finite $p$ moments for some $p > 2$. Let $\mu$ and $\sigma$ be the mean and the standard deviation of the weight. Then it was shown that [**BS05, BM05, Sui06**] for

$$0 < a < \frac{6}{7}\left(\frac{1}{2} - \frac{1}{p}\right), \tag{10.233}$$

we have

$$\lim_{M\to\infty} \mathbb{P}\left(\frac{L(M, M^a) - \mu(M + M^a) - 2\sigma M^{(1+a)/2}}{\sigma M^{1/2 - a/6}} \le x\right) = F(x). \tag{10.234}$$

In particular, if all moments of the weight are finite, then the limit theorem holds for $a < \frac{3}{7}$.

The result (10.234) can be understood as follows. Donsker's theorem implies that if $X_1, X_2, \dots$ are iid random variables, then

$$\frac{X_1 + \cdots + X_{[tM]} - \mu t M}{\sigma\sqrt{M}} \Rightarrow B_t, \qquad M \to \infty, \tag{10.235}$$

where $B_t$ is a standard Brownian motion. Hence for *fixed* $N$, the maximizing path consists of $N$ long horizontal parts and it is reasonable to expect that, as $M \to \infty$,

$$\frac{L(M,N) - \mu M}{\sigma\sqrt{M}} \Rightarrow \sup_{0 \le t_1 \le t_2 \le \cdots \le t_{N-1} \le 1} \left[\sum_{i=1}^{N} (B_{t_i}^{(i)} - B_{t_{i-1}}^{(i)})\right], \tag{10.236}$$

where $t_0 = 0$ and $t_N = 1$ by definition and $B_t^{(1)}, \dots, B_t^{(N)}$ are independent standard Brownian motions. This was proved in [**GW91**]. Note that this limit theorem holds for general weights with finite second moment. For a special choice of weights we can evaluate the limit more explicitly. Indeed, we have seen in Exercise 10.48 that for exponential weights, the left-hand side of (10.236) converges to a random variable $\xi(N)$ which is distributed as the largest eigenvalue of an $N \times N$ GUE matrix. Since the right-hand side of (10.236) is independent of the underlying random variables, we can conclude that

$$\frac{L(M,N) - \mu M}{\sigma\sqrt{M}} \Rightarrow \xi(N), \qquad M \to \infty. \tag{10.237}$$

This was first shown in [**GTW01, Bar01**]. From the result on GUE in (6.86) (there is a difference of normalization for the density function which accounts for a multiple of $\sqrt{2}$ in the formula below),

$$(\xi(N) - 2N)N^{1/6} \Rightarrow \chi_{TW}, \qquad N \to \infty, \tag{10.238}$$

where $\chi_{TW}$ is a Tracy-Widom random variable. Thus if we combine (10.237) and (10.238) and *assume* that the limits $M \to \infty$ and $N \to \infty$ are interchangeable, then we arrive at (10.234). This can be made rigorous as long as $N \to \infty$ at a suitably slow rate relative to $M$. The condition (10.233), in particular, is sufficient.

There is also a dynamical version of (10.236). If we take $M = [tm]$ and set $m \to \infty$ while fixing $N$, then (10.236) is changed to

$$\frac{L([tm],N) - \mu[tm]}{\sigma\sqrt{[tm]}} \Rightarrow \tilde{L}(t) \equiv \sup_{0=t_0 \le t_1 \le t_2 \le \cdots \le t_{N-1} \le t_N = t} \left[\sum_{i=1}^{N} (B_{t_i}^{(i)} - B_{t_{i-1}}^{(i)})\right] \tag{10.239}$$

for $t > 0$ where the convergence holds in distribution jointly for finitely many $t$'s. Hence if we consider $L(M,N)$ as a random process in time $M$, then its fluctuations converge to $\tilde{L}(t)$ in the finite-dimensional sense. Above we discussed that $\tilde{L}(1)$ is distributed as the largest eigenvalue of an $N \times N$ GUE matrix. By scaling, $\tilde{L}(t)$ for any fixed $t > 0$ is distributed as the largest eigenvalue of a scaled $N \times N$ GUE matrix. The joint distribution of $\tilde{L}(t)$ is precisely given by the top curve of the Brownian Dyson process that will be discussed in Chapter 13 (see Corollary 4.10 of [**BJ02**]).

## 10.7. Equivalent Models

Directed last passage percolation in the case of Schur measure is equivalent to other interesting models in probability. We discuss a few of them in this section (cf. Section 3.3).

**10.7.1. Tandem Queues.** Suppose that there are $N$ tellers and $M$ customers. Each customer is served by each of the $N$ tellers successively from teller 1 to teller $N$. Assume that initially all $M$ customers are in a queue waiting to be served by the first teller. The first customer is served by the first teller and then moves to the second teller. At that time the second customer comes forward to the first teller and is served immediately. If the second customer finishes his or her business before the first customer finishes his or her business with the second teller, the second customer will line up in the queue for the second teller, and so on. At any instance, only one customer can be served by a teller and all customers should be served by all tellers in order. We are interested in the total service time $Q(M, N)$ for all $M$ customers to be served by the $N$ tellers (see, for example, [**For10**]).

Let $a(i, j)$ be the service time for customer $j$ by teller $i$. We assume that $a(i, j)$ are independent random variables. Now consider the last customer, customer $M$. Immediately before the last customer starts to be served by teller $N$, either he/she advances from the queue behind the customer $M-1$ for the teller $N$ or he/she finished up with teller $N-1$ and moved to teller $N$ who was not serving anyone at that time. Therefore, we obtain the recurrence relation

$$Q(M, N) = \max\{Q(M-1, N), Q(M, N-1)\} + a(M, N), \tag{10.240}$$

$$Q(0, N) = Q(M, 0) = 0. \tag{10.241}$$

This is the same recurrence relation (10.109) for the last passage time $L(M, N)$ of the directed last passage percolation model. Since both $Q(M, N)$ and $L(M, N)$ satisfy the same boundary conditions, we see that $Q(M, N)$ and $L(M, N)$ have the same distribution.

A special case is when the service times are independent exponential random variables which depend on the tellers but not the customers. Namely, the parameter of the exponential random variable $a(i, j)$ is $\mu_j$. Then by Subsection 10.3.3, we find that $Q(M, N)$ is distributed as the largest eigenvalue of the sample covariance matrix $\Sigma^{1/2}YY^*\Sigma^{1/2}$ where $\Sigma = \mathrm{diag}(\mu_1, \ldots, \mu_N)$ and $Y$ is a $N \times M$ matrix with iid complex standard normal entries.

**10.7.2. TASEP.** The totally asymmetric simple exclusion process (TASEP) is one of the fundamental stochastic interacting particles systems (see, e.g., [**Lig99**]). At any time, each site $j \in \mathbb{Z}$ is either occupied by a particle or is empty. Let $\eta(t) = \{\eta_j(t)\}_{j\in\mathbb{Z}}$ be the *occupation variables* defined by $\eta_j(t) = 1$ if the site $j$ is occupied by a particle at time $t$ and $\eta_j(t) = 0$ if it is vacant. The stochastic updating rule is as follows. Particles jump to the neighboring site on the right and are allowed to do so only if the right neighboring site is empty. At the time instant when the right neighboring site is empty, a jump is performed after an exponential waiting time with mean 1. All jumps are independent of each other. An example of a configuration is

$$\begin{aligned} \eta(1) &= \cdots 1, 1, 0, 1, 0, \dots, \\ \eta(1.4) &= \cdots 1, 0, 1, 1, 0, \dots, \\ \eta(2.7) &= \cdots 1, 0, 1, 0, 1, \dots, \\ \eta(3.5) &= \cdots 0, 1, 1, 0, 1, \dots. \end{aligned} \tag{10.242}$$

Consider the particular initial condition (called the *step initial condition*)

$$\eta_j(0) = \begin{cases} 1, & j < 0, \\ 0, & j \geq 0. \end{cases} \tag{10.243}$$

Let $T(M,N)$ denote the first time that the $M$*th particle* (the particle that was at the site $-M$ at time 0) arrives at site $-M+N$, i.e., it made a total of $N$ jumps. Let $a(i,j)$ denote the waiting time for the $i$th particle to perform its $j$th jump once site $j-i$ is empty. By assumption, $a(i,j)$ are iid exponential random variables with parameter 1. For the $M$th particle to jump to the site $-M+N$, it should first arrive at the site $-M+N-1$, which takes $T(M,N-1)$ time. Also the site $-M+N$ should be vacant. This means that the $(M-1)$th particle which was at site $-M+1$ at time 0 should have made at least $N$ jumps due to the initial condition (10.243). This takes $T(M-1,N)$ units of time. Once the $M$th particle is at site $-M+N-1$ and the site $-M+N$ is empty, then it takes $a(M,N)$ time until the $M$th particle arrives at site $-M+N$. Therefore, we obtain the recurrence relation

$$\begin{aligned} T(M,N) &= \max\{T(M,N-1), T(M-1,N)\} + a(M,N), \\ T(M,0) &= T(0,N) = 0. \end{aligned} \tag{10.244}$$

This again implies that $T(M,N)$ is distributed as $L(M,N)$. Note that the special initial condition (10.243) was used in the derivation of this recursion. On the other hand, the fact that $a(i,j)$ are iid exponential was not used. The relations (10.244) hold for any (possibly dependent) $a(i,j)$'s.

We may also consider the integrated current defined as follows. For $k \geq 0$, let $\#(t,k)$ denote the number of particles that have jumped across from site $k-1$ to site $k$ by time $t$. The condition $\#(t,k) \geq n$ holds if and only if the $n$th particle is at site $k$ or its right at time $t$. Hence we have

$$\mathbb{P}(\#(t,k) \geq n) = \mathbb{P}(T(n,n+k) < t) = \mathbb{P}(L(n,n+k) < t). \tag{10.245}$$

For initial conditions other than (10.243), it is still possible to relate TASEP to a directed last passage percolation problem. For example, under the *flat initial condition* that $\eta_j(0) = 1$ if $j$ is even and $= 0$ if $j$ is odd, then one can show that TASEP is related to directed last passage percolation in the triangle with sites $(i, j)$, $i, j \leq M$ and $i + j \geq M$. On the other hand, if one considers TASEP only on the positive integer sites with the condition that at the origin a new particle is created at a certain rate if it is empty, then the associated directed last passage percolation model is defined on the triangle with sites $(i, j)$, $i, j \leq M$ and $j \leq i$. See, for example, [**BF08, BFS08**] and references therein for more discussion.

**10.7.3. Corner Growth Model.** Consider the directed last percolation model where the weights are nonnegative (see, for example, [**Joh00a**]). Define the random set

$$\Gamma(t) = \{(i, j) \in \mathbb{N}^2 : L(i, j) \leq t\}. \tag{10.246}$$

Note that since the weights are positive, $L(M, N)$ is larger than $L(M-1, N)$ and $L(M, N-1)$. Hence if $(M, N) \in \Gamma(t)$, then $(M-1, N), (M, N-1) \in \Gamma(t)$. Therefore, $\Gamma(t)$ looks like the left figure in Figure 10.11. The set $\Gamma(t)$ is a random growing set. The sites which can be added to $\Gamma(t)$ are marked as an $\times$ in the right figure in Figure 10.11. In other words, only the *corner sites* are allowed to grow.

For the special case when the weights are iid exponential, it is easy to check from the recurrence relation (10.109) for $L(M, N)$ that once a site $(i, j)$ becomes a corner site, it takes an exponential waiting time for the site to be added to the figure. In this case, the waiting times are all independent of each other. Hence in this case, at each instance, all possible corner sites are added into the figure independently of each other at the rate 1. As $t \to \infty$, the set $\Gamma(t)$ grows linear in time. The shape of $\Gamma(t)$ can be studied using (10.230). Namely, the set $\frac{1}{t}\Gamma(t)$ converges in probability to a deterministic set $\Gamma_\infty$ where

$$\Gamma_\infty = \{(x, y) \in \mathbb{R}_+^2 : (\sqrt{x} + \sqrt{y})^2 \leq 1, \ x, y \in [0, 1]\}. \tag{10.247}$$

**Figure 10.11.** $\Gamma(t)$

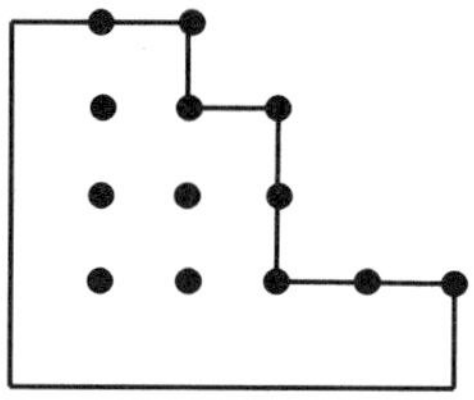

**Figure 10.12.** $\Gamma(t)$

The boundary of the random set $\Gamma(t)$ fluctuates around the curve $t\Gamma_\infty$ and along any fixed direction $\{(cx, x) : x \in \mathbb{R}_+\}$, $c > 0$, the fluctuations are of order $O(t^{1/3})$ and are distributed as $F(t)$, the Tracy-Widom distribution, after a proper scaling. More precisely, let $\overline{\Gamma(t)} = \Gamma(t) + [-1, 0]^2$ be the set in $\mathbb{R}^2_+$ which is obtained from $\Gamma(t)$ by filling in between the occupied sites as in Figure 10.12. For $c \in (0, \infty)$, define the random variable $\gamma_c(t)$ as

(10.248) $$\gamma_c(t) = \max\{u \geq 0 : (cu, u) \in \overline{\Gamma(t)}\}.$$

Then (see [**Joh00a**])

(10.249) $$\lim_{t\to\infty} \mathbb{P}\left(\frac{\gamma_c(t) - \frac{1}{(\sqrt{c}+1)^2}t}{c^{-1/6}(\sqrt{c}+1)^{-4/3}t^{1/3}} \geq -x\right) = F(x).$$

Chapter 11

# Determinantal Point Processes

As discussed in the Introduction, Theorem 1.2 on the limiting statistics of $\ell_N$ can be reinterpreted and generalized to the following result. Let $\mu_1, \ldots, \mu_k$ be the number of boxes in rows $1, \ldots, k$ of a Young diagram of size $N$ under Plancherel measure. Then, as $N \to \infty$, $\mu_1, \ldots, \mu_k$ behave statistically like the first $k$ eigenvalues $\lambda_1, \ldots, \lambda_k$ of a (large) GUE random matrix. This result relating two configurations of points on the line, $\{\mu_1, \ldots, \mu_k\}$ and $\{\lambda_1, \ldots, \lambda_k\}$, is best described in the language of point processes. A basic reference for the theory of point processes is [**DVJ03**]. The relationship between point processes and their correlation functions (see below) was investigated in detail by Lenard [**Len75a, Len75b, Len73**]. Recent expositions of Lenard's work, taking into account recent developments in random matrix theory, are [**Sos00, Joh06**].

We follow [**Joh06**] closely, often repeating certain arguments/explanations verbatim. Let $\Lambda$ be a complete separable metric space. A (boundedly finite) counting measure on $\Lambda$ is a measure $\xi$ for which $\xi(B)$ is a finite, nonnegative integer for each bounded Borel set $B \subset \Lambda$. Let $\mathcal{N}(\Lambda)$ denote the set of all counting measures $\xi$ on $\Lambda$. Typically, $\Lambda$ will be $\mathbb{R}$ or $\mathbb{Z}$, some subset of these, or the disjoint union of several copies of $\mathbb{R}$ or $\mathbb{Z}$. The smallest $\sigma$-algebra on $\mathcal{N}(\Lambda)$ for which the mapping $\xi \mapsto \xi(B)$ are measurable for all Borel sets $B \subset \Lambda$ is denoted by $\mathcal{F}_{\mathcal{N}(\Lambda)}$.

Clearly, any $\xi \in \mathcal{N}(\Lambda)$ can be written in the form $\xi = \sum_i \delta_{x_i}$ for a suitable countable set $\{x_1, x_2, \ldots\} \subset \Lambda$. This notation allows for multiple points, i.e., $x_i = x_j$ for some $i \neq j$. In this presentation we only consider $\xi$ for which $\xi(\{x\}) \leq 1$ for all $x \in \Lambda$; such $\xi$'s are called *simple*. A *point process* on $\Lambda$ is a probability space $(\mathcal{N}(\Lambda), \mathcal{F}_{\mathcal{N}(\Lambda)}, \mathbb{P})$ for some probability measure $\mathbb{P}$ on $(\mathcal{N}(\Lambda), \mathcal{F}_{\mathcal{N}(\Lambda)})$. In this presentation $\mathbb{P}(\xi \text{ is simple}) = 1$; such processes are called *simple processes*.

Basic information about point processes is carried by *correlation functions*, which will be defined below (see Definition 11.1). We begin by considering, for any $n \geq 1$, the map $X_n : \mathcal{N}(\Lambda) \to \mathcal{N}(\Lambda^n)$ defined by

$$\xi = \sum_i \delta_{x_i} \mapsto X_n(\xi) = \sum_{x_{i_1} \neq x_{i_2} \neq \cdots \neq x_{i_n}} \delta_{(x_{i_1}, \ldots, x_{i_n})}. \tag{11.1}$$

Note that $X_n(\xi)$ is boundedly finite because $X_n(\xi)(B \times \cdots \times B) \leq (\xi(B))^n < \infty$ for any bounded Borel set $B \subset \Lambda$.

We claim that $X_n$ is measurable as a map from $(\mathcal{N}(\Lambda), \mathcal{F}_{\mathcal{N}(\Lambda)})$ to $(\mathcal{N}(\Lambda^n), \mathcal{F}_{\mathcal{N}(\Lambda^n)})$. To prove this it is sufficient (we leave this as an exercise) to show that for any $k \in \{0, 1, \ldots\}$ and any bounded Borel sets $B_1, \ldots, B_n \subset \Lambda$, the set

$$\begin{aligned} &X_n^{-1}\{\xi^{(n)} \in \mathcal{N}(\Lambda^n) : \xi^{(n)}(B_1 \times \cdots \times B_n) = k\} \\ &\qquad = \{\xi : X_n(\xi)(B_1 \times \cdots \times B_n) = k\} \end{aligned} \tag{11.2}$$

lies in $\mathcal{F}_{\mathcal{N}(\Lambda)}$. For $\xi = \sum \delta_{x_i}$,

$$X_n(\xi)(B_1 \times \cdots \times B_n) = \sum_{i_1 \neq \cdots \neq i_n} \chi_{B_1}(x_{i_1}) \cdots \chi_{B_n}(x_{i_n}). \tag{11.3}$$

We leave it to the reader as an exercise to check that it is sufficient to consider the case where either $B_i = B_j$ or $B_i \cap B_j = \emptyset$ for any pair $i \neq j$. For such $B_i$'s, after rearrangement,

$$B_1 \times \cdots \times B_n = A_1^{n_1} \times \cdots \times A_\ell^{n_\ell} \tag{11.4}$$

for suitable disjoint bounded Borel sets $A_1, \ldots, A_\ell$, and $n_1 + \cdots + n_\ell = n$. As $\chi_{A_i}(x)\chi_{A_j}(x) = 0$ for $i \neq j$, it follows that (11.3) breaks up into a product of the form

$$\prod_{i=1}^{\ell} \left( \sum_{j_1 \neq \cdots \neq j_{n_1}} \chi_{A_i}(x_{j_1}) \cdots \chi_{A_i}(x_{j_{n_i}}) \right). \tag{11.5}$$

But, for any $1 \leq i \leq \ell$,

$$\sum_{j_1 \neq \cdots \neq j_{n_i}} \chi_{A_i}(x_{j_1}) \cdots \chi_{A_i}(x_{j_{n_i}}) = \xi(A_i)(\xi(A_i) - 1) \cdots (\xi(A_i) - n_i + 1). \tag{11.6}$$

Thus, for $A_1, \ldots, A_\ell$ disjoint,

$$X_n(\xi)(A_1^{n_1} \times \cdots \times A_\ell^{n_\ell}) = \prod_{i=1}^{\ell} \xi(A_i)(\xi(A_i) - 1) \cdots (\xi(A_i) - n_i + 1). \tag{11.7}$$

As $\{\xi : \xi(A_i) = m\}$ are measurable for all $m$, it follows that

$$X_n^{-1}\{\xi^{(n)} \in \mathcal{N}(\Lambda^n) : \xi^{(n)}(B_1 \times \cdots \times B_n) = k\} \in \mathcal{F}_{\mathcal{N}(\Lambda)}.$$

Therefore, $X_n$ is a measurable map from $(\mathcal{N}(\Lambda), \mathcal{F}_{\mathcal{N}(\Lambda)})$ to $(\mathcal{N}(\Lambda^n), \mathcal{F}_{\mathcal{N}(\Lambda^n)})$.

It follows that for a given probability measure $\mathbb{P}$ on $(\mathcal{N}(\Lambda), \mathcal{F}_{\mathcal{N}(\Lambda)})$, $X_n$ induces a point process on $\Lambda^n$ where

$$\mathbb{P}^{(n)}\{\Delta\} = \mathbb{P}\{X_n^{-1}(\Delta)\} \tag{11.8}$$

for any $\Delta \in \mathcal{F}_{\mathcal{N}(\Lambda^n)}$. Clearly, as $(\mathcal{N}(\Lambda), \mathcal{F}_{\mathcal{N}(\Lambda)}, \mathbb{P})$ is simple, so is $(\mathcal{N}(\Lambda^n), \mathcal{F}_{\mathcal{N}(\Lambda^n)}, \mathbb{P}^{(n)})$.

For a bounded Borel set $A \subset \Lambda^n$, define

$$M_n(A) \equiv \mathbb{E}[X_n(\xi)(A)]. \tag{11.9}$$

If $\eta$ is a random variable, then $\mathbb{E}\left[\eta(\eta-1)\cdots(\eta-k+1)\right]$ is the $k$th *factorial moment* of $\eta$. By formula (11.7), for disjoint bounded sets $A_1, \ldots, A_\ell$, $n_1 + \cdots + n_\ell = n$,

$$M_n(A_1^{n_1} \times \cdots \times A_\ell^{n_\ell}) = \mathbb{E}\left[\prod_{i=1}^{\ell} \xi(A_i)(\xi(A_i)-1)\cdots(\xi(A_i)-n_i+1)\right]. \tag{11.10}$$

For this reason $M_n$ is called the *factorial moment measure* for $(\mathcal{N}, \mathcal{F}_{\mathcal{N}(\Lambda)}, \mathbb{P})$.

Let $\lambda$ be a given (reference) measure on $\Lambda$. For example, if $\Lambda = \mathbb{R}$ of $\mathbb{Z}$, then we may take $\lambda$ to be Lebesgue measure or standard counting measure, respectively.

**Definition 11.1.** If $M_n$ is absolutely continuous with respect to $\lambda^n$ on $\Lambda^n$, i.e.,

$$M_n(A_1 \times \cdots \times A_n) = \int_{A_1 \times \cdots \times A_n} \rho_n(x_1, \ldots, x_n)\lambda(dx_1)\cdots\lambda(dx_n) \tag{11.11}$$

for all Borel sets $A_1, \ldots, A_n \subset \Lambda$, we call the symmetric nonnegative functions $\rho_n(x_1, \ldots, x_n)$ the $n$th *correlation function* or *correlation density* for $M_n$.

In particular, in the absolutely continuous case, we see from (11.10) and (11.11) that

$$\mathbb{E}[\xi(A)] = \int_A \rho_1(x_1)\lambda(dx_1). \tag{11.12}$$

In the case that $\Lambda = \mathbb{R}$ and $M_n$ is absolutely continuous with respect to Lebesgue measure, it follows that for distinct points $y_1, \ldots, y_n \in \mathbb{R}$,

$$\rho_n(y_1, \ldots, y_n) = \lim_{\Delta y_i \to 0} \frac{\mathbb{P}\{\text{one particle in each } (y_i, y_i + \Delta y_i)\}}{\Delta y_1 \cdots \Delta y_n}. \tag{11.13}$$

In the (simple) discrete case, say $\Lambda = \mathbb{Z}$, $M_n$ is of course always absolutely continuous with respect to counting measure and for distinct points $y_1, \ldots, y_n \in \mathbb{Z}$,

$$\rho_n(y_1, \ldots, y_n) = \mathbb{P}\{\text{there are particles at } y_1, \ldots, y_n\}. \tag{11.14}$$

Now suppose that $\phi : \Lambda \to \mathbb{C}$ is a bounded Borel function with support in a bounded Borel set $B$. Then, for $\xi = \sum \delta_{x_i}$, we have

$$\prod_i (1 + \phi(x_i)) = \sum_{n=0}^{\infty} \frac{1}{n!} \sum_{x_{i_1} \neq \cdots \neq x_{i_n}} \phi(x_{i_1}) \cdots \phi(x_{i_n}), \tag{11.15}$$

hence

$$\begin{aligned} \prod_i (1 + \phi(x_i)) &= \sum_{n=0}^{\infty} \frac{1}{n!} \sum_{x_1 \neq \cdots \neq x_{i_n}} \delta_{x_i}(\phi) \cdots \delta_{x_{i_n}}(\phi) \\ &= \sum_{n=0}^{\infty} \frac{1}{n!} \int_{\Lambda^n} \left( \prod_{i=1}^{n} \phi(y_i) \right) X_n(\xi)(d^n y). \end{aligned} \tag{11.16}$$

In particular, $\xi = \sum \delta_{x_i} \mapsto \prod_i (1 + \phi(x_i))$ is measurable, and if

$$\sum_{n=0}^{\infty} \frac{\|\phi\|_\infty^n}{n!} M_n(B^n) < \infty, \tag{11.17}$$

it follows by Fubini's theorem and definition (11.9) that

$$\mathbb{E}\left[ \prod_i (1 + \phi(x_i)) \right] = \sum_{n=0}^{\infty} \frac{1}{n!} \int \left( \prod_{i=1}^{n} \phi(y_i) \right) M_n(d^n y). \tag{11.18}$$

If, in addition, $M_n$ is absolutely continuous with respect to $\lambda^n$, then (11.19)

$$\mathbb{E}\left[ \prod_i (1 + \phi(x_i)) \right] = \sum_{n=0}^{\infty} \frac{1}{n!} \int \left( \prod_{i=1}^{n} \phi(y_i) \right) \rho_n(y_1, \ldots, y_n) \lambda^{\otimes n}(d^n y).$$

Formula (11.19) is useful when computing the gap probabilities (of Chapter 4). Let $B$ be a bounded Borel set, then for $\phi = -\chi_B$

$$\begin{aligned} \mathbb{P}(\text{ no particles in B }) &= \mathbb{E}\left[ \prod (1 - \chi_B(x_i)) \right] \\ &= \sum_{n=0}^{\infty} \frac{(-1)^n}{n!} \int_{B^n} \rho_n(y_1, \ldots, y_n) \lambda^{\otimes n}(d^n y). \end{aligned} \tag{11.20}$$

Below we will be interested in processes $\xi$ on $\mathbb{R}$ or possibly a discrete subset of $\mathbb{R}$, which have a rightmost particle. If there is a $t$ such that $\xi_{(t,\infty)} < \infty$, we say that $\xi$ has a *last particle*. Clearly, if $\xi$ has a last particle, then $\xi_{(t,\infty)} < \infty$ for all $t$ since $\xi(A) < \infty$ for all bounded Borel sets $A$. If $x_1(\xi) \leq \cdots \leq x_n(\xi)$ are the finitely many particles in $(t, \infty)$, we define

$x_{\max}(\xi) = x_n(\xi)$ to be the position of the last particle. If $\mathbb{E}[\xi_{(t,\infty)}] < \infty$ for some $t \in \mathbb{R}$, then $\xi$ has a last particle almost surely.

**Proposition 11.2.** *Consider a process $\xi$ on $\mathbb{R}$ or a subset of $\mathbb{R}$ all of whose correlation functions exist and assume that*

$$\sum_{n=0}^{\infty} \frac{1}{n!} \int_{(t,\infty)^n} \rho_n(x_1, \dots, x_n) \lambda^{\otimes n}(d^n x) < \infty \tag{11.21}$$

*for each $t \in \mathbb{R}$. Then the process has a last particle and*

$$\mathbb{P}\{x_{\max}(\xi) \leq t\} = \sum_{n=0}^{\infty} \frac{(-1)^n}{n!} \int_{(t,\infty)^n} \rho_n(x_1, \dots, x_n) \lambda^{\otimes n}(d^n x). \tag{11.22}$$

**Proof.** From the monotone convergence theorem, equation (11.12), and the condition (11.21),

$$\begin{aligned} \mathbb{E}[\xi_{(t,\infty)}] = \lim_{s\to\infty} \mathbb{E}[\xi_{(t,s)}] &= \lim_{s\to\infty} \int_{(t,s)} \rho_1(x)\lambda(dx) \\ &= \int_t^{\infty} \rho_1(x)\lambda(dx) < \infty. \end{aligned} \tag{11.23}$$

Hence the process has a last particle almost surely. But by formula (11.20),

$$\begin{aligned} \mathbb{P}\{ \text{ no particles in } (t,\infty)\} &= \lim_{s\to\infty} \mathbb{P}\{ \text{ no particles in } (t,s)\} \\ &= \lim_{s\to\infty} \sum_{n=0}^{\infty} \frac{(-1)^n}{n!} \int_{(t,s)^n} \rho_n(x_1, \dots, x_n) \lambda^{\otimes n}(d^n x) \\ &= \sum_{n=0}^{\infty} \frac{(-1)^n}{n!} \int_{(t,\infty)^n} \rho_n(x_1, \dots, x_n) \lambda^{\otimes n}(d^n x) \end{aligned} \tag{11.24}$$

by the dominated convergence theorem. □

We now consider some examples of point processes.

**Example 11.3** (Poisson process with intensity $r$). We assume that $r \in L^1(\mathbb{R})$ and that $r$ is nonnegative. The case where $r$ is locally $L^1$ is left to the reader as an exercise. Recall (see for example [**Dur96**]) that a Poisson process with intensity $r$ is constructed from a collection of independent random variables $(N, X_1, X_2, \dots)$ on a probability space $(\Omega, \mathcal{F}, \mathbb{P}_\Omega)$ where $N$ is a Poisson random variable with parameter $\int_{\mathbb{R}} r(x)dx$, i.e.,

$$\mathbb{P}_\Omega\{N = k\} = e^{-\int_{\mathbb{R}} r(x)dx} \frac{(\int_{\mathbb{R}} r(x)dx)^k}{k!}, \quad k \geq 0, \tag{11.25}$$

and the $X_i$'s are identically distributed on $\mathbb{R}$ with density $\frac{r(x)}{\int_{\mathbb{R}} r(x)dx}$. Consider the map $\Phi : \Omega \to \mathcal{N}(\mathbb{R})$ given by

$$\Phi(\omega) = \xi = \sum_{i=1}^{N(\omega)} \delta_{X_i(\omega)} \tag{11.26}$$

for $\omega \in \Omega$. It is easy to see that $\Phi$ is a measurable map, and we define $\mathbb{P}$ on $(\Omega, \mathcal{F})$ as the push forward of $\mathbb{P}_\Omega$, i.e.,

$$\mathbb{P}\{\mathcal{A}\} = \mathbb{P}_\Omega\{\Phi^{-1}(\mathcal{A})\}, \qquad \mathcal{A} \in \mathcal{F}_{\mathcal{N}(\mathbb{R})}. \tag{11.27}$$

This constructs a simple (exercise) point process $(\Omega, \mathcal{F}_{\mathcal{N}(\mathbb{R})}, \mathbb{P})$. Note that for any Borel set $B \subset \mathbb{R}$,

$$\begin{aligned} \mathbb{P}\{\xi_B = k\} =& \mathbb{P}_\Omega \left\{ |B \cap \{x_1(\omega), \ldots, x_{N(\omega)}(\omega)\}| = k \right\} \\ =& \sum_{n=k}^{\infty} \mathbb{P}_\Omega\{N = n\} \mathbb{P}_\Omega\{|B \cap \{x_1, \ldots, x_N)\}| = k\} \\ =& \sum_{n=k}^{\infty} e^{-s} \frac{s^n}{n!} \cdot \binom{n}{k} \left(\frac{s_B}{s}\right)^k \left(1 - \frac{s_B}{s}\right)^{n-k}, \end{aligned} \tag{11.28}$$

where $s_B = \int_B r(x)dx$ and $s = \int_{\mathbb{R}} r(x)dx$. Therefore,

$$\mathbb{P}\{\xi_B = k\} = e^{-\int_B r(x)dx} \frac{(\int_B r(x)dx)^k}{k!}. \tag{11.29}$$

Thus, $\xi_B$ is a Poisson random variable with parameter $\int_B r(x)dx$. A similar calculation (which we leave as an exercise) shows that $\xi_B$ and $\xi_C$ are independent if $B \cap C = \emptyset$. Observe now that if $A_1, \ldots, A_k$ are disjoint bounded Borel sets and $n_1 + \cdots + n_k = n$, then by (11.10), independence, and (11.29),

$$\begin{aligned} M_n(A_1^{n_1} \times \cdots \times A_k^{n_k}) =& \mathbb{E}\left[\prod_{i=1}^{k} \xi_{A_i}(\xi_{A_i} - 1) \cdots (\xi_{A_i} - n_i + 1)\right] \\ =& \prod_{i=1}^{k} \sum_{\ell=0}^{\infty} \ell(\ell - 1) \cdots (\ell - n_i + 1) e^{-s_{A_i}} \frac{s_{A_i}^\ell}{\ell!}. \end{aligned} \tag{11.30}$$

Since

$$\begin{aligned} \sum_{\ell=0}^{\infty} \ell(\ell - 1) \cdots (\ell - k + 1) \frac{s^\ell}{\ell!} &= \frac{d^k}{dt^k}\bigg|_{t=1} \sum_{\ell=0}^{\infty} \frac{(ts)^\ell}{\ell!} \\ &= \frac{d^k}{dt^k}\bigg|_{t=1} e^{ts} = s^k e^s, \end{aligned} \tag{11.31}$$

we obtain

$$M_n(A_1^{n_1} \times \cdots \times A_k^{n_k}) = \prod_{i=1}^{k} s_{A_i}^{n_i} = \int_{A_1^{n_1} \times \cdots \times A_k^{n_k}} \left(\prod_{i=1}^{n} r(x_i)\right) d^n x. \tag{11.32}$$

It follows that for all bounded Borel sets $B_1, \ldots, B_n$

$$M_n(B_1 \times \cdots \times B_n) = \int_{B_1 \times \cdots \times B_n} \left( \prod_{i=1}^{n} r(x_i) \right) d^n x, \tag{11.33}$$

and hence $M_n$ is absolutely continuous with respect to Lebesgue measure, and the correlation functions $\rho_n$ are given by

$$\rho_n(x_1, \ldots, x_n) = r(x_1) \cdots r(x_n), \qquad x_i \in \mathbb{R}. \tag{11.34}$$

**Example 11.4** (Symmetric probability densities: random matrix theory). If $u_N(x_1, \ldots, x_N)$ is a symmetric probability density on $\mathbb{R}^N$, then $(x_1, \ldots, x_N) \mapsto \xi = \sum_{i=1}^{N} \delta_{x_i}$ gives rise to a finite point process $(\mathcal{N}(\Lambda), \mathcal{F}_{\mathcal{N}(\Lambda)}, \mathbb{P})$ where for any bounded Borel set $B$, and any $k \in \{1, \ldots, N\}$

$$\mathbb{P}\{\xi(B) = k\} = \frac{N!}{k!(N-k)!} \int_{B^k \times (B^c)^{N-k}} u_N(x_1, \ldots, x_N) d^N x. \tag{11.35}$$

Now, from (11.15), for any $\phi$ with bounded support in a bounded Borel set $B$, we have

(11.36)
$$\begin{aligned}
\mathbb{E}\left[ \prod_{i=1}^{N} (1 + \phi(x_i)) \right] &= \sum_{n=0}^{N} \frac{1}{n!} \mathbb{E}\left[ \sum_{x_{i_1} \neq \cdots \neq x_{i_n}} \phi(x_{i_1}) \cdots \phi(x_{i_n}) \right] \\
&= \sum_{n=0}^{N} \frac{1}{n!} \frac{N!}{(N-n)!} \int_{\mathbb{R}^N} \phi(x_1) \cdots \phi(x_n) u_N(x_1, \ldots, x_N) d^N x \\
&= \sum_{n=0}^{\infty} \frac{1}{n!} \int_{\mathbb{R}^n} \prod_{i=1}^{n} \phi(x_i) \rho_n(x_1, \ldots, x_n) dx_1 \cdots dx_n,
\end{aligned}$$

where

(11.37)
$$\rho_n(x_1, \ldots, x_n) = \frac{N!}{(N-n)!} \int u_N(x_1, \ldots, x_n, x_{n+1}, \ldots, x_N) dx_{n+1} \cdots dx_N.$$

It follows that $M_n$ is absolutely continuous with respect to Lebesgue measure with correlation functions $\rho_n(x_1, \ldots, x_n)$ as in (11.37).

In particular, for a unitary ensemble of $N \times N$ Hermitian matrices $\{M\}$ with distribution

$$P^{(N)}(M) dM = \frac{1}{Z_N} e^{-\operatorname{tr} Q(M)} dM, \tag{11.38}$$

as in (6.8), the eigenvalues have distribution

$$p^{(N)}(x) dx = \frac{1}{Z_n} e^{-\sum_{i=1}^{N} Q(x_i)} \prod_{1 \le i < j \le N} (x_i - x_j)^2 d^N x \tag{11.39}$$

as in (6.13). In light of the above comment, the $x_i$'s give rise to a finite point process, $\{x_i\} \mapsto \xi = \sum_{i=1}^N \delta_{x_i}$, with correlation functions $\rho_j$ ($R_j$ with the notation of Subsection 6.3.4) given by (11.37) where $u_N(x_1, \dots, x_N) = p^{(N)}(x)$. By (6.70), we see that for $u_N(x_1, \dots, x_N) = p^{(N)}(x)$, the correlation functions can be expressed explicitly in determinantal form

$$\rho_n(x_1, \dots, x_n) = \det(K(x_i, x_j))_{1 \le i,j \le n}, \qquad x_i \in \mathbb{R}, \tag{11.40}$$

where $K(\lambda, \lambda')$ is given in (6.35). This leads us to the important class of so-called *determinantal point processes*, which we discuss below.

**Example 11.5** (Young diagrams). Recall from Subsection 1.1.4 that $\mathcal{Y}_N$ is the set of Young diagrams (equivalently partitions) of size $N$. A Young diagram can written as $\lambda = (\lambda_1, \dots, \lambda_{\ell(\lambda)})$, where $\sum_{i=1}^{\ell(\lambda)} \lambda_i = N$ and $\lambda_1 \ge \cdots \ge \lambda_{\ell(\lambda)} \ge 1$. Together with Plancherel measure $\mathbb{P}_N\{\lambda\} = \frac{F_\lambda^2}{N!}$ (see (1.37)), $\mathcal{Y}_N$ becomes a probability space. Now let $\mathcal{Y} = \bigcup_{N=1}^\infty \mathcal{Y}_N$ be the collection of all Young diagrams of arbitrary size. For any $t > 0$, $\mathcal{Y}$ carries a natural probability measure, the Poissonized Plancharel measure $\mathbb{P}^{P.P.}$ (see (2.5)), given by

$$\mathbb{P}^{P.P.}\{\lambda\} = e^{-t} \frac{|\lambda|^t}{|\lambda|!} \mathbb{P}_{|\lambda|}\{\lambda\} = e^{-t} \frac{|\lambda|^t F_\lambda^2}{(|\lambda|!)^2}, \tag{11.41}$$

where $|\lambda| = \sum_{i=1}^{\ell(\lambda)} \lambda_i$. Consider the map $\phi : \mathcal{Y} \to \mathcal{N}(\mathbb{Z})$

$$\lambda = (\lambda_1, \lambda_2, \dots) \mapsto \xi = \sum_{i=1}^{\ell(\lambda)} \delta_{\lambda_i - i}. \tag{11.42}$$

Then, the push forward of $\phi$ defines a finite point process $(\mathcal{N}(\mathbb{Z}), \mathcal{F}_{\mathcal{N}(\mathbb{Z})}, \mathbb{P})$ where for any bounded set $B \subset \mathbb{Z}$

$$\mathbb{P}\{\xi(B) = k\} = \mathbb{P}^{P.P.} \left\{ \lambda : \sum_{i=1}^{\ell(\lambda)} \chi_B(\lambda_i - i) = k \right\}. \tag{11.43}$$

The result (8.75) implies that the point process $(\mathcal{N}(\mathbb{Z}), \mathcal{F}_{\mathcal{N}(\mathbb{Z})}, \mathbb{P})$ has the correlation functions

$$\rho_n(x_1, \dots, x_n) = \det(\mathcal{K}(x_i, x_j))_{1 \le i,j \le n}, \qquad x_i \in \mathbb{Z}, \tag{11.44}$$

where $\mathcal{K}$ is defined in (8.76). Similarly, (10.199) and (10.200) imply that the Schur measure also gives rise to a determinantal point process.

We now define an important class of point processes called the determinantal point processes. The processes considered in Examples 11.4 and 11.5 are all determinantal point processes.

**Definition 11.6.** Consider a point process $\xi$ on a complete separable metric space $\Lambda$ with reference measure $\lambda$ all of whose correlation functions $\rho_n$ exist. If there is a function $K : \Lambda \times \Lambda \to \mathbb{C}$ such that

$$\rho_n(x_1, \ldots, x_n) = \det(K(x_i, x_j)_{1 \le i,j \le n} \tag{11.45}$$

for all $x_1, \ldots, x_n \in \Lambda$, $n \ge 1$, then we say that $\xi$ is a *determinantal point process* with correlation kernel $K$.

**Exercise 11.7.** Show that a determinantal point process is always simple.

The kernel $K(x, y)$ induces an operator $K$ on $L^2(\Lambda, \lambda)$ in the usual way:

$$Kf(x) = \int_\Lambda K(x, y) f(y) \lambda(dy). \tag{11.46}$$

Consider a determinantal process on $\Lambda$. Let $\phi \in L^\infty(\Lambda, \lambda)$ have support in a bounded Borel set $B$. Then by (11.19) and (11.45)
(11.47)

$$\mathbb{E}\left[\prod_j (1 + \phi(x_j))\right] = \sum_{n=0}^\infty \frac{1}{n!} \int_{B^n} \det(K(x_i, x_j)\phi(x_j))_{i,j=1}^n \lambda^{\otimes n}(d^n x)$$

provided that (see (11.17))

$$\sum_{n=0}^\infty \frac{\|\phi\|_\infty^n}{n!} \int_{B^n} \det(K(x_i, x_j))_{i,j=1}^n \lambda^{\otimes n}(d^n x) < \infty. \tag{11.48}$$

Suppose that either $\Lambda$ is an open subset of $\mathbb{R}$ or a subset of $\mathbb{Z}$. In the first case we suppose that $\lambda$ is Lebesgue measure and in the second case we suppose that $\lambda$ is counting measure.

In the first case, suppose that $K(x, y) = \overline{K(y, x)}$ for all $x, y \in B$ (i.e., $K(x, y)$ is Hermitian), and $K(x, y)$ induces a trace class operator on $L^2(B, dx)$ and is continuous on $B \times B$. Then (see Appendix A) the operator $K\phi$ given by the kernel $K(x, y)\phi(y)$ is a trace class operator for all $\phi \in L^\infty$ whose support lies in $B$ and the right-hand side of (11.47) is just the Fredholm expansion of $\det(I + K\phi)_{L^2(B)}$. Thus, if (11.48) holds (with $\lambda(dx) = dx$), then

$$\mathbb{E}\left[\prod_i (1 + \phi(x_i))\right] = \det(1 + K\phi)_{L^2(B)}. \tag{11.49}$$

If, in addition, $K$ induces a trace class operator on $L^2((t, \infty), dx)$ for any $t \in \mathbb{R}$, then the point process $\xi$ has a last particle almost surely and

$$\mathbb{P}\{x_{\max}(\xi) \le t\} = \det(1 - K)_{L^2(t,\infty)}. \tag{11.50}$$

We now verify that (11.48) holds under the assumptions on $K$. From the definition of the correlation function, $\rho_n(x_1, \ldots, x_n) \ge 0$ for all $n$ and for all

$x_1, \ldots, x_n \in \Lambda$. Hence, from (11.45) it follows by direct computation that for all $\epsilon > 0$ all the leading principal subdeterminants of $K_\epsilon^{(n)} = (K(x_i, x_j) + \epsilon\delta_{ij})_{i,j=1}^n$ are strictly positive, hence by standard linear algebra, $K_\epsilon^{(n)} > 0$ which implies $(K(x_i, x_j))_{i,j=1}^n$ is a positive semidefinite matrix. (Note that a Hermitian matrix may not be positive semidefinite even if all its leading principal subdeterminants are nonnegative; consider $\left(\begin{smallmatrix} 0 & 0 \\ 0 & -1 \end{smallmatrix}\right)$.) Hence, $0 \leq \det(K(x_i, x_j))_{i,j=1}^n \leq \prod_{j=1}^n K(x_j, x_j)$ by Hadamard's inequality. It follows that

$$\begin{aligned} \sum_{n=0}^{\infty} \frac{\|\phi\|_\infty^n}{n!} \int_B \det(K(x_i, x_j) d^n x \leq & \sum_{n=0}^{\infty} \frac{\|\phi\|_\infty^n}{n!} \left( \int_B K(x, x) dx \right)^n \\ = & \; e^{\|\phi\|_\infty \int_B K(x,x)dx} < \infty \end{aligned} \tag{11.51}$$

since $K$ is a a trace class operator on $L^2(B, dx)$ and, therefore, $\int_B K(x, x) dx = \operatorname{tr}(K)_{L^2(B)} < \infty$. Formulas (11.49) and (11.50) remain true in the discrete case: here, all we need is that $K(x, y) = \overline{K(y, x)}$ is trace class in $\ell_2(B)$ and $\ell_2(t, \infty)$, respectively.

As noted at the beginning of this section, Lenard [**Len75a, Len75b, Len73**] investigated the relationship between point processes and their correlation functions. It emerges in [**Len75a, Len75b, Len73**] that this relationship has many similarities to the classical moment problem which relates a measure to its moments. Lenard determined (see also [**Sos00**]) necessary and sufficient conditions for a collection of functions $\{\rho_n\}_{n \geq 1}$ to be the correlation functions for a (unique) point process. In the context of determinantal point processes, Soshnikov [**Sos00**] used results from [**Len75a, Len75b, Len73**] to prove the following results. Again, let $\Lambda$ be either an open subset of $\mathbb{R}$ where $\lambda$ is Lebesgue measure or a subset of $\mathbb{Z}$ where $\lambda$ is counting measure.

**Theorem 11.8.** *A Hermitian locally trace class operator $K$ on $L^2(\Lambda, \lambda)$ defines a determinantal point process with correlation functions $\rho_n(x_1, \ldots, x_n) = \det(K(x_i, x_j))_{i,j=1}^n$, $n \geq 0$, if and only if $0 \leq K \leq 1$. If the corresponding point process exists, it is unique.*

**Theorem 11.9.** *Let $(\mathcal{N}(\Lambda), \mathcal{F}_{\mathcal{N}(\Lambda)}, \mathbb{P})$, $(\mathcal{N}(\Lambda), \mathcal{F}_{\mathcal{N}(\Lambda)}, \mathbb{P}_n)$, $n = 1, 2, \ldots$, be determinantal point processes with associated Hermitian kernels $K$ and $K_n$. If $K_n \to K$ in trace norm in $L^2(B)$ for all bounded Borel sets $B \subset \Lambda$, then $\mathbb{P}_n \to \mathbb{P}$ on cylinder sets, i.e., for all $k_1, \ldots, k_n \in \mathbb{Z}_+$ and all disjoint Borel sets $B_1, \ldots, B_n$,*

$$\mathbb{P}_n \{\xi(B_1) = k_1, \ldots, \xi(B_n) = k_n\} \to \mathbb{P} \{\xi(B_1) = k_1, \ldots, \xi(B_n) = k_n\} \tag{11.52}$$

*as $n \to \infty$.*

**Example 11.10** (Sine kernel). Let $K(x,y) = \frac{\sin(\pi(x-y))}{\pi(x-y)}$, $x \neq y$ real, and let $B$ be a bounded Borel subset of $\Lambda = \mathbb{R}$. Then,

$$K(x,y) = \overline{K(y,x)} = \frac{1}{2\pi}\int_{-\pi}^{\pi} e^{i(x-y)s}ds = K_1K_2(x,y), \quad x,y \in B, \tag{11.53}$$

where $K_1(x,s) = \frac{1}{2\pi}e^{ixs}$ and $K_2(s,y) = e^{-iys}$ are Hilbert-Schmidt operators from $L^2((-\pi,\pi),ds) \to L^2(B,dx)$ and $L^2(B,dy) \to L^2((-\pi,\pi),ds)$, respectively. Also, for any $f \in L^2(B,dx)$

$$\begin{aligned} 0 \leq (f,Kf)_{L^2(B)} &= \int_{-\pi}^{\pi}\left|\int_{-\infty}^{\infty} e^{-iys}\chi_B(y)f(y)dy\right|^2 \frac{ds}{2\pi} \\ &\leq \int_{-\infty}^{\infty}\left|\widehat{(\chi_B f)}(s)\right|^2 ds \\ &= \int_{-\infty}^{\infty} |\chi_B(y)f(y)|^2 dy = (f,f)_{L^2(B)}. \end{aligned} \tag{11.54}$$

Thus, $0 \leq K \leq 1$ and $K$ induces a determinantal point process on $\Lambda = \mathbb{R}$. This process is called the sine process.

**Example 11.11** (Airy kernel process). For any $L \in \mathbb{R}$, let

$$K(x,y) = \int_0^{\infty} \mathrm{Ai}(x+t)\,\mathrm{Ai}(y+t)dt, \quad x,y \geq L, \tag{11.55}$$

where $\mathrm{Ai}(x)$ is the standard Airy function

$$\mathrm{Ai}(x) = \frac{1}{2\pi}\int_{-\infty}^{\infty} e^{i\left(\frac{s^3}{3}+xs\right)}ds, \tag{11.56}$$

where the improper integral exists for $x \in \mathbb{R}$. It was shown in Lemma 6.15 that $0 \leq K \leq 1$. Hence, $K$ defines a determinantal point process, the *Airy kernel process*. We note that the terminology *Airy process* is used for a different determinantal point process [**PS02**].

Chapter 12

# Tiling of the Aztec Diamond

Consider the Aztec diamond $\mathcal{A}_n$ of order $n$ which is defined (see Section 1.2) to be the union of all lattice squares $[m, m+1] \times [l, l+1]$ inside $\{(x, y) : |x| + |y| \leq n+1\}$ colored black and white. A *domino* is a closed $1 \times 2$ or $2 \times 1$ rectangle in $\mathbb{R}^2$ with corners in $\mathbb{Z}^2$. A tiling $\tau$ of a region $R \subset \mathbb{R}^2$ by dominos is a set of dominos whose interiors are disjoint and whose union is $R$. Let $T(\mathcal{A}_n)$ be the set of all domino tilings of the Aztec diamond $\mathcal{A}_n$. For example, for $n = 4$, the Aztec diamond has the form in Figure 12.1. An example of a tiling of $\mathcal{A}_4$ is given in Figure 12.2. This tiling, which we denote by $\tau$, will be used throughout this section to illustrate various constructions that arise in the analysis of $T(\mathcal{A}_n)$.

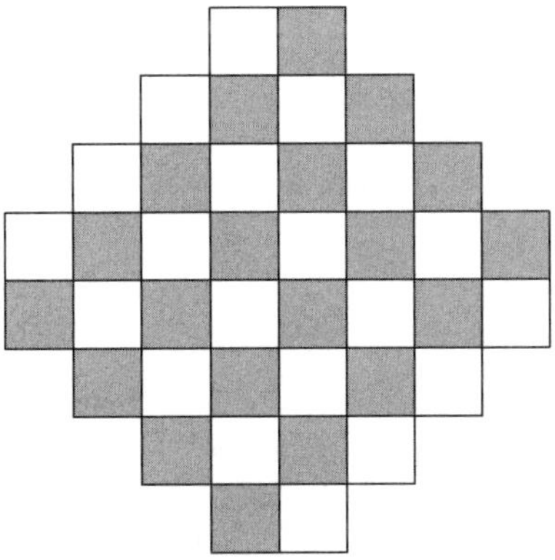

**Figure 12.1.** Aztec diamond $\mathcal{A}_4$

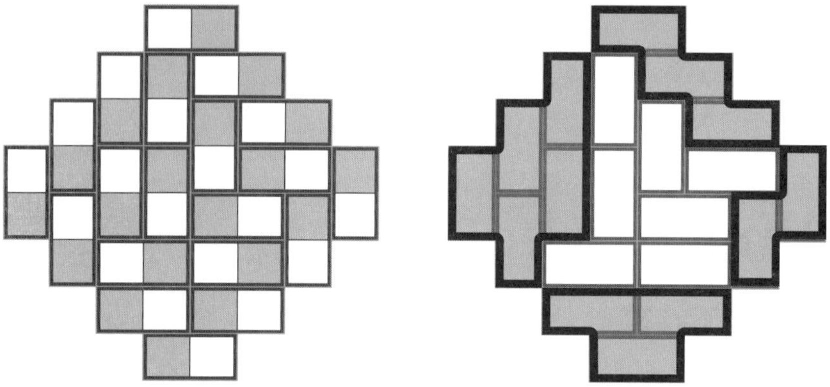

**Figure 12.2.** Tiling $\tau$ of $\mathcal{A}_4$ and the polar regions

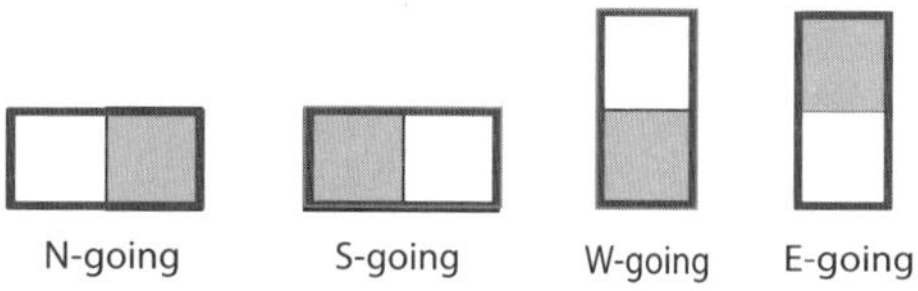


**Figure 12.3.** Orientations

**Note 12.1.** We will always assume that the squares are covered so that the lower-right boundary of $\mathcal{A}_n$ consists of white squares as in Figure 12.1, for example.

A domino in a tiling $\tau \in \mathcal{A}_n$ can have one of four possible orientations which are displayed in Figure 12.3. The reason for the notation will become clear later. We say that two dominos are *adjacent* if they share an edge and we say that a domino is *adjacent to the boundary* if it shares an edge with the boundary of $\mathcal{A}_n$. The *north polar region* of a tiling $\tau$ is defined to be the union of those north going dominos that are connected to the boundary by a sequence of adjacent north-going dominos. A simple argument (we leave this as an exercise) shows that the north-polar region, if it is nonempty, is connected and always covers the "top" two squares. The west-, east-, and south-going regions are defined similarly. The *temperate zone* is defined to be the complement of the union of the four polar regions in $\tau$. For $\tau$, the picture on the right in Figure 12.2 depicts these regions.

## 12.1. Nonintersecting Lattice Paths

**12.1.1. DR-Paths.** We now define a one-to-one mapping (in fact a bijection—see Theorem 12.4 below) from $T(\mathcal{A}_n)$ to families of nonintersecting lattice paths. Given a tiling we mark individual dominos according to the

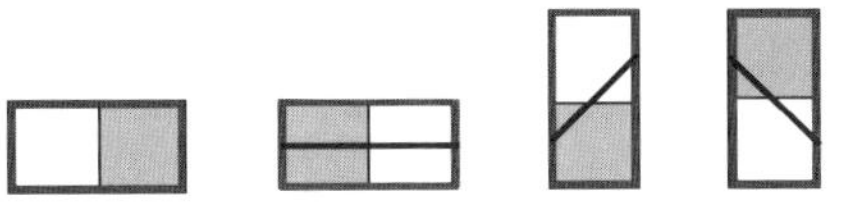


**Figure 12.4.** CS-I markings

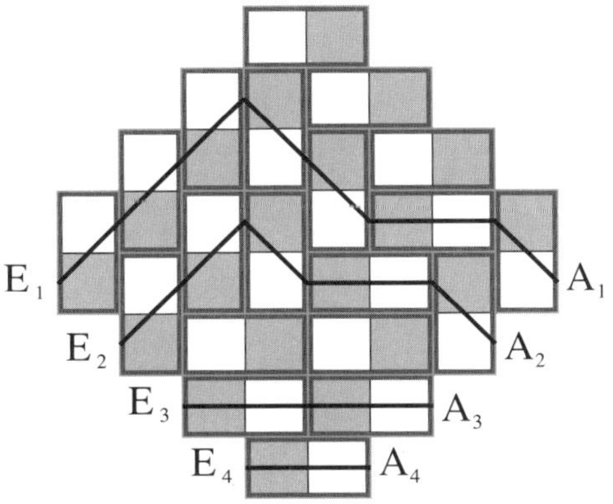


**Figure 12.5.** Type I DR-paths for $\tau$

rule in Figure 12.4 where $N$ denotes north-going, etc. Note that $N$-dominos are unmarked. Given a tiling $\tau$, draw lines on all the dominos as described.

**Lemma 12.2.** *This procedure gives $n$ nonintersecting paths starting at $A_j = (n+1-j, \frac{1}{2}-j)$ and ending at $E_j = (-n-1+j, \frac{1}{2}-j)$, $1 \le j \le n$. These paths are called DR-paths of type I by Johansson* [**Joh02**] *after Dana Randall (see* [**Sta99**, p. 277]*).*

For $\tau$, this marking is shown in Figure 12.5. The nonintersecting paths in Lemma 12.2 and their analogue, Lemma 12.3 below, constitute the basic algebraic tools needed to analyze the Aztec diamond.

**Proof of Lemma 12.2.** Lemma 12.2 is really a statement about the orbits of a *vector field.*

(i) (Analogue of local existence.) Consider any domino in $\mathcal{A}_n$. Then either the domino is north-going and there is no path through it, or it is a west/east/south domino. In all three cases, a path meets the left edge of the black square at a midpoint "o". For each of these cases, there are three possibilities for the adjacent domino as indicated in Figure 12.6. In each of these cases the path can clearly be continued to the left as indicated, ending at the midpoint of a black square.

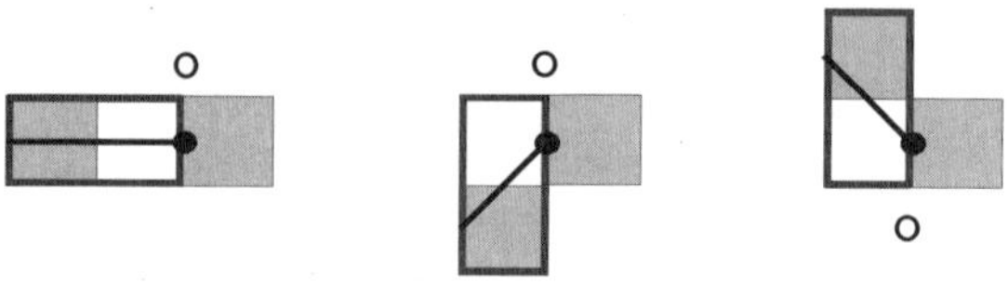


**Figure 12.6.** Three possible continuations of a path to the left

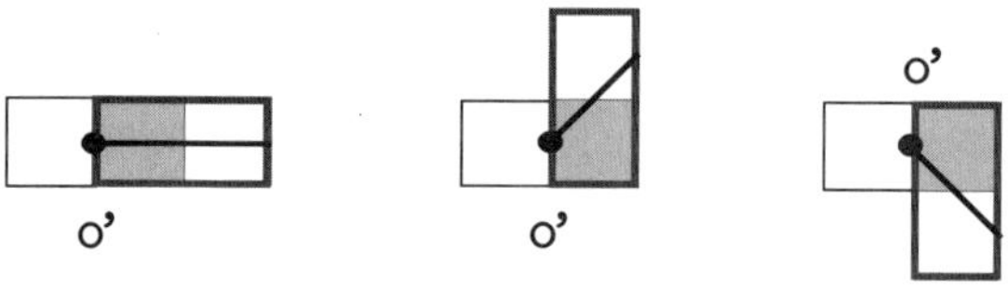


**Figure 12.7.** Three possible continuations of a path to the right

On the other hand, on the right the path ends up at the midpoint of a white square (o'). We see that there are three possibilities for the domino adjacent to (o'), as in Figure 12.7, and the paths can be continued to the right as indicated, ending up once again at the midpoint of a white square.

(ii) (Analogue of global existence.) The above argument shows that either a domino contains no path, or it contains a path and that path can be continued strictly to the left and to the right. This shows that once a path is started, it must continue until it reaches the boundary of $\mathcal{A}_n$ on the left and on the right.

(iii) (Analogue of orbits are disjoint.) The argument in (i) shows that the orbit through a point is uniquely determined: this is clear within a domino. On the other hand, the way in which the path continues to the left or the right is uniquely determined by the kind of domino lying to the left or right of the domino. Thus, orbits are either disjoint or identical.

Now, it follows from the fact that the leftmost (resp. rightmost) square in each row in the top half is white (resp. black) that no path can cross the northern boundary of $\mathcal{A}_n$. It follows in particular that the $n$ paths emanating from $A_j$, $1 \leq j \leq n$, must cross $\partial \mathcal{A}_n$ at points $E_j$, $1 \leq j \leq n$, in the third quadrant. Moreover, as the paths cannot intersect, we must have $A_j \to E_j$, $1 \leq j \leq n$. This establishes Lemma 12.2. □

We now introduce a system of coordinates, the *CS-I coordinate system*, which is particularly useful for the calculations that follow and is well adapted to the description of type I DR-paths. The origin $O_I$ of the system is

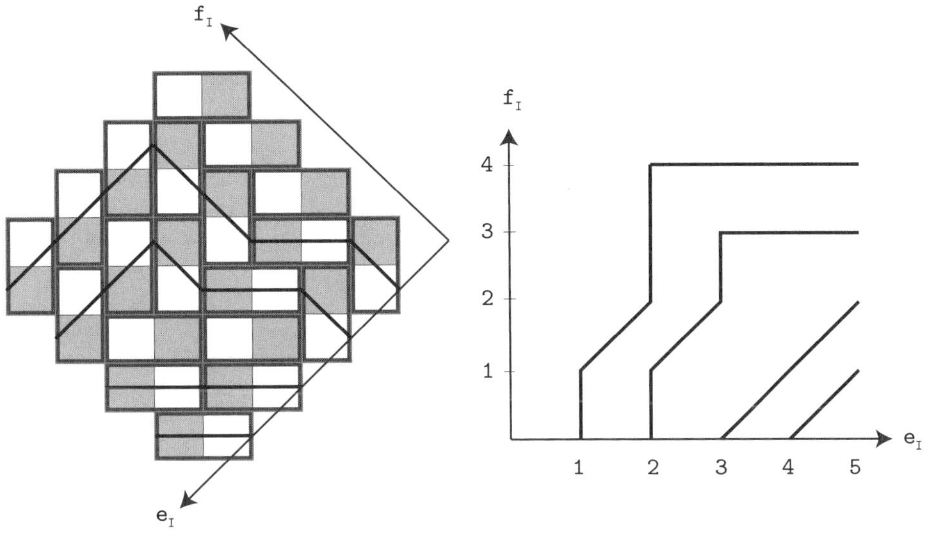


**Figure 12.8.** CS-I coordinates and lattice paths in lattice $L_I$

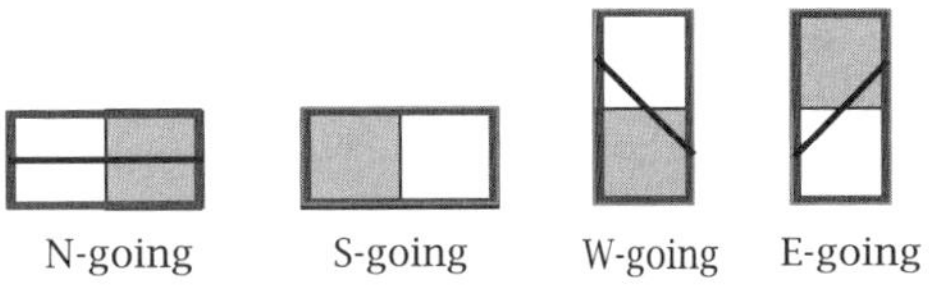


**Figure 12.9.** CS-II markings

the point $(n+1, \frac{1}{2})$ and the basis vectors are $e_I = (-1,-1)$ and $f_I = (-1,1)$. Note that the CS-I markings $S, E$, and $W$ for the dominos in Figure 12.4 correspond to $e_I + f_I$, $f_I$, and $e_I$, respectively. Let $L_I = \{je_I + kf_I : j, k \in \mathbb{Z}\}$ denote the integer lattice in CS-I. Clearly, type I DR-paths give rise to walks in $L_I$ that move up and to the right. For example, the type I DR-paths in $\tau$ give rise to the lattice paths in $L_I$ as shown in Figure 12.8.

A second family of nonintersecting lattice paths is defined as follows. Given a tiling, we mark individual dominos according to the rule in Figure 12.9 where $S$-dominos are now unmarked. Again, given a tiling $\tau$, draw lines on all the dominos as described. The proof of the following lemma is completely analogous to that of Lemma 12.2.

**Lemma 12.3.** *This procedure gives $n$ nonintersecting paths starting at $A'_j = (-n-1+j, j-\frac{1}{2})$ and ending at $E'_j = (n+1-j, j-\frac{1}{2})$, $1 \le j \le n$, the so-called DR-paths of type II.*

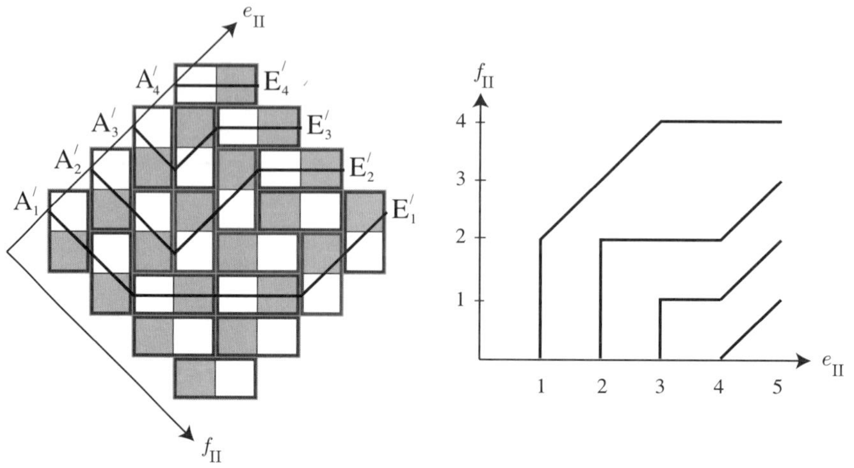


**Figure 12.10.** Type II DR-paths, CS-II coordinates, and lattice paths in $L_{II}$

For example, for $\tau$, we have the markings in Figure 12.10. In analogy to CS-I, we now introduce a system of coordinates that is well adapted to the description of type II DR-paths. The origin $O_{II}$ of the system is the point $(-n-1, -\frac{1}{2})$, and the basis vectors are $e_{II} = (1, 1)$ and $f_{II} = (1, -1)$. Note that the markings $N$, $E$, and $W$ correspond to $e_{II} + f_{II}$, $e_{II}$, and $f_{II}$, respectively. Let $L_{II} = \{je_{II} + kf_{II} : j, k \in \mathbb{Z}\}$ denote the integer lattice in CS-II. As before, type II DR-paths give rise to walks in $L_{II}$ that move to the right and up.

By a *family of nonintersecting lattice paths* $\mathcal{L}$ *in* $\mathcal{A}_n$ we mean a family of $n$ nonintersecting paths, $p_1, \ldots, p_n$, where $p_j \subset \mathcal{A}_n$ is a continuous piecewise linear curve from $A_j$ to $E_j$, $1 \le j \le n$, constructed by concatenating the vectors $e_I$, $f_I$, and $e_I + f_I$. Clearly, $\varphi$ is a map from $T(\mathcal{A}_n)$ to

$$\mathcal{F} \equiv \{\mathcal{L} : \mathcal{L} \text{ is a family of nonintersecting lattice paths in } \mathcal{A}_n\}.$$

**Theorem 12.4.** *$\varphi : T(\mathcal{A}_n) \to \mathcal{F}$ is a bijection.*

**Proof.** Suppose $\tau_1$ and $\tau_2$ are two tilings with $\varphi(\tau_1) = \varphi(\tau_2)$. If $\tau_1 \neq \tau_2$, then there exists in particular at least one white square, say $S$, which is covered by differently oriented dominos. But then it is clear from Figure 12.4 that $S \cap \varphi(\tau_1) \neq S \cap \varphi(\tau_2)$, which is a contradiction.

It remains to show that $\varphi$ is onto. Suppose that $\mathcal{L}$ is a family of nonintersecting lattice paths, $p_j : A_j \to E_j$. Then $\mathcal{L}$ induces a tiling of $\mathcal{A}_n$ in the following way. Consider $p_j$ for some $j \in \{1, \ldots, n\}$ which starts at $A_j$, which, in turn, lies at the midpoint of the eastern edge of a white square. Now, the endpoint of the first segment of $p_j$ must also lie on the eastern edge of a white square; hence, by induction this is true for the endpoints of each segment of each $p_j$. It follows that if $p_j$ intersects a white square, it must pass through the midpoint of the eastern edge of the square, and similarly, if it intersects a black square, it must pass through the midpoint of the western edge of the square. In particular, each square intersects with at most one path from the family of nonintersecting paths $p_j$, $1 \leq j \leq n$. Now for any $j$, $1 \leq k \leq n$, each segment $e_I, f_I$, or $e_I + f_I$ in $p_j$ intersects two squares. Associate a domino to this pair. Note that this association is unambiguous as each square intersects at most one $p_j$.

Let $\mathcal{B}$ be the union of the dominos obtained in this way. It remains to show that $\mathcal{A}_n \setminus \mathcal{B}$ can be tiled by horizontal dominos. Number the rows of $\mathcal{A}_n$ by $q = 1, \ldots, 2n$ starting from the top (row $q = 1$ has two boxes, etc.). Now fix a row $q = q'$.

**Lemma 12.5.** *Suppose $p_j$ intersects boxes $B_1, \ldots, B_{2l}$ in row $q'$ (necessarily there are an even number of them). Then the complement of $\bigcup_{i=1}^{2l} B_i$ in row $q'$ is a union of segments, each of which contains an even number of boxes, except possibly for two outside segments.*

**Example 12.6.** In Figure 12.5, for $p_2$, the complement of $\bigcup_{i=1}^{2\ell} B_i$ in row $q' = 6$ consists of an interior segment with four boxes, while in $q' = 5$ the complement consists of two singleton outside segments.

**Note 12.7.** Lemma 12.5 really implies the following. Every interior segment of the complement of $\bigcup B_j$ in row $q'$ contains an even number of elements.

**Proof of Lemma 12.5.** To prove Lemma 12.5, note that if we travel along $p_j$ from the right to the left, the CS-I markings imply that the path leaves a given row through a white box while it enters the row through a black box (see the left pictures in Figure 12.11). Let $p_j$ enter row $q'$ for the first time through a black box $B$ as in the right picture in Figure 12.11. Now suppose $p_j$ stays on $q'$ for a few steps, and then leaves $q'$ and then later returns; see Figure 12.11. Then we observe that when it leaves, it leaves through a white square as in $W$ in Figure 12.11, hence the next square $B'$ is black. But when

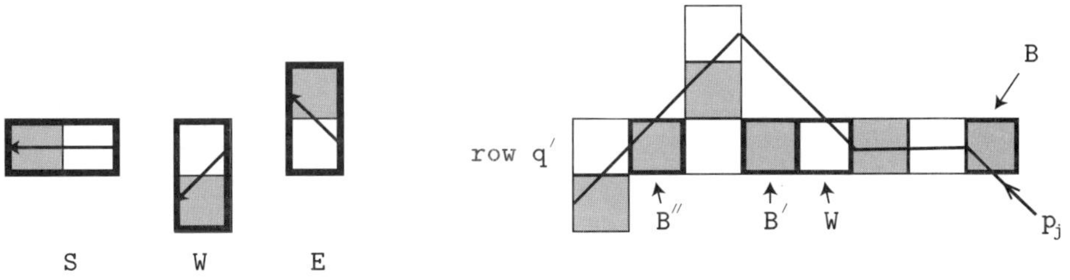


**Figure 12.11.** Example for the proof of Lemma 12.5

$p_j$ re-enters $q'$ it must do so through a black square ($B''$ above) and so there are necessarily an even number of squares in the segment between $W$ and $B''$. Repeating this argument, we prove Lemma 12.5. (The cases when $p_j$ starts (in a necessarily white square) in $q'$, or when $p_j$ never intersects with $q'$, are left to the reader.) Note that when $p_j$ eventually leaves $q'$, it must do so through a white square. So, it enters through a black square and leaves through a white square. □

We continue the proof of Theorem 12.4. Consider the region in $\mathcal{A}_n \setminus \mathcal{B}$ above $p_1$, which is denoted by $R_1$. Now the intersection of $R_1$ with any row $q$ is a union of at most two outside segments and some interior segments each of which, by Lemma 12.5, has an even number of boxes. Call these outside segments, if they exist, $s_1$ and $s_2$, where $s_1$ lies to the left of $s_2$. But $s_2$ must have a black box on the right and $s_1$ must have a white box on the left. This is because $R_1$ necessarily lies in the upper part of $\mathcal{A}_n$. As $p_1$ enters $q$ for the first time through a black box, it follows that $s_2$ must have an even number of boxes, and as it leaves $q$ for the last time through a white box, it follows that $s_1$ must also have an even number of boxes. Hence, all segments contain an even number of boxes, so $R_1$ may be tiled by horizontal boxes. This is true, in particular, if $q$ does not intersect with $p_1$ as in Figure 12.5 for $q = 1$.

Now consider the region $R_2 \subset \mathcal{A}_n \setminus \mathcal{B}$ between $p_1$ and $p_2$. Consider the intersection of $R_2$ with row $q'$, say. Typically, we may have the situation shown in Figure 12.12. Recall that a path $p_j$ enters a line through a black square and leaves through a white square. Segment $\Gamma_1$ is an interior segment for the intersection of $p_2$ with $q'$, and $\Gamma_3$ is an interior segment for the intersection of $p_1$ with $q'$. On the other hand, segments of type $\Gamma_2$, which arise from the intersection of $p_1$ and $p_2$ with $q'$ must necessarily have a black box on the right and a white box on the left. Hence, segments of type $\Gamma_2$ must also contain and even number of segments. It follows that each row in $R_2$ can be tiled with horizontal dominos. Continuing in this fashion, we see that all of the $\mathcal{A}_n \setminus \mathcal{B}$ can be tiled by horizontal dominos, all north-going. Thus we conclude that $\mathcal{L}$ induces a tiling, $\tau$, of $\mathcal{A}_n$.

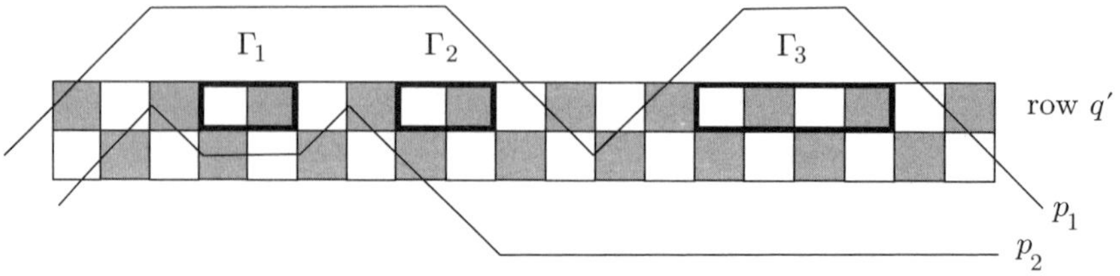


**Figure 12.12.** Example for the proof of Theorem 12.4

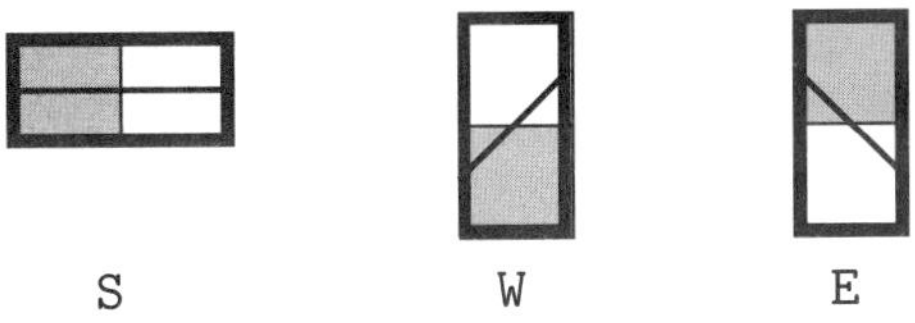


**Figure 12.13.** S/W/E-going markings

Finally, observe that $\mathcal{L}$ is the family of orbits for $\tau$, i.e., $\varphi(\tau) = \mathcal{L}$. To prove this, note that every segment of a path $p_j$ in $\mathcal{L}$ starts at the eastern edge of a white square, hence there are three possibilities; see Figure 12.13. Thus, these dominos are S/W/E for the tiling $\tau$ induced by $\mathcal{L}$. On the other hand, as noted above, all the dominos in $\mathcal{A}_n \setminus \mathcal{B}$ are north facing for $\tau$. Thus, $\varphi(\tau) = \mathcal{L}$, hence $\varphi$ is a bijection. This proves Theorem 12.4. □

**Note 12.8.** We remark that a consequence of Theorem 12.4 is that every family $\mathcal{L} \in \mathcal{F}$ is a family of type I DR-paths.

**Remark 12.9.** For the convenience of the reader, we now give a second, and very different proof that $\mathcal{L}$ induces a tiling. We must show that $\mathcal{A}_n \setminus \mathcal{B}$ can be tiled by horizontal tiles. Note that if a white square, say $W_1$, is not covered by the dominos in $\mathcal{B}$, then if the black square, say $B_1$, to its right lies in $\mathcal{A}_n$, $B_1$ must also not be covered by the dominos in $\mathcal{B}$. For if $B_1$ were covered, then there are three possibilities; see Figure 12.14 for the dominoes covering $B_1$. But the dominos must intersect a segment and the segment must be oriented as indicated in Figure 12.14. In all three cases the segment must be extended to the left in particular, and so $W_1$ must be covered. This is a contradiction. Thus, if $W_1$ is uncovered, so is $B_1$, and this gives rise to a north-facing domino. On the other hand, the only white squares in $\mathcal{A}_n$ which do not have a black square to the right are the squares forming the south-east boundary on $\mathcal{A}_n$. But each of these squares are intersected by one of the segments emanating from the $A_j$'s, and hence are covered by dominos.

Similarly, we see that if $B_2$ is an uncovered black square and if there is a white square $W_2$ to its left, then $W_2$ is also uncovered giving rise to a north-facing domino. Also, the only black dominos with no white neighbor to the

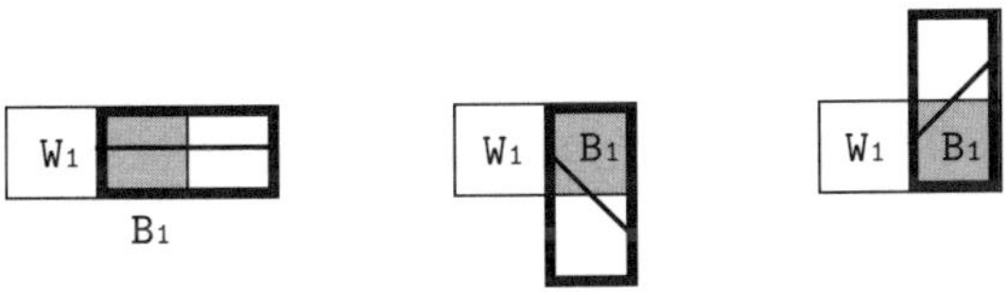


**Figure 12.14.** Example for Remark 12.9

left lie on the south-west boundary of $\mathcal{A}_n$, and these intersecting segments emanate from the $E_j$'s. Thus, $\mathcal{A}_n \setminus \mathcal{B}$ can be tiled with north-facing dominos.

By a *family of nonintersecting lattice paths* $\mathcal{L}'$ *in* $\mathcal{A}_n$ we mean a family of nonintersecting paths $p'_1, \ldots, p'_n$ where each $p'_j \subset \mathcal{A}_n$ is a continuous piecewise linear curve from $A'_j$ to $E'_j$, $1 \le j \le n$, constructed by concatenating the vectors $e_{II}$, $f_{II}$, and $e_{II} + f_{II}$ (see Figure 12.10). The following analogue of Theorem 12.4 is easily proved. Let

$$\mathcal{F}' = \{\mathcal{L}' : \mathcal{L}' \text{ be a family of nonintersecting family of lattice paths in } \mathcal{A}_n\}.$$

Clearly, there is an analogue $\varphi'$ of $\varphi$ mapping $T(\mathcal{A}_n)$ to $\mathcal{F}'$.

**Theorem 12.10.** $\varphi' : T(\mathcal{A}_n) \to \mathcal{L}'$ *is a bijection.*

**12.1.2. Zig-zag paths.** Another set of paths, the *zig-zag paths*, also play an important role in describing the statistics of $T(\mathcal{A}_n)$. For any $r \in \{1, 2, \ldots, n\}$, consider the sequence of $n+1$ white squares with top-left/bottom-right corners given by $Q^r_k = (-r+k, n+1-k-r)$, $k \in \{0, 1, \ldots, n+1\}$. For example, for $\mathcal{A}_4$ see the left picture in Figure 12.15. Given a tiling $\tau$, the *zig-zag path* $Z_r = Z_r(\tau)$ in $\mathcal{A}_n$, $1 \le r \le n$, is defined to be the unique path of edges that goes around the white squares with corners $Q^r_k$, $0 \le k \le n+1$, and does not cross a domino. Each move $Q^r_k \to Q^r_{k+1}$ in $Z_r(\tau)$ is either an ES-step or SE-step. For example, for $\tau$ the right picture in Figure 12.15 illustrates the zig-zag paths for different values of $r$.

As we will see below, for any $r \in \{1, \ldots, n\}$ and any $\tau \in T(\mathcal{A}_n)$, the zig-zag path $Z_r(\tau)$ contains *precisely* $r$ ES-steps. For example, $Z_2(\tau)$ in Figure 12.15 has two ES-steps. The statistics of the locations of the ES-steps are key objects in the study of tilings $\tau$ of $\mathcal{A}_n$.

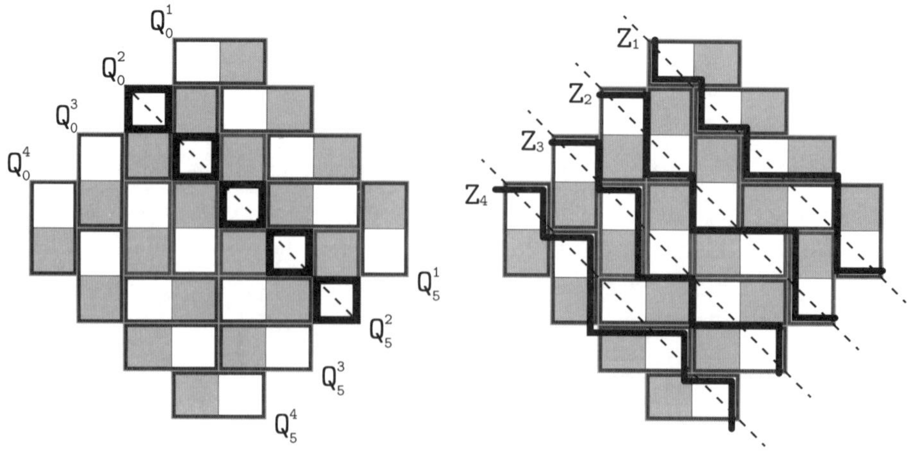


**Figure 12.15.** Zig-zag paths. Here $n = 4$.

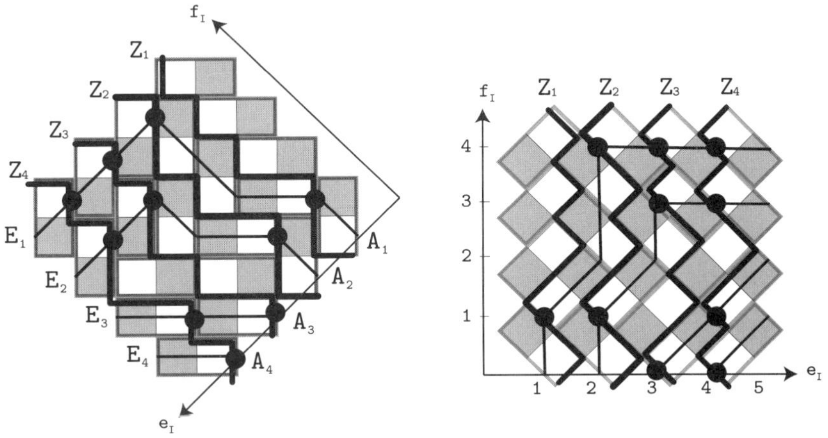


**Figure 12.16.** Black dots are the intersections $S_j^r$ of zig-zag paths and type I DR-paths. Their CS-I coordinates can be seen easily in the right picture. As an example, for $r = 3$, the intersection points are $S_1^3 = (3, h_3)$, $S_2^3 = (3, h_2)$, and $S_3^3 = (3, h_1)$, where $h_1 = 0$, $h_2 = 3$ and $h_3 = 4$.

We first show that every family $\mathcal{L} \in \mathcal{F}$ intersects a given $Z_r(\tau)$ in exactly $r$ points, and every family $\mathcal{L}' \in \mathcal{F}$ intersects $Z_r(\tau)$ in $n+1-r$ complementary points (see, for example, $r = 3$ in Figure 12.16).

**Lemma 12.11.** *Fix $r$, $1 \le r \le n$. For any $\tau \in T(\mathcal{A}_n)$, let $\mathcal{L} = p_1, \dots, p_n$ be its associated set of nonintersecting lattice paths of type DR-I. For $1 \le j \le n$, $p_j : A_j = (n+1-j, \frac{1}{2} - j) \to E_j = (-n-1+j, \frac{1}{2} - j)$. Let $Z_r(\tau)$ be the zig-zag path from $Q_0^r = (-r, n+1-r)$ to $Q_{n+1}^r = (-r+n+1, -r)$. Then $p_j$ intersects $Z_r(\tau)$ in precisely one point, $S_j^r$, $1 \le j \le r$, and does not intersect $Z_r(\tau)$ for $r+1 \le j \le n$. Moreover, $p_j$, $1 \le j \le r$, intersects $Z_r(\tau)$ at a point which lies at the midpoint of the right edge of a white square. The parts of the path $p_j$, $1 \le j \le r$, that lie above $Z_r(\tau)$ go from $A_j \to S_j^r$. In CS-I coordinates, $A_j = (j, 0)$, $S_j^r = (r, h_{r+1-j})$,[1] $1 \le j \le r$, where $0 \le h_1 < h_2 < \cdots < h_r \le n$, $h_i \in \mathbb{Z}_+$. The points $S_j^r$, $1 \le j \le r$, lie precisely in the ES-steps in $Z_r(\tau)$ and, conversely, each ES-step contains a point $S_j^r$. In particular, $Z_r(\tau)$ has precisely $r$ ES-steps.*

**Proof.** Note first that the formula $(x, y) = Xe_I + Yf_I + (n+1, \frac{1}{2})$ converts the original coordinates $(x, y)$ of a point into CS-I coordinates $(X, Y) \equiv CSI(x, y)$. Recall that the $X$ and $Y$ coordinates are nondecreasing along any path $p_j$. Observe that in CS-I coordinates the zig-zag path goes from $CSI(Q_{n+1}^r) = (r + \frac{1}{4}, -\frac{1}{4})$ to $CSI(Q_0^r) = (r + \frac{1}{4}, n + \frac{3}{4})$. See Figure 12.17 for the CS-I coordinates of the left picture in Figure 12.16. As $r < r + \frac{1}{4} < n+1$,

[1] Note that $h_i$ depends on $r$, i.e., $h_i = h_i^{(r)}$, etc.

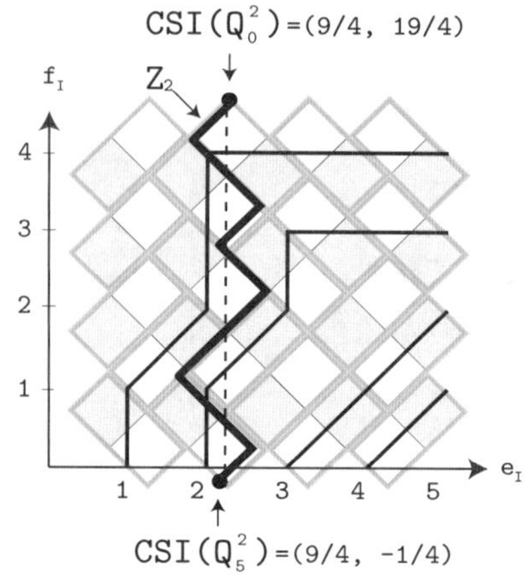


**Figure 12.17.** Zig-zag path in CS-I

it follows by continuity that $p_j$, $1 \leq j \leq r$, must intersect the straight line from $Q_{n+1}^r$ to $Q_0^r$ (the dashed line in Figure 12.17). Since the straight line does not intersect the interior of black squares, and since $p_j$ never crosses the corner of a square, $p_j$ must meet this straight line in the interior of a white square $W$. Now, in CS-I coordinates, $p_j$ is built out of segments as displayed in Figure 12.18. Clearly, $p_j$ cannot intersect a zig-zag path in an east-going tile. We are then left with two possibilities: south-going tiles and west-going tiles. In both cases, two facts are clear:

(i) An ES-step for $Z_r(\tau)$ around a white square in the original coordinates corresponds to a move to the left around a square in CS-I coordinates.

(ii) The line $p_j$ must meet $Z_r(\tau)$ at the midpoint (indicated by white circles in Figure 12.18) of the left-bottom edge of this white square in CS-I coordinates.

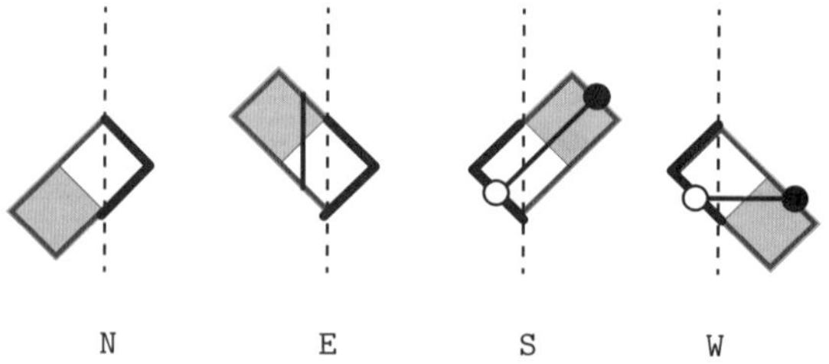


**Figure 12.18.** N/E/S/W-going markings in CS-I coordinates. Note that the zig-zag path (thick line segments) goes around the white boxes which intersect the dashed line.

Thus, $p_j$ meets $Z_r(\tau)$ precisely in those white squares around which the zig-zag path goes to the left in CS-I coordinates. It is also clear from Figure 12.18 that, conversely, if the zig-zag path goes around a white square to the left in CS-I coordinates, then some $p_j$ must intersect $Z_r(\tau)$ in that square. Now once $p_j$ intersects $Z_r(\tau)$, then it cannot intersect $Z_r(\tau)$ again. Indeed, after crossing $Z_r(\tau)$ and emerging at a point indicated by black circles in Figure 12.18, $p_j$ has a first CS-I coordinate equal to $r+1$. But, the first coordinate of $Z_r(\tau)$ is always less than or equal to $r+\frac{3}{4}$. Hence, $p_j$ cannot intersect $Z_r(\tau)$ again.

It is clear that the white circles in Figure 12.18 have first CS-I coordinate $r$. Thus, $S_j^\tau$ has CS-I coordinate $(r, h_{r+1-j})$ and $0 \le h_1 < \cdots < h_r \le n$. Indeed, if $h_i < h_j$ for $i > j$, then the two associated $p$'s would have to intersect, which is impossible.

Finally, as the first CS-I coordinate cannot decrease along a path $A_j \to E_j$, we see that for $r+1 \le j \le n$, the first coordinates are greater than or equal to $r+1$. But, as noted above, the first coordinate of $Z_r(\tau)$ is always less than $r+\frac{3}{4}$, and so $p_j$ for $r+1 \le j \le n$ never intersects $Z_r(\tau)$. This completes the proof of Lemma 12.11. □

The following result is proved similarly to Lemma 12.11; see also Figure 12.19.

**Lemma 12.12.** *Fix $r$, $1 \le r \le n$. For any $\tau \in T(\mathcal{A}_n)$, let $\mathcal{L}' = \{p'_1, \dots, p'_n\}$ be its associated lattice paths of type DR-II: For $1 \le j \le n$, $p'_j : A'_j = (-n-1+j, j-\frac{1}{2}) \to E'_j = (n+1-j, j-\frac{1}{2})$. Let $Z_r(\tau)$ be the zig-zag path from $Q_0^r = (-r, n+1-r) \to Q_{n+1}^r = (-r+n+1, -r)$, then $p'_j$ intersects $Z_r(\tau)$ in precisely one point $(S_j^\tau)'$, $1 \le j \le n+1-r$, and does not intersect $Z_r(\tau)$ for $n+1-r < j \le n$. Moreover, $p'_j$, $1 \le j \le n+1-r$, intersects $Z_r(\tau)$ at a point which lies at the midpoint of the left edge of a white square. The parts of the paths $p'_j$, $1 \le j \le n+1-r$, that lie below $Z_r(\tau)$ go from $A'_j \to (S_j^\tau)'$. In CS-II coordinates, $A'_j = (j, 0)$ and $S_j^\tau{}' = (n+1-r, n-k_j)$, where $0 \le k_1 \le k_2 \le \cdots \le k_{n+1-r} \le n$, $k_j \in \mathbb{Z}_+$. The points $(S_j^\tau)'$, $1 \le j \le n+1-r$, lie precisely in the SE-steps in $Z_r(\tau)$, and, conversely, each SE-step contains one point $(S_j^\tau)'$. In particular, $Z_r(\tau)$ has precisely $n+1-r$ SE-steps.*

The above lemmas show that each of the $r$ ES-steps in $Z_r(\tau)$ are intersected at the midpoints $\{S_j^\tau\}$ of their eastern boundaries by $\{p_j; 1 \le j \le r\}$, respectively; each of the $n+1-r$ SE-squares in $Z_r(\tau)$ are intersected at the midpoints $\{(S_j^\tau)'\}_{j=1}^{n+1-r}$ of their western boundaries by $\{p'_j\}_{j=1}^{n+1-r}$, respectively. Now the midpoints of the eastern (resp. western) boundaries of all the white squares in $Z_r(\tau)$ have second CS-I (respectively, CS-II) coordinates in

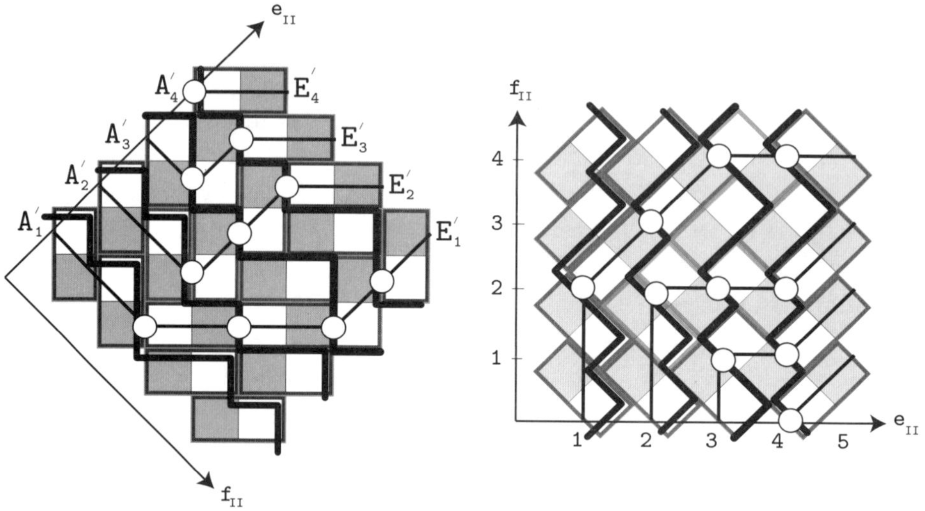


**Figure 12.19.** White dots are the intersections $S_j^r$ of zig-zag paths and type II DR-paths.

$\{0, 1, 2, \ldots, n\}$. Moreover, if the right endpoint of a white square has a second CS-I coordinate $\alpha$, say, then the left endpoint of the square has a second CS-II coordinate $n - \alpha$. Given the fact that each white square is either ES or SE, and given the definition of the $h_i$'s and $k_j$'s, we see that $\{h_1, \ldots, h_r\}$ and $\{k_1, \ldots, k_r\}$ are disjoint and $\{h_1, \ldots, h_r\} \cup \{k_1, \ldots, k_{n+1-r}\} = \{0, 1, \ldots, n\}$. We call $p(Z_r) = \{h_1, \ldots, h_r\}$ the *zig-zag (particle) configuration* and $h(Z_r) = \{k_1, \ldots, k_{n+1-r}\}$ the *zig-zag (hole) configuration.*[2] See Figure 12.20.

An important consequence of the previous considerations is that for any $\tau \in T(\mathcal{A}_n)$, if one follows the path $Z_r(\tau)$ from the top-left point to the bottom-right point, one goes around precisely $r$ white squares by ES-steps for $r$ times and $n - r$ white squares by SE-steps. This can also be proved in the following way by using the (domino) height function, $H$, associated with $\tau \in \mathcal{A}_n$ (see [**EKLP92a, EKLP92b**]).

$H$ is a map from the lattice points $\{u\}$ in $\mathcal{A}_n$ to $\mathbb{Z}$ and is defined as follows. Suppose $u$ and $v$ are two adjacent lattice points such that the edge connecting them is not covered by a domino. If the edge $u \to v$ has a black square to its right as one moves from $u$ to $v$, then $H(v) = H(u) - 1$, and if it has a black square to its left as one moves from $u$ to $v$, $H(v) = H(u) + 1$. See the left picture in Figure 12.21. This prescription fixes $H$ apart from an overall constant, which we specify by setting $H(n, 0) = 0$. We must check that this prescription is well defined. Now clearly for any vertex $u$, there

[2]Note again that $h_i$ depends on $r$, i.e., $h_i = h_i^{(r)}$, etc. The same remark also applies to $k_i$.

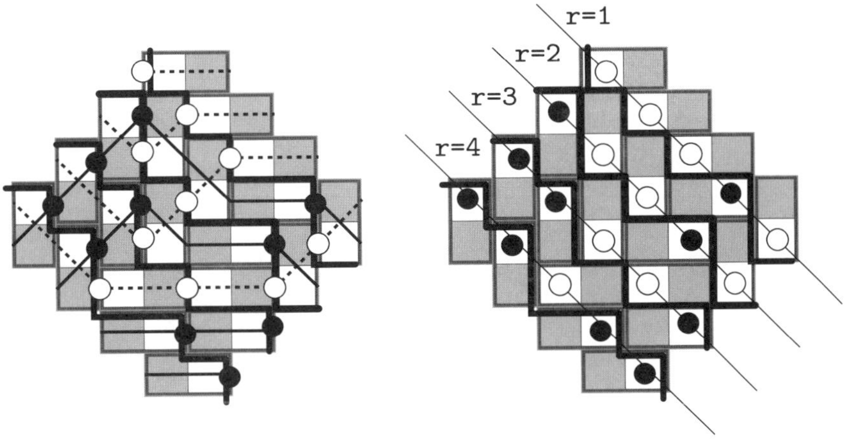


**Figure 12.20.** (Left) The black dots are $S_j^r$ and the white dots are $(S_j^r)'$. (Right) We moved the black dots to the left and the white dots to the right by a half-size of a square. It is easy to find $h_j$ and $k_j$ from this picture: read off the locations of dots along a diagonal straight line from the lower-bottom corner. As an example, for $r = 3$, we have $\{h_1, h_2, h_3\} = \{0, 3, 4\}$, $\{k_1, k_2\} = \{1, 2\}$.

exists a path

$$u_0 = (n, 0) \to u_1 \to \cdots \to u_n \to u_{n+1} = u,$$

where $u_i, u_{i+1}$ are adjacent and the edge $u_i \to u_{i+1}$ is not covered by a domino. Using the above rules for $H(u_{i+1}) - H(u_i) = \pm 1$, we then obtain $H(u)$. To check that $H(u)$ is well defined, we need to know that if $\{u_i\}$ is any sequence as above and $u_{n+1} = u_0$, then $\sum_{i=0}^{n}(H(u_{i+1}) - H(u_i)) = 0$. In addition, without loss of generality, we can assume that the $u_i$'s are distinct apart from $u_0 = u_{n+1}$. Now, the interior of the loop $u_0 \to \cdots \to u_0$ is tiled by the dominos in $\tau$. But by a standard topological argument, $\sum_{i=0}^{n}(H(u_{i+1}) - H(u_i))$ is the sum of the sum of the differences of the $H$'s around any domino $\sum_{i=0}^{5} H(c_{i+1}) - H(c_i)$, where the $c_i$'s are the corners of the two squares making up the domino. But three of the differences are $+1$ and the remaining three are $-1$. Thus, $\sum_{i=0}^{5}(H(c_{i+1}) - H(c_i)) = 0$. Hence, $H$ is well defined.

Now, by considering the outermost boundaries of the $\mathcal{A}_n$, we have $H(Q_{n+1}^r) = 2r - 1$ and $H(Q_0^r) = 2n - (2r - 1)$. On the other hand,

$$\begin{aligned} 2n + 2 - 4r = H(Q_0^r) - H(Q_{n+1}^r) &= \sum_{k=0}^{n}(H(Q_k^r) - H(Q_{k+1}^r)) \\ &= 2 \times (\#\text{SE-steps}) + (-2) \times (\#\text{ES-steps}) \\ &= -4 \times (\#\text{ES-steps}) + 2n + 2. \end{aligned}$$

So, the number of ES-steps equals $r$. See Figure 12.21 for an example.

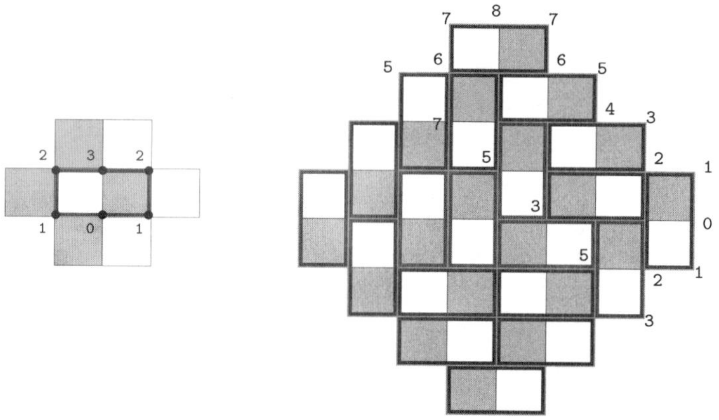

**Figure 12.21.** Height function

From the previous considerations we see that for any fixed $r$, $1 \leq r \leq n$, any $\tau \in T(\mathcal{A}_n)$ gives rise to a zig-zag path $Z_r(\tau)$ with $r$ ES-steps, a tiling $\tau^\uparrow$ of the part of $\mathcal{A}_n$ above $Z_r(\tau)$, and a tiling $\tau^\downarrow$ of the part of $\mathcal{A}_n$ below $Z_r(\tau)$. For any fixed $r$, $1 \leq r \leq n$, the mapping

$$T(\mathcal{A}_n) \ni \tau \to (Z_r(\tau), \tau^\uparrow, \tau^\downarrow) \tag{12.1}$$

is clearly a bijection onto

$$\begin{aligned} X_r \;\; = \;\; &\{(Z_r, \tau^\uparrow, \tau^\downarrow) : Z_r \text{ has } r \text{ ES-steps and } \tau^\uparrow \text{ (resp., } \tau^\downarrow) \text{ is} \\ &\quad \text{a tiling of the upper (resp., lower) part of } \mathcal{A}_n \setminus Z_r\}. \end{aligned}$$

In other words, for any $r \in \{1, \ldots, n\}$, $X_r$ is an equivalent and, as we will see, very useful description of $T(\mathcal{A}_n)$.

For a zig-zag path $Z_r$, let $A_r^\uparrow$ (resp. $A_r^\downarrow$) denote the upper (resp. lower) parts of $\mathcal{A}_n \setminus Z_r$.

**Proposition 12.13.**

(i) *$A_r^\uparrow$ can be tiled by dominos if and only if $Z_r$ has $r$ ES-steps.*

(ii) *$A_r^\downarrow$ can be tiled by dominos if and only if $Z_r$ has $r$ ES-steps or, equivalently, $n - 1 + r$ SE-steps.*

**Proof.** We prove (i) and leave the proof of (ii) to the reader. Let $\tau^\uparrow$ be a domino tiling of $A_r^\uparrow$. Then we can define a height function $H^\uparrow$ on the vertices in the connected set $A_r^\uparrow$ as in the above. The same argument now shows that $Z_r$ must have $r$ ES-steps.

Conversely, suppose $Z_r$ has $r$ ES-steps. In the notation of Lemma 12.11, let these squares be parametrized by the CS-I points $(r, h_j)$, $1 \leq j \leq r$, where $0 \leq h_1 < h_2 < \cdots < h_r \leq n$. For each $1 \leq j \leq r$ construct the unique lattice path $p_j$ from $(j, 0) \to (r, h_{r-j+1})$ which is the union of a vertical line

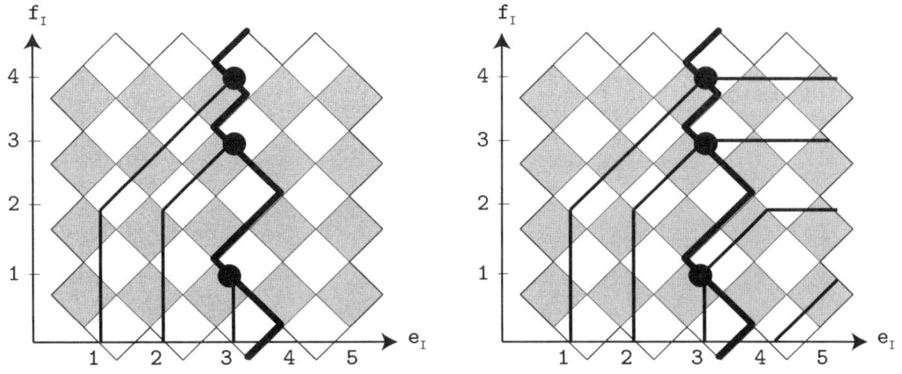


**Figure 12.22.** Zig-zag path $Z_3$ given by $h_1 = 1$, $h_2 = 3$, $h_3 = 4$, and a tiling of $A_3^{\uparrow}$.

from $(j, 0)$ and the 45° line from $(r, h_{r-j+1})$ as indicated by the left picture in Figure 12.22 (note that $p_r$ is always just the vertical line $(r, 0) \to (r, h_1)$). It is clear that the $p_j$'s, $1 \le j \le r$, do not intersect. Note that the construction of these paths relies on the fact that the points $S_j^r$ lie at the midpoints of the eastern edge (in the diamond $A_n$) of a white square (see for example the proof of Theorem 12.4).

We now show how to extend the paths to a family $\mathcal{L}$ of nonintersecting type I DR-paths. For $r+1 \le j \le n$, we define the paths $p_j$ to be the straight line at 45° in DR-I coordinates from $(j, 0)$ to $(n+1, n-j+1)$. For $1 \le j \le r$, $p_j$ continues as the straight line at 45° from $(r, h_j)$ until it $y'$ coordinate equals $n-r+j$. Then $p_j$ continues horizontally to $(r, n-r+j)$ (see the right picture in Figure 12.22). We leave it to the reader to verify that $\mathcal{L} = \{p_j\}_{j=1}^n$ is indeed a family of nonintersecting type I DR-paths. But then by Theorem 12.4, $\mathcal{L}$ induces a tiling $\tau = \varphi^{-1}(\mathcal{L})$ of $\mathcal{A}_n$. Then $\tau^{\uparrow}$ is a tiling of $A_r^{\uparrow}$. This proves (i). We leave the proof of the second to the reader. □

**Remark 12.14.** It follows of course from Proposition 12.13 that $A_r^{\uparrow}$ can be tiled by dominos if and only if $A_r^{\downarrow}$ can be tiled by dominos. Note that for $r = n+1$, $A_{r=n+1}^{\uparrow} = \mathcal{A}_n$ and $Z_{n+1}$ is the unique zig-zag path from $Q_0^{n+1} \to Q_{n+1}^{n+1}$ with $n+1$ ES-steps. Similarly, for $r = 0$, $A_{r=0}^{\downarrow} = \mathcal{A}_n$ and $Z_0$ is the unique zig-zag path from $Q_0^0 \to Q_{n+1}^0$ with no ES-steps. Thus the question of the tiling of $A_r^{\uparrow}$ and $A_r^{\downarrow}$ for $r = 1, \ldots, n$ is a natural extension of Theorems 12.4 and 12.10, to regions with more general boundaries.

The above arguments and results imply the following.

**Theorem 12.15.** *Fix $r$, $1 \le r \le n$, and let $Z_r$ be a zig-zag path with $r$ ES-steps. Then the tilings of $A_r^{\uparrow}$ are in one-to-one correspondence with families of nonintersecting lattice paths $p_j$, $1 \le j \le r$, of type DR-I that lie in $A_r^{\uparrow}$ and go from (CS-I) $A_j = (j, 0) \to S_j^r = (r, h_{r+1-j})$, $1 \le j \le r$, where $0 \le h_1 < \cdots < h_r \le n$ parametrizes $Z_r$ in CS-I coordinates. The tilings of $A_r^{\downarrow}$ are in one-to-one correspondence with families of nonintersecting lattice paths $p'_j$, $1 \le j \le n+1-r$, of type DR-II that lie in $A_r^{\downarrow}$ and go from (CS-II) $A'_j = (j, 0) \to S_j^r{\prime} = (n+1-r, n-k_j)$, $1 \le j \le n+1-r$, where $0 \le k_1 < \cdots < k_{n+1-r} \le n$ parametrizes $Z_r$ in CS-II coordinates.*

**Corollary 12.16.** *Fix $r$, $1 \le r \le n$. Then*

$$T(\mathcal{A}_n) \cong X_r \cong (Z_r, \{p_1, \ldots, p_r\}, \{p'_1, \ldots, p'_{n+1-r}\}), \tag{12.2}$$

*where $Z_r$ has $r$ ES-steps, $\{p_1, \ldots, p_r\}$ and $\{p'_1, \ldots, p'_{n+1-r}\}$ are families of nonintersecting paths for $A_r^{\uparrow}$ and $A_r^{\downarrow}$, respectively, as in Theorem* 12.15.

The latter equivalence in terms of nonintersecting paths will be particularly useful.

## 12.2. Density Function

Fix $\omega > 0$ and a positive integer $n$. For $\tau \in T(\mathcal{A}_n)$, let $v(\tau)$ be the number of vertical dominos in $\tau$. Introduce the probability measure on $T(\mathcal{A}_n)$ by

$$\mathbb{P}(\tau) = \frac{\omega^{v(\tau)}}{\sum\limits_{\tau' \in T(\mathcal{A}_n)} \omega^{v(\tau')}}. \tag{12.3}$$

In other words, we give weight 1 to horizontal dominos and weight $\omega$ to vertical dominos. Alternatively, in the equivalent path picture (see again Theorem 12.4), put weight 1 on steps $(1,1)$ in CS-I and CS-II coordinates and put weight $\omega$ on steps $(1,0)$ and $(0,1)$. Thus, the weight of $n$ given nonintersecting DR-paths is defined as the product of the weights on all steps. But this is clearly equal to $\omega^{v(\tau)}$ if $\tau$ is the tiling defined by the paths. For example, the tiling in Figure 12.8 has weight $w^{10}$.

### 12.2.1. Zig-zag Paths and Krawtchouk Ensemble.

Fix $r$, $1 \le r \le n$, and let $h = (h_1, \ldots, h_r)$ where $0 \le h_1 < h_2 < \cdots < h_r \le n$. Let $Z_r(\tau)$ denote the $r$th zig-zag path of tiling $\tau$. We will also, when there is no confusion, let $Z_r(\tau)$ denote the second CS-I coordinates of the points $S_j^r = (r, h_{r+i-j})$, $1 \le j \le r$, in Lemma 12.11. Thus, for example, for the tiling in Figure 12.16, $Z_3(\tau) = (0, 3, 4)$.

We want to compute

$$\mathbb{P}\left(Z_r(\tau)=h\right)=\frac{\sum\limits_{\tau\in T(\mathcal{A}_n);Z_r(\tau)=h}\omega^{v(\tau)}}{\sum\limits_{\tau'\in T(\mathcal{A}_n)}\omega^{v(\tau')}} \tag{12.4}$$

for $0\leq h_1<h_2<\cdots<h_r\leq n$. If $\tau=\tau^{\uparrow}\oplus\tau^{\downarrow}$ as in (12.1), then $\omega^{v(\tau)}=\omega^{v(\tau^{\uparrow})}\omega^{v(\tau^{\downarrow})}$, hence

$$\begin{aligned}&\sum_{\tau\in T(\mathcal{A}_n);Z_r(\tau)=h}\omega^{v(\tau)}\\&=\left(\sum_{\{\tau^{\downarrow}:Z_r(\tau)=Z_r(\tau^{\downarrow})=h\}}\omega^{v(\tau^{\downarrow})}\right)\left(\sum_{\{\tau^{\uparrow}:Z_r(\tau)=Z_r(\tau^{\uparrow})=h\}}\omega^{v(\tau^{\uparrow})}\right)\\&\equiv\omega[h]\omega'[h].\end{aligned}$$

Thus, equation (12.4) can be rewritten as

$$\mathbb{P}\left(Z_r(\tau)=h\right)=\frac{\omega[h]\omega'[h]}{\sum_{0\leq\tilde{h}_1<\tilde{h}_2<\cdots<\tilde{h}_r\leq n}\omega[\tilde{h}]\omega'[\tilde{h}]}. \tag{12.5}$$

By Theorem 12.15, $\omega[h]$ is the weight of all the type I DR-paths from $(j,0)\to(r,h_{r+1-j})$, $1\leq j\leq r$, in CS-I coordinates and $\omega'[h]$ is the weight if all the type II DR-paths from $(j,0)\to(n+1-r,n-k_j)$, $1\leq j\leq n+1-r$, in CS-II coordinates where

$$0\leq k_1<k_2<\cdots<k_{n+1-r}\leq n \text{ is the complement of } h \text{ in } \{0,1,\ldots,n\}.$$

**Remark 12.17.** In (12.5) above we used the relation

$$\sum_{\tau'\in T(\mathcal{A}_n)}\omega^{v(\tau')}=\sum_{0\leq\tilde{h}_1<\tilde{h}_2<\cdots<\tilde{h}_r\leq n}\omega[\tilde{h}]\omega'[\tilde{h}]. \tag{12.6}$$

When $\omega=1$, the left-hand side equals the number of elements in $T(\mathcal{A}_n)$.

We now compute $\omega[h]$ and $\omega'[h]$ using the Lindström-Gessel-Viennot method (see, e.g., [**GV85**]). First, we compute $\omega[h]$. Fix $h$ and consider the set $Y(h)$ of all paths $\{p\}$ with the following properties:

(a) each $p$ is composed of steps $(0,1)$, $(1,0)$, or $(1,1)$, and

(b) each $p$ starts at some $(j,0)$, $1\leq j\leq r$, and ends at one of the points $(r,h_i)$ for some $i$.

Fix $1\leq r\leq n$. A *family of paths* $\mathcal{L}=\mathcal{L}_{(r)}$ in $Y(h)$ is a set of $r$ paths, $\{p_1,\ldots,p_r\}$, where $p_j\in Y(h)$ for $1\leq j\leq r$ such that $p_j$ starts at $(j,0)$ and ends at $h_{\sigma(j)}$, $1\leq j\leq r$. for some permutation $\sigma=\sigma(\mathcal{L})\in S_r$. The *signed weight* of $\mathcal{L}$ is defined by $\omega^{f(\mathcal{L})}\operatorname{sgn}(\sigma(\mathcal{L}))$, where $f(\mathcal{L})$ denotes the number

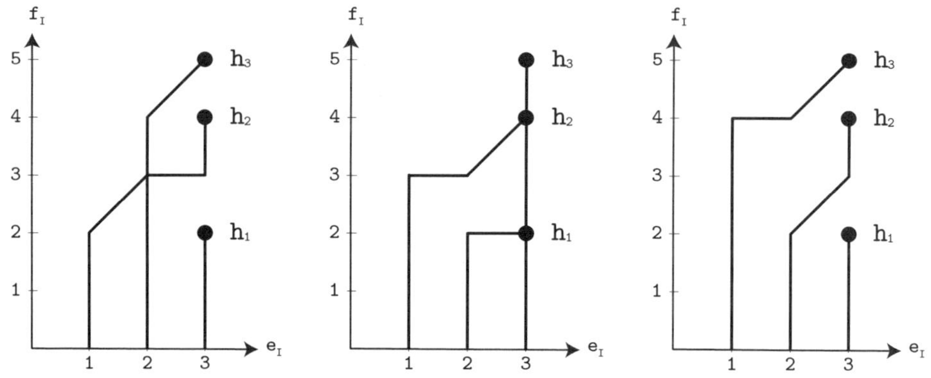


**Figure 12.23.** Examples of families of paths $\{p_1, p_2, p_3\}$ where $h_1 = 2$, $h_2 = 4$, and $h_3 = 5$. The corresponding permutation is (left) $\sigma = 231$ or $\sigma = 321$, (middle) $\sigma = 213$ or $\sigma = 231$ or $\sigma = 312$ or $\sigma = 321$, (right) $\sigma = 321$.

of horizontal or vertical steps in $\mathcal{L}$. For example, the leftmost picture in Figure 12.23 is an example of $\mathcal{L} = \{p_1, p_2, p_3\}$ such that $p_j$ ends at $h_{\sigma(j)}$ where $\sigma = \left(\begin{smallmatrix}1&2&3\\2&3&1\end{smallmatrix}\right)$. The signed weight of this family of paths is $w^{10}$. Note that this also can be thought of as the picture of the family $\mathcal{L}' = \{p_1', p_2', p_3'\}$ such that $p_j$ ends at $h_{\sigma'(j)}$ where $\sigma' = \left(\begin{smallmatrix}1&2&3\\3&2&1\end{smallmatrix}\right)$. The signed weight of this family of paths is $-w^{10}$.

Let $G(h) = \sum_{\mathcal{L} \in Y(h)} \omega^{f(\mathcal{L})} \operatorname{sgn}(\sigma(\mathcal{L}))$ and let

$$\mathcal{P} = \{\mathcal{L} \in Y(h) : \text{ at least two paths } p_i, p_j \text{ in } \mathcal{L} \text{ intersect}\}.$$

We now define an involution $\phi$ on $\mathcal{P}$ (cf. the evaluation of the number of the standard Young tableaux in Section 3.2 and the proof of the Jacobi-Trudi determinant formula in Subsection 10.1.3). For $\mathcal{L} \in \mathcal{P}$, let $i_1 = \min\{i : p_i \text{ intersects with some } p_j, j > i\}$. Let $i_2 = \min\{i > i_1 : p_i \text{ intersects with } p_{i_1}\}$. Necessarily, $p_{i_1}$ and $p_{i_2}$ intersect at a vertex in $\mathbb{Z}^2$. Let $q(\mathcal{L})$ be the first vertex along $p_{i_1}$, starting from $(i_1, 0)$ where $p_{i_1}$ intersects $p_{i_2}$. For $\mathcal{L} \in \mathcal{P}$, let $\mathcal{L}' = \phi(\mathcal{L}) = \{p_1', \ldots, p_r'\}$ be defined as follows:

$$p_{i_1}' \equiv (i_1, v) \xrightarrow{p_{i_1}} q(\mathcal{L}) \xrightarrow{p_{i_2}} h_{\sigma_{i_2}},$$

$$p_{i_2}' \equiv (i_2, 0) \xrightarrow{p_{i_2}} q(\mathcal{L}) \xrightarrow{p_{i_1}} h_{\sigma_{i_1}},$$

and $p_i' = p_i$ for $i \neq i_1, i_2$. Thus, up to $q(\mathcal{L})$, $p_{i_1}'$ and $p_{i_2}'$ agree with $p_{i_1}$ and $p_{i_2}$, respectively, but after $q(\mathcal{L})$ they switch. We see that $\mathcal{L}' \in \mathcal{P}$ and $\omega^{f(\mathcal{L}')} \operatorname{sgn}(\sigma(\mathcal{L}')) = -\omega^{f(\mathcal{L})} \operatorname{sgn}(\sigma(\mathcal{L}))$. Thus, $\phi$ is an involution on $\mathcal{P}$.

It follows that $G(h) = \sum_{\mathcal{L} \in Y(h); \mathcal{L} \text{ nonintersecting}} \omega^{f(\mathcal{L})} \operatorname{sgn}(\sigma(\mathcal{L}))$. But if $\mathcal{L}$ is nonintersecting, then necessarily $\sigma(\mathcal{L}) = \sigma^{\text{rev}}$, where $\sigma^{\text{rev}}(1) = r$,

$\sigma^{\text{rev}}(2) = r-1, \ldots, \sigma_r^{\text{rev}} = 1$. Thus

$$G(h) = \text{sgn}(\sigma^{\text{rev}}) \sum_{\mathcal{L}\in Y(h);\mathcal{L} \text{ nonintersecting}} \omega^{f(\mathcal{L})} = \text{sgn}(\sigma^{\text{rev}})\omega[h]. \tag{12.7}$$

On the other hand,

$$\begin{aligned} G(h) &= \sum_{\mathcal{L}'\in Y(h)} \text{sgn}(\sigma(\mathcal{L}'))\omega^{f(\mathcal{L}')} \\ &= \sum_{\sigma\in S_r} \text{sgn}(\sigma) \sum_{\mathcal{L}'\in Y(h);\text{sgn}(\mathcal{L})=\sigma} \omega^{f(\mathcal{L})} \\ &= \sum_{\sigma\in S_r} \text{sgn}(\sigma) \sum_{p_1,\ldots,p_r\in Y(h);p_j:(j,0)\to(r,h_{\sigma_j})} \omega(p_1)\cdots\omega(p_r), \end{aligned} \tag{12.8}$$

where $\omega(p) = \omega^{\#\{\text{horizontal and vertical steps in } p_j\}}$. Thus,

$$G(h) = \sum_{\sigma\in S_r} \text{sgn}(\sigma) W_{1,\sigma_1} W_{2,\sigma_2} \cdots W_{r,\sigma_r}, \tag{12.9}$$

where $W_{j,k} = \sum_{p:(j,0)\to(r,h_k)} \omega(p)$. Hence, by (12.7), we obtain $\text{sgn}(\sigma^{\text{rev}})\omega[h] = \det(W_{j,k})_{1\le j,k\le r}$ by (12.7) and (12.9). Now, by translation invariance, $W_{j,k} = \Omega(r-j, h_k)$, where $\Omega(n,m)$ is the total weight of all possible paths from $(0,0)\to(n,m)$. Recalling the definition of $\sigma^{\text{rev}}$, we conclude that

$$\omega[h] = \det(\Omega(i-1, h_j))_{1\le i,j\le r}. \tag{12.10}$$

Similarly, the total weight of all possible nonintersecting paths $(j,0)\to(n+1-r, n-k_j)$, $1\le j\le n+1-r$, is

$$\omega'[h] = \det(\Omega(i-1, h'_j))_{1\le i,j\le r'}, \tag{12.11}$$

where $r' = n+1-r$, $h'_j = n - k_{n+2-r-j}$, $1\le j\le n+1-r$, and $0\le h'_1 < \cdots < h'_{n+1-r}\le n$. Clearly, $\omega'[h]$ is of the same form as $\omega[h]$. This is the advantage of using both types of DR-paths.

Summarizing, we have:

**Lemma 12.18.** *Let $\Omega(n,m)$ be the total weight of all possible paths from $(0,0)\to(n,m)$ in the CS-I coordinates. For $h=(h_1,\ldots,h_r)$ such that $0\le h_1<\cdots<h_r\le n$, let $h'=(h'_1,\ldots,h'_{n+1-r})$ be defined by the conditions that $0\le h'_1<\cdots<h'_{n+1-r}\le n$ and $\{h_1,\ldots,h_r,h'_1,\ldots,h'_{n+1-r}\}=\{0,1,\ldots,n\}$. Then*

$$\omega[h] = \det(\Omega(i-1, h_j))_{1\le i,j\le r} \tag{12.12}$$

*and*

$$\omega'[h] = \det(\Omega(i-1, h'_j))_{1\le i,j\le r'}. \tag{12.13}$$

To evaluate $\Omega(n,m)$, let $p$ be a path from $(0,0)$ to $(n,m)$ with $a$ steps $(1,0)$, $b$ steps $(0,1)$, and $c$ steps $(1,1)$. Then $n = a+c$, $m = b+c$, and the number of paths with a given number of steps $a,b,c$ equals $\frac{(a+b+c)!}{a!b!c!} = \frac{(n+m-c)!}{(n-c)!(m-c)!c!}$. Note that $c$ can take any value between 0 and $\min\{m,n\}$. Thus,

$$\Omega(n,m) = \sum_{c=0}^{\min(m,n)} \frac{(n+m-c)!}{(n-c)!(m-c)!c!}\omega^{n+m-2c}, \tag{12.14}$$

which can be rewritten using the Pochhammer symbol $(a)_k = a(a+1)\cdots(a+k-1)$, $a_0 = 1$ as

$$\Omega(m,n) = \frac{\omega^{n+m}}{n!}\sum_{c=0}^{\infty}(m-c+1)_n(n-c+1)_c\frac{\omega^{-2c}}{c!}. \tag{12.15}$$

The determinant $\det(\Omega(i-1,h_j))_{1\le i,j\le r}$ can be evaluated as follows.

**Lemma 12.19.**
(12.16)

$$\omega[h] = \det(\Omega(i-1,h_j))_{1\le i,j\le r} = \left(\prod_{j=1}^{r}\frac{(1+\omega^2)^{j-1}}{(j-1)!\omega^{j-1}}\right)\Delta_r(h)\prod_{j=1}^{r}\omega^{h_j},$$

*where* $\Delta_r(h) \equiv \prod_{1\le j<i\le r}(h_i - h_j)$.

**Proof.** Using the fact that $\Omega(m,n)$ is symmetric in $n$ and $m$ (see (12.14)), we have by multilinearity of the determinant

$$\begin{aligned}
&\det(\Omega(i-1,h_j))_{1\le i,j\le r}\\
&= \det\left(\frac{\omega^{i-1+h_j}}{(i-1)!}\sum_{c=0}^{\infty}(h_j-c+1)_{i-1}(i-c)_c\frac{\omega^{-2c}}{c!}\right)_{i,j=1}^{r}\\
&= \left(\prod_{j=1}^{r}\frac{\omega^{j-1+h_j}}{(j-1)!}\right)\det\left(\sum_{c=0}^{\infty}(h_j-c+1)_{i-1}(i-c)_c\frac{\omega^{-2c}}{c!}\right)_{i,j=1}^{r}\\
&= \left(\prod_{j=1}^{r}\frac{\omega^{j-1+h_j}}{(j-1)!}\right)\sum_{c_1,\dots,c_r=0}^{\infty}\det((h_j-c_i+1)_{i-1})_{1\le i,j\le r}\prod_{i=1}^{r}\frac{\omega^{-2c_i}}{c_i!}(i-c_i)_{c_i}.
\end{aligned}$$

The last determinant is of the form $\det(q_i(h_j))_{1\le i,j\le r}$, where $q_i(h) = (h-c_i+1)_{i-1} = h^{i-1} +$ lower order terms. Thus, again by multilinearity of the determinant, we can use a sequence of row operations to show that

$\det(q_i(h_j))_{1\le i,j\le r} = \det(h_j^{i-1})_{1\le i,j\le r} = \Delta(h)$. Thus, the above becomes

$$\begin{aligned}&\prod_{j=1}^{r} \frac{\omega^{j-1+h_j}}{(j-1)!} \sum_{c_1,\dots,c_r=0}^{\infty} \Delta_r(h) \prod_{i=1}^{r} \frac{\omega^{-2c_i}}{c_i!}(i-c_i)_{c_i}\\ &= \Delta_r(h) \prod_{j=1}^{r} \frac{\omega^{j-1+h_j}}{(j-1)!} \prod_{i=1}^{r} \left( \sum_{c=0}^{\infty} (i-c)_c \frac{1}{c!\omega^{2c}} \right).\end{aligned}$$

Since

$$\sum_{c=0}^{\infty}(i-c)_c \frac{1}{c!\omega^{2c}} = \sum_{c=0}^{i-1} \frac{(i-1)!}{c!(i-1-c)!}\omega^{-2c} = (1+\omega^{-2})^{i-1},$$

we obtain (12.16). □

Similarly, we have the following lemma. Recall that $0 \le k_1 \le \cdots \le k_{n+1-r} \le n$ is the complement of $h$ in $\{0, 1, \dots, n\}$.

**Lemma 12.20.**

$$\omega'[h] = \left( \prod_{j=1}^{n+1-r} \frac{(1+\omega^2)^{j-1}}{(j-1)!\omega^{j-1}} \right) \Delta_{n+1-r}(k) \prod_{j=1}^{n+1-r} \omega^{n-k_j}. \tag{12.17}$$

**Proof.** From Lemma 12.19, we have

$$\omega'[h] = \det(\Omega(i-1, h'_j))_{1\le r,j\le r'} = \left( \prod_{j=1}^{r'} \frac{(1+\omega^2)^{j-1}}{(j-1)!\omega^{j-1}} \right) \Delta_{r'}(h') \prod_{j=1}^{r'} \omega^{h'_j}. \tag{12.18}$$

But, $r' = n+1-r$ and $h'_j = n - k_{n+2-r-j} = n - k_{r'+1-j}$, $1 \le j \le r'$. Thus, $\prod_{j=1}^{r'} \omega^{h'_j} = \prod_{j=1}^{n+1-r} \omega^{n-k_j}$ and

$$\Delta_{r'}(h') = \prod_{1\le j<i\le r'} (h'_i - h'_j) = \Delta_{r'}(k) = \Delta_{n+1-r}(k). \tag{12.19}$$

Inserting these relations into (12.18), we obtain (12.17). □

We rewrite (12.17) using the following result.

**Lemma 12.21.** *If $s_1 < \cdots < s_N$ and $t_1 < \cdots < t_M$ are disjoint sets of integers whose union is precisely $\{0, 1, \dots, N+M-1\}$, then*

$$\Delta_M(t) = \left( \prod_{j=1}^{N+M-1} j! \right) \left( \prod_{j=1}^{N} \frac{1}{s_j!(N+M-1-s_j)!} \right) \Delta_N(s). \tag{12.20}$$

**Proof.** Let $M+N=L$ and set $\{z_1,\dots,z_L\}=\{t_1,\dots,t_M,s_1,\dots,s_N\}$. Then $(\Delta_L(z))^2=(\Delta_M(t))^2(\Delta_N(s))^2\prod_{i,j}(t_i-s_j)^2$. On the other hand,

$$\prod_{\ell=1}^{N}\prod_{k=1;z_k\neq s_\ell}^{L}(s_\ell-z_k)^2=\prod_{i,\ell}(s_\ell-t_i)^2\prod_{j\neq\ell}(s_\ell-s_j)^2=\prod_{i,\ell}(s_\ell-t_i)^2(\Delta_N(s))^4. \tag{12.21}$$

Thus,

$$(\Delta_M(t))^2=\frac{(\Delta_L(z))^2}{\prod_{\ell=1}^{N}\prod_{k=1;z_k\neq s_\ell}^{L}(s_\ell-z_k)^2}(\Delta_N(s))^2. \tag{12.22}$$

Equation (12.22) is true for arbitrary disjoint sets $\{s_j\},\{t_i\}$. But if $\{s_1,\dots,s_N\}\cup\{t_1,\dots,t_M\}=\{0,1,\dots,N+M-1\}$, then $\{z_1,\dots,z_L\}=\{0,1,\dots,N+M-1\}$. Hence,

$$(\Delta_L(z))^2=\prod_{0\le j<i\le N+M-1}(i-j)^2=\left(\prod_{\ell=1}^{N+M-1}\ell!\right)^2. \tag{12.23}$$

Also,

$$\begin{aligned}\prod_{\ell=1}^{N}\prod_{k=1;z_k\neq s_\ell}^{L}(s_\ell-z_k)^2&=\prod_{\ell=1}^{N}\prod_{k=1;k\neq s_\ell+1}^{L}(s_\ell-k+1)^2\\&=\left(\prod_{\ell=1}^{N}s_\ell!(N+M-1-s_\ell)!\right)^2.\end{aligned} \tag{12.24}$$

Taking square roots and taking account of the signature, we obtain (12.20). $\square$

By (12.17) and (12.20),

$$\omega'[h]=\omega^{n(n+1)/2-nr}\prod_{j=1}^{n+1-r}\frac{(1+\omega^2)^{j-1}}{(j-1)!\omega^{j-1}}\prod_{j=1}^{n}j!\prod_{j=1}^{r}\frac{\omega^{h_j}}{h_j!(n-h_j)!}\Delta_r(h) \tag{12.25}$$

since $\prod_{j=1}^{n+1-r}\omega^{n-k_j}=\omega^{n(n+1)/2-nr}\prod_{j=1}^{r}\omega^{h_j}$. Together with (12.16) this yields

$$\begin{aligned}\omega[h]\omega'[h]=&\omega^{n(n+1)/2-nr}\prod_{j=1}^{r}\frac{(1+\omega^2)^{j-1}}{(j-1)!\omega^{j-1}}\prod_{j=1}^{n+1-r}\frac{(1+\omega^2)^{j-1}}{(j-1)!\omega^{j-1}}\prod_{j=1}^{n}j!\\&\times(\Delta_r(h))^2\prod_{j=1}^{r}\frac{\omega^{2h_j}}{h_j!(n-h_j)!}.\end{aligned} \tag{12.26}$$

Since $\sum_{j=1}^{r}(j-1)+\sum_{j=1}^{n+1-r}(j-1)=\frac{1}{2}n(n+1)-nr+r(r-1)$, this shows that

$$\omega[h]\omega'[h]=\frac{(1+\omega^2)^{\frac{n(n+1)}{2}-nr+r(r-1)}}{\omega^{r(r-1)}}\left(\prod_{j=0}^{r-1}\frac{(n-j)!}{j!}\right) \times(\Delta_r(h))^2\prod_{j=1}^{r}\frac{\omega^{2h_j}}{h_j!(n-h_j)!}. \tag{12.27}$$

Set

$$q=\frac{\omega^2}{1+\omega^2},\qquad p=\frac{1}{1+\omega^2} \tag{12.28}$$

so that $p+q=1$. Then $q^{h_j}p^{n-h_j}=\frac{\omega^{2h_j}}{(1+\omega^2)^n}$. It follows that

$$\omega[h]\omega'[h]=c_{r,n}(\omega)(\Delta_r(h))^2\prod_{j=1}^{r}\frac{n!}{h_j!(n-h_j)!}q^{h_j}p^{n-h_j} \tag{12.29}$$

for some appropriate constant

$$c_{r,n}(\omega)=\frac{(1+\omega^2)^{\frac{n(n+1)}{2}+r(r-1)}}{\omega^{r(r-1)}}\prod_{j=1}^{r-1}\frac{(n-j)!}{n!j!}. \tag{12.30}$$

Note that we inserted the term $n!$ inside the products in (12.29) and (12.30). We arrive at the following

**Theorem 12.22.** *Fix $r$, $1\le r\le n$, and let $h=(h_1,\ldots,h_r)$, where $0\le h_1<h_2<\cdots<h_r\le n$. Then*

$$\mathbb{P}\left(Z_r(\tau)=h\right)=\frac{1}{Z_{r,n}}(\Delta_r(h))^2\prod_{j=1}^{r}\frac{n!}{h_j!(n-h_j)!}q^{h_j}p^{n-h_j}, \tag{12.31}$$

*where $Z_{r,n}$ is the partition function given by*

$$Z_{r,n}=\sum_{0\le h_1<h_2<\cdots<h_r\le n}(\Delta_r(h))^2\prod_{j=1}^{r}\frac{n!}{h_j!(n-h_j)!}q^{h_j}p^{n-h_j}. \tag{12.32}$$

We will compute $Z_{r,n}$ explicitly in Subsection 12.2.3 below. Note that $Z_{r,n}$ is precisely the Heine formula for the $r\times r$ Hankel determinant $H_{r-1}(\mu)$ for the discrete weight $\mu(dx)=\frac{n!}{x!(n-x)!}q^xp^{n-x}$. This weight is classical and is known as the Krawtchouk weight, and the associated orthonormal polynomials $p_k(x)=\gamma_kx^k+\cdots$, $\gamma_k>0$, $0\le k\le n$, are known as the Krawtchouk polynomials (see, e.g., [**Sze75**]. The relationship of (12.31) to random matrix theory is clear, and (12.31) gives rise, in particular, to a determinantal point process described in Chapter 11. We say that the $\{h_j\}$

belong to the *Krawtchouk Ensemble.* For a determinantal point process, key statistics can be expressed explicitly in terms of a Fredholm determinant (see (11.49) and Proposition 12.27 below).

### 12.2.2. Some Properties of Krawtchouk Polynomials.

Let $p_k(x) = \gamma_k x^k + \cdots$, $\gamma_k > 0$, $0 \le k \le n$, be the orthonormal polynomials with respect to the discrete measure $\mu(dx)$ on $\{0, 1, \ldots, n\}$ with weight

$$w(x) = \binom{n}{x} q^x p^{n-x}, \qquad x \in \{0, 1, \ldots, n\}. \tag{12.33}$$

Here $p$ and $q$ are positive real numbers satisfying $p + q = 1$. Thus,

$$\sum_{x=0}^{n} p_k(x) p_j(x) w(x) = \delta_{kj}. \tag{12.34}$$

As noted above, the polynomials $\{p_k\}$ are the *Krawtchouk polynomials*, and $w(x)$ is called the *Krawtchouk weight.*

**Lemma 12.23.** *For* $0 \le k \le n$,

$$p_k(x) = \binom{n}{k}^{-\frac{1}{2}} (pq)^{-\frac{k}{2}} \sum_{\nu=0}^{k} (-1)^{k-\nu} \binom{n-x}{k-\nu} \binom{x}{\nu} q^{k-\nu} p^{\nu}. \tag{12.35}$$

Note that for $0 \le \nu < k$

(12.36)

$$\binom{n-x}{k-\nu} = \frac{(n-x)(n-x-1)\cdots(n-x-k+\nu+1)}{(k-\nu)!} = \frac{(-1)^{k-\nu}}{(k-\nu)!} x^{k-\nu} + \cdots$$

and $\binom{n-x}{k-\nu} = 1$ for $k = \nu$. Also for $\nu > 0$

$$\binom{x}{\nu} = \frac{x(x-1)\cdots(x-\nu+1)}{\nu!} = \frac{x^\nu}{\nu!} + \cdots \tag{12.37}$$

and $\binom{x}{\nu} = 1$ for $\nu = 0$. Thus we see that $p_k(x)$ is indeed a polynomial of order $k$ and the leading coefficient is

$$\gamma_k = \binom{n}{k}^{-\frac{1}{2}} (pq)^{-\frac{k}{2}} \left( \sum_{\nu=0}^{k} \frac{1}{(k-\nu)!\nu!} q^{k-\nu} p^{\nu} \right) = \binom{n}{k}^{-\frac{1}{2}} (pq)^{-\frac{k}{2}} \frac{1}{k!}, \tag{12.38}$$

for $0 \le k \le n$.

**Proof of Lemma 12.23.** Let $P_k(x)$ be the polynomial defined by the right-hand side of (12.35). It is enough to show that $P_k(x)$ satisfies the orthonormality conditions (12.34). Set (cf. [**Sze75**, page 36])

$$\begin{aligned} K(x,s) &= \sum_{k=0}^{n} \binom{n}{k}^{\frac{1}{2}} (pq)^{\frac{k}{2}} P_k(x) s^k \\ &= \sum_{k=0}^{n}\sum_{\nu=0}^{k} (-1)^{k-\nu}\binom{n-x}{k-\nu}\binom{x}{\nu} q^{k-\nu}p^{\nu}s^k \\ &= \sum_{\nu=0}^{n}\sum_{k=\nu}^{n} (-1)^{k-\nu}\binom{n-x}{k-\nu}\binom{x}{\nu} q^{k-\nu}p^{\nu}s^k. \end{aligned} \tag{12.39}$$

Observe that if $\nu > n$ or $k > n$, then the polynomial $\binom{n-x}{k-\nu}\binom{x}{\nu}$ vanishes for all $x \in \{0,1,\ldots,n\}$. Thus, for $x \in \{0,1,\ldots,n\}$,

$$\begin{aligned} K(x,s) &= \sum_{\nu=0}^{\infty} \binom{x}{\nu} p^{\nu}s^{\nu} \sum_{k=\nu}^{\infty} (-1)^{k-\nu}\binom{n-x}{k-\nu} q^{k-\nu}s^{k-\nu} \\ &= \sum_{\nu=0}^{\infty} \binom{x}{\nu} p^{\nu}s^{\nu} \sum_{\ell=0}^{\infty} (-1)^{\ell}\binom{n-x}{\ell} (qs)^{\ell} \\ &= (1+ps)^x(1-qs)^{n-x}, \end{aligned} \tag{12.40}$$

from which we see that

$$\begin{aligned} &\sum_{x=0}^{n} w(x)K(x,u)K(x,v) \\ &\quad= \sum_{x=0}^{n} \binom{n}{x} q^x p^{n-x}(1+pu)^x(1-qu)^{n-x}(1+pv)^x(1-qv)^{n-x} \\ &\quad= \big(q(1+pu)(1+pv)+p(1-qu)(1-qv)\big)^n = (1+qpuv)^n. \end{aligned} \tag{12.41}$$

Thus,

$$\sum_{x=0}^{n} w(x)\left(\sum_{k=0}^{n}\binom{n}{k}^{\frac{1}{2}}(pq)^{\frac{k}{2}}P_k(x)u^k\right)\left(\sum_{\ell=0}^{n}\binom{n}{\ell}^{\frac{1}{2}}(pq)^{\frac{\ell}{2}}P_\ell(x)v^\ell\right) = (1+qpuv)^n.$$

So,

$$\begin{aligned} &\sum_{k,\ell=0}^{n}\binom{n}{k}^{\frac{1}{2}}(pq)^{\frac{k}{2}}\binom{n}{\ell}^{\frac{1}{2}}(pq)^{\frac{\ell}{2}}u^k v^\ell\left(\sum_{x=0}^{n} w(x)P_k(x)P_\ell(x)\right) \\ &\quad= \sum_{m=0}^{n}\binom{n}{m}(qp)^m(uv)^m. \end{aligned}$$

Hence, comparing the coefficients of $u^k v^\ell$, we find that for any $k \neq \ell$,

$$\sum_{x=0}^{n} w(x) P_k(x) P_\ell(x) = 0,$$

and for $k = \ell$,

$$\sum_{x=0}^{n} w(x) (P_k(x))^2 (uv)^k = 1.$$

So $P_k(x) = p_k(x)$, and the claim is proved. □

**Remark 12.24.** The Krawtchouk polynomials also have the integral representation

$$p_k(x) = \binom{n}{k}^{-\frac{1}{2}} (pq)^{-\frac{k}{2}} \frac{1}{2\pi i} \int_{\mathcal{C}} \frac{(1+pz)^x (1-qz)^{n-x}}{z^k} \frac{dz}{z}, \tag{12.42}$$

where $\mathcal{C}$ is a (counter-clockwise) circle about the origin with radius $\leq \min\{p^{-1}, q^{-1}\}$. Indeed, the right-hand side equals

$$\begin{aligned} &\binom{n}{k}^{-\frac{1}{2}} (pq)^{-\frac{k}{2}} \frac{1}{2\pi i} \sum_{m=0}^{\infty} \sum_{\ell=0}^{\infty} \binom{x}{m} \binom{n-x}{\ell} p^m (-q)^\ell \int_{\mathcal{C}} \frac{z^m}{z^k} z^\ell \frac{dz}{z} \\ &= \binom{n}{k}^{-\frac{1}{2}} (pq)^{-\frac{k}{2}} \sum_{m+\ell=k} \binom{x}{m} \binom{n-x}{\ell} p^m (-q)^\ell \qquad (12.43) \\ &= \binom{n}{k}^{-\frac{1}{2}} (pq)^{-\frac{k}{2}} \sum_{m=0}^{k} \binom{x}{m} \binom{n-x}{k-m} (-q)^{k-m} p^m, \end{aligned}$$

which is $p_k(x)$ by Lemma 12.23.

### 12.2.3. The Partition Function and Edge Distribution in Terms of a Fredholm Determinant.

For $1 \leq r \leq n$, consider the Hankel determinant $H_{r-1}$ of size $r$ associated with the discrete Krawtchouk measure $\mu_q = \sum_{x=0}^{n} \binom{n}{x} q^x p^{n-x} \delta_x(s)$:

$$H_{r-1} = \det \left( \int s^{i+j} \mu(ds) \right)_{0 \leq i,j \leq r-1}. \tag{12.44}$$

By the Heine formula (5.7),

$$\begin{aligned} H_{r-1} &= \frac{1}{r!} \int_{\mathbb{R}^r} (\Delta_r(s))^2 \mu(ds_1) \cdots \mu(ds_r) \\ &= \int_{h_1 < \cdots < h_r} (\Delta_r(h))^2 \mu(dh_1) \cdots \mu(dh_r) \qquad (12.45) \\ &= Z_{r,n}, \end{aligned}$$

the partition function given by (12.32). By (5.12), $\frac{H_k}{H_{k-1}} = \gamma_k^{-2}$, where $\gamma_k$ is the leading coefficient of the orthonormal polynomial $p_k(x)$ with respect to the discrete Krawtchouk measure. Hence,

$$H_{r-1} = \prod_{k=0}^{r-1} \gamma_k^{-2}. \tag{12.46}$$

From (12.38)

$$\gamma_k^{-2} = \binom{n}{k}(pq)^k(k!)^2 = \frac{n!k!}{(n-k)!}(pq)^k. \tag{12.47}$$

Hence, we obtain:

**Proposition 12.25.**

$$Z_{r,n} = \prod_{k=0}^{r-1} \frac{n!k!}{(n-k)!}(pq)^k = (pq)^{\frac{r(r-1)}{2}} \prod_{k=0}^{r-1} \frac{n!k!}{(n-k)!}. \tag{12.48}$$

**Remark 12.26.** Formula (12.48) can be used to compute some basic statistics for tilings. Recall that $v(\tau)$ denotes the number of vertical dominos in $\tau \in T(\mathcal{A}_n)$. From (12.6), (12.29), and (12.32),

$$\sum_{\tau \in T(\mathcal{A}_n)} \omega^{v(\tau)} = \sum_{0 \le h_1 < \cdots < h_r \le n} \omega[h]\omega'[h] = c_{r,n}(\omega) Z_{r,n}. \tag{12.49}$$

By (12.30) and (12.48),

$$\sum_{\tau \in T(\mathcal{A}_n)} \omega^{v(\tau)} = \frac{(1+\omega^2)^{\frac{n(n+1)}{2}+r(r-1)}}{\omega^{r(r-1)}}(pq)^{\frac{r(r-1)}{2}} = (1+\omega^2)^{\frac{n(n+1)}{2}}, \tag{12.50}$$

since $(pq)^{1/2} = \frac{\omega}{1+\omega^2}$ from (12.28). As the number of squares in $\mathcal{A}_n$ is $2n(n+1)$, we have the basic bound $v(\tau) \le n(n+1)$. Thus,

$$\sum_{\tau \in \mathcal{A}_n} \omega^{v(\tau)} = \sum_{k=0}^{n(n+1)} \#\{\tau : v(\tau) = k\}\omega^k. \tag{12.51}$$

On the other hand, expanding out the right-hand side of (12.50) by the binomial theorem, we find

$$(1+\omega^2)^{\frac{n(n+1)}{2}} = \sum_{k=0}^{\frac{n(n+1)}{2}} \binom{\frac{n(n+1)}{2}}{k} \omega^{2k}. \tag{12.52}$$

We conclude that

$$\#\{\tau : v(\tau) \text{ is odd}\} = 0 \tag{12.53}$$

and

$$\#\{\tau : v(\tau) = 2k\} = \binom{\frac{n(n+1)}{2}}{k}, \qquad 0 \leq k \leq \frac{n(n+1)}{2}. \tag{12.54}$$

Since, with the probability measure (12.3) on $T(\mathcal{A}_n)$,

$$\mathbb{P}\left(v(\tau) = 2k\right) = \sum_{\tau \in T(\mathcal{A}_n); v(\tau)=\ell} \mathbb{P}(\tau) = \frac{\#\{\tau : v(\tau) = 2k\} \cdot \omega^{2k}}{(1+\omega^2)^{\frac{n(n+1)}{2}}}, \tag{12.55}$$

we find find that

$$\mathbb{P}\left(v(\tau) = 2k\right) = \frac{\binom{\frac{n(n+1)}{2}}{k}\omega^{2k}}{(1+\omega^2)^{\frac{n(n+1)}{2}}} = \binom{\frac{n(n+1)}{2}}{k} q^k p^{\frac{n(n+1)}{2}-k} \tag{12.56}$$

for $0 \leq k \leq \frac{n(n+1)}{2}$, where again $q = \frac{\omega^2}{1+\omega^2}$, $p = \frac{1}{1+\omega^2}$. Otherwise, $\mathbb{P}(v(\tau) = \ell)$ $= 0$. Hence $v(\tau)/2$ is binomially distributed with parameters $\frac{n(n+1)}{2}$ and $q$.

When $\omega = 1$, the probability measure (12.3) becomes the uniform measure on the set of tilings of $\mathcal{A}_n$. In this case, (12.50) implies that

$$\#T(\mathcal{A}_n) = 2^{\frac{n(n+1)}{2}}, \tag{12.57}$$

and (12.56) becomes

$$\mathbb{P}\left(v(\tau) = 2k\right) = 2^{-\frac{n(n+1)}{2}} \binom{\frac{n(n+1)}{2}}{k}, \qquad 0 \leq k \leq \frac{n(n+1)}{2}. \tag{12.58}$$

Results (12.50) (and so also (12.57)), (12.53), (12.56), and (12.58) are due to Elkies, Kuperberg, Larson, and Propp [**EKLP92a, EKLP92b**].

In the next section, we evaluate the statistics of $h_r$ as $n \to \infty$ and $\frac{r}{n} \to t \in (0,1)$. This gives the fluctuations of the arctic circle. As noted earlier, the crucial fact to note is that the distribution of the $h_r$'s in (12.31) is precisely of the same form that occurs in random matrix theory. The proofs in Chapter 6 are written down for measures which are absolutely continuous with respect to Lebesgue measure, but the results go through for general measures. In particular, the analogue of Proposition 6.6 for the discrete Krawtchouk measure, $\mu = \sum_{x=0}^{n} \binom{n}{x} \delta_x$, yields the following basic result. Let $\{p_j(x)\}_{j=0}^{n}$ be the Krawtchouk polynomials orthonormal with respect to $\mu$.

**Proposition 12.27.** *Let $q, p \geq 0$, $p + q = 1$, be given. Then,*

$$\mathbb{P}\left(\max_{1 \leq i \leq r} h_i \leq s\right) = \det\left(1 - K_r \chi_{(s,\infty)}\right), \tag{12.59}$$

*where* $K_r$ *acts on* $\ell_2(\{0,1,\ldots,n\})$ *with kernel*

$$K_r(x,y) = \sqrt{w(x)w(y)} \sum_{j=0}^{r-1} p_j(x)p_j(y). \tag{12.60}$$

We recall that the asymptotic analysis of a similar Fredholm determinant was carried out in Section 9.4, in which case the operator was the product of two operators. We showed that each of the operators converge in Hilbert-Schmidt norm and utilized the general result (9.90) on the convergence of Fredholm determinants.

In this chapter, we consider the asymptotics using a different approach (see [**Joh00a**] and [**Joh01a**]; see also [**Joh05a**]). Note that the Fredholm determinant $\det\left(1 - K_r\chi_{(s,\infty)}\right)$ has a Fredholm expansion (A.42):

$$\det(1 - K_r\chi_{(s,\infty)}) = \sum_{k=0}^{\infty} \frac{(-1)^k}{k!} \sum_{h\in\{[s]+1,\ldots,n\}^k} \det\left(K_r(h_i,h_j)\right)_{i,j=1}^{k}, \tag{12.61}$$

where $[s]$ is the integral part of $s$. We evaluate this series asymptotically as $n \to \infty$, where $\frac{r}{n} \to t \in (0,1)$.

## 12.3. Asymptotics

Our goal is to prove the following result of Johansson. For the case $p = q = 1/2$, see [**Joh05a**]. The case of general $p$ and $q$ can be inferred from [**CJY**], in which the authors employed a method different from that in [**Joh05a**], for evaluating the distribution. Finally, see [**JN06**] for the case when $\alpha = p$ (for the special case $p = q = 1/2$) when (12.62) fails and we get fluctuations that differ from (12.62). References [**Joh00a**] and [**Joh01a**] also contain important calculations concerning tilings of the Aztec diamond. These calculations are used repeatedly in [**Joh05a**] and also are used repeatedly in the proof of Theorem 12.28 below.

**Theorem 12.28.** *Let* $r = [n\alpha]$, $0 < \alpha < 1$. *Set* $\beta \equiv 1 - \alpha$. *Suppose that* $\alpha < p$. *Then,*

$$\lim_{n\to\infty} \mathbb{P}\left(\max_{1\le j\le r} h_j \le nB + \xi\rho n^{\frac{1}{3}}\right) = F(\xi), \tag{12.62}$$

*where* $F(\xi)$ *is the Tracy-Widom distribution* (6.97). *Here,*

$$B = B(\alpha) \equiv q\beta + p\alpha + 2\sqrt{pq\alpha\beta} = \left(\sqrt{q\beta} + \sqrt{p\alpha}\right)^2 \tag{12.63}$$

*and*

$$\rho = \rho(\alpha) = \left(\frac{pq}{\alpha\beta}\right)^{\frac{1}{6}} (\sqrt{q\beta} + \sqrt{p\alpha})^{\frac{2}{3}} (\sqrt{p\beta} - \sqrt{q\alpha})^{\frac{2}{3}}. \tag{12.64}$$

**Remark 12.29.** Note that $\alpha < p$ (resp. $\alpha > p$) if and only if $p\beta > q\alpha$ (resp. $p\beta < q\alpha$). Note also that $1 - B = (\sqrt{p\beta} - \sqrt{q\alpha})^2$ and so $0 < B < 1$ for $\alpha < p$.

For the case $\alpha > p$ or $\alpha = p$, see the discussion following Figure 12.26 below.

### 12.3.1. General Lemma for Convergence of Fredholm Determinant.

In order to prove (12.62), we will use the following lemma of Johansson ([**Joh00a**, Lemma 3.1]).

**Lemma 12.30.** *Let $b > 0$ be a constant and assume $\rho_N > 1$, $\rho_N \to \infty$, $\rho_N = o(N)$, as $N \to \infty$. Suppose furthermore that $K_N : \mathbb{R}_+ \times \mathbb{R}_+ \to \mathbb{R}$, $N \geq 1$, satisfies the following properties.*

(i) *Given any $M_1 \geq 1$, let $N(M_1) \geq 1$ be an integer such that $bN + \rho_N t \geq 1$ for all $N \geq N(M_1)$ and $t \geq -M_1$. Then there exists a constant $C = C(M_1) > 0$ such that*

$$\sum_{m=1}^{\infty} K_N(bN + \rho_N t + m, bN + \rho_N t + m) \leq C \tag{12.65}$$

*for all $N \geq N(M_1)$ and $t \geq -M_1$.*

(ii) *Given $\epsilon > 0$, there exists $L(\epsilon) > 1$ such that*

$$\sum_{m=1}^{\infty} K_N(bN + \rho_N L + m, bN + \rho_N L + m) \leq \epsilon \tag{12.66}$$

*for all $N \geq 1$ and all $L \in [L(\epsilon) - 1, L(\epsilon)]$.*

(iii) *Let $M_0 > 0$ be any given constant. Then,*

$$\lim_{N\to\infty} \rho_N K_N(bN + \rho_N \xi, bN + \rho_N \eta) = \mathbb{A}(\xi, \eta) \tag{12.67}$$

*uniformly for $\xi, \eta \in [-M_0, M_0]$, where $\mathbb{A}(\xi,\eta)$ is the Airy kernel (6.85).*

(iv) *The matrix $(K_N(x_i, x_j))_{i,j=1}^k$ is positive definite for any $x_i, x_j \in (0, \infty)$, $k \geq 1$.*

*Then, uniformly for $t$ in a compact set*

(12.68)
$$\lim_{N\to\infty} \sum_{k=0}^{N} \frac{(-1)^k}{k!} \sum_{h \in \mathbb{N}_0^k} \det\left(K_N(bN + \rho_N t + h_i, bN + \rho_N t + h_j)\right)_{i,j=1}^k = F(t),$$

*where $\mathbb{N}_0 = \mathbb{N} \cup \{0\}$.*

We will of course apply this result with $N = r \sim \alpha n$, $b = B(\alpha)$, and $\rho_N = \rho(\alpha)n^{\frac{1}{3}}$ to the Fredholm expansion (12.61). In order to verify (i)–(iv), we need to analyze the asymptotics of the Krawtchouk orthogonal polynomials, $\{p_j\}$.

We will need the standard matrix theory result due to Hadamard:

**Lemma 12.31.** *If $(A_{ij})_{i,j=1}^k$ is a positive definite real matrix, then*

$$\det A \leq \prod_{i=1}^{k} A_{ii}. \tag{12.69}$$

**Proof.** We follow [**Lax02**]. Replacing $A$ by $A + \epsilon I$ and letting $\epsilon \downarrow 0$, we see that it is sufficient to consider the case where $A$ is strictly positive definite. As $A > 0$, $A_{ii} > 0$ for all $i$. Let $D = \operatorname{diag}(\sqrt{A_{11}}, \ldots, \sqrt{A_{kk}})$ and set $B = D^{-1}AD^{-1}$. Then $B$ is positive definite and $B_{ii} = 1$ for all $i$. As $\det B = \frac{\det A}{(\det D)^2} = \frac{\det A}{\prod_{i=1}^k A_{ii}}$, the problem is reduced to showing that $\det B \leq 1$. Let $b_1, \ldots, b_k > 0$ be the eigenvalues of $B$. Then by the algebraic-geometric inequality

$$\prod_{i=1}^{k} b_i \leq \left(\frac{\sum_{i=1}^k b_i}{k}\right)^k,$$

which gives

$$\det B \leq \left(\frac{\operatorname{tr} B}{k}\right)^k.$$

But, since $B_{ii} = 1$, $\operatorname{tr} B = k$, which completes the proof. ☐

In the proof that follows, certain auxiliary quantities (namely, $\ell_1(\epsilon, M_1)$, $\ell(\epsilon, M_1)$, $L_0(\epsilon, M_1)$) will be needed. These quantities are chosen in the following way. Let $M_1 \geq 1$ and $\epsilon > 0$ be given as in Lemma 12.30. Let $\ell_1(\epsilon, M_1)$ be a positive integer such that

$$\sum_{k=\ell_1(\epsilon,M_1)+1}^{\infty} \frac{C^k}{k!} < \epsilon, \tag{12.70}$$

where $C = C(M_1)$ is the constant in $(i)$. Let $\ell(\epsilon, M_1)$ and $L_0(\epsilon, M_1)$ be any constants satisfying

$$\ell(\epsilon, M_1) \geq \ell_1(\epsilon, M_1), \qquad L_0(\epsilon, M_1) \geq L(\epsilon) + M_1, \tag{12.71}$$

respectively, where $L(\epsilon)$ is the constant in (ii). The constants $\ell(\epsilon, M_1)$ and $L_0(\epsilon, M_1)$ will be specified in the proof of Lemma 12.30 below (see the discussion following (12.85)). For convenience we write $\ell_1 = \ell_1(\epsilon, M_1)$, $\ell = \ell(\epsilon, M_1)$, $L_0 = L_0(\epsilon, M_1)$.

**Proof of Lemma 12.30.** Let $M_1 \geq 1$ and $\epsilon > 0$ be given. Assume $|t| \leq M_1 - 1 < \infty$. Then we have $a_N \equiv bN + \rho_N t \geq 1 > 0$ for $N \geq N(M_1)$. From (iv) and (12.69),

$$\det\left(K_N(x_i, x_j)\right)_{i,j=1}^k \leq \prod_{j=1}^N K_N(x_j, x_j). \tag{12.72}$$

Consequently,

$$\begin{aligned} &\sum_{h\in\mathbb{N}_0^k} \det\left(K_N(a_N + h_i, a_N + h_j)\right)_{i,j=1}^k \\ &\leq \sum_{h\in\mathbb{N}_0^k} \prod_{j=1}^N K_N(a_N + h_j, a_N + h_j) \\ &= \left(\sum_{h=0}^{\infty} K_N(a_N + h, a_N + h)\right)^k \leq C^k \end{aligned} \tag{12.73}$$

as $a_N + h = bN + \rho_N t + h = bN + \rho_N\left(t - \frac{1}{\rho_N}\right) + (h+1)$. It follows from (i), (12.73), and (12.70) that

$$\begin{aligned} &\left|\sum_{k=\ell+1}^N \frac{(-1)^k}{k!} \sum_{h\in\mathbb{N}_0^k} \det\left(K_N(a_N + h_i, a_N + h_j)\right)_{i,j=1}^k\right| \\ &\leq \sum_{k=\ell+1}^N \frac{C^k}{k!} \leq \epsilon \end{aligned} \tag{12.74}$$

for all $N > \max\{N(M_1), \ell\}$.

Now, for $L_0 = L_0(\epsilon, M_1) \geq L(\epsilon) + M_1 > 1$, we have by (i), (ii), and (12.72)

$$\begin{aligned} &\left|\left(\sum_{h\in\mathbb{N}_0^k} - \sum_{\substack{0\leq h_q\leq\rho_N L_0 \\ 1\leq q\leq k}}\right) \det\left(K_N(a_N + h_i, a_N + h_j)\right)_{i,j=1}^k\right| \\ &= \left|\sum_{h_j\geq\rho_N L_0 \text{for some } j} \det\left(K_N(a_N + h_i, a_N + h_j)\right)_{i,j=1}^k\right| \\ &\leq \sum_{j=1}^k \sum_{h_j\geq\rho_N L_0} \prod_{i=1}^k K_N(a_N + h_i, a_N + h_i) \\ &= k\left(\sum_{m=0}^{\infty} K_N(a_N + m, a_N + m)\right)^{k-1} \sum_{h_1>\rho_N L_0} K_N(a_N + h_1, a_N + h_1). \end{aligned}$$

Now,

$$a_N + h_1 = bN + \rho_N t + h_1 = bN + \rho_N t + \rho_N L_0 + \omega_N + m,$$

where $0 < \omega_N \leq 1$ and $m$ is a nonnegative integer. Clearly, the map $h_1 \mapsto m$ is one-to-one. Write

$$a_N + h_1 = bN + L\rho_N + \rho_N \left( L_0 - L + t + \frac{\omega_N}{\rho_N} \right) + m,$$

where we will choose $L$ between $L(\epsilon) - 1$ and $L(\epsilon)$, where $L(\epsilon)$ is given in (ii). Now,

$$\begin{aligned} L_0 - L + t + \frac{\omega_N}{\rho_N} &\geq L(\epsilon) + M_1 - L - M_1 + 1 + \frac{\omega_N}{\rho_N} \\ &= L(\epsilon) - L + 1 + \frac{\omega_N}{\rho_N} \geq 1 \end{aligned}$$

for $L(\epsilon) - 1 \leq L \leq L(\epsilon)$. It follows that as $L$ varies from $L(\epsilon) - 1$ to $L(\epsilon)$, $\rho_N \left( L_0 - L + t + \frac{\omega_N}{\rho_N} \right)$ varies over an interval $[y(t, N, \epsilon, M_1), y(t, N, \epsilon, M_1) + \rho_N]$, where $y(t, N, \epsilon, M_1) \geq \rho_N > 1$. Hence, we may choose $L(t, N, \epsilon, M_1) \in [L(\epsilon) - 1, L(\epsilon)]$ such that $\rho_N \left( L_0 - L + t + \frac{\omega_N}{\rho_N} \right) = k_N = k(t, N, \epsilon, M_1)$ for some positive integer $k_N$ (necessarily $k_N \geq 2$). We conclude that for each $h_1 > \rho_N L_0$, $a_N + h_1 = bN + L\rho_N + m'$ for some $m' = k_N + m \geq k_N \geq 2$. As $L = L(t, N, \epsilon, M_1) \in [L(\epsilon) - 1, L(\epsilon)]$ and is independent of $m$, and as the integers $m' = k_N + m \geq 2$ are distinct, the estimate (12.66) applies, and we obtain

$$\left| \left( \sum_{h \in \mathbb{N}_0^k} - \sum_{\substack{0 \leq h_q \leq \rho_N L_0 \\ 1 \leq q \leq k}} \right) \det \left( K_N(a_N + h_i, a_N + h_j) \right)_{i,j=1}^k \right| \leq k C^{k-1} \epsilon. \tag{12.75}$$

Let

$$D_N(t) = \sum_{k=0}^{N} \frac{(-1)^k}{k!} \sum_{h \in \mathbb{N}_0^k} \det \left( K_N(bN + \rho_N t + h_i, bN + \rho_N t + h_j) \right)_{i,j=1}^k.$$

Then by (12.74) and (12.75), for $\ell = \ell(\epsilon, M_1)$ as above, $N > \ell$, and $\mathcal{K}_N(\xi,\eta) \equiv \rho_N K_N(bN + \rho_N\xi, bN + \rho_N\eta)$,

(12.76)

$$\left| D_N(t) - \sum_{k=0}^{\ell} \frac{(-1)^k}{k!} \sum_{\substack{0\le h_q \le \rho_N L_0 \\ 1\le q\le k}} \det\left(\mathcal{K}_N(t + \frac{h_i}{\rho_N}, t + \frac{h_j}{\rho_N})\right)_{i,j=1}^k \frac{1}{\rho_N^k} \right|$$

$$\le \epsilon + \left| \sum_{k=0}^{\ell} \frac{(-1)^k}{k!} \left( \sum_{h\in\mathbb{N}_0^k} - \sum_{\substack{0\le h_q\le \rho_N L_0 \\ 1\le q\le k}} \right) \det\left(K_N(a_N + h_i, a_N + h_j)\right)_{i,j=1}^k \right|$$

$$\le \epsilon + \sum_{k=0}^{\ell} \frac{1}{k!} k C^{k-1}\epsilon \le (1 + e^C)\epsilon.$$

We now compare $\mathcal{K}_N$ and $\mathbb{A}$. Set $M_0 \equiv M_1 + L_0$, and note that for $0 \le h_i \le \rho_N L_0$,

$$\left| t + \frac{h_i}{\rho_N} \right| \le M_1 - 1 + L_0 < M_0.$$

Thus, by (iii),

$$\mathcal{K}_N\left(t + \frac{h_i}{\rho_N}, t + \frac{h_j}{\rho_N}\right) - \mathbb{A}\left(t + \frac{h_i}{\rho_N}, t + \frac{h_j}{\rho_N}\right) \to 0 \tag{12.77}$$

as $N \to \infty$ uniformly for $|t| \le M_1 - 1$ and $\max_{1\le i\le k} |h_i| \le \rho_N L_0$. Thus, there exists $N_1 = N_1(\epsilon) > \ell$ such that

$$\left| \det\left(\mathcal{K}_N\left(t + \frac{h_i}{\rho_N}, t + \frac{h_j}{\rho_N}\right)\right)_{i,j=1}^k - \det\left(\mathbb{A}\left(t + \frac{h_i}{\rho_N}, t + \frac{h_j}{\rho_N}\right)\right)_{i,j=1}^k \right|$$

$$\le \frac{\epsilon}{L_0^k} \tag{12.78}$$

for $0 \le k \le \ell$, $0 \le h_j \le \rho_N L_0$ and $N \ge N_1$. Thus, for $N > N_1$

$$\left| \sum_{k=0}^{\ell} \frac{(-1)^k}{k!} \sum_{\substack{0\le h_q\le \rho_N L_0 \\ 1\le q\le k}} \frac{1}{\rho_N^k} \left( \det\left(\mathcal{K}_N\left(t + \frac{h_i}{\rho_N}, t + \frac{h_j}{\rho_N}\right)\right)_{i,j=1}^k \right.\right.$$

$$\left.\left. - \det\left(\mathbb{A}\left(t + \frac{h_i}{\rho_N}, t + \frac{h_j}{\rho_N}\right)\right)_{i,j=1}^k \right) \right| \tag{12.79}$$

$$\le \sum_{k=0}^{\ell} \frac{1}{k!} \frac{\epsilon}{(\rho_N L_0)^k} (1 + \rho_N L_0)^k \le e^{\frac{1+\rho_N L_0}{\rho_N L_0}} \epsilon \le C'\epsilon,$$

where $C' = e^2$ since $\rho_N L_0 > 1$.

From (12.76) and (12.79), we conclude that for some constants $C'' = C''(M_1)$,
(12.80)

$$\left| D_N(t) - \sum_{k=0}^{\ell} \frac{(-1)^k}{k!} \sum_{\substack{0 \le h_q \le \rho_N L_0 \\ 1 \le q \le k}} \frac{1}{\rho_N^k} \det \left( \mathbb{A} \left( t + \frac{h_i}{\rho_N}, t + \frac{h_j}{\rho_N} \right) \right)_{i,j=1}^k \right| \le C'' \epsilon$$

for $N > N_1$ and $|t| \le M_1 - 1$. By the continuity of $\mathbb{A}(\xi, \eta)$ we see that there are constants $C''' > 0$ and $N_2 > N_1$ such that

$$\left| D_N(t) - \sum_{k=0}^{\ell} \frac{(-1)^k}{k!} \int_{[0,L_0]^k} \det \left( \mathbb{A} \left( t + \xi_i, t + \xi_j \right) \right)_{i,j=1}^k d^k \xi \right| \le C''' \epsilon \tag{12.81}$$

for $N > N_2$ and $|t| \le M_1 - 1$.

By (6.87),

$$\mathbb{A}(x, y) = \int_0^\infty \operatorname{Ai}(x+s) \operatorname{Ai}(y+s) ds, \tag{12.82}$$

from which we conclude in the standard way that $(\mathbb{A}(x_i, x_j))_{i,j=1}^k$ is positive definite for all $x_i, x_j$. In particular, $\mathbb{A}(x, x) \ge 0$. Also it is not difficult to obtain using the steepest-descent method (see (B.18) or simply consult [**AS64**]) that there is a constant $c > 0$ such that

$$\begin{cases} |\operatorname{Ai}(x)| \le \frac{c}{(1+x)^{\frac{1}{4}}} e^{-\frac{2}{3} x^{\frac{3}{2}}}, & x \ge 0, \\ |\operatorname{Ai}'(x)| \le c(1+x)^{\frac{1}{4}} e^{-\frac{2}{3} x^{\frac{3}{2}}}, & x \ge 0. \end{cases} \tag{12.83}$$

Now,

$$\begin{aligned} \mathbb{A}(x, x) &= \lim_{y \to x} \frac{\operatorname{Ai}(y) \operatorname{Ai}'(x) - \operatorname{Ai}'(y) \operatorname{Ai}(x)}{y - x} \\ &= \operatorname{Ai}'(x) \operatorname{Ai}'(x) - \operatorname{Ai}''(x) \operatorname{Ai}(x) = (\operatorname{Ai}'(x))^2 - x(\operatorname{Ai}(x))^2. \end{aligned} \tag{12.84}$$

Thus from (12.83) we see that for some $c', c''$ we have $0 \le \mathbb{A}(x, x) \le c'' e^{-c' x^{\frac{3}{2}}}$ for $x \ge 0$. Hence, as $\mathbb{A}(x, x)$ is continuous, we see that for any $M_1 > 0$

$$\int_{-M_1}^\infty \mathbb{A}(x, x) dx \le c'' \left[ \int_0^\infty e^{-c' x^{\frac{3}{2}}} dx + \int_{-M_1}^0 \mathbb{A}(x, x) dx \right] < c''''$$

for some constant $c''' > 0$. Also, given $\epsilon > 0$, there exists $L_0' = L_0'(\epsilon)$ such that

$$\int_0^\infty \mathbb{A}(x + L_0, x + L_0) dx = \int_{L_0'}^\infty \mathbb{A}(x, x) dx < \epsilon.$$

Using similar arguments to those that lead to (12.76), we find that given $\epsilon > 0$ and $M_1$, if $|t| \leq M_1 - 1$,

$$\left| \left( \sum_{k=0}^{\infty} \int_{[t,\infty)^k} - \sum_{k=0}^{\ell'} \int_{[t, L_0''+t]^k} \right) \frac{(-1)^k}{k!} \det\left(\mathbb{A}(\xi_i, \xi_j)\right)_{i,j=1}^k d^k\xi \right| \leq \epsilon \tag{12.85}$$

for $\ell'$ and $L_0''$ sufficiently large. Set

$$\ell = \ell(\epsilon, M_1) \equiv \max\{\ell_1(\epsilon, M_1), \ell'\} \tag{12.86}$$

and

$$L_0 = L_0(\epsilon, M_1) \equiv \max\{L(\epsilon) + M_1, L_0''\}. \tag{12.87}$$

Using (12.81) and (12.85), we immediately obtain (12.68), as

$$F(t) = \det\left(1 - \mathbb{A}_{(t,\infty)}\right) = \sum_{k=0}^{\infty} \frac{(-1)^k}{k!} \int_{(t,\infty)^k} \det\left(\mathbb{A}(\xi_i, \xi_j)\right)_{i,j=1}^k d^k\xi.$$

The convergence in (12.68) as $N \to \infty$ is uniform for $|t| \leq M_1 - 1$. This proves Lemma 12.30. □

It remains to apply Lemma 12.30 to the Krawtchouk situation for the Aztec diamond. However, in order to prove (12.62) using Lemma 12.30, we first need to take care of a technicality. Instead of the expression in (12.68), it turns out that we must control an expression of the form

(12.88)

$$P_N(t) = \sum_{k=0}^{N} \frac{(-1)^k}{k!} \sum_{h \in \{[bN+\rho_N t]+1, [bN+\rho_N t]+2, \dots\}^k} \det\left(K_N(h_i, h_j)\right)_{i,j=1}^k$$

$$= \sum_{k=0}^{N} \frac{(-1)^k}{k!} \sum_{h \in \mathbb{N}_0^k} \det\left(K_n([bN + \rho_N t + 1] + h_i, [bN + \rho_N t + 1] + h_j)\right)_{i,j=1}^k.$$

But,

$$[bN + \rho_N t + 1] = bN + \rho_N t + (1 - \omega_N') = bN + \rho_N t',$$

where $0 \leq \omega_N' < 1$ and $t' = t_N' = t + \frac{1-\omega_N'}{\rho_N}$. But the convergence in (12.68) is uniform on compact sets. So,

$$|P_N(t) - F(t)| \leq \left|P_N(t) - F(t_N')\right| + \left|F(t_N') - F(t)\right|,$$

which tends to zero as $N \to \infty$ by (12.68) and the continuity of $F(t)$.

In order to verify properties (i), (ii), (iii), and (iv) of Lemma 12.30 in the case of the Aztec diamond, we must clearly (see (12.59) and (12.60)) control the asymptotics of the Krawtchouk polynomials $\{p_k(x)\}$ in the so-called Plancherel-Rotach limit; i.e., for fixed $0 < \alpha < 1$, we need to control $p_{[n\alpha]}([bn + \rho_n t + 1] + h)$, $h \in \mathbb{N}_0$, as $n \to \infty$ for suitable $b, \rho_n$. Such control can be obtained (see [**Joh02**]) via a steepest-descent analysis of the integral representation (12.42) for the Krawtchouk polynomials, and in the pages that follow we will present this approach in detail (cf. Appendix B). In other situations, however, such as tilings of a hexagon by rhombi [**Joh02, BKMM07**], the role of the Krawtchouk polynomials is played by other orthogonal polynomials (e.g., for hexagon tilings, one considers the Hahn polynomials) for which no viable integral representation is known. Nevertheless, it is still possible to analyze the asymptotics of these polynomials in the appropriate Plancherel-Rotach regime, using Riemann-Hilbert/nonlinear steepest-descent techniques (see [**BKMM07**]).

**12.3.2. Asymptotics of the Kernel.** We need to verify the estimates (i)–(iv) in Lemma 12.30 for

$$\begin{aligned} K_r(x,y) &= \sqrt{w(x)w(y)}\sum_{j=0}^{r-1} p_j(x)p_j(y) \\ &= \sqrt{w(x)w(y)}\,\frac{\gamma_{r-1}}{\gamma_r}\frac{p_r(x)p_{r-1}(y)-p_{r-1}(x)p_r(y)}{x-y}, \end{aligned} \tag{12.89}$$

where $\{p_j(x)\}$ are the Krawtchouk polynomials. Recall formulae (12.38) and (12.42):

$$\gamma_k = \binom{n}{k}^{-\frac{1}{2}}(pq)^{-\frac{k}{2}}\frac{1}{k!}, \qquad 0 \le k \le n, \tag{12.90}$$

and

$$p_k(x) = \binom{n}{k}^{-\frac{1}{2}}\frac{(pq)^{-\frac{k}{2}}}{2\pi i}\int_{\mathcal{C}}\frac{(1+pz)^x(1-qz)^{n-x}}{z^{k+1}}dz, \tag{12.91}$$

where $\mathcal{C}$ is any loop around zero within the circle of radius $\min\{\frac{1}{p},\frac{1}{q}\}$.

Note that $\frac{(1+pz)^x(1-qz)^{n-x}}{z^{k+1}}$ is analytic in $0<|z|<\min\{\frac{1}{p},\frac{1}{q}\}$. Note also that although (12.91) was derived initially only for integer values of $x$, the formula remains true for all real $x$. Indeed, by a residue calculation at $z=0$, the right-hand side is easily seen to be a polynomial of degree $k$ for all $x$, and as it agrees with the left-hand side on more than $k$ points of $x$, namely at $x=0,1,\dots,n$, it must agree for all $x$.

We consider the case $\alpha<p$ (see Theorem 12.28). When $\alpha=p$ or $\alpha>p$, the situation is different (see the discussion following Figure 12.26 below). Let $M_1>1$ be fixed. We are interested in the asymptotics of $K_r(x,y)$ for

$$r = [\alpha n], \tag{12.92}$$

with a fixed $0<\alpha<p$, and for $x, y$ of the form

$$x = Bn+\xi\rho_n, \quad y = Bn+\eta\rho_n, \qquad \xi,\eta \ge -M_1, \tag{12.93}$$

where

$$B = \left(\sqrt{\beta q} + \sqrt{\alpha p}\right)^2 = \beta q + \alpha p + 2\sqrt{\alpha\beta pq} \tag{12.94}$$

and

$$\rho_n = \rho n^{\frac{1}{3}}, \qquad \rho = \left(\frac{pq}{\alpha\beta}\right)^{\frac{1}{6}} \left(\sqrt{q\beta} + \sqrt{p\alpha}\right)^{\frac{2}{3}} \left(\sqrt{p\beta} - \sqrt{q\alpha}\right)^{\frac{2}{3}} \tag{12.95}$$

as defined in (12.63) and (12.64). Here $\beta = 1-\alpha$. Recall from Remark 12.29, that $0 < B < 1$ for $\alpha < p$.

Set

$$f_{n,k}(z, x) = x\log(1 + pz) + (n - x)\log(1 - qz) - (k + 1)\log(z), \tag{12.96}$$

where we take the principal branch for $\log(z)$. Then

$$\int_C \frac{(1 + pz)^x(1 - qz)^{n-x}}{z^{k+1}} dz = \int_C e^{f_{n,k}(z,x)} dz. \tag{12.97}$$

We are interested in

$$k = r = [\alpha n] = \alpha n - \omega_n, \quad \text{where } 0 \leq \omega_n < 1. \tag{12.98}$$

Define

$$\begin{aligned} f_n(z, \xi) \equiv& f_{n,[\alpha n]}(z, Bn + \xi\rho_n) \\ =& n\left(B\log(1 + pz) + (1 - B)\log(1 - qz) - \alpha\log(z)\right) \\ &+ \rho_n\xi(\log(1 + pz) - \log(1 - qz)) + (\omega_n - 1)\log(z). \end{aligned} \tag{12.99}$$

We have

$$\begin{aligned} f_n'(z, \xi) =& n\left(\frac{Bp}{1 + pz} - \frac{(1 - B)q}{1 - qz} - \frac{\alpha}{z}\right) \\ &+ \rho_n\xi\left(\frac{p}{1 + pz} + \frac{q}{1 - qz}\right) + \frac{\omega_n - 1}{z}. \end{aligned} \tag{12.100}$$

Direct evaluation shows that

$$\frac{Bp}{1 + pz} - \frac{(1 - B)q}{1 - qz} - \frac{\alpha}{z} = \frac{-z^2 qp\beta + 2z\sqrt{\alpha\beta pq} - \alpha}{z(1 + pz)(1 - qz)}. \tag{12.101}$$

The numerator has a unique root given by

$$z_c \equiv \sqrt{\frac{\alpha}{\beta pq}}. \tag{12.102}$$

For convenience in what follows we will assume that in addition to $\alpha < p$, we also have

$$\alpha < q. \tag{12.103}$$

Then a simple calculation shows that

$$z_c < \min\{\frac{1}{p}, \frac{1}{q}\}, \tag{12.104}$$

and so we can, and do, choose $C$ to be a contour around the origin passing through $z_c$ and lying in $|z| < \min\{\frac{1}{p}, \frac{1}{q}\}$. For the case $q \le \alpha < p$, see Exercise 12.32 below.

From the above calculations we see that the leading order term in $f_n'(z,\xi)$ has the form

$$-\frac{qp\beta(z-z_c)^2}{z(1+pz)(1-qz)}. \tag{12.105}$$

Thus, the second derivative of the leading term of $f_n$ also vanishes at $z = z_c$. Moreover, the third derivative is

$$\frac{-2qp\beta}{z_c(1+pz_c)(1-qz_c)}$$

at $z_c$. Finally, the fourth derivative at $z_c$ is easily seen to be

$$\frac{6qp\beta}{z_c^2(1+pz_c)^2(1-qz_c)^2}\left(1+2(p-q)z_c-3pqz_c^2\right).$$

Thus, using $\frac{p}{1+pz_c}+\frac{q}{1-qz_c}=\frac{1}{(1+pz_c)(1-qz_c)}$ for the $n^{\frac{1}{3}}$ term, we obtain

$$\begin{aligned} f_n(z,\xi) = f_n(z_c,\xi) &+ n\left(\frac{-qp\beta}{3z_c(1+pz_c)(1-qz_c)}(z-z_c)^3\right.\\ &\left.+\frac{qp\beta(1+2(p-q)z_c-3pqz_c^2)}{4z_c^2(1+pz_c)^2(1-qz_c)^2}(z-z_c)^4+O((z-z_c)^5)\right)\\ &+n^{\frac{1}{3}}\xi\left(\frac{\rho}{(1+pz_c)(1-qz_c)}(z-z_c)+O((z-z_c)^2)\right)\\ &+(\omega_n-1)\frac{z-z_c}{z_c}+O((z-z_c)^2). \end{aligned} \tag{12.106}$$

Set

$$b = \frac{\rho}{(1+pz_c)(1-qz_c)} > 0. \tag{12.107}$$

Note here that $1-qz_c=\frac{\sqrt{p\beta}-\sqrt{q\alpha}}{\sqrt{p\beta}}>0$ by the condition $\alpha<q<p$. Using

$$(1+pz_c)(1-qz_c)=\frac{(\sqrt{q\beta}+\sqrt{p\alpha})(\sqrt{p\beta}-\sqrt{q\alpha})}{\beta\sqrt{pq}}=\frac{\alpha^{1/4}}{(qp\beta)^{3/4}}\rho^{3/2},$$

simple algebraic manipulations show that

$$b^3 = \frac{pq\beta}{(1+pz_c)(1-qz_c)z_c}. \tag{12.108}$$

Thus, we see that for $z$ near $z_c$,

(12.109)
$$\begin{aligned} f_n(z,\xi) =& f_n(z_c,\xi) + n\left(-\frac{b^3}{3}(z-z_c)^3 + \hat{a}(z-z_c)^4 + O((z-z_c)^5)\right) \\ &+ n^{\frac{1}{3}}\xi b(z-z_c)(1+O((z-z_c))) \\ &+ \frac{(\omega_n - 1)}{z_c}(z-z_c) + O((z-z_c)^2), \end{aligned}$$

where

$$\hat{a} \equiv \frac{pq\beta}{4z_c^2(1+pz_c)^2(1-qz_c)^2}\left(1 + 2(p-q)z_c - 3pqz_c^2\right). \tag{12.110}$$

For later purposes, observe that

$$b^3 - 2z_c\hat{a} = \frac{pq}{2z_c(1+pz_c)^2(1-qz_c)^2}. \tag{12.111}$$

In particular,

$$b^3 - 2z_c\hat{a} > 0. \tag{12.112}$$

Note the following: by (12.100) and (12.105),

$$f_n'(z,\xi) = -n\frac{qp\beta(z-z_c)^2}{z(1+pz)(1-qz)} + \frac{\rho_n\xi}{(1+pz)(1-qz)} + \frac{\omega_n - 1}{z}. \tag{12.113}$$

Hence on the circle $z = z_c e^{i\theta}$, $-\pi \le \theta \le \pi$,

$$\begin{aligned} \frac{d}{d\theta}\operatorname{Re}\left(f_n(z_c e^{i\theta},\xi)\right) &= \operatorname{Re}\left(f_n'(z_c e^{i\theta},\xi) z_c i e^{i\theta}\right) \\ &= \operatorname{Re}\left(\frac{-nqp\beta z_c^2(e^{i\theta}-1)^2 i}{(1+pz_c e^{i\theta})(1-qz_c e^{i\theta})}\right) + \operatorname{Re}\left(\frac{\rho_n\xi z_c i e^{i\theta}}{(1+pz_c e^{i\theta})(1-z_c q e^{i\theta})}\right). \end{aligned}$$

Since $\operatorname{Re}\left(ie^{i\theta}((1+pz_c e^{-i\theta})(1-z_c q e^{-i\theta}))\right) = -(1+pqz_c^2)\sin(\theta) = -\frac{1}{\beta}\sin(\theta)$,

$$\frac{d}{d\theta}\operatorname{Re}\left(f_n(z_c e^{i\theta},\xi)\right) = -\frac{\left(4npqz_c^2\sin^2(\frac{\theta}{2}) + \frac{\rho_n\xi z_c}{\beta}\right)\sin(\theta)}{|1+pz_c e^{i\theta}|^2\,|1-qz_c e^{i\theta}|^2}. \tag{12.114}$$

In particular, we see that to leading order, and to all orders if $\xi \ge 0$, that $\operatorname{Re}\left(f_n(z_c e^{i\theta},\xi)\right)$ attains its maximum on $\{|z| = z_c\}$ at $\theta = 0$, i.e., $z = z_c$.

Now, from (12.89) and (12.91) we have

(12.115)
$$\begin{aligned} K_r(x,y) =& \sqrt{w(x)w(y)}\frac{\gamma_{r-1}}{\gamma_r}\binom{n}{r}^{-\frac{1}{2}}\binom{n}{r-1}^{-\frac{1}{2}}(pq)^{-r+\frac{1}{2}} \\ &\times \frac{Q}{(2\pi i)^2(x-y)}, \end{aligned}$$

where

$$(12.116)\quad \begin{aligned} Q \equiv & \left(\int_{\mathcal{C}} \frac{(1+pz)^x(1-qz)^{n-x}}{z^{r+1}}dz\right)\left(\int_{\mathcal{C}} \frac{(1+pz)^y(1-qz)^{n-y}}{z^{r}}dz\right) \\ & - \left(\int_{\mathcal{C}} \frac{(1+pz)^x(1-qz)^{n-x}}{z^{r}}dz\right)\left(\int_{\mathcal{C}} \frac{(1+pz)^y(1-qz)^{n-y}}{z^{r+1}}dz\right). \end{aligned}$$

Note that

(12.117)

$$\begin{aligned} Q = & \left(\int_{\mathcal{C}} \frac{(1+pz)^x(1-qz)^{n-x}}{z^{r+1}}dz\right)\left(\int_{\mathcal{C}} \frac{(1+pz)^y(1-qz)^{n-y}}{z^{r+1}}(z-z_c)dz\right) \\ & - \left(\int_{\mathcal{C}} \frac{(1+pz)^x(1-qz)^{n-x}}{z^{r+1}}(z-z_c)dz\right)\left(\int_{\mathcal{C}} \frac{(1+pz)^y(1-qz)^{n-y}}{z^{r+1}}dz\right). \end{aligned}$$

The terms proportional to $z_c$, which are inserted into the integrands, clearly drop out. We have

(12.118)

$K_r(x, y)$

$$\begin{aligned} = \frac{X_n(x,y)}{x-y} \Bigg[ & \left(\int_{\mathcal{C}} e^{f_n(z,\xi)-f_n(z_c,\xi)}\frac{dz}{2\pi i}\right)\left(\int_{\mathcal{C}} e^{f_n(z,\eta)-f_n(z_c,\eta)}(z-z_c)\frac{dz}{2\pi i}\right) \\ & - \left(\int_{\mathcal{C}} e^{f_n(z,\xi)-f_n(z_c,\xi)}(z-z_c)\frac{dz}{2\pi i}\right)\left(\int_{\mathcal{C}} e^{f_n(z,\eta)-f_n(z_c,\eta)}\frac{dz}{2\pi i}\right)\Bigg], \end{aligned}$$

where

$$(12.119)\quad \begin{aligned} X_n(x,y) \equiv & \sqrt{w(x)w(y)}\ \frac{\gamma_{r-1}}{\gamma_r}\binom{n}{r}^{-\frac{1}{2}}\binom{n}{r-1}^{-\frac{1}{2}}(pq)^{-r+\frac{1}{2}} \\ & \times \frac{(1+pz_c)^{x+y}(1-qz_c)^{2n-x-y}}{z_c^{2r+2}}. \end{aligned}$$

Here $r = [\alpha n] = \alpha n - \omega_n$, $0 \le \omega_n < 1$, and $x = Bn + \rho_n\xi, y = Bn + \rho_n\eta$, where $\xi, \eta \ge -M_1$,

We now evaluate the prefactor $X_n(x, y)$ asymptotically. First, by (12.90),

$$(12.120)\quad \frac{\gamma_{r-1}}{\gamma_r}\binom{n}{r}^{-\frac{1}{2}}\binom{n}{r-1}^{-\frac{1}{2}}(pq)^{-r+\frac{1}{2}} = \frac{r!(n-r+1)!}{n!}(pq)^{1-r}.$$

Now, Sterling's approximation gives $n! = \sqrt{2\pi n}n^n e^{-n}e^{\frac{\theta}{12n}}$ for some $0 < \theta < 1$. We have $r = \alpha n - \omega_n$, $0 \le \omega_n < 1$. Thus

$$\begin{aligned} \frac{r!(n-r+1)!}{n!} & = \frac{\sqrt{2\pi}}{e}\frac{(\alpha n-\omega_n)^{\alpha n-\omega_n+\frac{1}{2}}(\beta n+1+\omega_n)^{\beta n+\omega_n+\frac{3}{2}}}{n^{n+\frac{1}{2}}}e^{o(1)} \\ & = \sqrt{2\pi}\alpha^{\alpha n+\frac{1}{2}}\beta^{\beta n+\frac{3}{2}}\left(\frac{\beta}{\alpha}\right)^{\omega_n} n^{\frac{3}{2}}e^{o(1)}. \end{aligned}$$

Thus, using $pq = \frac{\alpha}{z_c^2\beta}$,

$$(12.121)\qquad \frac{\gamma_{r-1}}{\gamma_r}\binom{n}{r}^{-\frac{1}{2}}\binom{n}{r-1}^{-\frac{1}{2}}(pq)^{-r+\frac{1}{2}} = \sqrt{2\pi}\alpha^{\frac{3}{2}}\beta^{n+\frac{1}{2}}n^{\frac{3}{2}}z_c^{2(r-1)}e^{o(1)}.$$

Next, using the identities

$$(12.122)\qquad (1-qz_c)^2 = \frac{1-B}{p\beta}, \qquad (1+pz_c)^2 = \frac{B}{q\beta}, \qquad z_c = \sqrt{\frac{\alpha}{\beta pq}},$$

we have

$$(12.123)\qquad \frac{(1+pz_c)^{x+y}(1-qz_c)^{2n-x-y}}{z_c^{2r+2}} = \frac{\left(\frac{B}{q\beta}\right)^{\frac{x+y}{2}}\left(\frac{1-B}{p\beta}\right)^{n-\frac{x+y}{2}}}{z_c^{2r+2}}.$$

Finally,

$$(12.124)\qquad \begin{aligned}\sqrt{w(x)w(y)} &= \sqrt{\binom{n}{x}q^x p^{n-x}\binom{n}{y}q^y p^{n-y}} \\ &= \frac{n!}{(x!(n-x)!y!(n-y)!)^{\frac{1}{2}}}q^{\frac{x+y}{2}}p^{n-\frac{x+y}{2}}.\end{aligned}$$

By (12.121), (12.123), and (12.124), we find

$$(12.125)\qquad X_n(x,y) = \frac{n!\sqrt{2\pi}\alpha^{\frac{3}{2}}\beta^{\frac{1}{2}}z_c^{-4}n^{\frac{3}{2}}B^{\frac{x+y}{2}}(1-B)^{n-\frac{x+y}{2}}e^{o(1)}}{(x!(n-x)!y!(n-y)!)^{\frac{1}{2}}}.$$

We now insert the formulae $x = Bn + \rho_n\xi$ and $y = Bn + \rho_n\eta$. Note that

$$(12.126)\qquad 0 < Bn - \rho_n M_1 \le x, y \le n.$$

In order to avoid the error term in

$$(n-x)! = \sqrt{2\pi}(n-x)^{n-x+\frac{1}{2}}e^{-(n-x)}e^{\frac{\theta}{n-x}},$$

which blows up at $n = x$, we instead use the expansion

(12.127)

$$\begin{aligned}(n-x)! = \frac{(n-x+1)!}{n-x+1} &= \sqrt{2\pi}\frac{(n-x+1)^{n-x+\frac{3}{2}}}{n-x+1}e^{-(n-x+1)}e^{\frac{\theta}{n-x+1}} \\ &= \sqrt{2\pi}(n-x+1)^{n-x+\frac{1}{2}}e^{-(n-x+1)}e^{o(1)},\end{aligned}$$

where the error term is now controlled for all $0 \le x \le n$. For $x, y$ in the range (12.126) we find

$$\begin{aligned} x!(n-x)!y!(n-y)! \\ &= (2\pi)^2 e^{-2} e^{-2n} \left(\frac{x}{nB}\right)^{x+\frac{1}{2}} \left(\frac{n-x+1}{n(1-B)}\right)^{n-x+\frac{1}{2}} \\ &\quad \times \left(\frac{y}{nB}\right)^{y+\frac{1}{2}} \left(\frac{n-y+1}{n(1-B)}\right)^{n-y+\frac{1}{2}} \\ &\quad \times (nB)^{x+y+1}(n(1-B))^{2n-(x+y)+1} e^{o(1)}. \end{aligned}$$

Inserting this into (12.125), we obtain

(12.128)

$$X_n(x,y) = n \frac{\frac{e\alpha^{\frac{3}{2}}\beta^{\frac{1}{2}}}{z_c^4 (B(1-B))^{\frac{1}{2}}} e^{o(1)}}{\left[\left(\frac{x}{nB}\right)^{x+\frac{1}{2}} \left(\frac{n-x+1}{n(1-B)}\right)^{n-x+\frac{1}{2}} \left(\frac{y}{nB}\right)^{y+\frac{1}{2}} \left(\frac{n-y+1}{n(1-B)}\right)^{n-y+\frac{1}{2}}\right]^{\frac{1}{2}}}.$$

Now from (12.108) and (12.122)

$$b^3 = \frac{pq\beta}{(1+pz_c)(1-qz_c)z_c} = \frac{(pq)^{\frac{3}{2}}\beta^2}{(B(1-B))^{\frac{1}{2}} z_c} = \frac{\alpha^{\frac{3}{2}}\beta^{\frac{1}{2}}}{(B(1-B))^{\frac{1}{2}} z_c^4}.$$

Thus,

(12.129)

$$X_n(x,y) = n \frac{eb^3 e^{o(1)}}{\left[\left(\frac{x}{nB}\right)^{x+\frac{1}{2}} \left(\frac{n-x+1}{n(1-B)}\right)^{n-x+\frac{1}{2}} \left(\frac{y}{nB}\right)^{y+\frac{1}{2}} \left(\frac{n-y+1}{n(1-B)}\right)^{n-y+\frac{1}{2}}\right]^{\frac{1}{2}}}.$$

Now, if $x = nB + s$ where $s = O(n^{\frac{1}{3}})$, then

$$\left(\frac{x}{nB}\right)^{x+\frac{1}{2}} = e^s e^{O\left(n^{-\frac{1}{3}}\right)}.$$

Similarly,

$$\left(\frac{n-x+1}{n(1-B)}\right)^{n-x+\frac{1}{2}} = e^{-(s-1)} e^{O\left(n^{-\frac{1}{3}}\right)}.$$

Hence,

$$\left(\frac{x}{nB}\right)^{x+\frac{1}{2}} \left(\frac{n-x}{n(1-B)}\right)^{n-x+\frac{1}{2}} = e \times e^{O\left(n^{-\frac{1}{3}}\right)}. \tag{12.130}$$

We conclude from (12.129) and (12.130) that for

$$x = nB + s, \quad y = nB + t, \qquad s, t \in O(n^{\frac{1}{3}}), \tag{12.131}$$

that

$$X_n(x,y) = nb^3 e^{o(1)}. \tag{12.132}$$

We also need a uniform estimate for

$$Bn \le x, y \le n. \tag{12.133}$$

We may write

$$x = Bn + s, \quad y = Bn + t, \qquad 0 \le s, t \le (1-B)n.$$

Then,

$$\left(\frac{x}{nB}\right)^{x+\frac{1}{2}} \left(\frac{n-x+1}{n(1-B)}\right)^{n-x+\frac{1}{2}}$$
$$= \left(1+\frac{s}{nB}\right)^{nB+s+\frac{1}{2}} \left(1-\frac{s-1}{n(1-B)}\right)^{n(1-B)-s+\frac{1}{2}}.$$

Set $u = \frac{s}{n}$. Then, $0 \le u \le 1-B$. Thus, for $x = Bn + un = n(B+u)$, $0 \le u \le 1-B$,

$$\left(\frac{x}{nB}\right)^{x+\frac{1}{2}} \left(\frac{n-x+1}{n(1-B)}\right)^{n-x+\frac{1}{2}} = e^{ng_n(u)},$$

where

$$g_n(u) = \left(B+u+\frac{1}{2n}\right)\log\left(1+\frac{u}{B}\right) + \left(1-B-u+\frac{1}{2n}\right)\log\left(1-\frac{u-\frac{1}{n}}{1-B}\right).$$

Now,

$$g_n(0) = \left(1-B+\frac{1}{2n}\right)\log\left(1+\frac{1}{n(1-B)}\right), \tag{12.134}$$

$$g_n'(0) = \frac{1}{2n}\left(\frac{1}{B}+\frac{1}{1-B+\frac{1}{n}}\right) - \log\left(1+\frac{1}{n(1-B)}\right), \tag{12.135}$$

and as $n \to \infty$,

$$g_n''(u) = \frac{1}{B+u} + \frac{1}{1-B-u+\frac{1}{n}} - \frac{1}{2n}\frac{1}{(B+u)^2} + \frac{1}{2n}\frac{1}{\left(1-B-u+\frac{1}{n}\right)^2}$$
$$= \frac{1}{B+u}\left(1-\frac{1}{2n(B+u)}\right) + \frac{1}{1-B-u+\frac{1}{n}}\left(1+\frac{1}{2n\left(1-B-u+\frac{1}{n}\right)}\right)$$
$$\ge 3$$

for all $0 \le u \le 1-B$. Thus,

$$g_n(u) \ge g_n(0) + g_n'(0)u + \frac{3}{2}u^3. \tag{12.136}$$

By (12.134) and (12.135), we see that $ng_n(0), ng_n'(0)$ are bounded as $n \to \infty$. Hence, we conclude that for some $0 < c < \infty$,

$$\left(\frac{x}{nB}\right)^{x+\frac{1}{2}} \left(\frac{n-x+1}{n(1-B)}\right)^{n-x+\frac{1}{2}} \geq e^{-c}, \qquad \text{for } nB \leq x \leq n, \tag{12.137}$$

and similarly for $nB \leq y \leq n$

$$\left(\frac{y}{nB}\right)^{y+\frac{1}{2}} \left(\frac{n-y+1}{n(1-B)}\right)^{n-y+\frac{1}{2}} \geq e^{-c}. \tag{12.138}$$

Thus, by (12.129), we find that $|X_n(x,y)|$ is bounded above by a constant times $nb^3$ uniformly for $Bn \leq x, y \leq n$

We summarize the above calculations for the prefactor $X_n(x,y)$:

(i) for $Bn - \rho_n M_1 \leq x, y \leq n$,

$$|X_n(x,y)| \leq \text{const} \cdot nb^3; \tag{12.139}$$

and

(ii) for $x, y = nB + O(n^{\frac{1}{3}})$,

$$X_n(x,y) = nb^3(1+o(1)) \tag{12.140}$$

as $n \to \infty$. Moreover, the limit is uniform for $x, y = nB + \rho_n\xi$, $|\xi| \leq M_1$.

We now consider the integrals in (12.118). Consider first

$$\int_{\mathcal{C}} e^{f_n(z,\xi) - f_n(z_c,\xi)} \frac{dz}{2\pi i}, \tag{12.141}$$

where $\mathcal{C}$ is a contour around the origin passing through $z_c$ and lying in $|z| < \min\{p^{-1}, q^{-1}\}$. Let $\delta > 0$ be small (see calculation preceeding (12.146) below) and fixed. Let

$$\begin{aligned} \mathcal{C}_\delta =& \{z = z_c e^{i\theta} : \delta < |\theta| \leq \pi\} \\ &\cup \{z = z_c(1 + t(e^{i\delta} - 1)) : 0 \leq t \leq 1\} \\ &\cup \{z = z_c(1 + (1-t)(e^{-i\delta} - 1)) : 0 \leq t \leq 1\} \\ =& \mathcal{C}_1 \cup \mathcal{C}_2 \cup \mathcal{C}_3. \end{aligned} \tag{12.142}$$

See Figure 12.24. We may take $\mathcal{C} = \mathcal{C}_\delta$ in (12.141). We first consider $\xi$ in a compact set, $|\xi| \leq M_1$. Now it follows from (12.114) that there exist constants $c_0, c_1, c_2 > 0$ such that for any $\theta$, $\delta < \theta \leq \pi$, $|\xi| \leq M_1$,

$$\operatorname{Re}\left(f_n(z_c e^{i\theta}, \xi)\right) \leq \operatorname{Re}\left(f_n(z_c, \xi)\right) - nc_1 + c_2 n^{\frac{1}{3}} \leq \operatorname{Re}\left(f_n(z_c, \xi)\right) - nc_0$$

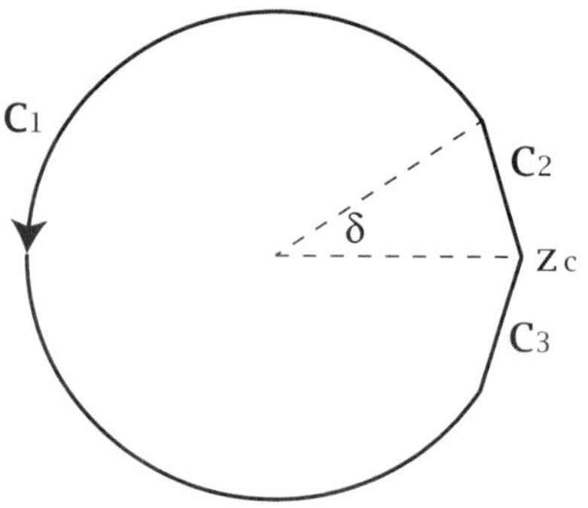


**Figure 12.24.** Contour $\mathcal{C}_\delta = \mathcal{C}_1 \cup \mathcal{C}_2 \cup \mathcal{C}_3$.

for some $c_0 > 0$ and $n$ sufficiently large. There is a similar estimate for $-\pi \le \theta < -\delta$. Hence, for $n \ge n(M_1, \delta)$,

$$\left| \int_{\mathcal{C}_1} e^{f_n(z,\xi) - f_n(z_c,\xi)} \frac{dz}{2\pi i} \right| \le z_c e^{-nc_0}. \tag{12.143}$$

Now for $z$ near $z_c$, i.e., $|z - z_c| \le \delta$, we have from (12.109)

$$\begin{aligned} f_n(z,\xi) =& f_n(z_c,\xi) + n\left( -\frac{b^3}{3}(z-z_c)^3 + \hat{a}(z-z_c)^4 + O(z-z_c)^5 \right) \\ &+ \xi n^{\frac{1}{3}} b(z-z_c)\left(1 + O(z-z_c)\right) + O(z-z_c). \end{aligned}$$

For $z \in \mathcal{C}_2$,

$$z = z_c + z_c t(e^{i\delta} - 1) = z_c + z_c t\left(2\sin\left(\frac{\delta}{2}\right)\right) e^{i\frac{\delta+\pi}{2}}$$

for $0 \le t \le 1$. Along $\mathcal{C}_2$, introduce the variable

$$u \equiv b n^{\frac{1}{3}}(z - z_c) = |u| e^{i\frac{\delta+\pi}{2}}, \tag{12.144}$$

where $0 \le |u| \le bn^{\frac{1}{3}} z_c 2\sin\left(\frac{\delta}{2}\right) \le bn^{\frac{1}{3}} z_c \delta$. We obtain

(12.145)

$$\begin{aligned} &\int_{\mathcal{C}_2} e^{f_n(z,\xi) - f_n(z_c,\xi)} \frac{dz}{2\pi i} \\ &= \frac{1}{bn^{\frac{1}{3}}} \int_0^{bn^{\frac{1}{3}} z_c(e^{i\delta}-1)} e^{-\frac{u^3}{3} + \frac{\hat{a}}{b^4}u^4 n^{-\frac{1}{3}} + O\left(u^5 n^{-\frac{2}{3}}\right)} e^{\xi u\left(1 + O\left(un^{-\frac{1}{3}}\right)\right)} e^{O\left(un^{-\frac{1}{3}}\right)} \frac{du}{2\pi i}. \end{aligned}$$

Along the line $u = |u|e^{i\frac{\delta+\pi}{2}}$, $0 \le |u| \le bn^{\frac{1}{3}} z_c 2\sin\left(\frac{\delta}{2}\right)$,

$$\begin{aligned}
&\operatorname{Re}\left(-\frac{u^3}{3} + \frac{\hat{a}}{b^4}\frac{u^4}{n^{\frac{1}{3}}} + O\left(\frac{u^5}{n^{\frac{2}{3}}}\right)\right) \\
&= \operatorname{Re}\left(-\frac{|u|^3}{3}e^{3i\frac{\delta+\pi}{2}} + \frac{\hat{a}}{b^4 n^{\frac{1}{3}}}|u|^4 e^{2i(\delta+\pi)} + O\left(\frac{|u|^5}{n^{\frac{2}{3}}}\right)\right) \\
&= |u|^3\left(-\frac{1}{3}\sin\left(\frac{3\delta}{2}\right) + \frac{\hat{a}}{b^4 n^{\frac{1}{3}}}|u|\cos(2\delta) + O\left(\frac{|u|^2}{n^{\frac{2}{3}}}\right)\right)
\end{aligned}$$

and hence

$$\begin{aligned}
\operatorname{Re}\left(-\frac{u^3}{3} + \frac{\hat{a}}{b^4}\frac{u^4}{n^{\frac{1}{3}}} + O\left(\frac{u^5}{n^{\frac{2}{3}}}\right)\right) &\le |u|^3\left(-\frac{\delta}{2} + \frac{\hat{a}}{b^4 n^{\frac{1}{3}}} bn^{\frac{1}{3}} z_c \delta + O(\delta^2)\right) \\
&= -\frac{\delta}{2}|u|^3\left(1 - 2\frac{\hat{a}}{b^3}z_c + O(\delta)\right).
\end{aligned}$$

But by (12.112), $\frac{2\hat{a}z_c}{b^3} < 1$; hence, for $\delta > 0$ sufficiently small,

$$\operatorname{Re}\left(-\frac{u^3}{3} + \frac{\hat{a}}{b^4}\frac{u^4}{n^{\frac{1}{3}}} + O\left(\frac{u^5}{n^{\frac{2}{3}}}\right)\right) \le -\hat{\delta}|u|^3 \tag{12.146}$$

for some $\hat{\delta} > 0$, for $u = |u|e^{i\frac{\delta+\pi}{2}}$, $0 \le |u| \le bn^{\frac{1}{3}} z_c 2\sin\left(\frac{\delta}{2}\right)$. Write the integral on the right-hand side of (12.145) as

$$\int_0^{bn^{\frac{1}{3}} z_c(e^{i\delta}-1)} \cdots = \int_0^{bn^{\mu} z_c(e^{i\delta}-1)} \cdots + \int_{bn^{\mu} z_c(e^{i\delta}-1)}^{bn^{\frac{1}{3}} z_c(e^{i\delta}-1)} \cdots \tag{12.147}$$

for any $0 < \mu < \frac{1}{3}$. From (12.146) we see that the second integral is bounded by

$$\left|\int_{bn^{\mu} z_c(e^{i\delta}-1)}^{bn^{\frac{1}{3}} z_c(e^{i\delta}-1)} \cdots\right| \le \int_{2bn^{\mu} z_c \sin\left(\frac{\delta}{2}\right)}^{2bn^{\frac{1}{3}} z_c \sin\left(\frac{\delta}{2}\right)} e^{-\hat{\delta}|u|^3} e^{M_1|u|c_1} e^{c_2}\frac{d|u|}{2\pi} \le c_3 e^{-\hat{\delta}_1 n^{3\mu}} \tag{12.148}$$

for suitable constants $\hat{\delta}_1, c_1, c_2, c_3 > 0$. Take

$$\mu < \frac{1}{12}. \tag{12.149}$$

Then for $|u| = O(n^{\mu})$,

$$\begin{cases}
|u|^4 n^{-\frac{1}{3}} = O(|n|^{4\mu-\frac{1}{3}}), \qquad |u|^5 n^{-\frac{2}{3}} = O(n^{5\mu-\frac{2}{3}}) = O(n^{-\frac{1}{4}}), \\
|u|^2 n^{-\frac{1}{3}} = O(n^{2\mu-\frac{1}{3}}) = O(n^{\frac{1}{6}-\frac{1}{3}}) = O(n^{-\frac{1}{6}}), \\
|u| n^{-\frac{1}{3}} = O(n^{\frac{1}{12}-\frac{1}{3}}) = O(n^{-\frac{1}{4}}).
\end{cases} \tag{12.150}$$

Hence, the first integral in (12.147) has the form

$$\int_0^{bn^\mu z_c(e^{i\delta}-1)} e^{-\frac{u^3}{3}+\xi u}\left(1+O\left(\frac{1}{n^{\frac{1}{3}-4\mu}}+O\left(\frac{1}{n^{1/4}}\right)+O\left(\frac{|\xi|}{n^{\frac{1}{6}}}\right)\right)\right)\frac{du}{2\pi i}$$
$$= \int_0^{bn^\mu z_c(e^{i\delta}-1)} e^{-\frac{u^3}{3}+\xi u}\frac{du}{2\pi i}$$
$$+O\left(\left(\int_0^{2bn^\mu z_c\sin\left(\frac{\delta}{2}\right)}\left|e^{-\frac{u^3}{3}+\xi u}\frac{d|u|}{2\pi}\right|\right)\left(O\left(\frac{1}{n^{\frac{1}{3}-4\mu}}\right)+O\left(\frac{M_1}{n^{\frac{1}{6}}}\right)\right)\right)$$

uniformly for $|\xi| \leq M_1$ as $n \to \infty$. We conclude that

$$\int_{C_2} e^{f_n(z,\xi)-f_n(z_c,\xi)}\frac{dz}{2\pi i} = \int_0^{\infty e^{i\frac{\delta+\pi}{2}}} e^{-\frac{u^3}{3}+\xi u}\frac{du}{2\pi i}+o(1) \tag{12.151}$$

uniformly for $|\xi| \leq M_1$. Similarly,

$$\int_{C_3} e^{f_n(z,\xi)-f_n(z_c,\xi)}\frac{dz}{2\pi i} = \int_{\infty e^{-i\frac{\delta+\pi}{2}}}^0 e^{-\frac{u^3}{3}+\xi u}\frac{du}{2\pi i}+o(1) \tag{12.152}$$

uniformly for $|\xi| \leq M_1$.

Summarizing, we see that as $n \to \infty$ (note that in (12.143), $n^{\frac{1}{3}}e^{-nc_0} \to 0$ as $n \to \infty$)

$$\begin{aligned}\int_C e^{f_n(z,\xi)-f_n(z_c,\xi)}\frac{dz}{2\pi i} &= \frac{1}{bn^{\frac{1}{3}}}\left(\int_{\infty e^{-i\frac{\delta+\pi}{2}}}^{\infty e^{i\frac{\delta+\pi}{2}}} e^{-\frac{u^3}{3}+\xi u}\frac{du}{2\pi i}+o(1)\right)\\ &= \frac{1}{bn^{\frac{1}{3}}}\left(\mathrm{Ai}(\xi)+o(1)\right)\end{aligned} \tag{12.153}$$

uniformly for $|\xi| \leq M_1$. Here we have used the fact that after changing variables $u \mapsto t = \frac{u}{i}$, the integral becomes $\int_{-\infty}^{\infty} e^{i\left(\frac{t^3}{3}+\xi t\right)}\frac{dt}{2\pi}$, which is the standard form for the Airy function, $\mathrm{Ai}(\xi)$ (see, e.g., [**AS64**]).

**Exercise 12.32.** Note from (12.102) that $\alpha < p$ if and only if $z_c < \frac{1}{q}$, and $\alpha < q$ if and only if $z_c < \frac{1}{p}$. Hence if $q \leq \alpha < p$, then $\frac{1}{p} \leq |z_c| < \frac{1}{q}$, and hence we can no longer take $\mathcal{C} = \mathcal{C}_\delta$ in (12.142). Instead, we define $\mathcal{C}$ as in Figure 12.25. It is straightforward to check using (12.100) that along the vertical line segment the function

$$Q(y) = \mathrm{Re}\left(f_n(-\frac{1}{2p}+iy,\xi)-(w_n-1)\log(-\frac{1}{2p}+iy)\right)$$

decreases as $y$ increases from 0 to $y_* = \mathrm{Im}(z_*)$, and decreases as $y$ decreases from 0 to $-y_*$. Hence $Q(y) \leq Q(y_*)$ for $-y_* \leq y \leq y_*$, and hence $\mathrm{Re}\left(f_n(z,\xi)\right) \leq \mathrm{Re}\left(f_n(z_*,\xi)\right)+C$ for a constant $C$ which is independent of $n$. This shows that the main contribution to the integral again comes from

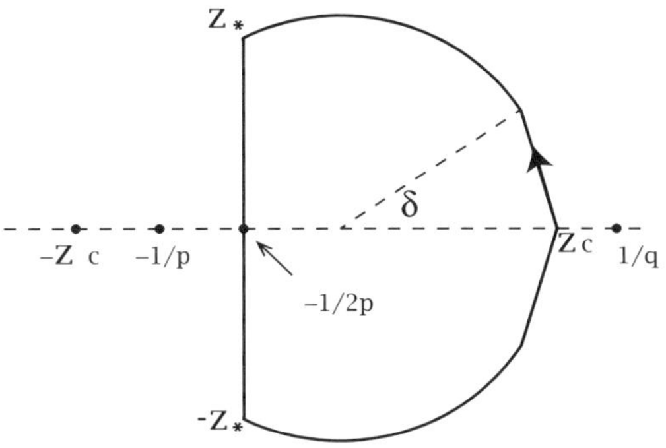


**Figure 12.25.** Contour $\mathcal{C}$ when $q \le \alpha < p$

a neighborhood of $z_c$ in the contour $\mathcal{C}$, and the same analysis as in the case $\alpha < p$ applies.

Now the other integrals in (12.118) are similar for $|\xi|, |\eta| \le M_1$. The integral with $z - z_c$ produces, after scaling, $z \mapsto u = bn^{\frac{1}{3}}(z - z_c)$ as in (12.144), an extra power $bn^{\frac{1}{3}}$ in the denominator. We obtain

(12.154)

$$\begin{aligned}\int_{\mathcal{C}} e^{f_n(z,\xi)-f_n(z_c,\xi)}(z-z_c)\frac{dz}{2\pi i z} &= \frac{1}{(bn^{\frac{1}{3}})^2}\left(\int_{\infty e^{-i\frac{\delta+\pi}{2}}}^{\infty e^{i\frac{\delta+\pi}{2}}} u e^{-\frac{u^3}{3}+\xi u}\frac{du}{2\pi i} + o(1)\right)\\ &= \frac{1}{(bn^{\frac{1}{3}})^2}\left(\mathrm{Ai}'(\xi) + o(1)\right)\end{aligned}$$

uniformly for $|\xi| \le M_1$.

Thus, inserting (12.140), (12.153), and (12.154) into (12.118), we see that for $\xi \ne \eta$ and $|\xi|, |\eta| \le M_1$,

(12.155)

$$\begin{aligned}&\rho_n K_{[\alpha n]}\left(Bn + \rho_n \xi, Bn + \rho_n \eta\right)\\ &= nb^3(1+o(1))\frac{1}{\xi - \eta}\frac{1}{bn^{\frac{1}{3}}}\frac{1}{(bn^{\frac{1}{3}})^2}\left(\mathrm{Ai}(\xi)\,\mathrm{Ai}'(\eta) - \mathrm{Ai}'(\xi)\,\mathrm{Ai}(\eta) + o(1)\right)\\ &\to \mathbb{A}(\xi, \eta).\end{aligned}$$

The convergence in (12.155) is in fact uniform for $|\xi|, |\eta| \le M_1$. This can be seen as follows. Set

(12.156) $$\Delta f_n(z, \xi) \equiv f_n(z, \xi) - f_n(z_c, \xi).$$

Using (12.118) and this notation, write $\rho_n K_{[\alpha N]}(x,y)$ for $x = Bn + \rho_n\xi$, $y = Bn + \rho_n\eta$, as

(12.157)
$$\begin{aligned}\rho_n K_{[\alpha n]}(x,y) = &\frac{X_n(x,y)}{\xi-\eta}\\ &\times\Bigg[\left(\int_C \left(e^{\Delta f_n(z,\xi)} - e^{\Delta f_n(z,\eta)}\right)\frac{dz}{2\pi i z}\right)\left(\int_C e^{\Delta f_n(z,\eta)}(z-z_c)\frac{dz}{2\pi i z}\right)\\ &-\left(\int_C \left(e^{\Delta f_n(z,\xi)} - e^{\Delta f_n(z,\eta)}\right)(z-z_c)\frac{dz}{2\pi i z}\right)\left(\int_C e^{\Delta f_n(z,\eta)}\frac{dz}{2\pi i z}\right)\Bigg].\end{aligned}$$

Clearly, the newly inserted terms cancel each other. Observe that

$$\begin{aligned}e^{\Delta f_n(z,\xi)} - e^{\Delta f_n(z,\eta)} &= \int_0^1 \frac{d}{dt} e^{\Delta f_n(z,\eta+t(\xi-\eta))}dt\\ &= (\xi-\eta)\rho_n \log\left(\frac{1+pz}{1-qz}\frac{1-qz_c}{1+pz_c}\right)\int_0^1 e^{\Delta f_n(z,\eta+t(\xi-\eta))}dt\end{aligned} \tag{12.158}$$

since

$$\begin{aligned}\Delta f_n(z,\xi) =& Bn\log\left(\frac{1+pz}{1+pz_c}\right) + (1-B)n\log\left(\frac{1-qz}{1-qz_c}\right)\\ &-([\alpha n]+1)\log\left(\frac{z}{z_c}\right) + \rho_n\xi\log\left(\frac{1+pz}{1-qz}\frac{1-qz_c}{1+pz_c}\right)\end{aligned} \tag{12.159}$$

by (12.99). Since $\frac{\rho_n}{(1+pz_c)(1-qz_c)} = n^{\frac{1}{3}}b$ by (12.107), we find that

$$\rho_n \log\left(\frac{1+pz}{1-qz}\frac{1-qz_c}{1+pz_c}\right) = bn^{\frac{1}{3}}(z-z_c)(1+O(z-z_c)). \tag{12.160}$$

Using (12.158) and (12.160) in (12.157), we obtain

(12.161)
$$\begin{aligned}\rho_n K_{[\alpha n]}(x,y) = &bn^{\frac{1}{3}}\int_0^1 dt X_n(x,y)\\ &\times\Bigg[\left(\int_C e^{\Delta f_n(z,\eta+t(\xi-\eta))}(z-z_c)(1+O(z-z_c))\frac{dz}{2\pi i z}\right)\left(\int_C e^{\Delta f_n(z,\eta)}(z-z_c)\frac{dz}{2\pi i z}\right)\\ &-\left(\int_C e^{\Delta f_n(z,\eta+t(\xi-\eta))}(z-z_c)^2(1+O(z-z_c))\frac{dz}{2\pi i z}\right)\left(\int_C e^{\Delta f_n(z,\eta)}\frac{dz}{2\pi i z}\right)\Bigg].\end{aligned}$$

Then the preceding methods show that uniformly for $|\xi|, |\eta| \le M_1$ as $n\to\infty$

$$\begin{aligned}&\rho_n K_{[\alpha n]}(nB+\rho_n\xi, nB+\rho_n\eta)\\ &\to \int_0^1 dt\left[\mathrm{Ai}'(\eta+t(\xi-\eta))\,\mathrm{Ai}'(\eta) - \mathrm{Ai}''(\eta+t(\xi-\eta))\,\mathrm{Ai}(\eta)\right].\end{aligned} \tag{12.162}$$

Note that if $\xi \neq \eta$, then this limit equals

$$\frac{1}{\xi-\eta}\int_0^1 dt\frac{d}{dt}\left[\operatorname{Ai}(\eta+t(\xi-\eta))\operatorname{Ai}'(\eta)-\operatorname{Ai}'(\eta+t(\xi-\eta))\operatorname{Ai}(\eta)\right]$$
$$=\frac{\operatorname{Ai}(\xi)\operatorname{Ai}'(\eta)-\operatorname{Ai}(\eta)\operatorname{Ai}'(\xi)}{\xi-\eta}=\mathbb{A}(\xi,\eta),$$

and if if $\xi=\eta$,

$$\left(\operatorname{Ai}'(\eta)\right)^2-\operatorname{Ai}''(\eta)\operatorname{Ai}(\eta)=\mathbb{A}(\eta,\eta).$$

This proves condition (iii) of Lemma 12.30 with the understanding that the convergence is uniform to $\mathbb{A}(\xi,\eta)$ in compact $\xi,\eta$ sets, and where $\mathbb{A}(\xi,\xi)\equiv \lim_{\eta\to\xi}\mathbb{A}(\xi,\eta)$.

In order to prove (i) and (ii) in Lemma 12.30, we must control $K_{[\alpha n]}(x,y)$ on the diagonal. Taking $y\to x=Bn+\rho_n\xi$ in (12.161) and using (12.158), we have

(12.163)
$$\begin{aligned}&\rho_n K_{[\alpha n]}(Bn+\rho_n\xi,Bn+\rho_n\xi)\\ =&X_n(x,x)\left[\int_{\mathcal{C}} e^{\Delta f_n(z,\xi)}\rho_n\log\left(\frac{1+pz}{1-qz}\frac{1-qz_c}{1+pz_c}\right)\frac{dz}{2\pi i}\int_{\mathcal{C}} e^{\Delta f_n(z,\xi)}(z-z_c)\frac{dz}{2\pi i}\right.\\ &\left.-\int_{\mathcal{C}} e^{\Delta f_n(z,\xi)}\rho_n\log\left(\frac{1+pz}{1-qz}\frac{1-qz_c}{1+pz_c}\right)(z-z_c)\frac{dz}{2\pi i}\int_{\mathcal{C}} e^{\Delta f_n(z,\xi)}\frac{dz}{2\pi i}\right].\end{aligned}$$

Again, the preceding methods show that

$$\lim_{n\to\infty}\rho_n K_{[\alpha n]}(Bn+\rho_n\xi,Bn+\rho_n\xi)=(\operatorname{Ai}'(\xi))^2-\operatorname{Ai}''(\xi)\operatorname{Ai}(\xi) \tag{12.164}$$

uniformly for $|\xi|\le M_1$. This of course also follows from (12.162), as already noted. As a corollary, we have

$$0\le K_{[\alpha n]}(Bn+\rho_n\xi,Bn+\rho_n\xi)\le\frac{C}{\rho_n},\qquad \text{for all } |\xi|\le M_1, \tag{12.165}$$

and for all $n\ge n(M_1)$ for some $n(M_1)$.

We now find an estimate for large $\xi$. Consider $\xi\ge 1$ satisfying $Bn+\rho_n\xi\le n$, i.e., $1\le\xi\le\frac{(1-B)n}{\rho_n}=O(n^{\frac{2}{3}})$. Consider the term

$$\int_{\mathcal{C}} e^{\Delta f_n(z,\xi)}\frac{dz}{2\pi i}=\int_{\mathcal{C}_1\cup\mathcal{C}_2\cup\mathcal{C}_3} e^{\Delta f_n(z,\xi)}\frac{dz}{2\pi i}$$

in (12.163), where $\mathcal{C}_1,\mathcal{C}_2,\mathcal{C}_3$ are as above. On $\mathcal{C}_1$, for $\delta<\theta\le\pi$, $\operatorname{Re}\left(\Delta f_n(z_c e^{i\theta},\xi)\right)\le -c\xi n^{\frac{1}{3}}$ for some $c>0$, by (12.114). Again, there is a similar estimate for $\pi\theta<-\delta$. Hence

$$\left|\int_{\mathcal{C}_1} e^{\Delta f_n(z,\xi)}\frac{dz}{2\pi i}\right|\le c_1' e^{-c_1\xi n^{\frac{1}{3}}},\qquad \xi\ge 1. \tag{12.166}$$

We now consider $\int_{C_2} e^{\Delta f_n(z,\xi)} \frac{dz}{2\pi i z}$. We need to estimate the term proportional to $\xi$ in (12.109) more carefully. This term arises from the expansion of

$$\log\left(\frac{1+pz}{1-qz}\frac{1-qz_c}{1+pz_c}\right)$$

near $z = z_c$ (cf. (12.101)). We note that

$$\frac{d}{dz}\log\left(\frac{1+pz}{1-qz}\frac{1-qz_c}{1+pz_c}\right)\bigg|_{z=z_c} = \frac{1}{(1+pz_c)(1-qz_c)}$$

and

$$\frac{d^2}{dz^2}\log\left(\frac{1+pz}{1-qz}\frac{1-qz_c}{1+pz_c}\right)\bigg|_{z=z_c} = \frac{q-p+2pqz_c}{(1+pz_c)^2(1-qz_c)^2}.$$

Thus, by (12.107)

$$\rho_n\xi\log\left(\frac{1+pz}{1-qz}\frac{1-qz_c}{1+pz_c}\right) = \xi\left(u+\hat{b}\frac{u^2}{bn^{\frac{1}{3}}}+O\left(\frac{u^3}{b^2n^{\frac{2}{3}}}\right)\right),$$

where $u \equiv bn^{\frac{1}{3}}(z-z_c)$ and $\hat{b} = \frac{1}{2}\frac{q-p+2pqz_c}{(1+pz_c)(1-qz_c)}$. We now obtain from (12.145)

$$\int_{C_2} e^{\Delta f_n(z,\xi)}\frac{dz}{2\pi i z} = \frac{1}{bn^{\frac{1}{3}}}\int_0^{bn^{\frac{1}{3}}z_c(e^{i\delta}-1)} e^{-\frac{u^3}{3}+\frac{\hat{a}}{b^4}u^4n^{-\frac{1}{3}}+O\left(u^5n^{-\frac{2}{3}}\right)} \times e^{\xi\left(u+\frac{\hat{b}}{b}u^2n^{-\frac{1}{3}}+O\left(u^3n^{-\frac{2}{3}}b^{-2}\right)\right)}e^{O\left(un^{-\frac{1}{3}}\right)}\frac{du}{2\pi i}. \tag{12.167}$$

Now observe that for $u = |u|ie^{i\frac{\delta}{2}}$, $|u| \le 2bn^{\frac{1}{3}}z_c\sin\left(\frac{\delta}{2}\right) \le bn^{\frac{1}{3}}z_c\delta$,

$$\begin{aligned}\mathrm{Re}\left(u+\frac{\hat{b}}{b}\frac{u^2}{n^{\frac{1}{3}}}+O\left(\frac{|u|^3}{n^{\frac{2}{3}}b^2}\right)\right) &= \mathrm{Re}\left(|u|ie^{i\frac{\delta}{2}}+\frac{\hat{b}}{b}\frac{|u|^2}{n^{\frac{1}{3}}}(-e^{i\delta})+O\left(\frac{|u|^3}{n^{\frac{2}{3}}b^2}\right)\right)\\ &= |u|\left(-\frac{\delta}{2}-\frac{\hat{b}}{b}\frac{|u|}{n^{\frac{1}{3}}}\cos\delta+O(\delta^2)\right).\end{aligned}$$

Now, a direct computation shows that

$$-\hat{b}z_c < \frac{1}{2}. \tag{12.168}$$

Write

$$\mathrm{Re}\left(u+\frac{\hat{b}}{b}\frac{u^2}{n^{\frac{1}{3}}}+O\left(\frac{u^3}{n^{\frac{2}{3}}b^2}\right)\right) = -\frac{\delta}{2}|u|\left(1+\frac{2\hat{b}z_c}{\delta b}\frac{|u|}{n^{\frac{1}{3}}z_c}\cos\theta+O(\delta)\right).$$

If $\hat{b} > 0$, then we clearly have

$$\mathrm{Re}\left(u+\frac{\hat{b}}{b}\frac{u^2}{n^{\frac{1}{3}}}+O\left(\frac{u^3}{n^{\frac{2}{3}}}\right)\right) \le -\frac{\delta}{4}|u|$$

for $\delta$ sufficiently small. If $\hat{b} < 0$, then, as $0 \leq \frac{|u|\cos\theta}{\delta b n^{\frac{1}{3}} z_c} \leq 1$,

$$1 + 2\hat{b} z_c \frac{|u|\cos\theta}{\delta b n^{\frac{1}{3}} z_c} \geq 1 + 2\hat{b} z_c > 0$$

by (12.168). Either way we see that

$$\operatorname{Re}\left(u + \frac{\hat{b}}{b}\frac{u^2}{n^{\frac{1}{3}}} + O\left(\frac{u^3}{n^{\frac{2}{3}} b^2}\right)\right) \leq -\delta'|u| \tag{12.169}$$

for some $\delta' > 0$. Now split the integral in (12.167) as follows:

$$\int_0^{bn^{\frac{1}{3}} z_c (e^{i\delta}-1)} = \int_0^{e^{i\delta}-1} + \int_{(e^{i\delta}-1)}^{bn^{\frac{1}{3}} z_c (e^{i\delta}-1)} . \tag{12.170}$$

Using (12.146), we now have

$$\begin{aligned} &\frac{1}{bn^{\frac{1}{3}}} \left| \int_{(e^{i\delta}-1)}^{bn^{\frac{1}{3}} z_c (e^{i\delta}-1)} \cdots \frac{du}{2\pi i} \right| \\ &\leq \frac{c_1}{n^{\frac{1}{3}}} \int_{2\sin\left(\frac{\delta}{2}\right)}^{2bn^{\frac{1}{3}} z_c \sin\left(\frac{\delta}{2}\right)} e^{-\hat{\delta}|u|^3} e^{-\delta'\xi|u|} d|u| \leq \frac{c_1}{n^{\frac{1}{3}}} e^{-c_2\xi} \end{aligned} \tag{12.171}$$

for some $c_1, c_2 > 0$. There is a similar estimate on the corresponding part of $\mathcal{C}_3$:

$$\frac{1}{bn^{\frac{1}{3}}} \left| \int_{z_c bn^{\frac{1}{3}} (e^{-i\delta}-1)}^{e^{-i\delta}-1} \cdots \frac{du}{2\pi i u} \right| \leq \frac{c_1}{n^{\frac{1}{3}}} e^{-c_2\xi}. \tag{12.172}$$

The remaining parts of $\mathcal{C}_2$ and $\mathcal{C}_3$ from (12.170) are combined to

$$\frac{1}{bn^{\frac{1}{3}}} \int_{e^{-i\delta}-1}^{e^{i\delta}-1} e^{-\frac{u^3}{3}+\cdots+} e^{\xi u \left(1+\frac{\hat{b}}{b} u n^{-\frac{1}{3}} + O\left(u^2 n^{-\frac{2}{3}}\right)\right)} e^{O\left(u n^{-\frac{1}{3}}\right)} \frac{du}{2\pi i}, \tag{12.173}$$

and by analyticity (note that $f_n(z,\xi)$ is analytic in a neighborhood of $z_c$ and the expansions in (12.173) are valid in this neighborhood) we can deform the contour to a straight line. Along this line, $\operatorname{Re}(u) = \cos\delta - 1 < 0$. It follows that the integral in (12.173) is bounded by

$$ce^{-c'\xi} \tag{12.174}$$

for some $c, c' > 0$, uniformly as $n \to \infty$. Thus from (12.166), (12.171), (12.172), and (12.174),

$$\left| \int_{\mathcal{C}} e^{\Delta f_n(z,\xi)} \frac{dz}{2\pi i} \right| \leq c \frac{e^{-c'\xi}}{n^{\frac{1}{3}}}. \tag{12.175}$$

There are similar estimates for all the integrals in (12.163) (integrals with factors $(z - z_c)$ give an extra power of $n^{-\frac{1}{3}}$, etc.), and so we obtain for $1 \leq \xi \leq \frac{(1-B)n}{\rho_n} = O(n^{\frac{2}{3}})$

$$0 \leq \rho_n K_{[\alpha n]}(Bn + \rho_n\xi, Bn + \rho_n\xi) \leq c\frac{X_n(x,x)}{n}e^{-c'\xi} \leq ce^{-c'\xi} \tag{12.176}$$

by (12.139).

It is now straightforward to prove (i), (ii) of Lemma 12.30. Fix $M_1 \geq 1$ and consider $t \geq -M_1$. Recall that $K_r(x,y)$ is only defined for $x, y \in \{0, 1, \ldots, n\}$. Now $Bn + \rho_n t + m \leq n$ holds if and only if $m \leq (1-B)n - \rho_n t \equiv m_t$. First we consider $1 \leq m \leq \rho_n(M_1 + 1)$. For such $m$, $-M_1 \leq t + \frac{m}{\rho_n} \leq M_1 + M_1 + 1 = 2M_1 + 1$. In this range

$$\begin{aligned}
&K_{[\alpha n]}(Bn + t\rho_n + m, Bn + t\rho_n + m) \\
&= K_{[\alpha n]}\left(Bn + \rho_n\left(t + \frac{m}{\rho_n}\right), Bn + \rho_n\left(t + \frac{m}{\rho_n}\right)\right) \leq \frac{C}{\rho_n}
\end{aligned}$$

by (12.165). Then

$$\begin{aligned}
&\sum_{1 \leq m \leq \rho_n(M_1+1)} K_{[\alpha n]}(Bn + t\rho_n + m, Bn + t\rho_n + m) \\
&\qquad \leq \frac{C}{\rho_n}\rho_n(M_1 + 1) = C(M_1 + 1)
\end{aligned} \tag{12.177}$$

for all $n \geq n(M_1)$ (such that $Bn + \rho_n t + 1 \geq 0$). For $m > \rho_n(M_1 + 1)$, we have

$$t + \frac{m}{\rho_n} \geq t + M_1 + 1 \geq 1.$$

Hence, by (12.176)

$$\begin{aligned}
&\sum_{\rho_n(M_1+1) < m \leq m_t} K_{[\alpha n]}(Bn + \rho_n t + m, Bn + \rho_n t + m) \\
&\leq \frac{c}{\rho_n} \sum_{\rho_n(M_1+1) < m \leq m_t} e^{-c'\left(t + \frac{m}{\rho_n}\right)} = ce^{-c't}\frac{e^{-\frac{c'}{\rho_n}(\rho_n(M_1-1)+1)}}{\rho_n(1 - e^{\frac{-c'}{\rho_n}})} \leq c''e^{-c't}
\end{aligned}$$

for all $n \geq n(M_1)$. Together with (12.177) this proves (i).

Now for $t \geq 1$, $t + \frac{m}{\rho_n} \geq 1$, so

$$\begin{aligned}
\sum_{1 \leq m \leq m_t} K_{[\alpha n]}(Bn + \rho_n t + m, Bn + \rho_n t + m) &\leq \frac{c}{\rho_n} \sum_{1 \leq m \leq m_t} e^{-c'\left(t + \frac{m}{\rho_n}\right)} \\
&\leq \frac{e^{-c't}}{\rho_n(e^{\frac{c'}{\rho_n}} - 1)} \leq c''e^{-c't}
\end{aligned}$$

as before. Thus, there exists an $L(\epsilon)$ such that $L(\epsilon) - 1 \le t = L \le L(\epsilon)$ implies

$$\sum_{1 \le m \le m_t} K_{[\alpha n]}(Bn + \rho_n L + m, Bn + \rho_n L + m) \le \epsilon.$$

This proves (ii).

Property (iv) is straightforward: $K_r(x,y) = \sum_{j=0}^{r-1} \psi_j(x)\psi_j(y)$, $\psi_j(x) = \sqrt{w(x)}p_j(x)$ is already positive definite. The verification of (i), (ii), (iii), (iv) in Lemma 12.30 in the Krawtchouk case is now complete. This in turn completes the proof of Theorem 12.28.

The analogous result for $\min_{1 \le j \le r} h'_j$ is:

**Theorem 12.33.** *Let $\alpha < q$ and set*

$$A = \beta q + \alpha p - 2\sqrt{pq\alpha\beta} = \left(\sqrt{q\beta} - \sqrt{p\alpha}\right)^2 \tag{12.178}$$

*and*

$$\rho' = \left(\frac{pq}{\alpha\beta}\right)^{\frac{1}{6}} \left((\sqrt{p\beta} + \sqrt{q\alpha})(\sqrt{q\beta} - \sqrt{p\alpha})\right)^{\frac{2}{3}}. \tag{12.179}$$

*Then,*

$$\lim_{n\to\infty} \mathbb{P}\left(\min_{1 \le j \le r} h'_j \ge nA - \xi\rho' n^{\frac{1}{3}}\right) = F(\xi). \tag{12.180}$$

The proof of (12.180) is similar to the proof of Theorem 12.28 and is left to the interested (and diligent!) reader.

As noted in the introduction, an *Arctic circle*, or more correctly, an *Arctic ellipse* emerges in the tilings as $n \to \infty$. After scaling by $\frac{1}{n}$, the ellipse has the form (see [**CEP96, Joh02**])

$$\frac{x^2}{p} + \frac{y^2}{q} = 1. \tag{12.181}$$

The ellipse, in the case $q < p$, is drawn in Figure 12.26 inscribed in the scaled Aztec diamond. The four *polar regions* lie outside the ellipse, and the *temperate region* lies inside.

The above ellipse can be seen from Theorem 12.28 as follows. Recall Figure 12.20. Note that along each line in the right picture labeled by $r$, $h_r = \max_{1 \le j \le r} h_j$ denotes the top-most location of a vertical tile. Hence the collection of such locations for all $r = 1, 2, \ldots, n$ includes the boundary of the *north polar region.* Theorem 12.28 implies that if $r = [\alpha n]$, then $\frac{h_r}{n} \approx B = B(\alpha)$, assuming that $\alpha < q$. If we scale the diamond so that it is bounded by four lines $y = \pm x \pm 1$, then this means that the top-most location of the vertical tile is at the location indicated by the black dot in

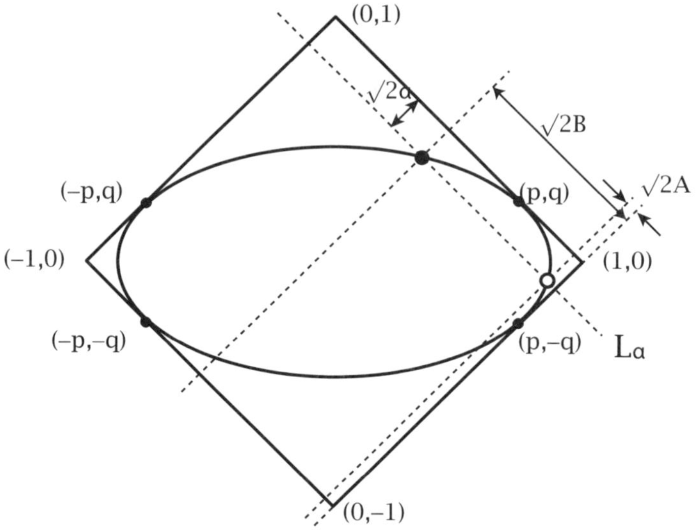


**Figure 12.26.** Arctic circle

Figure 12.26. Its rectangular coordinate is $(x, y) = (1 - B - \alpha, B - \alpha)$. Since $B = (\sqrt{q(1-\alpha)} + \sqrt{p\alpha})^2$, it follows that the black dot lies on the ellipse $\frac{x^2}{p} + \frac{y^2}{q} = 1$. Theorem 12.28 is proved under the assumption that $\alpha < q$, which corresponds to the case when the line $L_\alpha$ in Figure 12.26 is above the point $(p, -q)$. If the line $L_\alpha$ is below the point $(p, -q)$ but above the point $(-p, q)$, then we have $q < \alpha < p$. If the line $L_\alpha$ is below the point $(-p, q)$, then we have $p < \alpha$. The analysis for these cases is similar (cf. Exercise 12.32). Overall, the limit of $\frac{h_r}{n}$ gives rise to the solid curve from $(-1, 0)$ to $(p, q)$ through $(-p, q)$ in the left picture of Figure 12.27. The analysis for $h_1 = \min_{1\le j\le r} h_j$ leads us to the other solid curve from $(-p, -q)$ to $(1, 0)$ through $(p, -q)$. On the other hand, we can also consider $k_j$ in Figure 12.20. When $\alpha < q$, the lowest point $k_1 = \min_{1\le j\le n+1-r} k_j$ gives rise to the boundary of the *east polar region*. This location is indicated by the white dot in Figure 12.26. Theorem 12.33 implies that (note that $h'_j = k_j$) the white dot has the rectangular coordinate $(x, y) = (1 - B - \alpha, A - \alpha)$, where $A = (\sqrt{q(1-\alpha)} - \sqrt{p\alpha})^2$. This point is also on the ellipse. The analysis for all other cases implies the solid curve from $(0, -1)$ to $(p, q)$ through $(p, -q)$ in the right picture in Figure 12.27. The analysis for $k_{n+1-r} = \max_{1\le j\le n+1-r} k_j$ yields the other solid curve. The intersection of the domains indicated in the two pictures in Figure 12.27 is the region bounded by the ellipse.

Theorems 12.28 and 12.33 also show us that the fluctuations of the arc circle at a fixed location are given by the Tracy-Widom distribution $F(x)$. The joint fluctuations at multiple locations can also be evaluated. If the

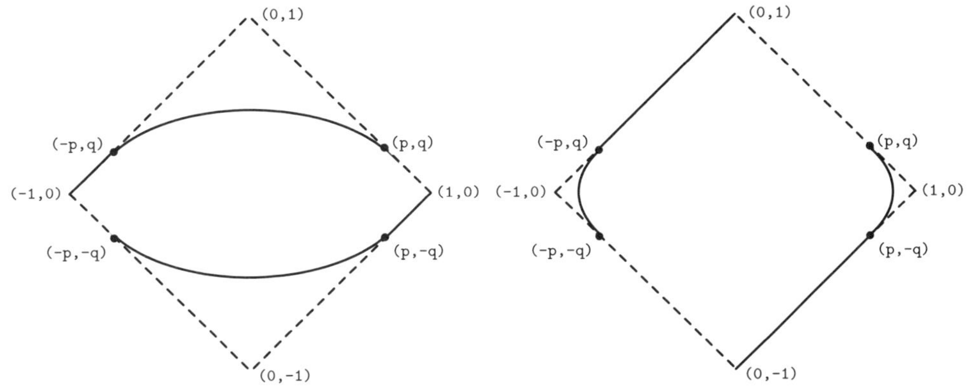


**Figure 12.27.** (Left) The two solid curves are the limit of $\frac{h_r}{n}$ and $\frac{h_1}{n}$ respectively. (Right) The two solid curves are the limit of $\frac{k_{n+1-r}}{n} = \frac{h'_{n+1-r}}{n}$ and $\frac{k_1}{n} = \frac{h'_1}{n}$ respectively.

locations are of order $n^{2/3}$ apart, then the fluctuations are given by the so-called Airy process; see [**Joh05a**]. The Airy process was first introduced by Prähofer and Spohn in the analysis of the polynuclear growth (PNG) model in [**PS02**] and is as fundamental as the Tracy-Widom distribution.

If we evaluate the Krawtchouk kernel for $h_j$, where $j$ is proportional to $n$, we can find the density of the vertical tiles inside the region given in the left picture of Figure 12.27; see [**Joh02**].

*Chapter 13*

# The Dyson Process and the Brownian Dyson Process

There is a very interesting dynamical version of random matrix theory due to Dyson [**Dys62**]. This dynamical version is obtained by assuming that the entries of the matrices (real symmetric or Hermitian) are Ornstein-Uhlenbeck processes instead of Gaussian random variables. This *matrix Ornstein-Uhlenbeck process* induces a diffusion on the eigenvalues of the matrices which is called the *Dyson process* (DP). Similarly, the *Brownian Dyson process* (BDP) is the diffusion induced on the eigenvalues by *matrix Brownian motion* for which the entries of the matrices are Brownian motions. (*Caveat*: The names for these processes are not standard and vary considerably from author to author.)

For the Hermitian case, the distribution of Hermitian Brownian Dyson process at a fixed time starting from the zero matrix is simply the distribution of a scaled version of the eigenvalues of the Gaussian unitary ensemble. Hence, we find, for example, that as the number of paths becomes large, the fixed-time distribution of the top path of the Hermitian Brownian Dyson process converges to the Tracy-Widom distribution. The same is true for the Hermitian Dyson process starting from a Gaussian matrix (or the zero matrix). Now it turns out that after a proper scaling, the process of the top path (of both Hermitian Brownian Dyson process and Hermitian Dyson process) itself also converges, and the limit is called the *Airy process* (see, for example, [**Joh03, Joh05b**]). The Airy process is a stationary process that is not Markovian. Its marginal distribution at a fixed time is, naturally, the Tracy-Widom distribution, and it was first introduced by Prähofer and Spohn in the study of random growth models [**PS02**]. See also [**CH14**] for various properties of the Airy process. The Airy process is as universal as the Tracy-Widom distribution itself in the sense that certain processes from last passage percolation, the PNG growth model, and random tilings, for example, give rise to the Airy process [**PS02, Joh03, Joh05a**]. The multi-time distributions of the Hermitian Brownian Dyson process and Hermitian Dyson process form determinantal point processes with explicit kernels as in the case of GUE [**Joh05b**], from which one can prove the convergence to the Airy process.

As we will see below, the Hermitian Dyson process and Hermitian Brownian Dyson process are equivalent to independent Ornstein-Uhlenbeck processes and Brownian motions *conditioned not to intersect for all time*, respectively. Hence the top path of nonintersecting Brownian motion converges to the Airy process, and in particular to the Tracy-Widom distribution at a fixed time. The analogous statement is true for Dyson process. We have already seen another example of such a nonintersecting process in Chapter 12; the tilings of an Aztec diamond are mapped to certain nonintersecting paths. Plancherel measure and Schur measure are also related to nonintersecting paths, as one can see in Section 3.2 and Subsection 10.1.3. From this perspective, the connection of the tiling of an Aztec diamond, and also the longest increasing subsequence problem, to random matrix theory is perhaps not as surprising as one might initially think.

In this chapter, we study a few fundamental properties of the Dyson process and the Brownian Dyson process. We first define the matrix Ornstein-Uhlenbeck process and matrix Brownian motion, and we then evaluate the induced processes on the eigenvalues, namely the Dyson process and the Brownian Dyson process. We then prove the noncolliding property for both processes. In the final section we prove the equivalence of the Hermitian Dyson process and independent Ornstein-Uhlenbeck processes conditioned not to intersect for all time. A similar result for Brownian motion is also proved. However, in this book we do not consider the determinantal structure of the Hermitian Dyson process and the Hermitian Brownian Dyson process, and also we do not consider the convergence to the Airy process; the interested reader is encouraged to to consult [**Joh05b**] and the references within.

We also mention that more recently the Brownian Dyson process and the Dyson process have played a fundamental analytical role in proving universality for Wigner ensembles (see, e.g., [**Joh01b, EPR$^+$10**]) and also for the adjacency matrices of random graphs (see, e.g., [**BHKY**] and the references therein).

## 13.1. Dyson Process

We can consider the Dyson process for either real symmetric matrices, Hermitian, or Hermitian self-dual matrices, i.e., $JM^TJ^T = M$ where $J = \operatorname{diag}(\sigma, \sigma, \ldots, \sigma)$, $\sigma = \left(\begin{smallmatrix} 0 & 1 \\ -1 & 0 \end{smallmatrix}\right)$. Dyson was led to these processes by his interest in the dynamics of Coulomb gases.

In this chapter, we discuss how to construct the Dyson process in the real symmetric case, and we leave the analogous constructions for the other two cases to the interested reader. We denote by $\mathcal{S}(n)$ the set of $n \times n$ real symmetric matrices. The various facts and results from stochastic analysis that are needed in the construction can be found in [**RY99**]. For the convenience of the reader some of these results are described in Appendix C.

Recall that the standard (scalar) Ornstein-Uhlenbeck process is a stationary process $H(t)$, $-\infty < t < \infty$, which solves the stochastic differential equation

$$dH(t) = -H(t)\,dt + c\,dB(t), \tag{13.1}$$

and whose single-time distribution is Gaussian. Here $c > 0$ and $B(t)$ is Brownian motion (see, e.g., [**RY99**]). We will abuse notation slightly and call any solution $H(t)$, $t \geq 0$, of (13.1) an Ornstein-Uhlenbeck process.

**Definition 13.1.** A *matrix Ornstein-Uhlenbeck process* (or matrix Ornstein-Uhlenbeck diffusion) $\{H(t)\}_{t\geq 0}$ is an $\mathcal{S}(n)$-valued process whose entries evolve according to the family of Ornstein-Uhlenbeck processes. For $k \leq j$,

$$dH_{kj}(t) = -H_{kj}(t)dt + c_{kj}dB_{kj}(t), \quad c_{kj} = 1 + (\sqrt{2} - 1)\delta_{kj}, \tag{13.2}$$

where $\{B_{kj}(t)\}_{k\leq j}$ are independent identically distributed Brownian motions, and we set $H_{kj}(t) = H_{jk}(t)$ for $k < j$.

For a given initial condition $H(0) \in \mathcal{S}(n)$, the solution $H_{kj}(t)$, $k \leq j$, of (13.2) for $t \geq 0$ forms an $n(n+1)/2$-dimensional Ito process (see Appendix C).

**Definition 13.2.** Let $\{H(t)\}_{t\geq 0}$ be a matrix Ornstein-Uhlenbeck process as in Definition 13.1. Then the process induced on the (ordered) eigenvalues $\lambda_1(t) \geq \lambda_2(t) \cdots \geq \lambda_n(t)$ of $H(t)$ is called the *Dyson process* for symmetric matrices.

The eigenvalues of a matrix Ornstein-Uhlenbeck process have the following fundamental property.

**Theorem 13.3.** *Let $H(t)$ be a matrix Ornstein-Uhlenbeck process with $H(0) \in \mathcal{S}(n)$. Then the eigenvalues of $H(t)$ are distinct for all $t > 0$ almost surely.*

Note that the theorem holds even if the initial matrix $H(0)$ has a nonsimple spectrum.

One of the main theorems in this chapter is that the Dyson process satisfies the following diffusion equation.

**Theorem 13.4.** *The Dyson process is given by the diffusion*

$$d\lambda_i(t) = \Big( -\lambda_i(t) + \sum_{j\neq i} \frac{1}{\lambda_i(t) - \lambda_j(t)} \Big) dt + \sqrt{2}\, dW_i(t), \quad t > 0, \tag{13.3}$$

*for $i = 1, \ldots, n$, where $\{W_i(t)\}_{i=1}^n$ are independent identically distributed Brownian motions.*

Note that Theorem 13.3 guarantees that $\lambda_i(t) \neq \lambda_j(t)$ for $i \neq j$, $t > 0$, almost surely and hence equation (13.3) is well defined for all $t > 0$ almost surely. We prove Theorem 13.4 in Section 13.3 below assuming Theorem 13.3. We then prove Theorem 13.3 in Section 13.5.

We note the following lemma.

**Lemma 13.5.** *The matrix Ornstein-Uhlenbeck process $H(t)$ has a (unique) invariant measure given by $e^{-\frac{1}{2}\operatorname{tr} H^2} dH$ on $\mathcal{S}(n)$.*

**Proof.** For each $j \leq k$, the invariant measure for the scalar Ornstein-Uhlenbeck process (13.2) is $p_{kj}(x) = \frac{1}{\sqrt{\pi c_{kj}^2}} e^{-x^2/c_{kj}^2}$. The invariant measure for the matrix Ornstein-Uhlenbeck process is thus given by the product measure of $p_{kj}$, $j \leq k$, which can be written as $e^{-\frac{1}{2}\operatorname{tr} H^2} dH$ on $\mathcal{S}(n)$. Uniqueness is standard (cf. the discussion following (13.23) below). □

In other words, the invariant measure for the matrix Ornstein-Uhlenbeck process is the Gaussian orthogonal ensemble (GOE); see, for example, [**Meh04**].

## 13.2. Brownian Dyson Process

We now introduce the Brownian motion version of the above processes.

**Definition 13.6.** A *matrix Brownian motion* $\tilde{H}(t), t \geq 0$, is an $\mathcal{S}(n)$-valued process whose entries evolve according to the family of Brownian motions. For $k \geq j$,

$$d\tilde{H}_{kj}(t) = c_{kj}\, dB_{kj}(t), \qquad c_{kj} = 1 + (\sqrt{2} - 1)\delta_{kj}, \tag{13.4}$$

where $B_{kj}$ are independent and identically distributed Brownian motions, and we set $\tilde{H}_{kj}(t) = \tilde{H}_{jk}(t)$ for $k < j$.

**Definition 13.7.** Let $\tilde{H}(t), t \geq 0$ be a matrix Brownian motion as in Definition 13.6. Then the process induced on the eigenvalues $\tilde{\lambda}_1(t) \geq \tilde{\lambda}_2(t) \geq \cdots \geq \tilde{\lambda}_n(t)$ of $\tilde{H}(t)$ is called *Brownian Dyson process* for symmetric matrices.

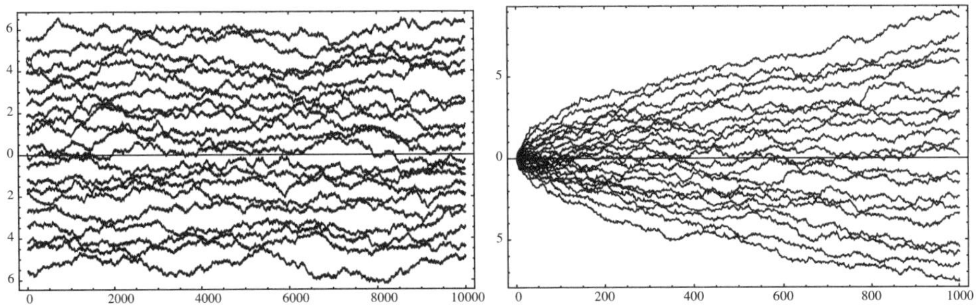


**Figure 13.1.** Simulation of the Dyson process (left) and Brownian Dyson process (right) when $n = 20$. The pictures are obtained using Mathematica by first generating a matrix Ornstein-Uhlenbeck process and a matrix Brownian motion, and then evaluating the eigenvalues of the matrices at each time.

The following theorems are the counterparts of Theorems 13.3 and 13.4.

**Theorem 13.8.** *Let $\tilde{H}(t)$ be a matrix Brownian motion with $\tilde{H}(0) \in \mathcal{S}(n)$. Then the eigenvalues of $\tilde{H}(t)$ are distinct for all $t > 0$ almost surely.*

**Theorem 13.9.** *Brownian Dyson process is given by the diffusion*

$$d\tilde{\lambda}_i(t) = \sum_{l \neq i} \frac{1}{\tilde{\lambda}_i(t) - \tilde{\lambda}_l(t)}\, dt + \sqrt{2}\, dW_i(t), \qquad t > 0, \tag{13.5}$$

*where again the $W_i(t)$'s are independent, identically distributed Brownian motions.*

Theorem 13.9 will be proved in Section 13.3 assuming Theorem 13.8, which will be proved in Section 13.4.

Note that the matrix Brownian process does not have an invariant measure. See Figure 13.1 for simulations of the Dyson process and Brownian Dyson process.

## 13.3. Derivation of the Dyson Process and the Brownian Dyson Process

In this section we prove Theorem 13.4 assuming Theorem 13.3 which states that the eigenvalues of the matrix Ornstein-Uhlenbeck process do not intersect for all $t > 0$. The proof of Theorem 13.9 assuming Theorem 13.8 is almost the same, and it is left to the reader as an exercise.

The proof proceeds by combining Ito's formula (see Appendix C) with the methods of classical perturbation theory. Theorem 13.3 is used here to justify the application of Ito's formula.

**Remark 13.10.** We note that the first rigorous derivation of the Brownian Dyson process is due to H. McKean [**McK69**].

Recall the following facts (see, e.g., [**RS78**]).

**Lemma 13.11.** *(i) For any $H \in \mathcal{S}(n)$ the eigenvalues of $H$ are real, well ordered by the min-max principle, $\lambda_1(H) \geq \cdots \geq \lambda_n(H)$, and are Lipschitz continuous functions of the entries of $H$.*

*(ii) If $\hat{\lambda}$ is a simple eigenvalue of $H$, i.e., $\dim(\mathrm{Ker}(H - \hat{\lambda})) = 1$, then $\hat{\lambda}$ depends locally analytically on the entries of $H$, i.e., for some $\epsilon > 0$ the map*

$$\{K \in \mathcal{S}(n) : \|K - H\| < \epsilon\} \ni K \to \lambda(K) \tag{13.6}$$

*satisfying $\lambda(H) = \hat{\lambda}$ is a real analytic function of the entries of $K$. Moreover, the normalized eigenvector $u(K)$ for $\lambda(K)$ can be chosen to be real analytic.*

Suppose that $H(t)$, $t \in [t_0, t_1]$, is a smooth $\mathcal{S}(n)$-valued map and $\lambda_1(t) > \cdots > \lambda_n(t)$ for all $t \in [t_0, t_1]$. In view of (ii), these assumptions guarantee that $\{\lambda_j(t)\}_{i=1}^n$ and the associated orthonormal eigenvectors $\{u_j(t)\}_{j=1}^n$ are smooth functions of $t \in [t_0, t_1]$.

The following calculation is standard in perturbation theory. Differentiating $H(t)u_j(t) = \lambda_j(t)u_j(t)$ with respect to $t$, we obtain

$$\frac{d}{dt}(H(t)u_j(t)) = \dot{H}(t)u_j(t) + H(t)\dot{u}_j(t) = \dot{\lambda}_j(t)u_j(t) + \lambda_j(t)\dot{u}_j(t). \tag{13.7}$$

Taking the inner product of equation (13.7) with $u_j(t)$ and noting that $\langle u_j(t), \dot{u}_j(t)\rangle = 0$ leads to the evolution equation for $\lambda_j(t)$:

$$\dot{\lambda}_j(t) = \langle u_j(t), \dot{H}(t)u_j(t)\rangle. \tag{13.8}$$

Here $\langle \cdot, \cdot \rangle$ denotes the standard inner product on $\mathbb{R}^n$. Similarly, taking the inner product of equation (13.7) with $u_k(t)$, using the orthogonality of the eigenvectors and the symmetry of $H$, we obtain the relation

$$\langle u_k(t), \dot{u}_j(t)\rangle = \frac{\langle u_k(t), \dot{H}(t)u_j(t)\rangle}{\lambda_j(t) - \lambda_k(t)}, \qquad k \neq j. \tag{13.9}$$

Hence, if we expand $\dot{u}_j(t)$ in terms of the eigenvectors $u_k(t)$, then the coefficient of $u_k(t)$'s for $k \neq j$ is given as above. On the other hand, the coefficient of $u_j(t)$ is zero since $\langle u_j(t), \dot{u}_j(t)\rangle = 0$. Thus, for any arbitrary smooth evolution $t \to H(t)$, the evolution equations for the eigenvalues and eigenvectors become

$$\dot{\lambda}_j(t) = \langle u_j(t), \dot{H}(t)u_j(t)\rangle, \tag{13.10}$$

$$\dot{u}_j(t) = \sum_{k \neq j} \frac{\langle u_k(t), \dot{H}(t)u_j(t)\rangle}{\lambda_j(t) - \lambda_k(t)} u_k(t) \tag{13.11}$$

provided $\lambda_1(t) > \cdots > \lambda_n(t)$, $0 \leq t \leq T$.

We use (13.10), (13.11) to compute the partial derivatives of the eigenvalues with respect to the matrix elements of $H$. Indeed, if $E_{kj}$ is the matrix that is 1 in the $(k, j)$ position and 0 elsewhere, the evolutions $t \mapsto H + t(E_{kj} + E_{jk})$, and $t \mapsto H + tE_{kk}$, (13.10) yield

$$\begin{aligned} \frac{\partial \lambda_i}{\partial H_{kj}} &= 2u_i^k u_i^j, \qquad k < j, \\ \frac{\partial \lambda_i}{\partial H_{kk}} &= (u_i^k)^2, \end{aligned} \tag{13.12}$$

respectively (provided the eigenvalues of $H$ are simple). Here, $u_i^k$ is the $k$th coordinate of $u_i$. Differentiation of (13.12) (we leave this as an exercise) leads to

$$\begin{aligned} \frac{\partial^2 \lambda_i}{\partial H_{kj}^2} &= 2\sum_{l\neq i} \frac{(u_l^k u_i^j + u_l^j u_i^k)^2}{\lambda_i - \lambda_l}, \qquad k < j, \\ \frac{\partial^2 \lambda_i}{\partial H_{kk}^2} &= 2\sum_{l\neq i} \frac{(u_l^k)^2 (u_i^k)^2}{\lambda_i - \lambda_l}. \end{aligned} \tag{13.13}$$

Now, let $H(t) = \{H_{kj}(t)\}$ be a matrix Ornstein-Uhlenbeck process (13.2) with $H(0) \in \mathcal{S}(n)$, and assume that Theorem 13.3 holds. We apply Ito's formula (see Theorem C.10) to the $n(n+1)/2$-dimensional process $H(t)$ and to the functions $f_i(H(t)) = \lambda_i(t)$, $i = 1, \ldots, n$. Due to Theorem 13.3 and Lemma 13.11 (ii), the functions $f_i(H)$ are real analytic at $H = H(t)$, and hence in particular $C^2$ functions of $H$, for all $t > 0$ almost surely. Also $f_i(H(t)) = \lambda_i(t)$, $i = 1, \ldots, n$, is also continuous in $t \geq 0$ by Lemma 13.11 (ii). In order to apply Ito's formula, however, there is a technical problem. Although $\lambda_i(H)$ is a $C^2$ function of $H$ in the neighborhood of $H(t)$, $t > 0$, $\lambda_i(H)$ is not globally a $C^2$ function of $H$. We address this problem as follows. Fix $k \geq 1$. Let $\psi_k(s)$ be a $C^\infty$ function on $(0, \infty)$ such that $\psi_k(s) = 1$ for $s \geq \frac{1}{k}$, $\psi_k(s) = 0$ for $0 < s < \frac{1}{2k}$, and $0 \leq \psi_k(s) \leq 1$ for all $s > 0$. Let

$$\hat{\lambda}_i(H) = \lambda_i(H) \prod_{j=1}^{n-1} \psi_k(\lambda_j(H) - \lambda_{j+1}(H)), \quad 1 \leq i \leq n. \tag{13.14}$$

Then, again by Lemma 13.11 (ii), $\hat{\lambda}_i(H)$ is globally $C^2$ for all $H \in \mathcal{S}(n)$, and we can apply Ito's formula to $\hat{\lambda}_i(H(t))$. Let $\epsilon > 0$ be given. With $k \geq 1$ fixed as above, let $\hat{\tau}_k = \hat{\tau}_{k,\epsilon}$ be the stopping time

$$\hat{\tau}_k = \inf\left\{ t \geq \epsilon : \min_{1\leq i\leq n-1} (\lambda_i(H(t)) - \lambda_{i+1}(H(t))) = \frac{1}{k} \right\}. \tag{13.15}$$

Now Ito's formula implies that for $t \geq \epsilon$

$$
\begin{aligned}
\hat{\lambda}_i(H(t)) =& \hat{\lambda}_i(H(\epsilon)) + \sum_{k\leq \ell} \int_\epsilon^t \frac{\partial \hat{\lambda}_i}{\partial H_{k\ell}} dH_{k\ell}(s) \\
&+ \frac{1}{2} \sum_{k\leq \ell}\sum_{q\leq p} \int_\epsilon^t \frac{\partial^2 \hat{\lambda}_i}{\partial H_{k\ell}\partial H_{qp}} d\langle H_{k\ell}, H_{qp}\rangle(s).
\end{aligned} \tag{13.16}
$$

But $\hat{\lambda}_i(H(t\wedge\hat{\tau}_k)) = \lambda_i(H(t\wedge\hat{\tau}_k))$ and so (13.16) is true at time $t\wedge\hat{\tau}_k$ with $\hat{\lambda}_i$ replaced by $\lambda_i$. This leads, as we show below, to the Dyson process (13.3) for the $\lambda_i(t)$, for $t < \hat{\tau}_k$, almost surely. But by Theorem 13.3, the eigenvalues of $H(\epsilon)$ are distinct almost surely, and hence $\lim_{k\to\infty} \hat{\tau}_k = \infty$ almost surely, and so the $\lambda_i$'s satisfy the Dyson process for all $t > \epsilon$, and hence for all $t > 0$.

The quadratic variation terms needed in Ito's formula are

$$
d\langle H_{kj}, H_{\ell m}\rangle_t = \begin{cases} c_{kj}^2\, dt, & \text{if } (k,j) = (\ell, m), \\ 0, & \text{otherwise.} \end{cases}
$$

The second equality is due to the fact that $\{B_{kj}\}_{k\leq j}$ are independent. Hence, the eigenvalues, $\{\lambda_i\}_{i=1}^n$, evolve as

$$
d\lambda_i(t) = \sum_{k\leq j} \frac{\partial \lambda_i}{\partial H_{kj}} dH_{kj}(t) + \frac{1}{2}\sum_{k\leq j} \frac{\partial^2\lambda_i}{\partial H_{kj}^2}(1+\delta_{kj})\, dt, \tag{13.17}
$$

where we have used $c_{kj}^2 = 1+\delta_{kj}$. Substituting (13.2), (13.12), and (13.13) into Ito's formula (13.17),

$$
\begin{aligned}
d\lambda_i =& \sum_{k=1}^n (u_i^k)^2(-H_{kk}dt + \sqrt{2}dB_{kk}) + 2\sum_{k<j} u_i^k u_i^j(-H_{kj}dt + dB_{kj}) \\
&+ 2\sum_{k=1}^n \sum_{l\neq i} \frac{(u_l^k)^2 (u_i^k)^2}{\lambda_i - \lambda_l} dt + \sum_{k<j}\sum_{l\neq i} \frac{(u_l^k u_i^j + u_l^j u_i^k)^2}{\lambda_i - \lambda_l} dt.
\end{aligned}
$$

The terms involving $dt$ can be simplified. First,

$$
\sum_{k=1}^n (u_i^k)^2(-H_{kk})dt + 2\sum_{k<j} u_i^k u_i^j (-H_{kj})dt = -\langle u_i, Hu_i\rangle dt = -\lambda_i dt.
$$

Second,

$$
\begin{aligned}
&2\sum_{k=1}^{n}\sum_{l\neq i}\frac{(u_l^k)^2(u_i^k)^2}{\lambda_i-\lambda_l}dt+\sum_{k<j}\sum_{l\neq i}\frac{(u_l^k u_i^j+u_l^j u_i^k)^2}{\lambda_i-\lambda_l}dt\\
&=\sum_{l\neq i}\left[2\sum_{k=1}^{n}(u_l^k)^2(u_i^k)^2+\sum_{k\neq j}(u_l^k)^2(u_i^j)^2+\sum_{k\neq j}u_l^k u_i^k u_l^j u_i^j\right]\frac{dt}{\lambda_i-\lambda_l}\\
&=\sum_{l\neq i}\left[\left(\sum_{k=1}^{n}(u_l^k)^2\right)\left(\sum_{j=1}^{n}(u_i^j)^2\right)+\left(\sum_{k=1}^{n}u_l^k u_i^k\right)\left(\sum_{j=1}^{n}u_l^j u_i^j\right)\right]\frac{dt}{\lambda_i-\lambda_l}\\
&=\sum_{l\neq i}\frac{dt}{\lambda_i-\lambda_l}
\end{aligned}
$$

due to the the orthonormality of the eigenvectors. Therefore,

$$
d\lambda_i=\left[-\lambda_i+\sum_{l\neq i}\frac{1}{\lambda_i-\lambda_l}\right]dt+\sqrt{2}\left[\sum_{k=1}^{n}(u_i^k)^2\,dB_{kk}+\sqrt{2}\sum_{k<j}u_i^k u_i^j\,dB_{kj}\right].
$$

We now apply Levy's characterization of Brownian motion (see Theorem C.11) to show that the last bracket is simply a Brownian increment for a Brownian motion $W_i$. Note that $\mathbb{E}\left[\int_0^t(u_i^k)^4ds\right]<\infty$, $\mathbb{E}\left[\int_0^t(u_i^k u_i^j)^2ds\right]<\infty$, and $\sum_{k=1}^n\int_0^t(u_i^k)^4ds+\sum_{k<j}\int_0^t(u_i^k u_i^j)^2ds=t$. These follow from the fact that $|u_i^\ell|\le 1$ for all $i,\ell$, and

$$
\sum_{k=1}^{n}(u_i^k)^4+2\sum_{k<j}(u_i^k)^2(u_i^j)^2=\left(\sum_{\ell=1}^{n}(u_i^\ell)^2\right)^2=1. \tag{13.18}
$$

Therefore,

$$
d\lambda_i(t)=\Bigg(-\lambda_i(t)+\sum_{j\neq i}\frac{1}{\lambda_i(t)-\lambda_j(t)}\Bigg)dt+\sqrt{2}dW_i(t), \tag{13.19}
$$

for $i=1,2,\ldots,n$. It remains to check that $\{W_i\}_{i=1}^n$ is a collection of independent Brownian motions. This is seen by evaluating the cross variation

$$
\begin{aligned}
\langle W_i,W_\ell\rangle_t&=\sum_{k=1}^{n}\int_0^t(u_i^k)^2(u_\ell^k)^2ds+2\sum_{k<j}\int_0^t u_i^k u_i^j u_\ell^k u_\ell^j ds\\
&=\int_0^t\left(\sum_{k=1}^{n}u_i^k u_\ell^k\right)^2 ds=0
\end{aligned} \tag{13.20}
$$

for $i\neq\ell$. Therefore, $\lambda_i$, $i=1,\ldots,n$, form the Dyson process (13.3). This completes the proof of Theorem 13.4 assuming that Theorem 13.3 holds.

**Remark 13.12.** As already noted, for general matrices $H \mapsto \lambda_i(H)$ is continuous but is not smooth at $H$ with double eigenvalues, and so in order to apply Ito's formula in the above argument we needed to know a priori that the eigenvalues of the matrix Ornstein-Uhlenbeck process are distinct for all $t > 0$ with probability 1. This is the content of Theorem 13.3, which we will prove in the next subsection. In [**AGZ10**], the authors discuss a different approach of proving Theorem 13.4 without proving Theorem 13.3 first. First they show[1] that the stochastic differential equation (13.3) for the Dyson process has a global solution with $\hat{\lambda}_1(t) > \cdots > \hat{\lambda}_n(t)$. Then they solve an associated stochastic differential equation for an orthogonal matrix $\hat{O}(t)$, the putative eigenvector matrix for $H(t)$. Finally, they note that $\hat{H}(t) \equiv \hat{O}(t)\hat{\Lambda}(t)\hat{O}(t)^T$, $\hat{\Lambda}(t) = \operatorname{diag}(\hat{\lambda}_1(t), \ldots, \hat{\lambda}_n)$, solves (13.2), and hence $\{\hat{\lambda}_i\}_{i=1}^n$ can be identified with the eigenvalues $\{\lambda_i(t)\}_{i=1}^n$. As a consequence, the eigenvalues are distinct and solve the Dyson process (13.3).

**Remark 13.13.** In the case of the Dyson process and Brownian Dyson process for Hermitian matrices, equations (13.3) and (13.5) are changed (see [**Dys62**]) to

$$d\lambda_i(t) = \Big( -\lambda_i(t) + 2\sum_{j\neq i} \frac{1}{\lambda_i(t) - \lambda_j(t)} \Big)\, dt + \sqrt{2}\, dW_i(t) \tag{13.21}$$

and

$$d\tilde{\lambda}_i(t) = 2\sum_{l\neq i} \frac{1}{\tilde{\lambda}_i(t) - \tilde{\lambda}_l(t)}\, dt + \sqrt{2}\, dW_i(t), \tag{13.22}$$

respectively. The difference from the symmetric matrix case is that the sum $\sum_{j\neq i} \frac{1}{\lambda_i(t)-\lambda_j(t)}$ in the drift term is multiplied by 2. This changes the behavior of the eigenvalue evolution in an important qualitative way. In particular, in the Brownian case for the symmetric matrix diffusion (13.5) the eigenvalues come arbitrarily close to each other without touching, whereas in the Hermitian Brownian case the eigenvalues remain a positive minimum distance apart. This situation is reminiscent and related to the simpler case of Brownian motion starting at some point $x \neq 0$ in two and three dimensions. In two dimensions Brownian motion comes arbitrarily close to, but never touches, the origin, whereas in three dimensions it remains a (realization dependent) positive distance away from the origin for all $t > 0$. Another way to describe the situation is as follows: Brownian motion comes arbitrarily close to, but does not touch, a codimension 2 subset, whereas it stays a positive distance away from a codimension 3 subset. One checks easily (we leave this as an exercise) that the symmetric matrices with at least two

[1] More precisely, they consider Brownian Dyson process (13.5), but their method also applies to (13.3) (see Exercise 4.3.9 in [**AGZ10**]).

coincident eigenvalues form a codimension 2 subset in the symmetric matrices, whereas the Hermitian matrices with at least two coincident eigenvalues form a codimension 3 subset in the Hermitian matrices. In the case of Dyson diffusion (13.3) there is a similar qualitative enhancement of repulsion in the Hermitian case over the symmetric case. Note that for $n = 2$, in particular, $(\lambda_1 - \lambda_2)^2 = (\tilde{H}_{11} - \tilde{H}_{22})^2 + 4(\tilde{H}_{12})^2$ in the real symmetric case, and $(\lambda_1 - \lambda_2)^2 = (\tilde{H}_{11} - \tilde{H}_{22})^2 + 4(\operatorname{Re}\tilde{H}_{12})^2 + 4(\operatorname{Im}\tilde{H}_{12})^2$ for the complex Hermitian case, and so the separation of the eigenvalues behaves precisely like Brownian motion in two and three dimensions, respectively. Some of these observations were communicated to us by S.R.S. Varadhan.

Since the unique invariant measure of the matrix Ornstein-Uhlenbeck process is the GOE measure $e^{-\frac{1}{2}\operatorname{tr} H^2}$, Dyson process (13.3) has a unique invariant measure given by the GOE eigenvalue distribution. More precisely, $\mu(d^n\lambda) = \frac{e^{-\frac{1}{2}\sum\lambda_i^2}}{Z_n}\prod_{i<j}|\lambda_i - \lambda_j|\, d^n\lambda$ is the unique probability measure such that

$$\mu(A) = \int p(s,t,x,A)\,\mu(d^n x) \tag{13.23}$$

holds for all Borel sets $A \subset W^{(n)}$ where

$$W^{(n)} \equiv \{x \in \mathbb{R}^n : x_1 > \cdots > x_n\} \tag{13.24}$$

and $t \geq s \geq 0$, where $p(s,t,x,A)$ is the time $t - s$ transition probability at time $s$ for the time homogeneous diffusion (13.3), i.e., $\partial_s p(s,t,x,y) + L_x p(s,t,x,y) = 0$ subject to the condition $p(t,t,x,y) = \delta_x(y)$ where $L_x = (-x + \nabla_x \log h)\cdot\nabla_x + \Delta_x$ with $h(x) = \prod_{1\leq i<j\leq n}|\lambda_j - \lambda_i|$ (see (C.2), (C.3), and (C.4)). The following computation shows directly that $\mu$ is an invariant measure. For any Borel set $A \subset W^{(n)}$,

$$\begin{aligned}\frac{\partial}{\partial s}\int p(s,t,x,A)\mu(d^n x) &= \int \frac{\partial}{\partial s}p(s,t,x,A)\mu(d^n x)\\ &= -\int L_x p(s,t,x,A)\mu(d^n x)\\ &= -\int p(s,t,x,A) L_x^*\mu(d^n x) = 0\end{aligned} \tag{13.25}$$

since $L^*\mu = 0$ by direct computation. Here we used the elementary identity for any smooth functions $p$ and $E$, $(L_x p)hE = pL_x^*(hE) + \nabla\cdot[h(E\nabla p - p\nabla E)]$ which shows that after integrating by parts the boundary terms drop out as $h = 0$ on $\partial W^{(n)}$. By contrast, the Brownian Dyson process (13.5) does not have an invariant measure. Uniqueness of the invariant measure, $\mu$, follows from the smoothness and positivity of $p(s,t,x,y)$ for $t > s$ and $x, y \in W^{(n)}$ and the fact that two distinct ergodic measures must be mutually singular; see, for example, [**Sin94, Wal82**]. For general information on properties of heat kernels of parabolic partial differential equations, see for example [**Kry95, SV79**].

It is natural to ask whether there are processes with invariant probability measures given by the general orthogonal ensembles $\frac{1}{Z_n}e^{-\operatorname{tr} V(H)}dH$ whose eigenvalue evolution is Markovian. Here, $V(x)$ is smooth and $V(x) \to \infty$ sufficiently fast as $x \to \infty$. This can be done as follows. Consider the process

$$dH_{kj}(t) = -\frac{1}{2}(V'(H(t)))_{kj}dt + c_{kj}dB_{kj}(t), \qquad k \leq j, \tag{13.26}$$

where again $c_{kj} = 1 + (\sqrt{2} - 1)\delta_{kj}$. Here $V(H)$ and $V'(H)$ are defined via the functional calculus for $H \in \mathcal{S}(n)$. Then, as above, one can show using the methods in [**AGZ10**] that the eigenvalues $\lambda_1(t) > \cdots > \lambda_n(t)$ of $H(t)$ satisfy the generalized Dyson process

$$d\lambda_i(t) = \left(-\frac{1}{2}V'(\lambda_i(t)) + \sum_{j\neq i} \frac{1}{\lambda_i(t) - \lambda_j(t)}\right) dt + \sqrt{2}dW_i(t) \tag{13.27}$$

for $i = 1, 2, \ldots, n$, where $\{W_i(t)\}_{i=1}^n$ are again independent standard Brownian motions. There is clearly an analogous evolution for the general unitary ensemble, with the extra factor 2 as in (13.21).

**Exercise 13.14.** Show that $\frac{e^{-\operatorname{tr}V(H)}}{Z_n}dH$ is an invariant probability measure for (13.26) and $\frac{e^{-\sum V(\lambda_i)}}{Z_n}\prod_{i<j}|\lambda_i - \lambda_j| d^n\lambda$ is an invariant probability measure for (13.27).

Note that if $V(H) = \frac{1}{2}H^2$, then (13.26) and (13.27) clearly reduce to (13.2) and (13.3), respectively.

**Remark 13.15.** In terms of the calculations in Chapter 11, we see that the Dyson process induces a dynamical point process on $\mathbb{R}$ by the map $(\lambda_1(t), \ldots, \lambda_n(t)) \mapsto \xi(t) = \sum_{i=1}^n \delta_{\lambda_i(t)}$.

In the next two sections, Sections 13.4 and 13.5, we prove Theorems 13.8 and 13.3, respectively. In Section 13.4, we prove Theorem 13.8 using a martingale argument, thereby completing the derivation of the Brownian Dyson process. This derivation can easily be adapted (we leave this as an exercise) to prove Theorem 13.3 for the Dyson process. We also note that Theorems 13.3 and 13.8 can be inferred from each other by invoking Girsanov's theorem [**RY99**]; we leave the details to the reader. However, there is a particularly elegant way of using the invariant measure to prove the noncolliding property of the eigenvalues of the matrix Ornstein-Uhlenbeck process, and we discuss it in Section 13.5 for a pedagogical purpose. Since matrix Brownian motion does not have an invariant measure, this argument does not apply, however, to Brownian Dyson process.[2]

## 13.4. Noncolliding Property of the Eigenvalues of Matrix Brownian Motion

We now prove Theorem 13.8 that the eigenvalues of matrix Brownian motion do not collide for all time $t > 0$. Let $\tilde{H}(t), t \geq 0$, be the matrix Brownian motion

$$d\tilde{H}_{ij} = c_{ij}\, dB_{ij}, \qquad c_{ij} = 1 + (\sqrt{2} - 1)\delta_{ij}. \tag{13.28}$$

Then the eigenvalues $\tilde{\lambda}_1(t) \geq \tilde{\lambda}_2(t) \geq \cdots \geq \tilde{\lambda}_n(t)$ of $\tilde{H}(t)$ are well defined, Lipschitz continuous functions (see Lemma 13.11), and the goal is to show that $\tilde{\lambda}_1(t) > \tilde{\lambda}_2(t) > \cdots > \tilde{\lambda}_n(t)$ for all $t > 0$.

The infinitesimal generator for matrix Brownian motion is

$$\tilde{L} = \sum_{i=1}^{n} \frac{\partial^2}{\partial \tilde{H}_{ii}^2} + \frac{1}{2} \sum_{1 \leq i < j \leq n} \frac{\partial^2}{\partial \tilde{H}_{ij}^2}. \tag{13.29}$$

For $C^2$ functions $f(H)$, Ito's formula for matrix Brownian motion takes the form (see (C.6))

(13.30)

$$\begin{aligned} f(\tilde{H}(t)) &= f(\tilde{H}(0)) + \int_0^t (\tilde{L}f)(\tilde{H}(s))ds \\ &\quad + \int_0^t \left[ \sqrt{2} \sum_{i=1}^{n} \frac{\partial f}{\partial \tilde{H}_{ii}}(\tilde{H}(s))dB_{ii}(s) + \sum_{1 \leq i < j \leq n} \frac{\partial f}{\partial \tilde{H}_{ij}}(\tilde{H}(s))dB_{ij}(s) \right]. \end{aligned}$$

We will apply Ito's formula to a certain well-chosen function and use a martingale argument.

We define two functions, $D(H)$ and $\sigma(H)$, for $H \in \mathcal{S}(n)$.

**Lemma 13.16.** *For an $n \times n$ (not necessarily symmetric) matrix $H$, let $\lambda_1, \ldots, \lambda_n$ be the eigenvalues of $H$. Set*

$$D = D(H) = \prod_{1 \leq i < j \leq n} (\lambda_i - \lambda_j)^2. \tag{13.31}$$

*Then $D = D(H)$ is a smooth function, in fact a polynomial, in the entries $H_{ij}$ of $H$.*

[2]The method can be applied, however, after an appropriate compactification procedure. (Private communication from S.R.S. Varadhan.)

**Proof.** Observe that $D$ is a symmetric polynomial in the $\lambda_i$'s. Hence $D$ can be expressed in terms of (the linear combinations of the products of) the power-sum symmetric functions $p_k(\lambda_1,\dots,\lambda_n)=\sum_{i=1}^n \lambda_i^k$ (see for example, Theorem 10.11). But $p_k(\lambda_1,\dots,\lambda_n)=\operatorname{tr}(H^k)$. Hence $D$ is a polynomial in the entries of $H$. □

There are many other proofs of the above lemma. (See, for example, Exercise 13.18 below.)

For a symmetric matrix $H$, define $\sigma=\sigma(H)\geq 0$ by

$$\sigma^2=\sum_{i=1}^n \lambda_i^2=\operatorname{tr}(H^2)=\sum_{i=1}^n H_{ii}^2+2\sum_{i<j}H_{ij}^2. \tag{13.32}$$

Then for a matrix Brownian motion $\tilde{H}(t)$,

$$\lim_{t\to\infty}\sigma(\tilde{H}(t))=\infty \quad \text{almost surely.} \tag{13.33}$$

We will prove Theorem 13.8 first assuming that

$$D(\tilde{H}(0))>0. \tag{13.34}$$

An argument at the end of this section will show how we can prove the general case. Choose $k,K>0$ sufficiently large so that

$$\frac{1}{k}<D(\tilde{H}(0)) \quad \text{and} \quad \sigma(\tilde{H}(0))<K. \tag{13.35}$$

Let

$$\tau_k=\inf\{t\in\mathbb{R}_+ : D(\tilde{H}(t))=\frac{1}{k}\} \tag{13.36}$$

and

$$T_K=\inf\{t\in\mathbb{R}_+ : \sigma(\tilde{H}(t))=K\}. \tag{13.37}$$

Note that by (13.33), $T_K<\infty$ almost surely. However, $\tau_k$ may be infinite with positive probability.

For $\epsilon>0$, let

$$f_\epsilon(H)=\log(D(H)+\epsilon), \qquad H\in\mathcal{S}(n). \tag{13.38}$$

As $D(H)\geq 0$, $f_\epsilon(H)$ is a smooth function of $H$. By (13.30) we have for given $t>0$

$$f_\epsilon(\tilde{H}(t\wedge\tau_k\wedge T_K))-f_\epsilon(\tilde{H}(0))=I+II, \tag{13.39}$$

where

(13.40)

$$I=\int_0^{t\wedge\tau_k\wedge T_K}\left[\sqrt{2}\sum_{i=1}^n\frac{\partial f_\epsilon}{\partial\tilde{H}_{ii}}(\tilde{H}(s))dB_{ii}(s)+\sum_{1\leq i<j\leq n}\frac{\partial f_\epsilon}{\partial\tilde{H}_{ij}}(\tilde{H}(s))dB_{ij}(s)\right]$$

and

$$II = \int_0^{t\wedge\tau_k\wedge T_K} (\tilde{L}f_\epsilon)(\tilde{H}(s))ds. \tag{13.41}$$

Now as $t\wedge\tau_k\wedge T_K \le \tau_k$, $D(\tilde{H}(s)) > 0$ for $0 \le s \le t\wedge\tau_k\wedge T_K$, and hence $\log D(H)$ is a smooth function of $H$ for $H$ near $H(s)$ for any $0 \le s \le t\wedge\tau_k\wedge T_K$. Note that if $\tau_k = \infty$, this is still clearly true.

We claim $\tilde{L}\log D(\tilde{H}(s)) = 0$ for $0 \le s \le t\wedge\tau_k\wedge T_K$. To see this, note that as the $\lambda_i$'s are distinct for $H$ near $H(s)$, we can change coordinates $H \to \{\{\lambda_i\}_{i=1}^n, \text{eigenvectors}\}$ and the *radial* part of $\tilde{L}$ is given by (see Lemma 13.17 below)

$$L^h = \frac{\nabla_\lambda h}{h}\cdot\nabla_\lambda + \Delta_\lambda, \tag{13.42}$$

where

$$\nabla_\lambda = \left(\frac{\partial}{\partial\lambda_1}, \dots, \frac{\partial}{\partial\lambda_n}\right), \qquad \Delta_\lambda = \sum_{i=1}^n \frac{\partial^2}{\partial\lambda_i^2}, \tag{13.43}$$

and $h = h(\lambda) = \prod_{i<j}(\lambda_i - \lambda_j)$, $h(\lambda)^2 = D$. Now $\Delta_\lambda h = 0$. One way to see this is to note that $h = \det(T_{ij})_{1\le i,j\le n}$ where $T_{ij} = \lambda_j^{n-i}$. By density it is enough to consider the case $h = h(\lambda) \ne 0$. We have $\Delta_\lambda h = \sum_{j=1}^n \hat{h}_j$, where $\hat{h}_j$ is the same as $h$ except the $j$th column is replaced by $((n-1)(n-2)\lambda_j^{n-3}, (n-2)(n-3)\lambda_j^{n-4}, \dots, 2, 0, 0)^T$, which is the same as $((n-1)(n-2)T_{3j}, (n-2)(n-3)T_{4j}, \dots, 2T_{nj}, 0, 0)^T$. Then, by Cramer's rule,

$$\begin{aligned}\frac{\Delta_\lambda h}{h} = \sum_{j=1}^n \frac{\hat{h}_j}{h} &= \sum_{j=1}^n \Big((n-1)(n-2)T_{3j}(T^{-1})_{j1} \\ &\qquad + (n-2)(n-3)T_{4j}(T^{-1})_{j2} + \cdots + 2T_{nj}(T^{-1})_{j(n-2)}\Big) \\ &= (n-1)(n-2)\delta_{31} + (n-2)(n-3)\delta_{42} + \cdots + 2\delta_{n(n-2)} = 0.\end{aligned}$$

Thus $\Delta_\lambda h = 0$. It follows that $L^h(\log h) = \frac{\Delta_\lambda h}{h} = 0$ and hence we have $L^h(\log D) = 0$. But $\log D = \sum_{i<j}\log(\lambda_i - \lambda_j)^2$ is a radial function, i.e., it only depends on the $\lambda_i$'s. Hence $\tilde{L}\log D(\tilde{H}(s)) = L^h \log D(\tilde{H}(s)) = 0$ for $0 \le s \le t\wedge\tau_k\wedge T_K$.

Thus

$$\sum_{i=1}^n \frac{\partial^2 \log D}{\partial H_{ii}^2} + \frac{1}{2}\sum_{1\le i<j\le n} \frac{\partial^2 \log D}{\partial H_{ij}^2} = \frac{\tilde{\Delta}D}{D} - \left(\frac{\tilde{\nabla}D}{D}\right)^2 = 0, \tag{13.44}$$

where

$$\tilde{\Delta} = \tilde{L} = \sum_{i=1}^{n} \frac{\partial^2}{\partial H_{ii}^2} + \frac{1}{2} \sum_{1 \le i < j \le n} \frac{\partial^2}{\partial H_{ij}^2} \tag{13.45}$$

and

$$\tilde{\nabla} = \left( \frac{\partial}{\partial H_{11}}, \ldots, \frac{\partial}{\partial H_{nn}}, \frac{1}{\sqrt{2}} \frac{\partial}{\partial H_{12}}, \ldots, \frac{1}{\sqrt{2}} \frac{\partial}{\partial H_{n,n-1}} \right). \tag{13.46}$$

This implies that

$$\tilde{\Delta} D(\tilde{H}(s)) = \frac{(\tilde{\nabla} D(\tilde{H}(s)))^2}{D(\tilde{H}(s))} \tag{13.47}$$

for $0 \le s \le t \wedge \tau_k \wedge T_K$.

The same calculation shows that for $0 \le s \le t \wedge \tau_k \wedge T_K$,

$$\tilde{L} f_\epsilon(\tilde{H}(s)) = \tilde{L} \log(D + \epsilon) = \frac{\tilde{\Delta} D}{D + \epsilon} - \left( \frac{\tilde{\nabla} D}{D + \epsilon} \right)^2 = \frac{\epsilon (\tilde{\nabla} D)^2}{(D + \epsilon)^2 D} \tag{13.48}$$

using (13.47) for the last equality. Hence as $0 \le s \le t \wedge \tau_k \wedge T_K \le \tau_k$, $D(\tilde{H}(s)) \ge \frac{1}{k}$, and so

$$0 \le \tilde{L} f_\epsilon(H(s)) \le k^3 \epsilon (\tilde{\nabla} D)^2. \tag{13.49}$$

But as $0 \le s \le t \wedge \tau_k \wedge T_K \le T_K$,

$$\operatorname{tr} \tilde{H}(s)^2 \le K, \tag{13.50}$$

and so $(\tilde{\nabla} D)^2$, which is a polynomial in the $\tilde{H}_{ij}$'s, is bounded by some polynomial in $K$. Thus for $0 \le s \le t \wedge \tau_k \wedge T_K$,

$$0 \le \tilde{L} f_\epsilon(\tilde{H}(s)) \le \epsilon C(k, K), \tag{13.51}$$

where $C(k, K)$ is a polynomial in $k$ and $K$.

Now the stochastic integral $I = I(t)$ in (13.40) is a sum of integrals of the form $\int_0^t 1_{s<\tau_k \wedge T_K} G_s dW_s$ with $\int_0^T |1_{s<\tau_k \wedge T_K} G_s|^2 ds < \infty$ almost surely (indeed, $1_{s<\tau_k \wedge T_K} G_s$ is bound in $s \in [0, T]$ almost surely) for all $T > 0$, and hence is a local martingale (see Appendix C). Therefore, there exist stopping times $\hat{\tau}_q$ such that $\hat{\tau}_q \to \infty$ almost surely as $q \to \infty$ and $I(t \wedge \hat{\tau}_q)$ is a martingale. This implies that $\mathbb{E}[I(t \wedge \hat{\tau}_q)] = \mathbb{E}[I(0)] = 0$. Hence (13.39) implies that for any $q$ and any $t > 0$,

(13.52)

$$\mathbb{E}[f_\epsilon(\tilde{H}(t \wedge \hat{\tau}_q \wedge \tau_k \wedge T_K))] = f_\epsilon(\tilde{H}(0)) + \mathbb{E}\left[ \int_0^{t \wedge \hat{\tau}_q \wedge \tau_k \wedge T_K} \tilde{L} f_\epsilon(\tilde{H}(s)) ds \right].$$

But from (13.51)

$$0 \le \mathbb{E}\left[\int_0^{t\wedge\hat{\tau}_q\wedge\tau_k\wedge T_K} \tilde{L}f_\epsilon(\tilde{H}(s))ds\right] \le \epsilon C(k,K)t \to 0 \tag{13.53}$$

as $\epsilon \downarrow 0$. Thus letting $\epsilon \downarrow 0$ in (13.52), we obtain

$$\mathbb{E}[\log D(\tilde{H}(t\wedge\hat{\tau}_q\wedge\tau_k\wedge T_K))] = \log D(\tilde{H}(0)) \tag{13.54}$$

(note that $\log D(\tilde{H}(t\wedge\hat{\tau}_q\wedge\tau_k\wedge T_K))$ is uniformly bounded so that the dominated convergence theorem applies). Letting $q\to\infty$, we find that, as $\hat{\tau}_q\to\infty$ almost surely, $\mathbb{E}[\log D(\tilde{H}(t\wedge\tau_k\wedge T_K))] = \log D(\tilde{H}(0))$. Therefore, we obtain

$$\begin{aligned}
&\log D(\tilde{H}(0))\\
&= \log(k^{-1})\,\mathbb{P}(\tau_k < t\wedge T_K) + \log D(\tilde{H}(T_K))\,\mathbb{P}(T_K < t\wedge\tau_k)\\
&\quad + \mathbb{E}\left[\log D(\tilde{H}(t))\chi_{t<\tau_k\wedge T_K}\right].
\end{aligned} \tag{13.55}$$

Now $|\log D(\tilde{H}(t))|$ is uniformly bounded for $t<\tau_k\wedge T_K$. Therefore, as $\mathbb{P}(t<\tau_k\wedge T_K)\to 0$ as $t\to\infty$, by (13.33) we find that

$$\lim_{t\to\infty}\mathbb{E}\left[\log D(\tilde{H}(t))\chi_{t<\tau_k\wedge T_K}\right] = 0. \tag{13.56}$$

Since

$$\lim_{t\to\infty}\mathbb{P}(\tau_k < t\wedge T_K) = \mathbb{P}(\tau_k < T_K) \tag{13.57}$$

and

$$\lim_{t\to\infty}\mathbb{P}(T_K < t\wedge\tau_k) = \mathbb{P}(T_K<\tau_k), \tag{13.58}$$

we have from (13.55)

$$\log D(\tilde{H}(0)) = -(\log k)\,\mathbb{P}(\tau_k<T_K) + \log D(\tilde{H}(T_K))\,\mathbb{P}(T_K\le\tau_k). \tag{13.59}$$

As $D(\tilde{H}(T_K)) \le C(K)$ for some $C(K)>0$, we find that

$$\log D(\tilde{H}(0)) \le -(\log k)\,\mathbb{P}(\tau_k<T_K) + (\log C(K))(1-\mathbb{P}(\tau_k<T_K)) \tag{13.60}$$

which implies

$$\mathbb{P}(\tau_k<T_K) \le \frac{\log C(K) - \log D(H(0))}{\log k + \log C(K)}. \tag{13.61}$$

In particular,

$$\lim_{k\to\infty}\mathbb{P}(\tau_k<T_K) = 0. \tag{13.62}$$

Let

$$\tau = \inf\{s : D(\tilde{H}(s)) = 0\}. \tag{13.63}$$

But then as $\tau_k \leq \tau$ for all $k$ sufficiently large, we have $\mathbb{P}(\tau < T_K) = 0$ for all $K$ sufficiently large. Now let $M$ be any positive number. Then $\mathbb{P}(\tau < M \wedge T_K) \leq \mathbb{P}(\tau < T_K) = 0$. As the paths $\tilde{H}(t)$ are continuous, $M' = \max_{0\leq t\leq M} \sigma(\tilde{H}(t)) < \infty$. Hence for any $K > M'$, $T_K > M$, and so $\lim_{K\to\infty} M \wedge T_K = M$. Thus $\mathbb{P}(\tau < M) = \lim_{K\to\infty} \mathbb{P}(\tau < M \wedge T_K) = 0$. As $M > 0$ is arbitrary, we conclude that $\mathbb{P}(\tau = \infty) = 1$ and hence if $\tilde{H}(t)$ starts at $\tilde{H}(0)$ with $D(\tilde{H}(0)) > 0$, then $D(\tilde{H}(t)) > 0$ for all $t > 0$.

This completes the proof in the case that $D(\tilde{H}(0)) > 0$. For general $\tilde{H}(0)$ we follow an argument in [**AGZ10**]. With $\tilde{H}(0)$ given, the transition probability density is given by $p(0, t, \tilde{H}(0), H) = c_n(t)e^{-\frac{1}{4t}\operatorname{tr}(H-\tilde{H}(0))^2}$, $t > 0$, where $c_n(t) = (4\pi t)^{-n/2}(2\pi t)^{-n(n-1)/4}$. Let $\mathcal{B} = \{H \in \mathcal{S}(n) : D(H) > 0\} = \{H \in \mathcal{S}(n) : H \text{ has simple spectrum}\} \subset \mathbb{R}^{n(n+1)/2}$. As in the Hermitian case (cf. (6.9)), $\mathcal{B}$ is open and dense in $\mathbb{R}^{n(n+1)/2}$ and has full measure, i.e., the Lebesgue measure of $\mathbb{R}^{n(n+1)/2} \setminus \mathcal{B}$ is zero. It follows that

$$p(0, t, \tilde{H}(0), \mathcal{B}) = 1 \text{ for any } t > 0. \tag{13.64}$$

Let $t_1 > t_2 > \cdots > 0$ be a sequence of times such that $t_q \downarrow 0$ as $q \to \infty$. By the Markov property and (13.64), we have for $\tilde{H}(0)$ fixed,

$$\begin{aligned}
&\mathbb{P}(D(\tilde{H}(t)) > 0,\, t > 0)\\
&= \lim_{q\to\infty} \mathbb{P}(D(\tilde{H}(t)) > 0,\, t \geq t_q)\\
&= \lim_{q\to\infty} \int p(0, t_q, \tilde{H}(0), H)\, \mathbb{P}(D(\tilde{H}(t)) > 0,\, t \geq t_q | \tilde{H}(t_q) = H) dH\\
&= \lim_{q\to\infty} \int_{\mathcal{B}} p(0, t_q, \tilde{H}(0), H)\, \mathbb{P}(D(\tilde{H}(t)) > 0,\, t \geq t_q | \tilde{H}(t_q) = H) dH.
\end{aligned} \tag{13.65}$$

But for $H \in \mathcal{B}$, $\mathbb{P}(D(\tilde{H}(t)) > 0,\, t \geq t_q | \tilde{H}(t_q) = H) = 1$ by the result just proved. Hence $\mathbb{P}(D(\tilde{H}(t)) > 0,\, t > 0) = \lim_{q\to\infty} \int_{\mathcal{B}} p(0, t_q, \tilde{H}(0), H) dH = 1$. This proves Theorem 13.8 for general $\tilde{H}(0) \in \mathcal{S}(n)$.

In the above proof, we used the fact that the radial part of $\tilde{L}$ is given by (13.42). The proof of the following lemma is somewhat lengthy but straightforward and is left to the reader.

**Lemma 13.17.** *Let*

$$\tilde{L} = \sum_{i=1}^{n} \frac{\partial^2}{\partial H_{ii}^2} + \frac{1}{2} \sum_{1\leq i<j\leq n} \frac{\partial^2}{\partial H_{ij}^2} \tag{13.66}$$

*be a differential operator on matrices $H$. On the set $\Sigma$ of real symmetric matrices $\{H\}$ with simple spectrum, we have*

$$\tilde{L}f = (\nabla_\lambda \log h) \cdot \nabla_\lambda f + \Delta_\lambda f \tag{13.67}$$

*for all smooth functions $f$ of the eigenvalues $\{\lambda_i\}$ of $H$, where $h = \prod_{i<j}(\lambda_i - \lambda_j)$, $\nabla_\lambda = (\partial_{\lambda_1}, \ldots, \partial_{\lambda_n})$, and $\Delta_\lambda = \sum_{i=1}^n \frac{\partial^2}{\partial \lambda_i^2}$.*

The next exercise concerns a different proof of Lemma 13.16 which asserts that $D = D(H) = \prod_{1 \le i < j \le n}(\lambda_i - \lambda_j)^2$ is a smooth function in the entries of $H$.

**Exercise 13.18.** Let $\Gamma_H = (H \otimes I - I \otimes H)^2$ act on the skew 2-tensors $\Lambda_2 = \{\sum_{i,j} a_{ij} u_i \wedge u_j\}$. If $H$ has $n$ distinct eigenvalues $\{\lambda_i\}$ with associated eigenvectors $v_i$, $Hv_i = \lambda_i v_i$, then $\Gamma_H(v_i \wedge v_j) = (\lambda_i - \lambda_j)^2 v_i \wedge v_j$ and as $\{v_i \wedge v_j\}_{i<j}$ spans the space $\Lambda_2$, we see that $D = \det \Gamma_H$, which is clearly a polynomial in the $H_{ij}$'s. If some of the eigenvalues of $H$ are not distinct, then perturb $H$ to $H'$ where $H'$ has simple spectrum and hence a complete set of eigenvectors. Show that $D' = \det \Gamma_{H'}$ and then $H' \to H$.

## 13.5. Noncolliding Property of the Eigenvalues of the Matrix Ornstein-Uhlenbeck Process

In this section we prove Theorem 13.3 which states that the eigenvalues of the matrix Ornstein-Uhlenbeck process do not collide for all time $t > 0$. As mentioned before, one can prove this fact by the method of Section 13.4 or by appealing to Girsanov's theorem. However, there is a different proof using invariant measures and, in the interests of advertising different approaches, we present it in this section. This method is due to S.R.S. Varadhan and, as the reader will see, is particularly elegant and informative.

Consider a matrix Ornstein-Uhlenbeck process $H(t)$ on the set $\mathcal{S}(n)$ of $n \times n$ symmetric matrices:

$$dH_{ij} = -H_{ij}\,dt + c_{ij}\,dB_{ij}, \qquad c_{ij} = 1 + (\sqrt{2} - 1)\delta_{ij} \tag{13.68}$$

for $i \le j$. We show first, as in the case of Brownian Dyson process, that the eigenvalues are nonintersecting for almost all paths $H(t)$ if the eigenvalues of the initial condition $H(0)$ are all distinct. Recall (see Lemma 13.5) that the matrix Ornstein-Uhlenbeck process has invariant measure

$$d\mu(H) = \frac{1}{Z} e^{-\frac{1}{2}\operatorname{tr} H^2} dH, \qquad \text{where } Z = \int e^{-\frac{1}{2}\operatorname{tr} H^2} dH. \tag{13.69}$$

The basic idea is as follows:

(i) There exists a broad class of measurable functions $F : \mathcal{S}(n) \to \mathbb{R}$ with the property that $F(H(t))$ is continuous for almost all paths $\{H(t) : t \ge 0\}$.

(ii) There exists, in particular, a function $F$ in the class such that $F(H) = \infty$ if the eigenvalues are not distinct.

It follows then that for almost all paths $\{H(t)\}$ the eigenvalues $\{\lambda_i(H(t))\}$ are distinct for all $t \geq 0$.

We consider stationary Markov process $P = \{X(t), t \in \mathbb{R}\}$ taking values in a state space $\Sigma$ which is a locally compact Polish space, i.e., a locally compact space which is a complete, separable metric space. The space $\Sigma$ is equipped with the Borel sets and the invariant probability measure for the process is denoted by $d\mu$. We denote the generator of the process by $L$, and the associated Markov semigroup $S(t) = e^{tL}$, $t \geq 0$, is a strongly continuous positivity-preserving contraction mapping on $C_b$, the set of bounded continuous functions on $\Sigma$,

$$S(t + t') = S(t)S(t'), \quad t, t' > 0, \qquad S(0) = id, \tag{13.70}$$

$$t \mapsto S(t)f \text{ is continuous from } \mathbb{R} \text{ to } C_b, \tag{13.71}$$

$$(S(t)f)(x) \geq 0, \quad \text{if } f(x) \geq 0, \tag{13.72}$$

and

$$\|S(t)f\|_{L^\infty}(x) \leq \|f\|_{L^\infty} \tag{13.73}$$

for all $f \in C_b$. We also have

$$(S(t)1)(x) = 1(x). \tag{13.74}$$

We assume in addition that $S(t)$ is symmetric with respect to $d\mu$, i.e., for all $t \geq 0$,

$$\int f(x)(S(t)g)(x)d\mu(x) = \int (S(t)f)(x)g(x)d\mu(x) \tag{13.75}$$

for all $f, g \in C_b$. Note that as $S(t)1 = 1$, (13.75) is consistent with, and in fact implied by, the invariance of $d\mu$,

$$\int (S(t)g)(x)d\mu(x) = \int g(x)d\mu(x) \tag{13.76}$$

for all $g \in C_b$.

The symmetry of $S(t)$ implies that $\{X(t), t \in \mathbb{R}\}$ is *reversible* (see, e.g., [**FS86, Lig99**]), i.e., $\{X(t), t \in \mathbb{R}\}$ and $\{X(-t), t \in \mathbb{R}\}$ have the same finite dimensional distribution functions. More precisely, if $t_1 < t_2 < \cdots < t_m$, then the joint distribution functions of $X_{t_1}, \ldots, X_{t_m}$ and $X_{-t_m}, \ldots, X_{-t_1}$ are the same. This follows directly from (13.76) and the Markov property for $\{X(t), t \in \mathbb{R}\}$ (we leave this as an exercise).

We record the following properties of locally compact, Polish spaces $\Sigma$ (see, e.g., [**Roy88**]):

(i) $\Sigma$ is $\sigma$-compact.

(ii) Any finite Borel measure $\nu$ on a complete, separable metric space $X$ is regular; i.e., for any Borel set $E \subset X$, we have

$$\nu(E) = \inf\{\nu(O) : E \subset O,\, O \text{ open}\} \tag{13.77}$$

and

$$\nu(E) = \sup\{\nu(K) : K \subset E,\, K \text{ compact}\}. \tag{13.78}$$

By regularity and Uryshon's lemma, the characteristic function of any Borel set can be approximated in $L^p(d\mu)$, $1 \le p < \infty$, by a bounded, continuous function. It follows that $C_b$ is dense in $L^p(d\mu)$ for any $1 \le p < \infty$.

For all $t \ge 0$, $S(t)$ extends to a semigroup which is a contraction on $L^p(d\mu)$ for any $1 \le p \le \infty$ and which is strongly continuous for $1 \le p < \infty$. Indeed, for any $f, g \in C_b \subset L^1(d\mu)$,

$$\left| \int f(x)(S(t)g)(x)d\mu(x) \right| = \left| \int (S(t)f)(x)g(x)d\mu(x) \right| \le \|f\|_{L^\infty} \|g\|_{L^1}.$$

By density, there exist functions $f_n \in C_b$, $\|f_n\|_{L^\infty} \le 1$ such that $f_n(x) \to \operatorname{sgn}(S(t)g)(x)$ pointwise a.e., and we conclude that $\|S(t)g\|_{L^1(d\mu)} \le \|g\|_{L^1(d\mu)}$ for $g \in C_b$ and so $S(t)$ extends to a contraction on $L^1(d\mu)$. Clearly, (13.75) continues to hold for $g \in L^1(d\mu)$ and $f \in C_b$. But $L^\infty(d\mu) \subset L^1(d\mu)$, and so for $g \in L^\infty(d\mu)$ and $f \in C_b$,

$$\left| \int f(x)(S(t)g)(x)d\mu(x) \right| = \left| \int (S(t)f)(x)g(x)d\mu(x) \right| \le \|f\|_{L^1} \|g\|_{L^\infty},$$

and as $L^\infty(d\mu)$ is the dual of $L^1(d\mu)$, it follows that $\|S(t)g\|_{L^\infty} \le \|g\|_{L^\infty}$. Finally, by interpolation, we conclude that $\|S(t)f\|_{L^p(d\mu)} \le \|f\|_{L^p(d\mu)}$. Strong continuity of $S(t)$ for $1 \le p < \infty$ now follows from the density of $C_b$ in $L^p(d\mu)$. We conclude, in particular, that for $p = 2$, the generator $L$ of the process, $S(t) = e^{tL}$, is a negative definite self-adjoint operator in $L^2(d\mu)$.

The following result is due to Kipnis and Varadhan [**KV86**] (see Lemmas 1.4 and 1.12). Recall (see [**RS80**]) that if $-L$ is a positive definite operator in a Hilbert space $\mathcal{H}$ and if $d\mu_f(\lambda)$ is the spectral measure for $L$ in the cyclic subspace generated by $f$, then $f$ lies in the *form domain* $Q(L)$ of $L$ if $Q_L(f) \equiv \int_0^\infty \lambda d\mu_f(\lambda) < \infty$. If $f \in D(L) = \{f : \int_0^\infty \lambda^2 d\mu_f(\lambda) < \infty\}$, then clearly $f \in Q(L)$ and $Q_L(f) = (f, -Lf)$.

**Lemma 13.19.** *Let $L$ be the generator of a stationary Markov process $P$ with continuous paths, $P = \{X(t), t \in \mathbb{R}\}$, which is reversible with respect to an invariant probability measure $d\mu$ on a locally compact Polish space $\Sigma$ as above. Let $f$ be a function in the form domain $Q(L) \subset L^2(d\mu)$ of $L$. Then* *for any $T > 0$ and any $\ell > 0$, $f(X(t))$ is almost surely continuous on $[0, T]$ and*

$$\mathbb{P}\left(\sup_{0\le t\le T} |f(X(t))| \ge \ell\right) \le \frac{3}{\ell}\left(\|f\|_{L^2}^2 + TQ_L(f)\right)^{1/2}. \tag{13.79}$$

**Proof.** Assume first that $f \in C_b$. Discretize time, $T = Nh$. We show that

$$\mathbb{P}\left(\sup_{0\le n\le N} |f(X(nh))| \ge \ell\right) \le \frac{e}{\ell}\left(\|f\|_{L^2}^2 + TQ_L(f)\right)^{1/2} \tag{13.80}$$

uniformly in $h$. Let $G$ be the closed set $G = \{x : |f(x)| \ge \ell\}$ and define the stopping time $\tau = \inf\{n : X(nh) \in G\}$. Clearly,

$$\mathbb{P}\left(\sup_{0\le n\le N} |f(X(nh))| \ge \ell\right) = \mathbb{P}(\tau \le N). \tag{13.81}$$

We have the Chebyshev estimate

$$\mathbb{P}(\tau \le N) \le e^{\lambda N} \int \mathbb{E}_x[e^{-\lambda\tau}]d\mu(x) \tag{13.82}$$

for any $\lambda > 0$. Let $\pi_h = p(h, xdx)$ be the transition probability for the discrete-time chain,

$$(S(h)f)(x) = (e^{hL}f)(x) = (\pi_h f)(x) = \int_X f(y)p(h, x, dy). \tag{13.83}$$

Then as $\tau$ decreases, for $x \in G^c$, by one unit after a time step $t = 0 \to t = h$, $\phi(x) \equiv \mathbb{E}_x[e^{-\lambda\tau}]$ is the solution of

$$\phi(x) = e^{-\lambda}(\pi_h\phi)(x) \quad \text{for } x \in G^c \tag{13.84}$$

with

$$\phi(x) = 1 \quad \text{for } x \in G. \tag{13.85}$$

As $S(h)$ is self-adjoint and $\|e^{-\lambda}\pi_h\|_{L^2(d\mu)} = \|e^{-\lambda}S(h)\|_{L^2(d\mu)} \le e^{-\lambda} < 1$, standard arguments show that $\phi$ is the unique solution of (13.84), (13.85) and is the (unique) minimizer of $((1 - e^{-\lambda}\pi_h)\psi, \psi)_{L^2(d\mu)}$ subject to $\psi = 1$ on $G$. Since $G = \{x : |f(x)| \ge \ell\}$, $f_\ell = \frac{1}{\ell}(|f| \wedge \ell)$ is an admissible choice for $\psi$. Therefore

$$((1 - e^{-\lambda}\pi_h)\phi, \phi) \le ((1 - e^{-\lambda}\pi_h)f_\ell, f_\ell) \le \frac{1}{\ell^2}((1 - e^{-\lambda}\pi_h)f, f). \tag{13.86}$$

The second inequality follows from the fact that $S(t) = e^{tL}$ is a contraction in $L^p(d\mu)$ for all $1 \le p \le \infty$ and for all $t \ge 0$ (see, for example, (the proof of) the second Beurling-Deny criterion in [**RS78**]). Rewrite (13.86) in the form

$$\begin{aligned}&((1-e^{-\lambda})\phi,\phi)+e^{-\lambda}((1-\pi_h)\phi,\phi)\\&\le\frac{1}{\ell^2}\left[((1-e^{-\lambda})f,f)+e^{-\lambda}((1-\pi_h)f,f)\right].\end{aligned}\tag{13.87}$$

Divide by $1-e^{-\lambda}$ and set $\lambda=\frac{h}{T}$ so that $\lambda N=1$. As $(\phi,(1-\pi_h)\phi)\ge 0$, we obtain

$$\begin{aligned}\|\phi\|_{L^2}^2&\le\frac{1}{\ell^2}\left[\|f\|_{L^2}^2+\frac{1}{e^{h/T}-1}((1-\pi_h)f,f)\right]\\&\le\frac{1}{\ell^2}\left[\|f\|_{L^2}^2+\frac{T}{h}((1-\pi_h)f,f)\right].\end{aligned}\tag{13.88}$$

We conclude that for all bounded, continuous functions $f$ and for all $h>0$,

$$\begin{aligned}\mathbb{P}\left(\sup_{0\le n\le N}|f(X(nh))|\ge\ell\right)&=\mathbb{P}(\tau\le N)\le e\|\phi\|_{L^1}\le e\|\phi\|_{L^2}^{1/2}\\&\le\frac{e}{\ell}\left[\|f\|_{L^2}^2+\frac{T}{h}((1-\pi_h)f,f)\right]^{1/2}.\end{aligned}\tag{13.89}$$

Now if $f$ lies in the form domain $Q(L)$ of $L$, then by the spectral resolution of $L$, $L=-\int_0^\infty\lambda dE_\lambda$, we have

$$\begin{aligned}0&\le\frac{1}{h}((1-\pi_h)f,f)\\&=\int_0^\infty\frac{1-e^{-h\mu}}{h}d(f,E_\mu f)\le\int_0^\infty\mu d(f,E_\mu f)=Q_L(f)<\infty.\end{aligned}\tag{13.90}$$

It follows that if $f\in C_b\cap Q(L)$, then

$$\mathbb{P}\left(\sup_{0\le n\le N}|f(X(nh))|\ge\ell\right)\le\frac{e}{\ell}\left[\|f\|_{L^2}^2+TQ_L(f)\right]^{1/2}\tag{13.91}$$

uniformly for $h>0$. Choosing $h$ dyadically $h=2^{-d}$, it follows that

$$\mathbb{P}\left(\sup_{0\le t\le T}|f(X(t))|\ge\ell\right)\le\frac{e}{\ell}\left[\|f\|_{L^2}^2+TQ_L(f)\right]^{1/2}\tag{13.92}$$

for $f\in C_b\cap Q(L)$.

Now $C_b\cap D(L)$ is dense in $Q(L)$. Indeed, if $f$ lies in the domain of $L$, then choose $g_n\in C_b$ such that $g_n\to g\equiv(-L+1)f$ in $L^2(d\mu)$. As $-L\ge 0$, $(-L+1)^{-1}$ exists in $L^2(d\mu)$ and

$$f_n=(-L+1)^{-1}g_n=\int_0^\infty e^{(L-1)s}g_n ds=\lim_{\epsilon\downarrow 0}\int_\epsilon^{1/\epsilon}e^{(L-1)s}g_n ds.\tag{13.93}$$

Approximating $f_{n,\epsilon} \equiv \int_\epsilon^{1/\epsilon} e^{(L-1)s} g_n ds$ by Riemann sums, and using the fact that $S(t) = e^{tL}$ is a strongly continuous semigroup in $C_b$, we conclude that $f_{n,\epsilon}(x)$ is continuous (and bounded). Moreover, $f_{n,\epsilon}(x) = \frac{1}{-L+1}[(e^{(L-t)\epsilon} - e^{(L-t)/\epsilon})g]$ clearly also lies in the domain of $L$ and as $\epsilon \downarrow 0$ we have $(-L+1)f_{n,\epsilon} \to (-L+1)f_n$. As $(-L+1)f_n \to (-L+1)f = g$, it follows that $C_b \cap D(L)$ is dense in the domain of $L$, and hence in $Q(L)$, by the spectral theorem. Given $f \in Q(L)$, there exist $f_j \in (C_b \cap D(L)) \subset (C_b \cap Q(L))$ such that $f_j \to f$ in $Q(L)$ and such that

$$[\|f_{j+1} - f_j\|_{L^2}^2 + TQ_L(f_{j+1} - f_j)]^{1/2} \le \frac{1}{j^4}, \quad j \ge 1. \tag{13.94}$$

It follows then from (13.91) that

$$\sum_{j=1}^{\infty} \mathbb{P}\left( \sup_{0 \le t \le T} |f_{j+1}(X(t)) - f_j(X(t))| \ge \frac{1}{j^2} \right) < \infty, \tag{13.95}$$

and hence by the usual Borel-Cantelli argument, for almost all paths $X(t)$, $f(X(t)) = \lim_{j\to\infty} f_j(X(t))$ converges uniformly for $0 \le t \le T$ and hence $f(X(t))$ is continuous. Finally, by Fatou's lemma, (13.92) remains true for $f \in Q(L)$. □

Now let $H(t)$, $t \in \mathbb{R}$, denote the matrix Ornstein-Uhlenbeck process. The generator for the process has the form

$$L = \sum_{i=1}^{n} \frac{\partial^2}{\partial H_{ii}^2} + \frac{1}{2} \sum_{1\le i<j\le n} \frac{\partial^2}{\partial H_{ij}^2} - \sum_{1\le i\le j\le n} H_{ij} \frac{\partial}{\partial H_{ij}}. \tag{13.96}$$

Direct calculation shows that

$$e^{-\frac{1}{4}\operatorname{tr} H^2} L e^{\frac{1}{4}\operatorname{tr} H^2}$$

$$= \sum_{i=1}^{n} \left( \frac{\partial^2}{\partial H_{ii}^2} + \frac{1}{2} - \frac{1}{4} H_{ii}^2 \right) + \sum_{1\le i<j\le n} \left( \frac{1}{2} \frac{\partial^2}{\partial H_{ij}^2} + \frac{1}{2} - \frac{1}{2} H_{ij}^2 \right) \equiv \mathcal{L}.$$

Now, as $\mathcal{L}$ is a direct sum of harmonic oscillators, $\mathcal{L}$ has a complete orthonormal set of eigenvectors in $L^2(\mathbb{R}^{n(n+1)/2})$ of the form

$$\phi_{\vec{k}} = \prod_{j=1}^{n} \mathcal{H}_{k_j}(H_{jj}) \prod_{1\le i<j\le n} \mathcal{H}_{k_{ij}}(H_{ii}) e^{-\frac{1}{4}\operatorname{tr} H^2},$$

where the $\mathcal{H}_k(x)$'s are Hermite polynomials. It follows that we can find finite linear combinations $\psi$ of $\phi_{\vec{k}}$'s such that $(-\mathcal{L}+1)\psi$ are dense in $L^2(\mathbb{R}^{n(n+1)/2})$. Let $\chi \in C_0^\infty(\mathbb{R}^{n(n+1)/2})$ with $\chi(x) = 1$ for $\|x\| \le 1$ and $\chi(x) = 0$ for $\|x\| \ge 2$. For $q = 1, 2, \ldots$, set $\chi_q(x) = \chi(x/q)$. A simple calculation now shows that for functions $\psi$ as above, the functions $\tilde{\psi}(H) = \chi_q(H)\psi(H)e^{\frac{1}{4}\operatorname{tr} H^2} \in C_0^\infty(\mathbb{R}^{n(n+1)/2})$ have the property that $\{(-\mathcal{L}+1)\tilde{\psi}\}$ are dense in $L^2(d\mu) = L^2(e^{-\frac{1}{2}\operatorname{tr} H^2} dH)$. By general theory, it follows that $L$ is essentially self-adjoint on $C_0^\infty(\mathbb{R}^{n(n+1)/2})$. It follows then, by uniqueness, that $L$ is the Friedrichs extension of $L \restriction C_0^\infty(\mathbb{R}^{n(n+1)/2})$. It follows, in particular, that $Q(L)$, the form domain of $L$, is precisely the set of functions $f$ in $L^2(d\mu)$ with

$$Df = \left(\frac{\partial f}{\partial H_{11}}, \ldots, \frac{\partial f}{\partial H_{nn}}, \frac{1}{\sqrt{2}}\frac{\partial f}{\partial H_{12}}, \ldots, \frac{1}{\sqrt{2}}\frac{\partial f}{\partial H_{n-1,n}}\right)^T \in L^2(d\mu)$$

and

$$\begin{aligned} Q_L(f) &= \int |Df(H)|^2 d\mu(H) \\ &= \int \left(\sum_{i=1}^n \left(\frac{\partial f}{\partial H_{ii}}\right)^2 + \frac{1}{2}\sum_{1\le i<j\le n}\left(\frac{\partial f}{\partial H_{ij}}\right)^2\right) d\mu(H). \end{aligned} \tag{13.97}$$

It is an exercise left to the reader that

$$f(H) \equiv \sum_{1\le i<j\le n} \log\left(-\log\left(\min\{\lambda_i(H)-\lambda_j(H), 1/e\}\right)\right) \in Q(L). \tag{13.98}$$

Clearly, $f(H) = \infty$ if and only if two eigenvalues of $H$ coincide. But by Lemma 13.19, for any $T > 0$, $f(H(t))$ is continuous for $0 \le t \le T$ almost surely, and hence, almost surely, the eigenvalues of $H(t)$ do not coincide for $t \ge 0$.

More precisely, we have shown that (note (13.69))

$$\int \mathbb{P}(D(H(t)) > 0, t \ge 0 | H(0) = H) e^{-\frac{1}{2}\operatorname{tr} H^2}\frac{dH}{Z} = 1 \tag{13.99}$$

and hence for Lebesgue almost all $H$.

$$\mathbb{P}(D(H(t)) > 0, t \ge 0 | H(0) = H) = 1 \tag{13.100}$$

or by translation invariance, for any $s > 0$

$$\mathbb{P}(D(H(t)) > 0, t \ge s | H(0) = H) = 1 \tag{13.101}$$

for Lebesgue almost all $H$.

**Exercise 13.20.** Show that the transition probability for the Ornstein-Uhlenbeck process is given by

$$p^{\mathrm{OU}}(0, t, H(0), H) = c_n^{\mathrm{OU}} e^{-\frac{\operatorname{tr}(H - e^{-t}H(0))^2}{2(1-e^{-2t})}}, \qquad t > 0, \tag{13.102}$$

where $c_n^{\mathrm{OU}} = (2\pi(1-e^{-2t}))^{-n/2}(\pi(1-e^{-2t}))^{-n(n-1)/4}$.

Now as in (13.65), for any fixed $H(0) \in \mathcal{S}(n)$ and $t_q \downarrow 0$,

$$\begin{aligned} \mathbb{P}(D(H(t)) > 0, t > 0) &= \lim_{q\to\infty} \mathbb{P}(D(H(t)) > 0, t \ge t_q) \\ &= \lim_{q\to\infty} \int p^{\mathrm{OU}}(0, t_q, H(0), H)\, \mathbb{P}(D(H(t)) > 0, t \ge t_q | H) dH = 1 \end{aligned} \tag{13.103}$$

by (13.101) and (13.102). This proves Theorem 13.3 for general $H(0) \in \mathcal{S}(n)$.

## 13.6. Nonintersecting Processes

For *Hermitian matrices*, Brownian Dyson process is given by

$$d\tilde{\lambda}_i(t) = \sum_{l \neq i} \frac{1}{\tilde{\lambda}_i(t) - \tilde{\lambda}_l(t)} dt + d\tilde{W}_i(t), \quad i = 1, \ldots, n, \tag{13.104}$$

and the Dyson process is given by

$$d\lambda_i(t) = \left( -\lambda_i(t) + \sum_{j \neq i} \frac{1}{\lambda_i(t) - \lambda_j(t)} \right) dt + dW_i(t), \quad i = 1, \ldots, n. \tag{13.105}$$

These processes are the Hermitian analogues of Brownian Dyson process (13.5) and Dyson process (13.3) for symmetric matrices, and we call them *Hermitian Brownian Dyson process* and *Hermitian Dyson process*, respectively.

**Remark 13.21.** Note that the equations (13.104) and (13.105) are the same as the equations described in Remark 13.13 above, after rescaling $\lambda_i \to \sqrt{2}\lambda_i$. Equations (13.104) and (13.105) are slightly more standard when dealing with conditioned diffusions.

We show in this subsection that the above two processes can be considered as nonintersecting diffusions.[3] More precisely, for Hermitian Brownian Dyson process, we consider the transition density $q_{t,T}(x, y)$ for Brownian motion conditioned not to intersect for $0 \leq t \leq T$. Taking the limit as $T \to \infty$, we obtain a semigroup $\Gamma_t^{\mathrm{Br}}$ with kernel $p_t^{\mathrm{Br}}(x, y) \equiv \lim_{T\to\infty} q_{t,T}(x, t)$. We then show that $p_t^{\mathrm{Br}}(x, y)$ is the transition density for Hermitian Brownian Dyson process. The connection between the Hermitian Dyson process and nonintersecting Ornstein-Uhlenbeck processes is obtained in a similar way. We begin with the case of Hermitian Brownian Dyson process, which is due to D. Grabiner [**Gra99**] (see also [**Bia09**]).

**13.6.1. Nonintersecting Brownian Motion and Hermitian Brownian Dyson process.** Let $p(s, t, x, y)$ denote the transition probability density from $(s, x)$ to $(t, y)$, $0 \leq s < t$, for standard one-dimensional Brownian motion conditioned to be at $x$ at time $s$, i.e., $p(s, t, x, y) = \frac{1}{\sqrt{2\pi(t-s)}} e^{-\frac{(x-y)^2}{2(t-s)}}$. Let

$$W^{(n)} \equiv \{x \in \mathbb{R}^n : x_1 > \cdots > x_n\}. \tag{13.106}$$

[3]The authors are indebted to Ofer Zeitouni for help and many suggestions regarding the proofs in this section.

For $x, y \in W^{(n)}$ define

$$p^{\mathrm{Br}}(s,t,x,y) = \det\left(p(s,t,x_i,y_j)\right)_{i,j=1}^{n} \frac{h(y)}{h(x)}, \tag{13.107}$$

where

$$h(x) = \prod_{i<j}(x_i - x_j)$$

is the Vandermonde determinant. For $t > 0$, set

$$p_t^{\mathrm{Br}}(x,y) = p^{\mathrm{Br}}(0,t,x,y) = \det\left[\frac{1}{\sqrt{2\pi t}} e^{-\frac{(x_i-y_j)^2}{2t}}\right]_{i,j=1}^{n} \frac{h(y)}{h(x)} \tag{13.108}$$

for $x, y \in W^{(n)}$.

We prove the following three results in this section.

(1) For $t > 0$, $p_t^{\mathrm{Br}}(x,y)$ is the transition density function from $x \in W^{(n)}$ to $y \in W^{(n)}$ of $n$ independent Brownian motions conditioned not to intersect for all time; see Proposition 13.22.

(2) The transition density $p_t^{\mathrm{Br}}(x,y)$ generates a strongly continuous semigroup $\Gamma_t^{\mathrm{Br}}$; see Proposition 13.23.

(3) $\Gamma_t^{\mathrm{Br}}$ is also the semigroup for Hermitian Brownian Dyson process; see Theorem 13.27.

Thus Hermitian Brownian Dyson process can be interpreted as independent nonintersecting Brownian motions.

By Theorem 3.37 (Karlin-McGregor formula),

$$P(s,t,x,y) \equiv \det(p(s,y,x_i,y_j))_{i,j=1}^{n}, \qquad 0 \le s < t,$$

is the probability density that $n$ independent Brownian motions $N_u$ conditioned on $N_s = x = (x_i)_{i=1}^n \in W^{(n)}$ satisfy $N_t = y = (y_i)_{i=1}^n \in W^{(n)}$ and do not intersect for $s \le u \le t$. By the Markov property, $P(s,t,x,y)$ satisfies the equation

$$P(s,t,x,y) = \int_{W^{(n)}} P(s,r,x,z)P(r,t,z,y)\, d^n z$$

for $x, y \in W^{(n)}$ and $0 \le s < r < t < \infty$. Hence $p^{\mathrm{Br}}(s,t,x,y)$ defined in (13.107) satisfies the Chapman-Kolmogorov equation for $x, y \in W^{(n)}$

$$p^{\mathrm{Br}}(s,t,x,y) = \int_{W^{(n)}} p^{\mathrm{Br}}(s,r,x,z)p^{\mathrm{Br}}(r,t,z,y)\, d^n z \tag{13.109}$$

for $0 \le s < r < t < \infty$.

We now show that $p_t^{\mathrm{Br}}(x,y)$ is the transition density of Brownian motions conditioned not to intersect in the following sense. Given $x, y \in W^{(n)}$ and

$0 < t < T < \infty$, set

$$
\begin{aligned}
q_{t,T}(x, y) &= \mathbb{P}^x \left\{ N_t = y \middle| N_s \in W^{(n)}, \forall s \in [0, T] \right\} \\
&= \frac{\mathbb{P}^x \left\{ N_t = y; N_s \in W^{(n)}, \forall s \in [0, T] \right\}}{\mathbb{P}^x \left\{ N_s \in W^{(n)}, \forall s \in [0, T] \right\}},
\end{aligned} \tag{13.110}
$$

where we abused the notation $\mathbb{P}^x$ to denote both probability density and probability for $n$ independent Brownian motions $N_s$ with $N_0 = x$. Hence $q_{t,T}(x, y)$ is the probability density that $n$ independent Brownian motions starting at $x$ at time 0 reach $y$ at time $t$, conditioned on the event that the Brownian motions do not intersect for all times in $[0, T]$. The limit of $q_{t,T}(x, y)$ as $T \to \infty$ can be interpreted as the same density conditioned on the event that the Brownian motions do not intersect for all time $[0, \infty)$. We show that $p_t^{\mathrm{Br}}(x, y)$ is precisely this limit.

**Proposition 13.22.** *We have*

$$
p_t^{\mathrm{Br}}(x, y) = \lim_{T \to \infty} q_{t,T}(x, y), \qquad x, y \in W^{(n)}. \tag{13.111}
$$

**Proof.** By the Markov property,

$$
\begin{aligned}
&q_{t,T}(x, y) \\
&\quad = \frac{\mathbb{P}^x \left\{ N_t = y; N_s \in W^{(n)}, \forall s \in [0, t] \right\} \mathbb{P}^y \left\{ N_s \in W^{(n)}, \forall s \in [0, T - t] \right\}}{\mathbb{P}^x \left\{ N_s \in W^{(n)}, \forall s \in [0, T] \right\}}.
\end{aligned}
$$

Each of these probabilities can be evaluated by the Karlin-McGregor formula (see Theorem 3.37), and we find

$$
q_{t,T}(x, y) = \frac{\det \left( p_t(x_i, y_j) \right) \int_{W^{(n)}} \det \left( p_{T-t}(y_i, z_j) \right) d^n z}{\int_{W^{(n)}} \det \left( p_T(x_i, z_j) \right) d^n z}, \tag{13.112}
$$

where $p_t(a, b) \equiv p(0, t, a, b) = \frac{1}{\sqrt{2\pi t}} e^{-\frac{1}{2t}(b-a)^2}$.

Note that

$$
\det(p_T(x_i, z_j)) = \frac{1}{(2\pi T)^{\frac{n}{2}}} e^{-\frac{1}{2T} \sum_{i=1}^n (x_i^2 + z_i^2)} \det \left( e^{\frac{x_i z_j}{T}} \right). \tag{13.113}
$$

Andréief's formula (Lemma 6.2) implies that

$$
\begin{aligned}
\det \left( e^{\frac{x_i z_j}{T}} \right) &= \det \left( \sum_{\ell=0}^{\infty} \frac{x_i^\ell z_j^\ell}{T^\ell \ell!} \right) \\
&= \frac{1}{n!} \sum_{\ell_1, \ldots, \ell_n = 0}^{\infty} \det \left( x_i^{\ell_j} \right) \det \left( z_i^{\ell_j} \right) \prod_{j=1}^n \frac{1}{T^{\ell_j} \ell_j!} \\
&= \sum_{0 \le \ell_1 < \cdots < \ell_n} \det \left( x_i^{\ell_j} \right) \det \left( z_i^{\ell_j} \right) \prod_{j=1}^n \frac{1}{T^{\ell_j} \ell_j!}.
\end{aligned} \tag{13.114}
$$

Singling out the term $(\ell_1, \ell_2, \dots, \ell_n) = (0, 1, \dots, n-1)$, which satisfies $\sum_{j=1}^n \ell_j = n(n-1)/2$, in the sum, we obtain for $x \in W^{(n)}$

$$
\begin{aligned}
&\int_{W^{(n)}} \det(p_T(x_i, z_j))\, d^n z \\
&\quad = \frac{e^{-\frac{1}{2T}\sum_{i=1}^n x_i^2}}{(2\pi T)^{n/2}} h(x) \int_{W^{(n)}} h(z) \frac{e^{-\frac{1}{2T}\sum_{i=1}^n z_i^2}}{T^{n(n-1)/2}\prod_{j=0}^{n-1} j!}\, d^n z + E,
\end{aligned}
\tag{13.115}
$$

where

$$
E = \frac{e^{-\frac{1}{2T}\sum_{i=1}^n x_i^2}}{(2\pi T)^{n/2}} \sum_{\substack{0 \le \ell_1 < \cdots < \ell_n \\ \sum_{j=1}^n \ell_j \ge \frac{n(n-1)}{2}+1}} \det(x_i^{\ell_j}) \int_{W^{(n)}} \det(z_i^{\ell_j}) \frac{e^{-\frac{1}{2T}\sum_{i=1}^n z_i^2}}{\prod_{j=1}^n T^{\ell_j}\ell_j!}\, d^n z.
$$

By Hadamard's inequality

$$
\left|\det(x_i^{\ell_j})\right| \le \prod_{j=1}^n \left(\sum_{i=1}^n x_i^{2\ell_j}\right)^{1/2} \le \prod_{j=1}^n \|x\|^{\ell_j},
$$

where $\|x\| = (\sum_{i=1}^n x_i^2)^{1/2}$. Hence

$$
\begin{aligned}
|E| &\le \frac{e^{-\frac{1}{2T}\|x\|^2}}{(2\pi T)^{n/2}} \sum_{\substack{0 \le \ell_1 < \cdots < \ell_n \\ \sum_{j=1}^n \ell_j \ge \frac{n(n-1)}{2}+1}} \int_{W^{(n)}} \prod_{j=1}^n \left[\frac{1}{\ell_j!}\left(\frac{\|x\|\|z\|}{T}\right)^{\ell_j}\right] e^{-\frac{1}{2T}\|z\|^2}\, d^n z \\
&= \frac{e^{-\frac{1}{2T}\|x\|^2}}{(2\pi)^{n/2}} \sum_{\substack{0 \le \ell_1 < \cdots < \ell_n \\ \sum_{j=1}^n \ell_j \ge \frac{n(n-1)}{2}+1}} \int_{W^{(n)}} \prod_{j=1}^n \left[\frac{1}{\ell_j!}\left(\frac{\|x\|\|u\|}{T^{1/2}}\right)^{\ell_j}\right] e^{-\frac{1}{2}\|u\|^2}\, d^n u.
\end{aligned}
$$

For any $0 < \gamma < 1$ and $T \ge 1$,

$$
\begin{aligned}
&\sum_{\substack{0 \le \ell_1 < \cdots < \ell_n \\ \sum_{j=1}^n \ell_j \ge \frac{n(n-1)}{2}+1}} \prod_{j=1}^n \left[\frac{1}{\ell_j!}\left(\frac{\|x\|\|u\|}{T^{1/2}}\right)^{\ell_j}\right] \\
&\quad = \sum_{\substack{0 \le \ell_1 < \cdots < \ell_n \\ \sum_{j=1}^n \ell_j \ge \frac{n(n-1)}{2}+1}} \frac{1}{T^{\frac{1-\gamma}{2}\sum_{j=1}^n \ell_j}} \prod_{j=1}^n \left[\frac{1}{\ell_j!}\left(\frac{\|x\|\|u\|}{T^{\gamma/2}}\right)^{\ell_j}\right] \\
&\quad \le \frac{e^{n\frac{\|x\|\|u\|}{T^{\gamma/2}}}}{T^{\frac{1-\gamma}{2}(\frac{n(n-1)}{2}+1)}} \le \frac{e^{\frac{n^2}{T^\gamma}\|x\|^2 + \frac{1}{4}\|u\|^2}}{T^{(1-\gamma)(\frac{n(n-1)}{4}+\frac{1}{2})}}.
\end{aligned}
$$

Therefore, for any $0 < \gamma < 1$ and $T \ge 1$,

$$
|E| \le \frac{e^{-\frac{1}{2}\|x\|^2(\frac{1}{T} - \frac{2n^2}{T^\gamma})}}{(2\pi)^{n/2} T^{(1-\gamma)(\frac{n(n-1)}{4}+\frac{1}{2})}} \int_{\mathbb{R}^n} e^{-\frac{1}{4}\|u\|^2}\, d^n u.
$$

Thus

(13.116)
$$\int_{W^{(n)}} \det(p_T(x_i, z_j))\, d^n z = \frac{e^{-\frac{1}{2T}\sum_{i=1}^n x_i^2}}{(2\pi)^{n/2} T^{\frac{n(n-1)}{4}} \prod_{j=0}^{n-1} j!} h(x) \int_{W^{(n)}} h(u) e^{-\frac{1}{2}\sum_{i=1}^n u_i^2}\, d^n u + E,$$

where

(13.117)
$$|E| \leq c_n \frac{e^{-\frac{1}{2T}\|x\|^2} e^{\frac{n^2}{T^\gamma}\|x\|^2}}{T^{(1-\gamma)(\frac{n(n-1)}{4}+\frac{1}{2})}}$$

for some constant $c_n > 0$. Choose $0 < \gamma < 1$ such that

(13.118)
$$\frac{(1-\gamma)}{2}\left(\frac{n(n-1)}{2}+1\right) \geq \frac{n(n-1)}{4} + \frac{1}{4}.$$

Thus for $T \geq 1$,

(13.119)
$$|E| \leq c_n \frac{e^{-\frac{1}{2T}\|x\|^2}}{T^{\frac{n(n-1)}{4}}} \frac{e^{\frac{n^2}{T^\gamma}\|x\|^2}}{T^{\frac{1}{4}}},$$

and hence,

(13.120)
$$\int_{W^{(n)}} \det(p_T(x_i, z_j))\, d^n z = \frac{e^{-\frac{1}{2T}\sum_{i=1}^n x_i^2}}{(2\pi)^{n/2} T^{\frac{n(n-1)}{4}} \prod_{j=0}^{n-1} j!} h(x) \int_{W^{(n)}} h(u) e^{-\frac{1}{2}\sum_{i=1}^n u_i^2}\, d^n u \left(1 + O(T^{-1/4})\right).$$

It now follows from (13.112) that for $x, y \in W^{(n)}$ and $t > 0$, as $T \to \infty$,

(13.121)
$$q_{t,T}(x,y) \to \det(p_t(x_i, y_j))_{i,j=1}^n \frac{h(y)}{h(x)} = p_t^{\mathrm{Br}}(x,y),$$

as desired. □

Before we show that $p_t^{\mathrm{Br}}(x,y)$ generates a continuous semigroup, we first show that $p_t^{\mathrm{Br}}(x,\cdot)$ is a probability density for fixed $x \in W^{(n)}$ and $t > 0$. This follows from

(13.122)
$$\int_{W^{(n)}} q_{t,T}(x,y)\, d^n y = 1, \qquad x \in W^{(n)},$$

by using the dominated convergence theorem as $T \to \infty$, if we show that for fixed $t > 0$ and $x \in W^{(n)}$, $|q_{t,T}(x,y)|$ is bounded by a function that is integrable in $y \in W^{(n)}$ uniformly in $T$. Note formula (13.112) for $q_{t,T}(x,y)$. Estimating the terms in the sum (13.114) as in the proof of Proposition 13.22, we obtain from (13.113),

$$
(13.123)\quad \begin{aligned} |\det(p_t(x,y))| &\le \frac{1}{(2\pi t)^{n/2}} e^{-\frac{1}{2t}(\|x\|^2+\|y\|^2)} e^{\frac{n}{t}\|x\|\|y\|} \\ &\le \frac{1}{(2\pi t)^{n/2}} e^{-\frac{1}{2t}\|x\|^2+\frac{4n^2}{t}\|x\|^2} e^{-\frac{1}{4t}\|y\|^2} \end{aligned}
$$

for fixed $t>0$ and $x,y\in W^{(n)}$. It also follows then from (13.116), (13.119) that

(13.124)

$$
\int_{W^{(n)}} \det\left(p_{T-t}(y_i,z_j)\right) d^n z \le \frac{e^{-\frac{1}{2(T-t)}\|y\|^2}}{(T-t)^{\frac{n(n-1)}{4}}} \left( \tilde{c}_n h(y) + c_n \frac{e^{\frac{n^2\|y\|^2}{(T-t)^\gamma}}}{(T-t)^\gamma} \right)
$$

for positive constants $\tilde{c}_n$ and $c_n$, where $\gamma\in(0,1)$ is a constant that satisfies (13.118). Then from (13.112), (13.120), and (13.123), it follows that for fixed $t>0$ and $x\in W^{(n)}$ there is a positive constant $c_n(x,t)$ such that

$$
(13.125)\quad |q_{t,T}(x,y)| \le c_n(x,t) e^{-\frac{1}{8t}\|y\|^2}
$$

for all $y\in W^{(n)}$, as $T\to\infty$. We conclude from (13.122) that for $x\in W^{(n)}$ and $t>0$

$$
(13.126)\quad \int_{W^{(n)}} p_t^{\mathrm{Br}}(x,y)\, d^n y = 1
$$

so that $p_t^{\mathrm{Br}}(x,\cdot)$ is indeed a probability density.

From the definition (13.108), $p_t^{\mathrm{Br}}(x,y)$ is clearly continuous for $x\in W^{(n)}$ and $y\in \overline{W^{(n)}}$, the closure of $W^{(n)}$. We now show that $p_t^{\mathrm{Br}}(x,y)$ is in fact continuous, and bounded, for $y$ and $x$, up to the boundary, for $t>0$. From (13.113) and (13.114),

$$
(13.127)\quad \begin{aligned} &p_t^{\mathrm{Br}}(x,y) \\ &= \frac{e^{-\frac{1}{2t}(\|x\|^2+\|y\|^2)}}{(2\pi t)^{n/2}} h(y) \sum_{0\le \ell_1<\cdots<\ell_n} \frac{\det(x_i^{\ell_j})}{h(x)} \det(y_i^{\ell_j}) \prod_{j=1}^n \frac{1}{t^{\ell_j}\ell_j!}. \end{aligned}
$$

We observe that, up to a sign, the term $\frac{\det(x_i^{\ell_j})}{h(x)}$ is equal to a Schur function $s_\lambda(x)$ (see (10.81)) with $\lambda_j=\ell_{n-j+1}+j-n$, $1\le j\le n$, and hence is a polynomial in the $x_j$'s. Thus each of the summands in (13.127) extends continuously to the boundary in $\overline{W^{(n)}}\times\overline{W^{(n)}}$. Therefore, the continuity of $p_t^{\mathrm{Br}}(x,y)$ follows if we show that the series is majored by a summable series uniformly for $(x,y)$ in any compact subset of $\overline{W^{(n)}}\times\overline{W^{(n)}}$. By (10.118) and (10.121), we have for $\lambda_j = \ell_{n-j+1} + j - n$,

$$\left|\frac{\det(x_i^{\ell_j})}{h(x)}\right| = |s_\lambda(x)| \le s_\lambda(\|x\|, \dots, \|x\|)$$
$$\le \|x\|^{\sum_{j=1}^n \lambda_j} \prod_{1\le i<j\le n} \frac{\lambda_i - \lambda_j + j - i}{j-i}$$
$$= \|x\|^{\sum_{j=1}^n \ell_j - \frac{n(n-1)}{2}} \frac{\det(\ell_j^{i-1})_{i,j=1}^n}{\prod_{k=0}^{n-1} k!}.$$

But by Hadamard's inequality,

$$\left|\det(\ell_j^{i-1})\right| \le \prod_{j=1}^n (\sum_{m=0}^{n-1} \ell_j^{2m})^{1/2} \le \prod_{j=1}^n e^{n\ell_j} \tag{13.128}$$

and also $|\det(y_i^{\ell_j})| \le \|y\|^{\sum_{j=1}^n \ell_j}$. We thus find that the summand in (13.127) is bounded by

$$\frac{\|x\|^{-n(n-1)/2}}{\prod_{k=0}^{n-1} k!} \prod_{j=1}^n \frac{1}{\ell_j!} \left(\frac{\|x\|\|y\|e^n}{t}\right)^{\ell_j}$$

which is summable for all $x, y \in \mathbb{R}^n$ since the summation condition implies that $\ell_j \ge j-1$. Setting $m_j = \ell_j - j + 1$,

$$\sum_{0\le \ell_1 < \cdots < \ell_n} \frac{\|x\|^{-n(n-1)/2}}{\prod_{k=0}^{n-1} k!} \prod_{j=1}^n \frac{1}{\ell_j!} \left(\frac{\|x\|\|y\|e^n}{t}\right)^{\ell_j}$$
$$= \frac{\|y\|^{n(n-1)/2} e^{n^2(n-1)/2}}{t^{n(n-1)/2} \prod_{k=0}^{n-1} k!} \sum_{0\le m_1 \le \cdots \le m_n} \prod_{j=1}^n \frac{1}{(m_j + j - 1)!} \left(\frac{\|x\|\|y\|e^n}{t}\right)^{m_j}$$
$$\le \frac{\|y\|^{n(n-1)/2} e^{n^2(n-1)/2}}{t^{n(n-1)/2} \prod_{k=0}^{n-1} k!} e^{n\|x\|\|y\|e^n/t}.$$

Therefore we find that the series (13.127) converges uniformly in any compact subset of $\overline{W^{(n)}} \times \overline{W^{(n)}}$, and therefore $p_t^{\mathrm{Br}}(x,y)$ is continuous in $\overline{W^{(n)}} \times \overline{W^{(n)}}$. The above estimates imply that

$$p_t^{\mathrm{Br}}(x,y) \le \frac{e^{-\frac{1}{2t}(\|x\|^2 + \|y\|^2)}}{(2\pi t)^{n/2}} h(y) \frac{\|y\|^{n(n-1)/2} e^{n^2(n-1)/2}}{t^{n(n-1)/2} \prod_{k=0}^{n-1} k!} e^{n\|x\|\|y\|e^n/t}. \tag{13.129}$$

Therefore, $p_t^{\mathrm{Br}}(x,y)$ is bounded in $\overline{W^{(n)}} \times \overline{W^{(n)}}$ for each $t > 0$.

The above bound implies, in particular, that (13.126) remains true for all $x \in \overline{W^{(n)}}$ and any $t > 0$:

$$\int_{W^{(n)}} p_t^{\mathrm{Br}}(x,y) d^n y = 1, \qquad x \in \overline{W^{(n)}}, \quad t > 0. \tag{13.130}$$

We now show that $p_t^{\mathrm{Br}}(x,y)$ generates a continuous semigroup on

$$C_\infty \equiv \{f : f(x) \text{ is a continuous real-valued function on } \overline{W^{(n)}} \text{ such that } \lim_{x\to\infty} f(x) = 0\}$$

equipped with the supreme norm $\|f\|_\infty = \sup_{x\in\overline{W^{(n)}}} |f(x)|$. For $f \in C_\infty$, set

$$\Gamma_t^{\mathrm{Br}} f(x) = \int_{W^{(n)}} p_t^{\mathrm{Br}}(x,y) f(y) d^n y, \qquad t > 0, \tag{13.131}$$

for $x \in W^{(n)}$.

**Proposition 13.23.** *The operator $\Gamma_t^{\mathrm{Br}}$ generates a strongly continuous semigroup on $C_\infty$.*

**Proof.** For $f \in C_\infty$, by the bound (13.129), $\Gamma_t^{\mathrm{Br}} f(x)$ is a continuous function on $W^{(n)}$ which extends continuously to the boundary. In order to check that $\Gamma_t^{\mathrm{Br}} f(x) \to 0$ as $x \to \infty$, note that for any $\epsilon > 0$, there exists a constant $M$ such that $|f(y)| < \epsilon$ for $|y| \geq M$, $y \in W^{(n)}$. Hence by (13.130)

$$\Gamma_t^{\mathrm{Br}} f(x) = \int_{\substack{W^{(n)} \\ |y|<M}} p_t^{\mathrm{Br}}(x,y) f(y) d^n y + O(\epsilon). \tag{13.132}$$

On the other hand, again by the bound (13.129), the first term in (13.132) goes to zero as $|x| \to \infty$. We conclude that $\Gamma_t^{\mathrm{Br}} f(x) \in C_\infty$, and so for all $t > 0$,

$$\Gamma_t^{\mathrm{Br}} : C_\infty \to C_\infty. \tag{13.133}$$

In addition, using (13.130),

$$\|\Gamma_t^{\mathrm{Br}}\| = 1. \tag{13.134}$$

It also follows from (13.109) and the time invariance of $p^{\mathrm{Br}}$, $p^{\mathrm{Br}}(0,t,x,y) = p^{\mathrm{Br}}(s,s+t,x,y)$, that for $s,t > 0$,

$$\Gamma_t^{\mathrm{Br}} \circ \Gamma_s^{\mathrm{Br}} f = \Gamma_{t+s}^{\mathrm{Br}} f, \qquad f \in C_\infty. \tag{13.135}$$

By (13.135) it is now sufficient to show that for any $f \in C_\infty$,

$$\lim_{t\to 0} \|\Gamma_t^{\mathrm{Br}} f - f\|_\infty = 0. \tag{13.136}$$

The bounds that we have already derived on $p_t^{\mathrm{Br}}(x,y)$ are sufficient to control $\lim_{t\to\infty}(\Gamma_t^{\mathrm{Br}} f)(x)$ uniformly for points $x \in W^{(n)}$ at a fixed distance from the boundary $\partial W^{(n)}$. Extra care is needed to control the limit uniformly for points $x \to \partial W^{(n)}$.

**Lemma 13.24.** *Let $K, \epsilon > 0$ be given satisfying $\epsilon^2 < 4K$. For $x, y \in W^{(n)}$, $\|x\|, \|y\| \leq K$, $\|x - y\| \geq \epsilon\sqrt{3(n-1)}$,*

$$p_t^{\mathrm{Br}}(x,y) \leq \frac{C_n}{t^{m_n/2}} e^{-\frac{\epsilon^2}{2t}} \tag{13.137}$$

*for some integer $m_n \geq 1$ and some constant $C_n = C_n(\epsilon, K)$.*

**Proof.** Set $\delta \equiv \frac{\epsilon^2}{4K} < 1$. We proceed by induction on $n$. Let $n = 2$. Suppose $0 < x_1 - x_2 < \delta$. Then

$$\begin{aligned} p_t^{\mathrm{Br}}(x,y) &= \frac{1}{2\pi t} \det \begin{bmatrix} e^{-\frac{(x_1-y_1)^2}{2t}} & e^{-\frac{(x_1-y_2)^2}{2t}} \\ e^{-\frac{(x_2-y_1)^2}{2t}} & e^{-\frac{(x_2-y_2)^2}{2t}} \end{bmatrix} \frac{y_1 - y_2}{x_1 - x_2} \\ &= \frac{y_1 - y_2}{2\pi t} \det \begin{bmatrix} \frac{e^{-\frac{(x_1-y_1)^2}{2t}} - e^{-\frac{(x_2-y_1)^2}{2t}}}{x_1 - x_2} & \frac{e^{-\frac{(x_1-y_2)^2}{2t}} - e^{-\frac{(x_2-y_2)^2}{2t}}}{x_1 - x_2} \\ e^{-\frac{(x_2-y_1)^2}{2t}} & e^{-\frac{(x_2-y_2)^2}{2t}} \end{bmatrix}. \end{aligned}$$

The $(1, j)$ entry of the last determinant is given by, for $j = 1, 2$,

$$\frac{1}{x_1 - x_2} \int_0^1 \left( \frac{d}{ds} e^{-\frac{(x(s)-y_j)^2}{2t}} \right) ds = -\frac{1}{t} \int_0^1 (x(s) - y_j) e^{-\frac{(x(s)-y_j)^2}{2t}} ds,$$

where $x(s) = x_2 + s(x_1 - x_2)$. Hence

$$\begin{aligned} p_t^{\mathrm{Br}}(x,y) &= -\frac{y_1 - y_2}{2\pi t^2} \\ &\quad \times \int_0^1 \det \begin{bmatrix} (x(s) - y_1) e^{-\frac{(x(s)-y_1)^2}{2t}} & (x(s) - y_2) e^{-\frac{(x(s)-y_2)^2}{2t}} \\ e^{-\frac{(x_2-y_1)^2}{2t}} & e^{-\frac{(x_2-y_2)^2}{2t}} \end{bmatrix} ds. \end{aligned}$$

We expand the determinant as a sum of two terms and estimate the absolute value of each term. For the exponent of the term that comes from the diagonal entries, note that

$$\begin{aligned} &(x(s) - y_1)^2 + (x_2 - y_2)^2 \\ &= (x_1 - y_1)^2 + (x(s) - x_1)(x(s) + x_1 - 2y_1) + (x_2 - y_2)^2 \\ &= \|x - y\|^2 + (x(s) - x_1)(x(s) + x_1 - 2y_1) \geq 3\epsilon^2 - \epsilon^2 = 2\epsilon^2 > \epsilon^2 \end{aligned}$$

since

$$|(x(s) - x_1)(x(s) + x_1 - 2y_1)| \leq |x_2 - x_1|(4K) \leq \delta(4K) = \epsilon^2.$$

Similarly, for the exponent of the term that comes from the off-diagonal entries, we use the simple and general inequality that if $a > a'$ and $b > b'$, then

$$(a - b')^2 + (a' - b)^2 \geq (a - b)^2 + (a' - b')^2, \tag{13.138}$$

and we find that

$$(x(s) - y_2)^2 + (x_2 - y_1)^2 \geq (x(s) - y_1)^2 + (x_2 - y_2)^2 > \epsilon^2.$$

Hence we find that

$$p_t^{\mathrm{Br}}(x, y) \leq \frac{2K^2}{\pi t^2} e^{-\frac{\epsilon^2}{2t}} \tag{13.139}$$

for $0 < x_1 - x_2 < \delta = \frac{\epsilon^2}{4K}$. On the other hand, if $x_1 - x_2 \geq \delta$, then

$$p_t^{\mathrm{Br}}(x, y) \leq \frac{K}{\pi \delta t} e^{-\frac{\epsilon^2}{2t}}. \tag{13.140}$$

Thus (13.137) is true for $n = 2$.

Now assume that (13.137) is true for $n$. We must show that (13.137) is true for $n+1$. It is more convenient to use the notation $x=(x_0, x_1, \ldots, x_n)$ and $y=(y_0, y_1, \ldots, y_n)$ instead of $x=(x_1, x_2, \ldots, x_{n+1})$ and $y=(y_1, y_2, \ldots, y_{n+1})$, and we use this convention from this point on until the end of the proof of this lemma. Let

$$f(u) = f(u; y_0, \ldots, y_n) = (e^{-\frac{(u-y_0)^2}{2t}}, \ldots, e^{-\frac{(u-y_n)^2}{2t}}).$$

Then

$$H(x, y) \equiv \det \left( e^{-\frac{(x_i - y_j)^2}{2t}} \right)_{i,j=0}^{n} = \det \begin{bmatrix} f(x_0) \\ f(x_1) \\ \vdots \\ f(x_n) \end{bmatrix}$$

and

$$p_t^{\mathrm{Br}}(x, y) = \frac{1}{(2\pi t)^{n/2}} H(x, y) \frac{h(y)}{\prod_{0 \leq q < p \leq n} x_q - x_p}.$$

Set $\delta_n = \frac{\delta}{n} = \frac{\epsilon^2}{4Kn} < 1$, and assume that $\min_{0 \leq i \leq n-1}(x_i - x_{i+1}) \leq \delta_n$. Let $i(x)$ be the first integer $i$, $0 \leq i \leq n-1$, such that $x_i - x_{i+1} > \delta_n$ if there is such an $i$. Otherwise set $i(x) = n$. Then $0 \leq x_0 - x_i \leq i(x)\delta_n \leq n\delta_n$ for $i \leq i(x)$ and $x_0 - x_i > \delta_n$ for $i > i(x)$.

For $x, y \in W^{(n)}$, $\|x\|, \|y\| \leq K$, consider $H(x_0) = H(x_0, x_1, \ldots, x_n, y_0, \ldots, y_n)$ as a function of $x_0$ with the remaining variables fixed. We have $H(x_i) = 0$ for $i = 1, \ldots, n$. Now

$$H(x_0) = \det \begin{bmatrix} f(x_0) - f(x_1) \\ f(x_1) \\ \vdots \\ f(x_n) \end{bmatrix} = (x_0 - x_1) \det \begin{bmatrix} \int_0^1 f'(x_{01}(s)) ds \\ f(x_1) \\ \vdots \\ f(x_n) \end{bmatrix},$$

where $x_{01}(s) = x_1 + s(x_0 - x_1)$, $0 \le s \le 1$. Note that $x_1 \le x_{01}(s) \le x_0$. Now as $x_0 > x_1 > x_2$ and $H(x_1) = H(x_2) = 0$, we see that

$$\det \begin{bmatrix} \int_0^1 f'(x_{21}(s))ds \\ f(x_1) \\ \vdots \\ f(x_n) \end{bmatrix} = 0,$$

where $x_{21}(s) = x_1 + s(x_2 - x_1)$. Hence, adding 0 to $H(x_0)$, we find

(13.141)

$$H(x_0) = (x_0 - x_1) \det \begin{bmatrix} \int_0^1 (f'(x_{01}(s_1)) - f'(x_{21}(s_1)))\, ds_1 \\ f(x_1) \\ \vdots \\ f(x_n) \end{bmatrix}$$

$$= (x_0 - x_1)(x_0 - x_2) \det \begin{bmatrix} \int_0^1 s_1 ds_1 \int_0^1 ds_2 f''(x_{012}(s_1, s_2)) \\ f(x_1) \\ \vdots \\ f(x_n) \end{bmatrix},$$

where $x_{012}(s_1, s_2) = x_{21}(s_1) + s_2(x_{01}(s_1) - x_{21}(s_1)) = s_1 s_2 x_0 + \hat{x}_{12}$, where $\hat{x}_{12}$ depends on $s_1, s_2, x_1$, and $x_2$. Note that $x_2 \le x_{012}(s_2) \le x_0$. Now $H(x_3) = 0$, and as $x_0 > x_1 > x_2 > x_3$, we see that

$$\det \begin{bmatrix} \int_0^1 s_1 ds_1 \int_0^1 ds_2 f''(x_{312}(s_1, s_2)) \\ f(x_1) \\ \vdots \\ f(x_n) \end{bmatrix} = 0, \tag{13.142}$$

where $x_{312}(s_1, s_2) = s_1 s_2 x_3 + \hat{x}_{12}$. Subtracting (13.142) from the determinant in (13.141), and expressing the difference of the second derivatives in terms of an integral of the third derivative, and continuing in this way, we find

$$H(x_0) = \prod_{j=1}^{i(x)} (x_0 - x_j) \int_0^1 s_1^{i(x)-1} ds_1 \int_0^1 s_2^{i(x)-2} ds_2 \cdots \int_0^1 ds_{i(x)} \times \det \begin{bmatrix} f^{(i(x))}(x_{01\cdots i(x)}(s_1, \ldots, s_{i(x)})) \\ f(x_1) \\ \vdots \\ f(x_n) \end{bmatrix}, \tag{13.143}$$

where $x_{01\cdots i(x)}(s) = s_1 s_2 \cdots s_{i(x)} x_0 + \hat{x}_{12\cdots i(x)}$ and $\hat{x}_{12\cdots i(x)}$ depends on $s_1, \ldots, s_{i(x)}, x_1, \ldots, x_{i(x)}$, and $x_{i(x)} \le x_{01\cdots i(x)}(s) \le x_0$.

Now $f(x_0)$ has $n+1$ components $f_i(x_0) = e^{-\frac{(x_0-y_i)^2}{2t}}$, $0 \le i \le n$, and hence for any $k$, $f_i^{(k)}(x_0) = P_k(x_0 - y_i, \frac{1}{t})e^{-\frac{(x_0-y_i)^2}{2t}}$ where $P_k(u,v)$ is a polynomial in $u$ and $v$; $P_k(x_0 - y_i, \frac{1}{t}) = (-1)^k \frac{(x_0-y_i)^k}{t^k} + \cdots$. It follows that

(13.144)

$$p_t^{\mathrm{Br}}(x,y) = \frac{h(y)}{(2\pi t)^{n/2}} \frac{\prod_{j=1}^{i(x)}(x_0 - x_j)}{\prod_{j=1}^{n}(x_0 - x_j)} \sum_{m=0}^{n} (-1)^m \frac{\hat{H}_m(x_1,\ldots,x_n)}{\prod_{1\le q<p\le n}(x_q - x_p)}$$
$$\times \int_0^1 s_1^{i(x)-1} ds_1 \int_0^1 s_2^{i(x)-2} ds_2 \cdots \int_0^1 ds_{i(x)} f_m^{(i(x))}(x_{01\cdots i(x)}(s)),$$

where $\hat{H}_m(x_1,\ldots,x_n)$ is the $n\times n$ determinant obtained by removing the first row and $m$th column from $H(x,y)$. The $(k,\ell)$ entry in $\hat{H}_m$ has the form

$$\begin{cases} e^{-\frac{(x_k-y_{\ell-1})^2}{2t}}, & 1\le k\le n, \quad 1\le \ell\le m, \\ e^{-\frac{(x_k-y_\ell)^2}{2t}}, & 1\le k\le n, \quad m+1\le \ell\le n. \end{cases}$$

In particular, the diagonal entries are given by

$$e^{-\frac{(x_1-y_0)^2}{2t}},\ldots,e^{-\frac{(x_m-y_{m-1})^2}{2t}}, e^{-\frac{(x_{m+1}-y_{m+1})^2}{2t}},\ldots,e^{-\frac{(x_n-y_n)^2}{2t}}.$$

Now using the argument in (13.138) repeatedly, we have, for $\|x-y\| \ge \epsilon\sqrt{3n}$,

$$\begin{aligned}
&(x_{01\cdots i(x)} - y_m)^2 + (x_1 - y_0)^2 + \cdots + (x_m - y_{m-1})^2 \\
&\quad + (x_{m+1} - y_{m+1})^2 + \cdots + (x_n - y_n)^2 \\
&= \big[(x_0 - y_m)^2 + (x_1 - y_0)^2 + \cdots + (x_m - y_{m-1})^2 \\
&\quad + (x_{m+1} - y_{m+1})^2 + \cdots + (x_n - y_n)^2\big] \\
&\quad + (x_{01\cdots i(x)} - y_m)^2 - (x_0 - y_m)^2 \\
&\ge \sum_{i=0}^{n} (x_i - y_i)^2 + (x_{01\cdots i(x)} - x_0)(x_{01\cdots i(x)} + x_0 - 2y_m) \\
&\ge \sum_{i=0}^{n} (x_i - y_i)^2 - 4n\delta_n K \ge 3n\epsilon^2 - \epsilon^2 = 3(n-1)\epsilon^2 + 2\epsilon^2
\end{aligned}$$

as $0 \le x_{01\cdots i(x)} - x_0 \le x_{i(x)} - x_0 \le i(x)\delta \le n\delta_n$. If

$(x_1-y_0)^2+\cdots+(x_m-y_m)^2+(x_{m+1}-y_{m+1})^2+\cdots+(x_n-y_n)^2 \ge 3(n-1)\epsilon^2$,

then by the induction hypothesis

$$\left| \frac{\hat{H}_m(x_1,\ldots,x_n)\prod'_{1\le q<p\le n}(y_q - y_p)}{(2\pi t)^{n/2}\prod_{1\le q<p\le n}(x_q - x_p)} \right| \le \frac{c_n}{t^{m_n/2}} e^{-\frac{\epsilon^2}{2t}}. \tag{13.145}$$

Here $\prod'_{1\le q<p\le n}(y_q - y_p)$ indicates that $p, q \ne m$.

On the other hand, if $(x_1 - y_0)^2 + \cdots + (x_m - y_m)^2 + (x_{m+1} - y_{m+1})^2 + \cdots + (x_n - y_n)^2 < 3(n-1)\epsilon^2$, then $|x_{01\cdots i(x)} - y_m|^2 \geq 2\epsilon^2 \geq \epsilon^2$ and so

$$e^{-\frac{(x_{01\cdots i(x)} - y_m)^2}{2t}} \leq e^{-\frac{\epsilon^2}{2t}}. \tag{13.146}$$

Furthermore, we see from (13.143) that

$$\begin{aligned} H(x_0) = \prod_{j=1}^{i(x)} (x_0 - x_j) \int_0^1 s_1^{n-1} ds_1 \int_0^1 s_2^{n-2} ds_2 \cdots \int_0^1 ds_n \\ \times \det \begin{bmatrix} f^{(n)}(x_{01\cdots n}(s_1, \ldots, s_n)) \\ f(x_1) \\ \vdots \\ f(x_n) \end{bmatrix} \end{aligned} \tag{13.147}$$

from which it follows as in (13.144) that $H(x_0)$ is a sum of $n+1$ terms of the form $q_m(x, y, t) \hat{H}_m(x_1, \ldots, x_n) \prod_{j=1}^n (x_0 - x_j)$, where for some positive constant $c_m$ and some positive integer $\ell_m$, $|q_m(x, y, t)| \leq \frac{c_m}{t^{\ell_m/2}}$ for $\|x\|, \|y\| \leq K$, $t > 0$. But then applying (13.147) to $\hat{H}_m(x_1, \ldots, x_n)$ we see that $\hat{H}_m(x_1, \ldots, x_n)$ is a sum of $n$ terms of the form $q_{mp}(x, y, t) \hat{H}_{mp}(x_2, \ldots, x_n) \times \prod_{j=2}^n (x_1 - x_j)$ where now for a suitable positive constant $c_{mp}$ and positive integer $\ell_{mp}$, $|q_{mp}(x, y, t)| \leq \frac{c_{mp}}{t^{\ell_{mp}/2}}$ for $\|x\|, \|y\| \leq K$, $t > 0$. Continuing in this way we eventually conclude that

$$\hat{H}_m(x_1, \ldots, x_n) = \hat{q}_m(x, y, t) \prod_{1 \leq q < p \leq n} (x_q - x_p), \tag{13.148}$$

where for some positive constant $\hat{c}_m$ and some positive integer $\hat{m}$, $|\hat{q}_m(x, y, t)| \leq \frac{\hat{c}_m}{t^{\hat{m}/2}}$ for $\|x\|, \|y\| \leq K$, $t > 0$. Thus, together with (13.146), we recover (13.145). As $\prod_{j=i(x)+1}^n (x_0 - x_j) \geq \delta_n^2 \left(\frac{\epsilon^2}{4Kn}\right)^n$, (13.137) follows from (13.144) for $n+1$. This completes the induction and proves Lemma 13.24. □

Observe that for $\|x\| \geq 4ne^n \|y\|$,

$$\|x\|^2 - 2ne^n \|x\| \|y\| = \|x\| (\|x\| - 2ne^n \|y\|) \geq \frac{1}{2} \|x\|^2.$$

Hence we have from (13.129)

**Lemma 13.25.** *Let $\hat{K} > 1$. Suppose that $x, y \in \overline{W^{(n)}}$ and $\|y\| \leq \hat{K}$, $\|x\| \geq 5ne^n \hat{K}$. Then for $0 < t < 1$*

$$p_t^{\mathrm{Br}}(x, y) \leq \frac{C_{n,\hat{K}}}{t^{n^2/2}} e^{-\frac{1}{4t}\|x\|^2}. \tag{13.149}$$

*Similarly, if $x, y \in \overline{W^{(n)}}$ and $\|x\| \le \hat{K}$, $\|y\| \ge 5ne^n\hat{K}$, then taking into account the polynomial terms in (13.129), we have for $0 < t < 1$*

$$p_t^{\mathrm{Br}}(x,y) \le \frac{C_{n,\hat{K}}}{t^{n^2/2}} e^{-\frac{1}{4t}\|y\|^2}. \tag{13.150}$$

**Remark 13.26.** From (13.148) and (13.144), we see that for any $K$, there exist a constant $c_n'$ and a positive integer $m_n'$, such that for $\|x\|, \|y\| \le K$,

$$p_t^{\mathrm{Br}}(x,y) \le \frac{c_n'}{t^{m_n'/2}}. \tag{13.151}$$

We now prove (13.136). Let $f \in C_\infty$ and $\eta > 0$ be given. For any $\hat{K} > 1$, let $\varphi_{\hat{K}}(x)$ be a continuous function in $\overline{W^{(n)}}$ such that $0 \le \varphi_{\hat{K}}(x) \le 1$, $\varphi_{\hat{K}}(x) = 1$ for $\|x\| \le \hat{K} - 1$, $\varphi_{\hat{K}}(x) = 0$ for $\|x\| \ge \hat{K}$. Choose $\hat{K} > 1$ such that

$$|f(x)| \le \eta \qquad \text{for } \|x\| \ge \hat{K} - 1. \tag{13.152}$$

Set

$$K = 5ne^n\hat{K}. \tag{13.153}$$

As $f$ is uniformly continuous in $\overline{W^{(n)}}$, there exists a constant $\mu > 0$ such that for all $x, y \in \overline{W^{(n)}}$,

$$|f(x) - f(y)| < \eta \qquad \text{if } \|x - y\| < \mu. \tag{13.154}$$

Set

$$\epsilon = \frac{1}{\sqrt{3(n-1)}}\mu. \tag{13.155}$$

Without loss of generality we may assume that $\epsilon < K/\sqrt{2}$.

Now by (13.130), for any $x \in \overline{W^{(n)}}$,

$$\begin{aligned} \Gamma_t^{\mathrm{Br}} f(x) - f(x) &= \int_{W^{(n)}} p_t^{\mathrm{Br}}(x,y)(f(y) - f(x))d^n y \\ &= I + I_1 + I_2 + I_3, \end{aligned} \tag{13.156}$$

where

$$I = I(x) = \int_{W^{(n)}, \|x-y\| < \mu} p_t^{\mathrm{Br}}(x,y)(f(y) - f(x))d^n y,$$

$$I_1 = I_1(x) = \int_{W^{(n)}, \|x-y\| \ge \mu} p_t^{\mathrm{Br}}(x,y)(f(y) - f(x))\varphi_{\hat{K}}(y)d^n y,$$

$$I_2 = I_2(x) = \int_{W^{(n)}, \|x-y\| \ge \mu} p_t^{\mathrm{Br}}(x,y)(f(y) - f(x))\varphi_{\hat{K}}(x)(1 - \varphi_{\hat{K}}(y))d^n y,$$

and

$$I_3 = I_3(x) = \int_{W^{(n)}, \|x-y\| \geq \mu} \times p_t^{\mathrm{Br}}(x,y)(f(y) - f(x))(1 - \varphi_{\hat{K}}(x))(1 - \varphi_{\hat{K}}(y))d^n y.$$

We have

$$|I_3(x)| \leq \max_{\|y\| \geq \hat{K}-1} |f(y)| + \max_{\|x\| \geq \hat{K}-1} |f(x)| \leq 2\eta. \tag{13.157}$$

Also,

$$|I_1(x)| \leq 2\|f\|_\infty \int_{W^{(n)},\, \|x-y\| \geq \mu,\, \|y\| \leq \hat{K}} p_t^{\mathrm{Br}}(x,y)d^n y.$$

Using Lemma 13.24 when $\|x\| \leq K$ and using Lemma 13.25 when $\|x\| > K$, we find that

$$|I_1(x)| \leq \frac{C_0'}{t^{m_0'/2}} e^{-\frac{\epsilon^2}{4t}}$$

for some $C_0', m_0' \geq n$ where $\epsilon$ is defined in (13.155).

Finally,

$$|I_2(x)| \leq \chi_{\|x\| \leq \hat{K}} 2\|f\|_\infty \int_{W^{(n)},\, \|x-y\| \geq \mu} p_t^{\mathrm{Br}}(x,y)d^n y. \tag{13.158}$$

Splitting up the $y$-integration in $I_2$ into $\{\|y\| < K\}$ and $\{\|y\| \geq K\}$, we obtain as in the case $I_1$,

$$|I_2(x)| \leq \frac{C_0'}{t^{m_0'/2}} e^{-\frac{\epsilon^2}{4t}}. \tag{13.159}$$

Assembling these estimates, we find that for all $w \in W^{(n)}$ and $0 < t < 1$,

$$\left|\Gamma_t^{\mathrm{Br}} f(x) - f(x)\right| \leq I(x) + O\left(\frac{e^{-\frac{\epsilon^2}{2t}}}{t^{m_0'/2}}\right) + O(\eta).$$

But $I(x) \leq \eta$, and as $\eta > 0$ is arbitrary, we conclude that

$$\lim_{t \downarrow 0} \Gamma_t^{\mathrm{Br}} f(x) = f(x)$$

in $C_\infty(\overline{W^{(n)}})$. Thus $t \to \Gamma_t^{\mathrm{Br}}$ is a strongly continuous semigroup in $C_\infty(\overline{W^{(n)}})$, and Proposition 13.23 is proved. □

We finally prove that $\Gamma_t^{\mathrm{Br}}$ is also the semigroup for Hermitian Brownian Dyson process.

**Theorem 13.27.** *Let $\Gamma_t^{\mathrm{Br}}$ be the semigroup generated by $p_t^{\mathrm{Br}}(x,y)$. Then $\Gamma_t^{\mathrm{Br}}$ is also the semigroup for Hermitian Brownian Dyson process, i.e., if $f \in C_\infty$, then for $x \in \overline{W^{(n)}}$*

$$\Gamma_t^{\mathrm{Br}} f(x) = \mathbb{E}^x[f(x(t))], \qquad t > 0, \tag{13.160}$$

*where $x(t) = \{\lambda_k(t)\}_{k=1,\dots,n}$ is Hermitian Brownian Dyson process with $x(0) = x$.*

**Proof.** For $w \in W^{(n)}$, $t > 0$ set

$$u(w,t) \equiv (\Gamma_t^{\mathrm{Br}} f)(w) = \int_{W^{(n)}} p_t^{\mathrm{Br}}(w,y) f(y) d^n y.$$

As $R \equiv \det(p_t(w_i, y_j))_{i,j=1}^n$ solves the heat equation, $\frac{\partial R}{\partial t} = \frac{1}{2}\Delta R$, and as $u(w,t)$ is smooth on $W^{(n)} \times \{t > 0\}$,

$$\frac{\partial u}{\partial t}(w,t) = \int \frac{1}{h(w)}(\Delta_w R) h(y) f(y) d^n y = \mathcal{L}^{\mathrm{Br}} u(w,t), \tag{13.161}$$

where $h(w) = \prod_{i<j}(h_i - h_j)$ and

$$\mathcal{L}^{\mathrm{Br}} = \frac{1}{2h(w)} \Delta_w h(w) = \frac{1}{2}\Delta_w + (\nabla_w \log h)(w) \cdot \nabla_w \tag{13.162}$$

as $\Delta_w h = 0$ (see the discussion preceding (13.44)). Note that $\mathcal{L}^{\mathrm{Br}}$ is also the generator for Hermitian Brownian Dyson process (cf. Appendix C).

Let $T > 0$ be given. We apply Ito's formula to $J(x(t),t)$ where $J(y,t) = u(y, T-t)$, $0 \le t \le T$. As $J$ is not $C^2$ on $\overline{W^{(n)}} \times [0,T]$, we need to use a stopping time argument as in (13.16), for example, as follows.

For $k \ge 1$, set

$$\hat{J}(y,t) = \chi_k(y) J(y,t), \qquad y \in \overline{W^{(n)}},$$

where $\chi_k(y)$ is smooth, $0 \le \chi_k(y) \le 1$, $\chi_k(y) = 1$ for $h(y) > \frac{1}{k}$ and $\chi_k(y) = 0$ for $h(y) \le \frac{1}{2k}$. Then, for any (small) $\epsilon > 0$, $\hat{J}(y,t)$ is a smooth function on $\overline{W^{(n)}} \times [0, T-\epsilon]$, and by Ito's formula

$$\begin{aligned} \hat{J}(x(t),t) =& \hat{J}(x(\epsilon),\epsilon) + \int_\epsilon^t \sum_{j=1}^n \frac{\partial \hat{J}}{\partial y_j}(x(s),s) dW_j(s) \\ &+ \int_\epsilon^t \left[ \frac{\partial \hat{J}}{\partial s}(x(s),s) + (\mathcal{L}^{\mathrm{Br}} \hat{J})(x(s),s) \right] ds \end{aligned} \tag{13.163}$$

for $\epsilon \le t \le T-\epsilon$. In particular (13.163) is valid if we replace $t$ by $t \wedge \tau_k$, where $\tau_k$ is the stopping time given by $\tau_k = \inf\{s : \epsilon \le s \le T-\epsilon,\ h(x(s)) \le \frac{1}{k}\}$. Now $\hat{J}(x(t\wedge\tau_k), t\wedge\tau_k) = J(x(t\wedge\tau_k), t\wedge\tau_k)$ and so (13.163) remains true with $\hat{J}(x(t),t)$ replaced by $J(x(t),t)$ for time $t \wedge \tau_k$. But as $x(s) \in W^{(n)}$

almost surely for $s > 0$ by Theorem 13.8, $\frac{\partial J}{\partial s}(x(s), s) + (\mathcal{L}^{\mathrm{Br}} J)(x(s), s) = 0$ by (13.161), and so

$$J(x(t \wedge \tau_k), t \wedge \tau_k) = J(x(\epsilon), \epsilon) + M,$$

where

$$M = \int_{\epsilon}^{t\wedge\tau_k} \sum_{j=1}^{n} \frac{\partial J}{\partial y_j}(x(s), s) dW_j(s)$$

is a martingale. Hence

$$\mathbb{E}^x[u(x(t \wedge \tau_k), T - t \wedge \tau_k)] = \mathbb{E}^x[u(x(\epsilon), T - \epsilon)].$$

But for any $t > 0$, $t \wedge \tau_k \to t$ as $l \to \infty$ almost surely, and so as $u(y, t)$ is uniformly bounded, $\mathbb{E}^x[u(x(t), T - t)] = \mathbb{E}^x[u(x(\epsilon), \epsilon)]$ for $\epsilon \leq t \leq T - \epsilon$. In particular,

$$\mathbb{E}^x[u(x(T - \epsilon), \epsilon)] = \mathbb{E}^x[u(x(\epsilon), T - \epsilon)].$$

Now as $\Gamma_t^{\mathrm{Br}}$ generates a continuous semigroup on $C_\infty = C_\infty(\overline{W^{(n)}})$, $u(y, t)$ is continuous in $t \in [0, \infty)$, uniformly in $y$, and we obtain finally

$$\mathbb{E}^x[f(x(T))] = \mathbb{E}^x[u(x(T), 0)] = \mathbb{E}^x[u(x, T)] = u(x, T) = \Gamma_T^{\mathrm{Br}} f(x),$$

which establishes (13.160). □

**Remark 13.28.** The underlying analytical issue that is addressed in Theorem 13.27 is the following. Let $\tilde{\mathcal{L}}$ be the infinitesimal generator of the semigroup for Hermitian Brownian Dyson process

$$\mathbb{E}^x[f(x(t))] = (e^{t\tilde{\mathcal{L}}} f)(x), \qquad x \in \overline{W^{(n)}}, \quad t \geq 0,$$

and $f \in C_\infty$. We would like to identify $\tilde{\mathcal{L}}$ and its domain $D(\tilde{\mathcal{L}})$ so that we can solve the equation

$$\frac{\partial u}{\partial t}(x, t) = \tilde{\mathcal{L}} u(x, t), \qquad u(x, 0) \in D(\tilde{\mathcal{L}})$$

uniquely for $t > 0$, and hence determine the transition density $\tilde{p}_t(x, y)$ for Hermitian Brownian Dyson process. By general theory we know that $C_0^\infty(W^{(n)}) \subset D(\tilde{\mathcal{L}})$ and

$$(\tilde{\mathcal{L}} f)(x) = \frac{1}{2h(x)} \Delta(hf)(x) = \frac{1}{2}\Delta f + \sum_{j \neq i} \frac{1}{x_i - x_j} \frac{\partial f}{\partial x_i} \tag{13.164}$$

for $f \in C_0^\infty(W^{(n)})$. However, because of the singularities when $x_i = x_j$, it is not at all clear how to extend $\tilde{\mathcal{L}}$ from $C_0^\infty(W^{(n)})$ to $D(\tilde{\mathcal{L}})$. What Theorem 13.27 tells us is that

$$e^{t\tilde{\mathcal{L}}} = \Gamma_t^{\mathrm{Br}}$$

and

$$\tilde{p}_t(x,y) = p_t^{\mathrm{Br}}(x,y).$$

Also, by general theory,

$$D(\tilde{\mathcal{L}}) = \{f \in C_\infty : \lim_{t\to 0} \frac{(\Gamma_t^{\mathrm{Br}} - 1)f}{t} \equiv \tilde{\mathcal{L}} f \text{ exists}\}.$$

**13.6.2. Nonintersecting Ornstein-Uhlenbeck Process and the Hermitian Dyson process.** The Ornstein-Uhlenbeck processes conditioned not to intersect for all time can be handled similarly to the above case of nonintersecting Brownian motion. For $x, y \in \mathbb{R}$ and $t > 0$, let $\tilde{p}_t(x,y)$ be the time $t$-transition density for the Ornstein-Uhlenbeck process given by the stochastic differential equation

$$dz_t = -z_t dt + dB_t, \tag{13.165}$$

where $B_t$ is a standard one-dimensional Brownian motion. Explicitly, $\tilde{p}_t(x,y) = \frac{1}{\sqrt{\pi(1-e^{-2t})}} e^{-\frac{1}{(1-e^{-2t})}(y-xe^{-t})^2}$ (see Exercise 13.20). As in the previous subsection, we note that the time $t$-transition density is sufficient to describe all features of the process since the stochastic differential equation (13.165) is time homogeneous.

For $x, y \in W^{(n)}$ and $t > 0$, we define a time $t$-transition probability density

$$p_t^{\mathrm{OU}}(x,y) \equiv e^{\frac{n(n-1)}{2}t} \det\left(\tilde{p}_t(x,y)\right)_{i,j=1}^n \frac{h(y)}{h(x)}. \tag{13.166}$$

As in (13.110) et seq., $p_t^{\mathrm{OU}}(x,y)$ can be obtained by considering the transition density for the Ornstein-Uhlenbeck process conditioned not to intersect for $0 \le t \le T$, and then taking the limit as $T \to \infty$.

For $f \in C_\infty$, define

$$\Gamma_t^{\mathrm{OU}} f(x) = \int_{W^{(n)}} p_t^{\mathrm{OU}}(x,y) f(y) d^n y, \qquad x \in W^{(n)}, \quad t > 0. \tag{13.167}$$

As in the case of Hermitian Brownian Dyson process, one shows that $\Gamma_t^{\mathrm{OU}}$ extends to a strongly continuous semigroup on $C_\infty$. Then one shows that $\Gamma_t^{\mathrm{OU}}$ is the semigroup for the Hermitian Dyson process $\{X(t)\}_{t\ge 0}$ on Hermitian matrices, i.e., if $f \in C_\infty$, then

$$\Gamma_t^{\mathrm{OU}} f(x) = \mathbb{E}^x[f(X(T))], \qquad T > 0, \tag{13.168}$$

where $X(0) = x \in \overline{W^{(n)}}$. The proofs of the above results are very similar to the case of Hermitian Brownian Dyson process, and the details are left to the reader. Thus we say that Hermitian Dyson process (13.105) is the same as that of the Ornstein-Uhlenbeck process *conditioned not to intersect for all time.*

**Remark 13.29.** As shown above, both the Dyson process and Brownian Dyson process for the spectrum of Hermitian matrices ($\beta = 2$) can be realized as diffusions conditioned not to intersect for all time. There is no known analogue of these results for the orthogonal ($\beta = 1$) or symplectic ($\beta = 4$) ensembles.

# Appendix A

# Theory of Trace Class Operators and Fredholm Determinants

This appendix contains a summary of the basic properties of trace class operators and Fredholm determinants. We present few proofs; all details and some history of the subject can be found in [**Sim05c, RS80, RS79, RS78**].

Recall that a bounded linear operator $T$ from a Banach space $X$ to a Banach space $Y$ is *compact* if $T$ takes bounded sets in $X$ to precompact sets in $Y$. We restrict our attention to the case where $X = Y = \mathcal{H}$, a separable Hilbert space with bounded linear operators $\mathcal{L}(\mathcal{H})$. Denote the inner product on $\mathcal{H}$ by $(\cdot, \cdot)$. Let $\mathcal{B}_\infty(\mathcal{H})$ denote the compact operators in $\mathcal{H}$: $\mathcal{B}_\infty(\mathcal{H})$ is a closed linear subspace of $\mathcal{L}(\mathcal{H})$ with the usual operator norm, $\|A\| = \sup_{0 \neq x \in \mathcal{H}} \frac{\|Ax\|}{\|x\|}$.

Let $B = B^* \geq 0$ be a positive, self-adjoint operator in $\mathcal{L}(\mathcal{H})$. Let $\{u_n\}_{n\geq 1}$ be an orthonormal basis for $\mathcal{H}$. Then the quantitiy $\sum_{n=1}^\infty (u_n, Bu_n)$, finite or infinite, is independent of the basis, $\{u_n\}_{n\geq 1}$. For $A \in \mathcal{L}(\mathcal{H})$, let $|A| = \sqrt{A^*A} \geq 0$ denote its modulus. We say that $A \in \mathcal{B}_1(\mathcal{H})$, the (Schatten) class of *trace class* operators, if

(A.1) $$\|A\|_1 \equiv \sum_{n=1}^\infty (u_n, |A|u_n) < \infty$$

for one, hence all, orthonormal bases, $\{u_n\}_{n\geq 1}$. The quantity $\|\cdot\|_1$ provides a norm on $\mathcal{B}_1(\mathcal{H})$ and $(\mathcal{B}_1(\mathcal{H}), \|\cdot\|_1)$ is a Banach space. If $A \in \mathcal{B}_1(\mathcal{H})$, we define the trace of $A$ by

$$\text{(A.2)} \qquad \operatorname{tr}(A) \equiv \sum_{n=1}^{\infty} (u_n, Au_n).$$

As before the sum is independent of the orthonormal basis $\{u_n\}_{n\geq 1}$, and we find

$$\text{(A.3)} \qquad |\operatorname{tr}(A)| \leq \operatorname{tr}(|A|) = \|A\|_1.$$

If $A = (a_{ij})_{i,j=1}^n$ is an $n\times n$ matrix, then (A.2) clearly agrees with the matrix definition of a trace, viz. $\sum_{i=1}^n a_{ii}$.

Trace class operators are automatically compact. This implies, in particular, that $\operatorname{spec}(A) \setminus \{0\}$ consists of a discrete set $\{\lambda_k\}_{k\geq 1}$ of eigenvalues, each of finite (algebraic, hence geometric) multiplicity, and which can accumulate only at 0. Recall that the algebraic multiplicity of $\lambda_k$ is given by the dimension of the range of the spectral projection $P_{\lambda_k} = \oint_{\mathcal{C}_k} \frac{1}{s-A}\frac{ds}{2\pi i}$, where $\mathcal{C}_k$ is a suitably small contour enclosing $\lambda_k$. If $A$ is an $n \times n$ matrix with eigenvalues $\{\lambda_k\}_{k=1}^n$, then $\operatorname{tr}(A) = \sum_{k=1}^n \lambda_k$. It is a fundamental result of Lidskii that the same is true for trace class operators $A$,

$$\text{(A.4)} \qquad \operatorname{tr}(A) = \sum_{k\geq 1} \lambda_k, \qquad A \in \mathcal{B}_1(\mathcal{H}),$$

where the sum is taken over the (nonzero) eigenvalues of $A$, counted with algebraic multiplicity. The fact that the sequence $\{\lambda_k\}$ is summable and

$$\text{(A.5)} \qquad \sum_{k\geq 1} |\lambda_k| \leq \|A\|_1$$

is an earlier result due essentially to Schur, Lalesco, and Weyl.

We say that $A \in \mathcal{B}_2(\mathcal{H})$, the (Schatten) class of *Hilbert-Schmidt* operators if

$$\text{(A.6)} \qquad \|A\|_2 \equiv (\operatorname{tr}(|A|^2))^{1/2} < \infty.$$

The quantity $\|\cdot\|_2$ provides a norm on $\mathcal{B}_2(\mathcal{H})$ and $(\mathcal{B}_2(\mathcal{H}), \|\cdot\|_2)$ is a Banach space (in fact, a Hilbert space).

The spaces $\mathcal{B}_p(\mathcal{H})$, $p = 1, 2, \infty$ are $*$-ideals in $\mathcal{L}(\mathcal{H})$, i.e., in addition to being linear subspaces, the following is true: for $p = 1, 2, \infty$,

- if $A \in \mathcal{B}_p(\mathcal{H})$ and $B \in \mathcal{L}(\mathcal{H})$, then $AB \in \mathcal{B}_p(\mathcal{H})$ and $BA \in \mathcal{B}_p(\mathcal{H})$ and $\|AB\|_p, \|BA\|_p \leq \|B\|\|A\|_p$;
- if $A \in \mathcal{B}_p(\mathcal{H})$, then $A^* \in \mathcal{B}_p(\mathcal{H})$.

Also, if $A \in \mathcal{B}_1(\mathcal{H})$, then

$$\operatorname{tr}(A^*) = \overline{\operatorname{tr}(A)}, \tag{A.7}$$

$$\operatorname{tr}(AB) = \operatorname{tr}(BA) \text{ for all } B \in \mathcal{L}(\mathcal{H}). \tag{A.8}$$

As $|A|^2 = A^*A$, it follows that

$$\mathcal{B}_1(\mathcal{H}) \subset \mathcal{B}_2(\mathcal{H}) \subset \mathcal{B}_\infty(\mathcal{H}) \tag{A.9}$$

and

$$\|A\| \leq \|A\|_2 \leq \|A\|_1. \tag{A.10}$$

Most importantly, we have the following result:

$$A \in \mathcal{B}_1(\mathcal{H}) \text{ if and only if } A = BC \tag{A.11}$$

for some $B, C \in \mathcal{B}_2(\mathcal{H})$ and $\|A\|_1 \leq \|B\|_2 \|C\|_2$.

Finally, the following property is very useful. Let $A \in \mathcal{B}_p(\mathcal{H})$, $p = 1, 2$, and let $C_n, D_n \in \mathcal{L}(\mathcal{H})$, $n \geq 1$.

(A.12) If $C_n \to C$, $D_n^* \to D^*$ strongly in $\mathcal{H}$, then $C_n A D_n$ converges in $(\mathcal{B}_p(\mathcal{H}), \|\cdot\|_p)$ and $\lim_{n\to\infty} C_n A D_n = CAD$.

The square root in the definition of $|A|$ makes it difficult in general to relate the properties of $|A|$ to the properties of $A$, and this complicates the determination whether a particular given operator lies in $\mathcal{B}_1(\mathcal{H})$ (see (A.22) et seq. below). On the other hand, $|A|^2 = A^*A$ is directly related to $A$ (and $A^*$). So the determination of whether $A \in \mathcal{B}_2$ is a relatively simple matter. Indeed, the following result shows that if $\mathcal{H}$ is (separable and) of the form $L^2(M, d\mu)$ for some measure space $(M, d\mu)$, $A \in \mathcal{L}(\mathcal{H})$ is Hilbert-Schmidt if and only if there is a (kernel) function $K \in L^2(M \times M, \mu \otimes \mu)$ such that

$$Af(x) = \int_M K(x, y) f(y) \mu(dy), \qquad f \in \mathcal{H}. \tag{A.13}$$

Moreover,

$$\|A\|_2^2 = \int_M \int_M |K(x, y)|^2 \mu(dx) \mu(dy). \tag{A.14}$$

Given an operator $A \in \mathcal{L}(\mathcal{H})$, what techniques can one use to show that $A \in \mathcal{B}_p(\mathcal{H})$ for $p = 1, 2, \infty$? Note first:

- Every finite rank operator $A$, i.e., $A(f) = \sum_{i=1}^n (x_i, f) y_i$ for some $x_i, y_j \in \mathcal{H}$, is trace class. A fortiori, $A \in \mathcal{B}_p$, $p = 2, \infty$, and we have $\|A\| \leq \|A\|_2 \leq \|A\|_1 \leq \sum_{i=1}^n \|x_i\| \|y_i\|$. This is a consequence of the triangle inequality and the fact that each rank 1 operator $(x_i, \cdot) y_i \in \mathcal{B}_1(\mathcal{H})$ and satisfies $\|(x_i, \cdot) y_i\|_1 = \|x_i\| \|y_i\|$. In particular, it follows that if $\mathcal{H} = \ell_2^+(\mathbb{N})$ and $A(u)_i = \sum_{i=1}^{\infty} a_{ij} u_j$ for some semi-infinite matrix $(a_{ij})_{i,j\geq 1}$ with $\sum |a_{ij}| < \infty$, then $A \in \mathcal{B}_1(\mathcal{H})$ and $\|A\|_1 \leq \sum_{i,j\geq 1} |a_{ij}|$.

- As $\mathcal{B}_\infty(\mathcal{H})$ is $\|\cdot\|$-closed in $\mathcal{L}(\mathcal{H})$, it follows that every limit of finite rank operators in $\mathcal{L}(\mathcal{H})$ is compact. For separable $\mathcal{H}$, the converse is true, i.e., if $A \in \mathcal{B}_\infty(\mathcal{H})$, then there exists a sequence of finite rank operators $A_n \in \mathcal{L}(\mathcal{H})$ such that $\lim_{n\to\infty} \|A - A_n\| = 0$.

If $A$ has a kernel $K(x,y)$ as in (A.13), the finiteness of $\|K\|_{L^2(M\times M,\mu\otimes\mu)}$ provides a useful and direct method to determine whether an operator is Hilbert-Schmidt, and a fortiori compact or even just bounded. In particular, for given functions $f, g : \mathbb{R}^n \to \mathbb{C}$, let $A = f(x)g(s)$ denote the product of multiplication by $f$ in $x$-space and $g$ in Fourier $s$-space, i.e.,

$$Ah(x) = f(x) \int_{\mathbb{R}^n} g(s)\hat{h}(s) e^{ix\cdot s} \frac{ds}{(2\pi)^{\frac{n}{2}}}, \quad h \in L^2(\mathbb{R}^n), \tag{A.15}$$

where $\hat{h}(s) = \int_{\mathbb{R}^n} h(x) e^{-ix\cdot s} \frac{ds}{(2\pi)^{\frac{n}{2}}}$. If $g \in L^2(\mathbb{R}^n)$, then $A$ has kernel

$$K(x,y) = \frac{1}{(2\pi)^{\frac{n}{2}}} f(x)\check{g}(x-y), \tag{A.16}$$

where $\check{g}$ is the inverse Fourier transform of $g$. Hence, if $f, g \in L^2(\mathbb{R}^n)$, then for $\mathcal{H} = L^2(\mathbb{R}^n, dx)$, $A \in \mathcal{B}_2(\mathcal{H})$ and

$$\|A\|_2 = \left( \int \int_{\mathbb{R}^n\times\mathbb{R}^n} |K(x,y)|^2 dxdy \right)^{\frac{1}{2}} = \frac{1}{(2\pi)^{n/2}} \|f\|_{L^2} \|g\|_{L^2}. \tag{A.17}$$

If $f, g \in L^\infty(\mathbb{R}^n)$, then from (A.15) $Ah = f(g\hat{h})^\vee$, so by the unitarity of the Fourier transform and its inverse in $L^2(\mathbb{R}^n)$,

$$\begin{aligned} \|Ah\|_{L^2} &= \|f(g\hat{h})^\vee\|_{L^2} \leq \|f\|_\infty \|(g\hat{h})^\vee\|_{L^2} \\ &= \|f\|_\infty \|\check{gh}\|_{L^2} \leq \|f\|_\infty \|g\|_\infty \|\check{h}\|_{L^2} \\ &= \|f\|_\infty \|g\|_\infty \|h\|_{L^2}, \end{aligned}$$

so, $A \in \mathcal{L}(\mathcal{H})$ and

$$\|A\| \leq \|f\|_\infty \|g\|_\infty. \tag{A.18}$$

If $f, g \in L^\infty(\mathbb{R}^n)$ and $\lim_{x\to\infty} f(x) = 0$, $\lim_{s\to\infty} g(s) = 0$, then $A = f(x)g(s) \in \mathcal{B}_\infty(\mathcal{H})$. Indeed, for $R > 0$, let $\chi_R$ be the characteristic function of the ball $\{x \in \mathbb{R}^n : |x| < R\}$. Then for any $R$,

$$\begin{aligned} A = {} & \chi_R(x) f(x) \chi_R(s) g(s) + \chi_R(x) f(x)(1 - \chi_R(s)) g(s) \\ & + (1 - \chi_R(x)) f(x) g(s). \end{aligned} \tag{A.19}$$

The first operator is Hilbert-Schmidt by (A.17) and the last two operators have arbitrarily small norm as $R \to \infty$ by (A.18). Thus $A$ is a norm limit of compact operators, hence compact.

The above calculations have the following important implications for the Laplacian $-\frac{d^2}{dx^2}$. As the Laplacian corresponds to multiplication by $s^2$ in Fourier space, we have, in particular,

$$A \equiv \frac{1}{(x^2+1)^\alpha} \frac{1}{(1-\frac{d^2}{dx^2})^\beta} \in \mathcal{B}_\infty(L^2(\mathbb{R}, dx)) \tag{A.20}$$

for any $\alpha, \beta > 0$ and

$$A = \frac{1}{(x^2+1)^\alpha} \frac{1}{(1-\frac{d^2}{dx^2})^\beta} \in \mathcal{B}_2(L^2(\mathbb{R}, dx)) \tag{A.21}$$

for any $\alpha, \beta > \frac{1}{4}$. Of course, these results immediately generalize to $\mathbb{R}^n$ (we leave this as an exercise).

As noted earlier, it is a more difficult matter to determine when a given operator $A$ is trace class. Mostly one shows that $A \in \mathcal{B}_1(\mathcal{H})$ by showing that $A = BC$ where $B$ and $C$ are Hilbert-Schmidt operators. For example, the operator $\mathbb{S}_\theta$ on $L^2(-\theta, \theta)$ in (6.79), has kernel $\mathbb{S}(x,y) = \frac{\sin(\pi(x-y))}{\pi(x-y)} = \frac{1}{2}\int_{-1}^1 e^{i\pi x t} e^{-i\pi y t} dt = (BC)(x,y)$, where $C$ is the Hilbert-Schmidt operator from $L^2(-\theta,\theta)$ to $L^2(-1,1)$ with kernel $C(t,y) = \frac{1}{\sqrt{2}} e^{-i\pi t y}$, and $B = C^*$. Hence $\mathbb{S}_\theta$ is trace class. In a similar way we obtain the following very useful result. For $\delta > 0$, let

$$L^2_\delta = \{f : \|f\|_\delta = \|(1+|x|^2)^{\frac{\delta}{2}} f(x)\|_{L^2} < \infty\}. \tag{A.22}$$

Consider $A = f(x)g(s)$ with $f, g \in L^2_\delta(\mathbb{R}^n)$ for some $\delta > \frac{n}{2}$. Then $A \in \mathcal{B}_1(L^2(\mathbb{R}^n))$ and

$$\|A\|_1 \le c\|f\|_\delta \|g\|_\delta \tag{A.23}$$

for some $c = c(n, \delta)$. To prove this result, we follow [**RS79**, p. 48] and write $A = BC$ where

$$B = f(x)(1+|s|^2)^{-\frac{\delta}{2}}(1+|x|^2)^{\frac{\delta}{2}}, \quad C = (1+|x|^2)^{-\frac{\delta}{2}}(1+|s|^2)^{\frac{\delta}{2}} g(s). \tag{A.24}$$

Then, $C \in \mathcal{B}_2(L^2(\mathbb{R}^n))$ by (A.17). Let $h$ be $(2\pi)^{-\frac{n}{2}}$ times the Fourier transform of $(1+|s|^2)^{-\frac{\delta}{2}} \in L^2(\mathbb{R}^n)$. Then, $B$ is an integral operator with kernel $f(x)h(x-y)(1+|y|^2)^{\frac{\delta}{2}}$. To prove that $B \in \mathcal{B}_2(L^2(\mathbb{R}^n))$ it is clearly enough to show that

$$\int |h(x-y)|^2 (1+|y|^2)^\delta dy \le c(1+|x|^2)^\delta. \tag{A.25}$$

Now, $(1+|s|^2)^{-\frac{\delta}{2}}$ has an analytic continuation $H(s)$ to $\{|\operatorname{Im}(s)| < 1\}$ and $\int_{\mathbb{R}^n} |H(s+ik)|^2 ds < \infty$ for $|\operatorname{Re}(k)| < 1$. It follows by the Paley-Wiener principle that $\int_{\mathbb{R}^n} e^{a|x|}|h(x)|^2dx<\infty$ for all sufficiently small $a>0$, and hence $\int_{\mathbb{R}^n}(1+|x|^2)^\delta|h(x)|^2dx<\infty$. As $(1+|y|^2)^\delta \le 2^\delta(1+|x-y|^2)^\delta(1+|x|^2)^\delta$, (A.25) now follows, which proves that $B \in \mathcal{B}_2(L^2(\mathbb{R}^n))$.

Consider an operator $A$ in $\mathcal{H} = L^2(0,1)$ with kernel $k(x,y)$ that is differentiable with respect to $x$ such that $k(x,y)$ and $k_x(x,y)$ are continuous on $[0,1]\times[0,1]$. Then $A$ is trace class. Indeed,

$$k(x,y) = k(0,y) + \int_0^x k_x(t,y)dt, \tag{A.26}$$

which shows that $A$ is the sum of a rank 1 operator, $h \mapsto \int_0^1 k(0,y)h(y)dy$ and a product $BC$ of operators, where $B$ has kernel $H(x,t)$, the characteristic function of the set $0\le t\le x\le 1$ in $[0,1]\times[0,1]$, and $C$ has kernel $k_x(t,y)$. Clearly, $B, C \in \mathcal{B}_2(L^2(0,1))$, and we conclude that $A$ is a trace class operator. Note, in particular, that this argument provides an alternative proof that the sine kernel operator $\mathbb{S}_\theta$ is trace class.

The following result is well known. Let $A$ be an operator in $\mathcal{H} = L^2(a,b)$, $-\infty<a<b<\infty$, with kernel $k(x,y)$. If $A\in\mathcal{B}_1(\mathcal{H})$ and $k(x,y)$ is continuous on $[a,b]\times[a,b]$, then

$$\operatorname{tr}(A) = \int_a^b k(x,x)dx. \tag{A.27}$$

For example, for the sine kernel operator $\mathbb{S}_\theta$ above, $\mathbb{S}_\theta(x,x) = 1$ and so $\operatorname{tr}\mathbb{S}_\theta = \int_{-\theta}^{\theta} 1dx = 2\theta$. If $A\in\mathcal{B}_1(L^2(\mathbb{R}))$ and $k(x,y)$ is continuous, then (A.27) is also true provided we interpret the right-hand side as an improper Riemann integral, $\lim_{c\to\infty}\int_{-c}^{c} k(x,x)dx$.

**Remark A.1** (A note of caution)**.** If $k(x,y)$ is simply continuous on $[a,b]\times[a,b]$, then we cannot conclude in general that $A\in\mathcal{B}_1(\mathcal{H})$. Indeed, let $g(t)$ be a continuous function on $\mathbb{R}$ with period $2\pi$ and Fourier coefficients $\{g_n\}_{n\in\mathbb{Z}}$ that are *not* summable (*Exercise*: Construct such a function $g$.) Then $(Ah)(x)\equiv\int_0^{2\pi} g(x-y)h(y)dy$ defines an operator with a continuous kernel $k(x,y) = g(x-y)$ on $[0,2\pi]\times[0,2\pi]$. But, by Fourier theory $A$ is unitarily equivalent to multiplication by $\sqrt{2\pi}g_n$ on $\ell_2(\mathbb{Z})$. In particular, $A$ has eigenvalues $2\pi g_n$, $n\in\mathbb{Z}$. If $A$ were trace class the $g_n$'s would be summable by (A.5). But this contradicts the construction of $g$, so $A\notin\mathcal{B}_1(L^2(0,2\pi))$.

As noted earlier, the square root in the definition of $|A|$ makes it difficult in general to relate the properties of $|A|$ to those of $A$. In particular, for trace class operators $A_n, A$, it is in general a complicated matter to determine when $A_n\to A$ in trace norm. However, the situation is simplified if $A_n, A$ are self-adjoint and satisfy $0\le A_n, A\le 1$. We have:

**Lemma A.2** (See [**Sim05c**, Theorem 2.20]). *For trace class operators $A_n$,$A$, $0 \leq A_n, A \leq 1$, in a separable Hilbert space $\mathcal{H}$, a necessary and sufficient condition for the convergence of $A_n \to A$ in trace norm is that*

$$(f, A_n f) \to (f, Af) \tag{A.28}$$

*as $n \to \infty$ for all $f \in \mathcal{H}$ and*

$$\operatorname{tr}(A_n) \to \operatorname{tr}(A). \tag{A.29}$$

We now consider Fredholm determinants. If $A$ is an $n \times n$ matrix with eigenvalues $\{\lambda_k\}_{k=1}^n$, then

$$\begin{aligned} \det(1+A) &= \prod_{k=1}^{n}(1+\lambda_k) \\ &= 1 + \sum_{k=1}^{n} \lambda_k + \sum_{1 \leq k_1 < k_2 \leq n} \lambda_{k_1}\lambda_{k_2} + \cdots \\ &\quad + \sum_{1 \leq k_1 < \cdots < k_{n-1} \leq n} \lambda_{k_1} \cdots \lambda_{k_{n-1}} + \lambda_1 \cdots \lambda_n. \end{aligned} \tag{A.30}$$

Now, $\sum_{k=1}^n \lambda_k = \operatorname{tr}(A)$, $\sum_{1\leq k_1<k_2\leq n} \lambda_{k_1}\lambda_{k_2} = \operatorname{tr}(\bigwedge^2(A))$ where $A \wedge A$ denotes the tensor product $A \otimes A$ acting in the antisymmetric tensor space $\bigwedge^2(\mathbb{C}^n)$, and more generally $\sum_{1\leq k_1<\cdots<k_\ell\leq n} \lambda_{k_1}\cdots\lambda_{k_\ell} = \operatorname{tr}(\bigwedge^\ell(A))$, where $\bigwedge^\ell(A) = A \otimes \cdots \otimes A$ acting in $\bigwedge^\ell(\mathbb{C}^n)$. Thus,

$$\det(1+A) = \sum_{j=0}^{n} \operatorname{tr}(\bigwedge^{j}(A)). \tag{A.31}$$

In order to extend the above definition to operators $A$ in a (separable) Hilbert space $\mathcal{H}$, one would need, at the very least, that the eigenvalues, $\{\lambda_k\}$ of $A$ are summable. In view of (A.5) we are led to consider operators in $\mathcal{B}_1(\mathcal{H})$. A simple calculation shows that if $A \in \mathcal{B}_1(\mathcal{H})$, then $\bigwedge^\ell(A)$ is a trace class operator on $\bigwedge^j \mathcal{H}$ with

$$\|\bigwedge^{j}(A)\|_1 \leq \frac{1}{j!}\|A\|_1^j. \tag{A.32}$$

For $A \in \mathcal{B}_1(\mathcal{H})$ we define

$$\det(1+A) \equiv \sum_{j=0}^{\infty} \operatorname{tr}(\bigwedge^{j}(A)). \tag{A.33}$$

By (A.32), the sum in (A.33) is convergent. The following properties of $\det(1+A)$ are basic.

(i) The map $A \mapsto \det(1+A)$ is Lipshitz continuous in $\mathcal{B}_1(\mathcal{H})$.

(ii) $|\det(1+A) - \det(1+B)| \le \|A-B\|_1 e^{\|A\|_1 + \|B\|_1 + 1}$.

(iii) By (A.12) and the property (ii), if $A \in \mathcal{B}_1(\mathcal{H})$ and $C_n \to C$, $D_n^* \to D^*$ strongly in $\mathcal{H}$, then as $n \to \infty$, $\det(1 + C_n A D_n) \to \det(1 + CAD)$.

(iv) Let $\mathcal{H}_1, \mathcal{H}_2$ be two separable Hilbert spaces. Let $A, A_n : \mathcal{H}_2 \to \mathcal{H}_1$ and $B, B_n : \mathcal{H}_1 \to \mathcal{H}_2$, $n = 1, 2, \ldots$, be Hilbert-Schmidt operators (i.e., $|A|^2 = A^*A \in \mathcal{B}_1(\mathcal{H}_2)$ etc.). If $A_n \to A$ and $B_n \to B$ in $\mathcal{B}_2(\mathcal{H}_2)$, $\mathcal{B}_2(\mathcal{H}_1)$ respectively, then as $n \to \infty$, $\det(1 + A_n B_n) \to \det(1 + AB)$.

(v) For $A \in \mathcal{B}_1(\mathcal{H})$, $\det(1 + zA)$ is an entire function of $z$ satisfying $|\det(1 + zA)| \le e^{|z| \|A\|_1}$.

The next set of properties shows that $\det(1 + A)$ retains many of the properties of the finite-dimensional case.

(vi) For $A, B \in \mathcal{B}_1(\mathcal{H})$, $\det(1 + A + B + AB) = \det(1 + A)\det(1 + B)$.

(vii) For $A \in \mathcal{B}_1(\mathcal{H})$, $\det(1 + A) \neq 0$ if and only if $1 + A$ is invertible.

(viii) By Lidskii's formula (A.4) and the definition (A.33), $\det(1 + A) = \prod_{k=1}^{\infty}(1 + \lambda_k)$ where $\{\lambda_k\}$ are the (nonzero) eigenvalues of $A \in \mathcal{B}_1(\mathcal{H})$, counting multiplicity.

(ix) Let $A \in \mathcal{B}_1(\mathcal{H})$ and suppose in addition that $\|A\| < 1$. Then $\log(1 + A) = \sum_{k=1}^{\infty} \frac{(-1)^{k+1}}{k} A^k$ is trace class and

$$\|\log(1 + A)\|_1 \le \|A\|_1 \sum_{k=1}^{\infty} \frac{\|A\|^{k-1}}{k} < \infty. \tag{A.34}$$

We have by Lidskii's formula (A.4),

$$\begin{aligned} \operatorname{tr}\log(1 + A) &= \sum_{k=1}^{\infty} \frac{(-1)^{k+1}}{k} \operatorname{tr}(A^k) = \sum_{k=1}^{\infty} \frac{(-1)^{k+1}}{k} \sum_{i=1}^{\infty} \lambda_i^k \\ &= \sum_{i=1}^{\infty} \log(1 + \lambda_i) = \log\det(1 + A). \end{aligned} \tag{A.35}$$

Thus for $A \in \mathcal{B}_1(\mathcal{H})$ and $\mu$ sufficiently small

$$\log\det(1 + \mu A) = \operatorname{tr}\log(1 + \mu A) \tag{A.36}$$

and

$$\det(1 + \mu A) = \exp\left(\sum_{k=1}^{\infty} \frac{(-1)^{k+1}}{k} \mu^k \operatorname{tr}(A^k)\right). \tag{A.37}$$

(x) Let $A$ be a trace class operator all of whose eigenvalues are real and less than 1. Then, by Lidskii's determinant formula and the inequality $0 \le 1 - x \le e^{-x}$ for $x \le 1$, we have

$$0 \le \det(1 - A) = \prod_{k=1}^{\infty}(1 - \lambda_k) \le \prod_{k=1}^{\infty} e^{-\lambda_k} = e^{-\operatorname{tr}(A)}. \tag{A.38}$$

(xi) Let $\mathcal{H}_1$ and $\mathcal{H}_2$ be two Hilbert spaces. If $A : \mathcal{H}_1 \to \mathcal{H}_2$ and $B : \mathcal{H}_2 \to \mathcal{H}_1$ are bounded operators such that $AB \in \mathcal{B}_1(\mathcal{H}_2)$ and $BA \in \mathcal{B}_1(\mathcal{H}_1)$, then

$$\det(1 - AB) = \det(1 - BA). \tag{A.39}$$

Finally we have the following classical result. Let $A$ be a trace class operator in $\mathcal{H} = L^2(a,b)$, $-\infty < a < b < \infty$, with a continuous kernel $k(x,y)$. Then by (A.27) $\operatorname{tr}(A) = \int_a^b k(x,x)dx$. Direct evaluation shows that $\bigwedge^\ell(A)$ has continuous kernel

$$k(x_1, \ldots, x_\ell, y_1, \ldots, y_\ell) = \frac{1}{\ell!} \det\left(k(x_i, y_j)\right)_{i,j=1}^{\ell} \tag{A.40}$$

for any $\ell \ge 1$, hence by the $\ell$-dimensional analogue of (A.27),

$$\operatorname{tr}(\bigwedge^{\ell}(A)) = \frac{1}{\ell!} \int_a^b \cdots \int_a^b \det(k(x_i, x_j))_{i,j=1}^{\ell} dx_1 \cdots dx_\ell. \tag{A.41}$$

Thus we have the Fredholm expansion

$$\det(1 + A) = \sum_{\ell=0}^{\infty} \frac{1}{\ell!} \int_a^b \cdots \int_a^b \det(k(x_i, x_j))_{i,j=1}^{\ell} dx_1 \cdots dx_\ell \tag{A.42}$$

for $A \in \mathcal{B}_1(L^2(a,b))$ with continuous kernel $k(x,y)$.

*Appendix B*

# Steepest-descent Method for the Asymptotic Evaluation of Integrals in the Complex Plane

In this appendix we describe briefly the steepest-descent method to evaluate integrals of the form

$$I(t) = \int_C e^{th(z)} g(z) dz, \qquad t \in \mathbb{R}, \tag{B.1}$$

as $t \to \infty$. Here $C$ is a contour in $\mathbb{C}$ and $h(z)$, $g(z)$ are given functions. Our goal is simply to illustrate how the methods work in contexts that are relevant to the text. A systematic exposition of the method, together with the associated method of stationary phase, can be found, for example, in [**Erd56, Mil02**].

First consider the case when $h(z)$ is *real* on $C$. This situation is analyzed by the *method of Laplace*, which shows that the leading order behavior of $I(t)$ as $t \to \infty$ comes from the points on $C$, where $h(z)$ is a maximum. These maximum points (called critical points) could be endpoints or interior points of $C$. For example, if $C = [0, \infty)$ and $h(z) = -z$, then the maximum of $h(z)$

is achieved at the endpoint $z = 0$ and integrating by parts,

$$I(t) = \int_0^\infty e^{-tx} g(x)dx = \frac{g(0)}{t} + o(t^{-1}), \tag{B.2}$$

assuming that $g$ is bounded and has an integrable derivative. Alternatively, if $h(z)$ achieves its maximum at an interior point $z_0$, say, of $C$, then again the leading behavior will arise from $z_0$, but the order of decay will be slower than $t^{-1}$. For example, consider the Gamma function

$$\Gamma(t) = \int_0^\infty e^{-x} x^{t-1} dx, \qquad t > 0. \tag{B.3}$$

We have

$$\Gamma(t+1) = \int_0^\infty e^{-x} x^t dx = t^{t+1} \int_0^\infty e^{-ts} s^t ds = t^{t+1} \int_0^\infty e^{th(s)} ds, \tag{B.4}$$

where

$$h(s) = -s + \log s. \tag{B.5}$$

Now $h(s)$ attains its maximum $h(1) = -1$ at $s = 1$. For $\epsilon$ small and positive, and $|s-1| < \epsilon$, $h(s) = -1 - \frac{1}{2}(s-1)^2 g(s)$, where $g(s)$ is smooth in the neighborhood $(1-\epsilon, 1+\epsilon)$ of $s = 1$ and $g(1) = 1$. Write

$$\Gamma(t+1) = t^{t+1} \int_{\mathbb{R}_+ \setminus (1-\epsilon, 1+\epsilon)} e^{th(s)} ds + t^{t+1} e^{-t} \int_{1-\epsilon}^{1+\epsilon} e^{-\frac{t}{2}(s-1)^2 g(s)} ds. \tag{B.6}$$

Now

$$\begin{aligned} &\int_{\mathbb{R}_+ \setminus (1-\epsilon, 1+\epsilon)} e^{th(s)} ds \\ &= \int_0^{1-\epsilon} e^{th(s)} ds + \int_{1+\epsilon}^{e} e^{th(s)} ds + \int_e^\infty e^{th(s)} ds \\ &\le (1-\epsilon) e^{th(1-\epsilon)} + (e - 1 - \epsilon) e^{th(1+\epsilon)} + \int_e^\infty e^{t\frac{h(e)}{e} s} ds \end{aligned} \tag{B.7}$$

as $h(s)$ is monotonically increasing (resp. decreasing) in $(0,1)$ (resp. $(1,\infty)$), and $\frac{h(s)}{s}$ is monotonically decreasing in $(e, \infty)$. Thus

$$\int_{\mathbb{R}_+ \setminus (1-\epsilon, 1+\epsilon)} e^{th(s)} ds \le c_\epsilon e^{-\gamma_\epsilon t} \tag{B.8}$$

for some constants $\gamma_\epsilon > 1$ and $c_\epsilon > 0$. On the other hand, changing variables $\xi = (s-1)\sqrt{\frac{g(s)}{2}}$, $|s-1| < \epsilon$, we find

$$\int_{1-\epsilon}^{1+\epsilon} e^{-\frac{t}{2}(s-1)^2 g(s)} ds = \int_{-\epsilon\sqrt{\frac{g(1-\epsilon)}{2}}}^{\epsilon\sqrt{\frac{g(1+\epsilon)}{2}}} e^{-t\xi^2} k(\xi) d\xi, \tag{B.9}$$

where $k(\xi)$ is smooth and positive, and $k(0) = \sqrt{2}$. Letting $t \to \infty$, a standard Gaussian calculation shows that

$$\text{(B.10)} \qquad \int_{-\epsilon\sqrt{\frac{g(1-\epsilon)}{2}}}^{\epsilon\sqrt{\frac{g(1+\epsilon)}{2}}} e^{-t\xi^2} k(\xi) d\xi = \sqrt{\frac{2\pi}{t}} + o(t^{-1/2}).$$

Thus as $t \to \infty$,

$$\text{(B.11)} \qquad \Gamma(t+1) = \sqrt{2\pi} t^{t+1/2} e^{-t}(1 + o(1)),$$

which is Stirling's well-known formula. Note that, as indicated above, the decay in (B.10) is slower than in (B.2).

If $h(z)$ is *purely imaginary*, then the *method of stationary phase* of Stokes and Kelvin (see [**Erd56, Mil02**]) again shows that the main contribution to $I(t)$ as $t \to \infty$ comes from the critical point $z_0$ (now called a point of stationary phase), $h'(z_0) = 0$. Although the integral *localizes* to the point $z_0$ where $h'(z_0) = 0$, the mechanism is completely different from the mechanism in the case of the method of Laplace. In the case of Laplace, $|e^{th(z)}|$ is simply maximal near $z_0$. In the case when $h(z) \in i\mathbb{R}$, however, $|e^{th(z)}| = 1$ and the contribution to $I(t)$ as $t \to \infty$ from any point $z \in C$ is canceled out by the rapid oscillation of the integral near $z$. Such cancelations are minimal at points where $|h(z)|$ changes at slower rate: in particular, the main contribution to $I(t)$ comes from points $z_0$ where $h'(z_0) = 0$.

Thus in two distinct situations, $h(z) \in \mathbb{R}$ or $h(z) \in i\mathbb{R}$, the integral $I(t)$ localizes to a neighborhood of a point on the contour as $t \to \infty$. The situation is very different if $h$ has *nontrivial real and imaginary parts.* For example, consider

$$\text{(B.12)} \qquad I_\epsilon(t) = \int_{-\infty}^{\infty} e^{-t(x+i\epsilon)^2}, \qquad \epsilon \neq 0.$$

This classical integral equals $\sqrt{\frac{\pi}{t}}$, independent of $\epsilon$, from which we see that $I_\epsilon(t)$ does not localize to any point $z \in C = \mathbb{R}$ as $t \to \infty$. When $h(z)$ and $g(z)$ in (B.1) are *analytic* in some region in $\mathbb{C}$ containing $C$, we may try to deform the integral to a new contour $C'$ on which $\operatorname{Im} h(z)$ is constant, $\operatorname{Im} h(z) = c$. Then

$$\text{(B.13)} \qquad I(t) = \int_{C'} e^{th(z)} g(z) dz = e^{itc} \int_{C'} e^{t \operatorname{Re} h(z)} g(z) dz,$$

and the Laplace's method can now be applied. This deformation method is known as the *steepest-descent method* and was introduced by Riemann.

Note first that as $\operatorname{Im} h(z) = c$ on $C'$, the derivative of $\operatorname{Im} h(z)$ along $C'$ is zero. On the other hand, if $\operatorname{Re} h(z)$, $z \in C'$, is maximal at some point $z_0 \in C'$, then the derivative of $\operatorname{Re} h(z)$ is also zero at $z_0$ along $C'$. By the Cauchy-Riemann equation, we conclude that $h'(z_0) = 0$. Thus any such contour $C'$ must pass through critical points $z = z_0$, $h'(z_0) = 0$. Secondly, note that a critical point $z_0$ for $h$ never corresponds to a local maximum of $\operatorname{Re} h(z)$ at $z = z_0$ in $\mathbb{C}$. Indeed, by the Cauchy-Riemann equation, $\operatorname{Re} h(z)$ is a harmonic function of $x = \operatorname{Re}(z)$, $y = \operatorname{Im}(y)$, and hence $z_0$ is always a saddle point for $\operatorname{Re} h(z)$. For this reason, the steepest-descent method is sometimes called the saddle-point method. Finally, as $\operatorname{Im}(h(z) - h(z_0)) = 0$ in $C'$, it follows, again by the Cauchy-Riemann equation, that at $z_0$, $\operatorname{Re}(h(z) - h(z_0))$ changes maximally in the direction of on $C'$. Thus $C'$ is a *curve of steepest descent* for $\operatorname{Re} h(z)$ near $z_0$.

We illustrate the steepest-descent method for the Airy function (see, e.g., [**AS64**])

(B.14) $$\operatorname{Ai}(x) = \frac{1}{2\pi} \int_C e^{i(\frac{1}{3}s^3 + xs)} ds.$$

Here $C$ is a contour which passes from $\infty e^{i\theta_1}$ to $\infty e^{i\theta_2}$, where $\theta_1 \in (2\pi/3, \pi)$ and $\theta_2 \in (0, \pi/3)$. With this choice of $C$, the integral converges and a simple calculation shows that $\operatorname{Ai}(x)$ (indeed) solves the Airy equation $\operatorname{Ai}''(x) = x \operatorname{Ai}(x)$. We are interested in evaluating $\operatorname{Ai}(x)$ as $x \to \infty$, $x \in \mathbb{R}$.

First we scale $s \mapsto z = x^{-1/2}s$ and set $t = x^{3/2}$. We find, after using Cauchy's theorem, that with the same contour $C$,

(B.15) $$I(t) = \operatorname{Ai}(t^{2/3}) = \frac{t^{1/3}}{2\pi} \int_C e^{th(z)} dz, \qquad h(z) = i(\frac{1}{3}z^3 + z).$$

The contour $C'$ of steepest descent that contains a point $z_0 \in C'$ such that $\operatorname{Re} h(z)$, $z \in C'$, is maximum at $z = z_0$, must pass the critical point of $h$, $h'(z) = i(z^2 + 1) = 0$, i.e.,

(B.16) $$z = \pm z_0 = \pm i.$$

Now if we write $z = x + iy$, then $\operatorname{Im} h(z) = \frac{1}{3}x^3 - xy^2 + x$. At $z = \pm z_0$, $\operatorname{Im} h(\pm z_0) = 0$ and the points where $\operatorname{Im} h(z) = \operatorname{Im} h(\pm z_0) = 0$ are given by the vertical line $x = 0$ or the hyperbola $y^2 = 1 + \frac{1}{3}x^2$; see Figure B.1. As $\tan \frac{\pi}{6} = -\tan \frac{5\pi}{6} = \frac{1}{\sqrt{3}}$, we may choose $C'$ to be the part of the hyperbola lying in the upper half-plane, $y = \sqrt{1 + \frac{1}{3}x^2}$, $x \in \mathbb{R}$. On this contour $h(z)$ is real and $\operatorname{Re} h(x+iy) = h(x+iy) = 2y - \frac{8}{3}y^3$. As $\frac{d}{dy} \operatorname{Re} h(x+iy) = 2 - 8y^2 < 0$ for $y \geq 1$, we see that $z = i$ is a maximum point for $\operatorname{Re} h(z)$ on $C'$. (Note that the curve $\operatorname{Re} z = x$ through $z_0 = i$ is a curve of *steepest ascent* for $\operatorname{Re} h(z)$.) As $h(z)$ is real on $C'$, the Laplace's method applies and $I(t)$ localizes on $C'$ as $t \to \infty$ to a small neighborhood of $z_0 = i$, where we find $\operatorname{Re} h(x + iy) = h(x + iy) = -\frac{2}{3} - x^2 + O(x^4)$. We obtain for $\epsilon$ small and

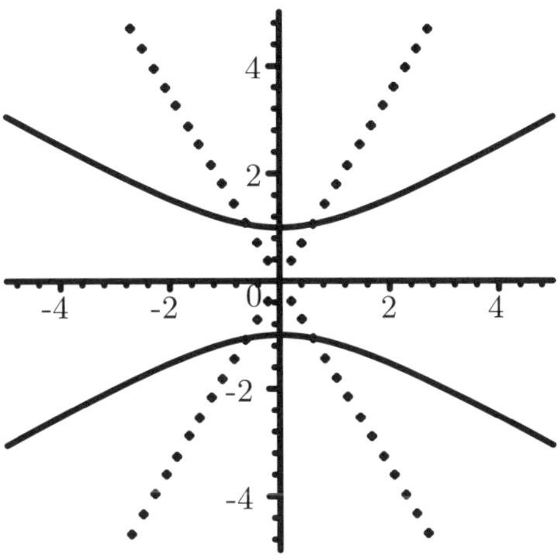


**Figure B.1.** Steepest-descent curves for Airy function

positive,

$$I(t) \sim \frac{t^{1/3}}{2\pi} \int_{-\epsilon}^{\epsilon} e^{t(-\frac{2}{3}-x^2+O(x^4))}(1+O(x))dx. \tag{B.17}$$

Gaussian integration then leads to the formula

$$\mathrm{Ai}(x) = I(t) \sim \frac{1}{2\sqrt{\pi}t^{1/6}} e^{-\frac{2}{3}t} = \frac{1}{2\sqrt{\pi}x^{1/4}} e^{-\frac{2}{3}x^{3/2}} \tag{B.18}$$

as $x \to +\infty$ (see [**AS64**]).

If $x \to -\infty$, the critical points are given by $\pm z_0 = \pm 1$, and both of these points contribute to the leading asymptotics of $\mathrm{Ai}(x)$.

**Exercise B.1.** Evaluate $\mathrm{Ai}(x)$ as $x \to -\infty$.

Finally, it is clear that we do not need to choose $C'$ precisely as the curve of steepest descent. If $C'$ agrees with the steepest-descent curve in a neighborhood $C'_\epsilon$ of the critical point $z_0$ and the contribution of the integral is of smaller exponential order on $C \setminus C'_\epsilon$, then the method of steepest descent still goes through. We may call such a curve a *curve of steep descent*. Moreover, in other situations, $C'$ does not even need to agree precisely with the steepest-descent curve near $z_0$. This is true, in particular, in the case of the Krawtchouk polynomials in Chapter 12 (see Figure 12.24 and also Figures 12.25).

*Appendix C*

# Basic Results of Stochastic Calculus

In this appendix we define several notions which are used in stochastic calculus and list (without proof) a few results which are used in the text. All notions and proofs below can be found, for example, in [**RY99**]. We will focus only on results for processes with continuous paths.

We will always assume that $(\Omega, \mathcal{F}, \{\mathcal{F}_t\}_{t\geq 0}, \mathbb{P})$ is a filtered probability space with a right continuous filtration: $(\Omega, \mathcal{F}, \mathbb{P})$ is a probability space and $\{\mathcal{F}_t\}_{t\geq 0}$ is an increasing family of $\sigma$-subalgebras of $\mathcal{F}$ such that $\bigcap_{\epsilon>0} \mathcal{F}_{t+\epsilon} = \mathcal{F}_t$ for all $t \geq 0$.

**Definition C.1.** A *continuous path stochastic process* $\{X_t\}_{t\geq 0}$ is a family of $\mathcal{F}$-measurable map from $(\Omega, \mathcal{F})$ into $(C([0,\infty)), \mathcal{B})$ where $\mathcal{B}$ is the Borel $\sigma$-algebra on $C([0,\infty))$. The process $\{X_t\}_{t\geq 0}$ is *adapted* to the filtration $\{\mathcal{F}_t\}_{t\geq 0}$ if $X_t$ is $\mathcal{F}_t$-measurable for every $t \geq 0$. The *natural filtration* associated to a stochastic process, $X = \{X_t\}$, is denoted by $\{\mathcal{F}_t^X\}_{t\geq 0}$ and is defined to be $\mathcal{F}_t^X \equiv \sigma\left(\sigma(X_s : s \leq t) \cup \mathcal{N}\right)$ where $\mathcal{N}$ consists of the subsets $A \in \mathcal{F}$ with $\mathbb{P}(A) = 0$. For brevity, in the remainder of this appendix we will simply call $\{X_t\}_{t\geq 0}$ a stochastic process.

**Definition C.2.** An $\{\mathcal{F}_t\}$-*stopping time* $\tau$ is a nonnegative random variable satisfying $\{\tau \leq t\} \in \mathcal{F}_t$ for all $t \geq 0$. Given an $\{\mathcal{F}_t\}$-*stopping time* $\tau$, the stopped $\sigma$-algebra $\mathcal{F}_\tau$ is defined as

$$\mathcal{F}_\tau \equiv \{A \in \mathcal{F} : A \cap \{\tau \leq t\} \in \mathcal{F}_t \text{ for all } t \geq 0\}.$$

**Definition C.3.** Let $\mu$ be a probability measure on $\left(\mathbb{R}^d, \mathcal{B}\left(\mathbb{R}^d\right)\right)$ ($\mathbb{R}^d$ endowed with the Borel $\sigma$-algebra). An adapted $d$-dimensional process $X = \{X_t\}_{t\geq 0}$ is said to be a *Markov process* with initial distribution $\mu$ if

- $\mathbb{P}^\mu\{X_0 \in A\} = \mu(A)$ for all $A \in \mathcal{B}\left(\mathbb{R}^d\right)$ and
- for $s, t > 0$ and $A \in \mathcal{B}(\mathbb{R}^d)$, $\mathbb{P}^\mu\{X_{t+s} \in A|\mathcal{F}_s\} = \mathbb{P}^\mu\{X_{t+s} \in A|X_s\}$, $\mathbb{P}^\mu$-a.s.

$X$ is a *strong Markov process* if for any stopping time $\tau$ of $\{\mathcal{F}_t\}_{t\geq 0}$ and $A \in \mathcal{B}(\mathbb{R}^d)$,

- $\mathbb{P}^\mu\{X_{t+\tau} \in A|\mathcal{F}_\tau\} = \mathbb{P}^\mu\{X_{t+\tau} \in A|X_\tau\}$, $\mathbb{P}^\mu$-a.s on $\{\tau < \infty\}$.

**Brownian motion.** Brownian motion is the simplest and most important continuous time stochastic process with continuous paths. One of many equivalent definitions of Brownian motion is as follows.

**Definition C.4.** A *Brownian motion* is a real-valued, continuous stochastic process $\{X_t\}_{t\geq 0}$ with independent stationary increments. In other words:

- Continuity: $\mathbb{P}$-a.s. the map $t \mapsto X_t$ is continuous.
- Independent increments: If $s \leq t$, $X_t - X_s$ is independent of $\mathcal{F}_s^X$.
- Stationary increments: If $s \leq t$, $X_t - X_s$ and $X_{t-s} - X_0$ have the same distribution.

A Brownian motion is called *standard* if $X_0 = 0\ \mathbb{P} - a.s.$ and $\mathbb{E}(X_t) = 0$, $\mathbb{E}(X_t^2) = t$ for all $t \geq 0$.

Brownian motion is a strong Markov process. The following theorem describes basic properties of Brownian motion.

**Theorem C.5.** *If $\{X_t\}_{t\geq 0}$ is a Brownian motion, then $X_t - X_0$ is a normal random variable with mean $rt$ and variance $\sigma^2 t$, where $r$ and $\sigma$ are real constants.*

**Stochastic integrals.** We briefly introduce Ito's stochastic integral with respect to Brownian motion. Our description is not the most general but will suit our needs. For a general treatment of stochastic integration see, for example, [**RY99**]. We begin with a Hilbert space of integrands $\mathcal{H}$ and an appropriate dense subspace $\mathcal{H}'$ on which we first define the integral. We then note that the stochastic integral can be extended to the entire space $\mathcal{H}$.

Let $W_t$ be a standard Brownian motion, and let $\mathcal{F}_t = \mathcal{F}_t^W$ be the natural filtration associated to $W$.

**Definition C.6.** Fix $T > 0$ and define

$$\mathcal{H} \equiv \left\{ \{H_t\}_{t\in[0,T]} : H_t \text{ is adapted to } \{\mathcal{F}_t\} \text{ and } \mathbb{E}\left[\int_0^T |H_t|^2 dt\right] < \infty \right\}.$$

The *simple stochastic processes* are defined to be the set of processes, $\{H_t\} \in \mathcal{H}$, such that $H_t = \sum_{n=1}^N \phi_i \chi_{(t_{i-1},t_i]}(t)$ for some $N < \infty$ and $0 = t_0 < t_1 < t_2 < \cdots < t_N = T$ where $\phi_i$ is a $\mathcal{F}_{t_{i-1}}$ measurable random variable. Denote the subspace of simple processes by $\mathcal{H}'$.

The *Ito integral* of $\{H_t\} \in \mathcal{H}'$ with respect to $W$ where $H_t = \sum_{n=1}^N \phi_i \chi_{(t_{i-1},t_i]}(t)$ is given by

$$I(H)(t) \equiv \sum_{n=1}^{N} \phi_{t_{i-1}\wedge t}(W_{t_i\wedge t} - W_{t_{i-1}\wedge t}), \text{ for } t \in [0,T].$$

This integral extends (see for example [**RY99**]) to the entire space $\mathcal{H}$. We denote the stochastic integral of $H \in \mathcal{H}$ with respect to $W$ by $I(H)(t) = \int_0^t H_s dW_s$. The limiting procedure which leads to the stochastic integral above involves convergence in $L^2(\Omega \times [0,T])$. The space $\mathcal{H}$ of integrands can be enlarged by weakening the restriction $\mathbb{E}[\int_0^T |H_t|^2 dt] < \infty$ to $\mathbb{P}\left(\int_0^T |H_t|^2 dt < \infty\right) = 1$ in which case $L^2$ convergence is replaced by convergence in probability. We note (see [**RY99**]) that the limit so constructed is almost surely a continuous path stochastic process. Two basic properties of the stochastic integral are contained in the next theorem.

**Theorem C.7.** *For any $H \in \mathcal{H}$, the following relations hold:*

- $\mathbb{E}\left[\left(\int_0^T H_t dW_t\right)^2\right] = \mathbb{E}\left[\int_0^T |H_t|^2 dt\right]$,
- $\mathbb{E}\left[\sup_{t\in[0,T]} \left|\int_0^t H_s dW_s\right|^2\right] \le 4\mathbb{E}\left[\int_0^T |H_t|^2 dt\right]$.

**Ito's formula**. We now define the notion of an Ito process and state Ito's formula, first in one dimension and then in $n$ dimensions.

**Definition C.8.** Assume that $\{\mathcal{F}_t\}$ is a filtration and $\{W_t\}$ is an $\mathcal{F}_t$-Brownian motion. $\{X_t\}_{t\ge 0}$ is an *Ito process* if it can be written as

$$X_t = X_0 + \int_0^t K_s ds + \int_0^t H_s dW_s, \text{ for all } t \ge 0, \ \mathbb{P}\text{-a.s.},$$

where

- $X_0$ is $\mathcal{F}_0$-measurable,
- $\{K_t\}$ and $\{H_t\}$ are $\mathcal{F}_t$ adapted processes,
- $\mathbb{P}$-a.s. $\int_0^T |K_s| ds < \infty$ for all $T > 0$,
- $\mathbb{P}$-a.s. $\int_0^T |H_s|^2 ds < \infty$ for all $T > 0$.

**Theorem C.9** (One-dimensional Ito's formula). *Let $\{X_t\}$ be an Ito process*

$$X_t = X_0 + \int_0^t K_s ds + \int_0^t H_s dW_s,$$

*and let $f : \mathbb{R} \to \mathbb{R}$ be a twice continuously differentiable function. Then*

$$f(X_t) = f(X_0) + \int_0^t f'(X_s)dX_s + \frac{1}{2}\int_0^t f''(X_s)d\langle X, X\rangle_s,$$

*where*

$$\langle X, X\rangle_t = \int_0^t |H_s|^2 ds$$

*is the* quadratic variation process *of $X_t$ and*

$$\int_0^t f'(X_s)dX_s = \int_0^t f'(X_s)K_s ds + \int_0^t f'(X_s)H_s dW_s.$$

**Theorem C.10** ($n$-dimensional Ito formula). *Let $(X_t^1, \ldots, X_t^n)$ be an $n$-dimensional Ito process*

$$X_t^i = X_0^i + \int_0^t K_s^i ds + \sum_{j=1}^{p} \int_0^t H_s^{i,j} dW_s^j,$$

*where $(W_1, \ldots, W_p)$ is a $p$-dimensional Brownian motion (i.e., the $p$ coordinates are independent standard Brownian motions) and $K_s^i, H_s^{i,j}$ satisfy analogous conditions to those in the definition of an Ito process above. If $f : \mathbb{R} \times \mathbb{R}^n \to \mathbb{R}$ is once continuously differentiable in the first coordinate and has continuous second-order partial derivatives in the remaining $n$ coordinates, then*

$$\begin{aligned} f(t, X_t^1, \ldots, X_t^n) =& f(0, X_0^1, \ldots, X_0^n) + \int_0^t \frac{\partial f}{\partial s}(s, X_s^1, \ldots, X_s^n)ds \\ &+ \sum_{i=1}^{n} \int_0^t \frac{\partial f}{\partial x_i}(s, X_s^1, \ldots, X_s^n)dX_s^i \\ &+ \frac{1}{2}\sum_{i,j=1}^{n} \int_0^t \frac{\partial^2 f}{\partial x_i \partial x_j}(s, X_s^1, \ldots, X_s^n)d\langle X^i, X^j\rangle_s, \end{aligned}$$

*where*

- $dX_s^i = K_s^i ds + \sum_{j=1}^{p} H_s^{i,j} dW_s^j$,
- $d\langle X^i, X^j\rangle_s = \sum_{m=1}^{p} H_s^{i,m} H_s^{j,m} ds$.

A linear combination of stochastic integrals (considered as a function of time) with respect to independent standard Brownian motions may, under certain conditions, be a standard Brownian motion. The next theorem provides a sufficient condition.

**Theorem C.11** (Levy's characterization of Brownian motion). *Let $W_t^{(1)}, \dots, W_t^{(n)}$ be $n$ independent one-dimensional standard Brownian motions. Let $h^{(1)}, \dots, h^{(n)} \in \mathcal{H}$, where $\mathcal{H}$ is defined as above for the case of the $n$-dimensional Brownian motion $W = (W^{(1)}, \dots, W^{(n)})$. Let $Z_t = \sum_{k=1}^n \int_0^t h_s^{(k)} dW_s^{(k)}$. If*

$$\sum_{k=1}^n \int_0^t \left(h_s^{(k)}\right)^2 ds = t$$

*for all $t \geq 0$, $\mathbb{P}$-a.s., then $Z_t$ is a standard Brownian motion. Moreover, if $Z_t^i = \sum_{k=1}^n \int_0^t h_s^{(i,k)} dW_s^{(k)}$, $i = 1, \dots, m$, and*

$$\sum_{k=1}^n \int_0^t h_s^{(i,k)} h_s^{(j,k)} ds = t\delta_{ij}$$

*for all $t \geq 0$, $\mathbb{P}$-a.s., then $Z_t^i$ are independent standard Brownian motions.*

**Martingale.** We now give the definition of a martingale and a local martingale.

**Definition C.12.** A stochastic process $\{X_t\}_{t \geq 0}$ is an $\{\mathcal{F}_t\}$-*martingale* if for every $0 \leq s \leq t < \infty$ the equality $\mathbb{E}[X_t|\mathcal{F}_s] = X_s$ holds $\mathbb{P}$-a.s. The process is a *submartingale (supermartingale)* if the equality is replaced by $\geq$ ($\leq$, respectively).

**Theorem C.13** (Optional sampling theorem). *If $\{X_t\}_{t \geq 0}$ is a martingale with respect to the filtration $\{\mathcal{F}_t\}_{t \geq 0}$ and if $\tau_1$ and $\tau_2$ are two stopping times such that $0 \leq \tau_1 \leq \tau_2 \leq K < \infty$, then $\mathbb{E}[X_{\tau_2}|\mathcal{F}_{\tau_1}] = X_{\tau_1}$ $\mathbb{P}$-a.s. This statement in the cases of submartingales and supermartingales is true if the equality is replaced with the inequalities $\geq$ and $\leq$, respectively.*

**Definition C.14.** A process, $\{X_t\}_{t \geq 0}$, is an $\{\mathcal{F}_t\}$-*local martingale* if there exists an increasing sequence of $\{\mathcal{F}_t\}$-stopping times, $0 \leq \tau_1 \leq \tau_2 \leq \cdots \leq \tau_n \leq \cdots$, such that $\lim_{n \to \infty} \tau_n = \infty$ a.s. and each of the processes $\{X_{t \wedge \tau_n}\}_{t \geq 0}$ is an $\{\mathcal{F}_t\}$-martingale where $t \wedge \tau_n$ denotes the minimum of $t$ and $\tau_n$.

Stochastic integrals, $\int_0^t H_s dW_s$ with $\int_0^T |H_s|^2 ds < \infty$ almost surely for all $T > 0$, are local martingales. If a local martingale is bounded for all $t$, then it is a martingale; see for example [**RY99**]. A standard example of a local martingale which is not a martingale is given by $X_t = \frac{1}{|B_t^{(3)}|}$ where $B_t^{(3)} = (B_t^1, B_t^2, B_t^3)$ is a three-dimensional Brownian motion (see, for example, [**KS91**, Exercise 3.36]).

**Stochastic differential equations, diffusions, and generators.** Let $W = (W^1, \dots, W^p)$ be an $\mathbb{R}^p$-valued $\mathcal{F}_t$-Brownian motion, let $b : \mathbb{R}^+ \times \mathbb{R}^n \to \mathbb{R}^n$ be a smooth $\mathbb{R}^n$-valued function in all coordinates, let $\sigma : \mathbb{R}^+ \times \mathbb{R}^n \to \mathbb{R}^{n \times p}$

be a smooth $n \times p$ matrix-valued function in all coordinates, and let $Z = (Z^1, \dots, Z^n)$ be a $\mathcal{F}_0$-measurable $\mathbb{R}^n$-valued random variable. The integral equation

$$X_t = Z + \int_0^t b(s, X_s)ds + \int_0^t \sigma(s, X_s)dW_s \tag{C.1}$$

is called a stochastic differential equation and is often written in shorthand as

$$dX_t = b(t, X_t)dt + \sigma(t, X_t)dW_t, \qquad X_0 = Z. \tag{C.2}$$

A solution to (C.1) on the time interval $[0, T]$ is an $\mathbb{R}^n$-valued process $\{X_t\}_{t \in [0,T]}$ which is adapted to $\{\mathcal{F}_t\}$ and such that $\mathbb{P}$-a.s. for any $t \in [0, T]$ and $i \in \{1, \dots, n\}$

$$X_t^i = Z^i + \int_0^t b^i(s, X_s)ds + \sum_{j=1}^p \int_0^t \sigma_{ij}(s, X_s)dW_s^j.$$

Questions of existence and uniqueness of solutions are addressed in [**RY99**]. Existence and uniqueness of solutions can be shown for each of the stochastic differential equations which arise in this text.

Solutions to stochastic differential equations of the form (C.1) are called *diffusion processes*. The generator of a diffusion process is a second-order differential operator $L_{t,x}$ defined as follows: If $f : \mathbb{R}^n \to \mathbb{R}$ has continuous second-order partial derivatives in all coordinates, then

$$\lim_{h \downarrow 0} \frac{\mathbb{E}[f(X_{t+h})|X_t = x] - f(x)}{h} = (L_{t,x}f)(x),$$

where

$$(L_{t,x}f)(x) \equiv \frac{1}{2} \sum_{i,j=1}^n a_{ij}(t, x) \frac{\partial^2 f}{\partial x_i \partial x_j}(x) + \sum_{j=1}^n b_j(t, x) \frac{\partial f}{\partial x_j}(x), \tag{C.3}$$

and $a(t, x) = \sigma(t, x)\sigma^*(t, x)$ where $\sigma^*$ is the transpose of $\sigma$. For $t > s \geq 0$, the transition probability $p(s, t, x, y)$ from position $x$ at time $s$ to position $y$ at time $t$ for the (time homogeneous) diffusion $dX_t^i = b^i(t, X_t)dt + \sum_{j=1}^p \sigma_{ij}(x, X_t)dW_t^i$ is the (unique) solution of the Kolmogorov backward equation

$$\partial_s p(s, t, x, y) + L_{s,x}p(s, t, x, y) = 0 \tag{C.4}$$

and is also a solution of the Kolmogorov forward equation

$$\partial_t p(s, t, x, y) - L_{t,y}^* p(s, t, x, y) = 0 \tag{C.5}$$

subject to the condition $p(t, t, x, y) = \delta_x(y)$.

We remark that a diffusion is an Ito process. Ito's formula, Theorem C.10, can be written as

(C.6)
$$f(X_t) = f(X_0) + \sum_{i=1}^{n} \sum_{j=1}^{p} \int_0^t \sigma_{ij}(s, X_s) \frac{\partial f}{\partial x_i}(X_s) dW_s^j + \int_0^t (L_{s,x} f)(X_s) ds,$$

when $\{X_t\}_{t \geq 0}$ is a diffusion process with generator $L_{t,x}$.

# Bibliography


[ABDF11] G. Akemann, J. Baik, and P. Di Francesco (eds.), *The Oxford handbook of random matrix theory*, Oxford University Press, Oxford, 2011. MR2920518 (2012m:60007)

[AD95] D. Aldous and P. Diaconis, *Hammersley's interacting particle process and longest increasing subsequences*, Probab. Theory Related Fields **103** (1995), no. 2, 199–213. MR1355056 (96k:60017)

[AD99] ———, *Longest increasing subsequences: from patience sorting to the Baik-Deift-Johansson theorem*, Bull. Amer. Math. Soc. (N.S.) **36** (1999), no. 4, 413–432. MR1694204 (2000g:60013)

[AD14] A. Auffinger and M. Damron, *A simplified proof of the relation between scaling exponents in first-passage percolation*, Ann. Probab. **42** (2014), no. 3, 1197–1211. MR3189069

[AG94] G. S. Ammar and W. B. Gragg, *Schur flows for orthogonal Hessenberg matrices*, Hamiltonian and gradient flows, algorithms and control, Fields Inst. Commun., vol. 3, Amer. Math. Soc., Providence, RI, 1994, pp. 27–34. MR1297983 (95m:15013)

[AGZ10] G. W. Anderson, A. Guionnet, and O. Zeitouni, *An introduction to random matrices*, Cambridge Studies in Advanced Mathematics, vol. 118, Cambridge University Press, Cambridge, 2010. MR2760897 (2011m:60016)

[AL75] M. J. Ablowitz and J. F. Ladik, *Nonlinear differential-difference equations*, J. Mathematical Phys. **16** (1975), 598–603. MR0377223 (51 #13396)

[AL76] ———, *Nonlinear differential-difference equations and Fourier analysis*, J. Mathematical Phys. **17** (1976), no. 6, 1011–1018. MR0427867 (55 #897)

[And83] C. Andréief, *Note sur une relation les intégrales définies des produits des fonctions*, Mém. de la Soc. Sci. Bordeaux **2** (1883), 1–14.

[And87] D. André, *Solution directe du problème résolu par M. Bertrand*, Comptes Rendus Acad. Sci. Paris. **105** (1887), 436–437.

[AS64] M. Abramowitz and I. A. Stegun, *Handbook of mathematical functions with formulas, graphs, and mathematical tables*, National Bureau of Standards Applied Mathematics Series, vol. 55, U.S. Government Printing Office, Washington, D.C., 1964. MR0167642 (29 #4914)

[AvM03] M. Adler and P. van Moerbeke, *Recursion relations for unitary integrals, combinatorics and the Toeplitz lattice*, Comm. Math. Phys. **237** (2003), no. 3, 397–440. MR1993333 (2004g:37109)

[Bac14] E. Bachmat, *Mathematical adventures in performance analysis: From storage systems, through airplane boarding, to express line queues*, Birkhauser, 2014, Modeling and Simulation in Science, Engineering and Technology.

[Bai03] J. Baik, *Riemann-Hilbert problems for last passage percolation*, Recent developments in integrable systems and Riemann-Hilbert problems (Birmingham, AL, 2000), Contemp. Math., vol. 326, Amer. Math. Soc., Providence, RI, 2003, pp. 1–21. MR1989002 (2004h:60147)

[Bar01] Y. Baryshnikov, *GUEs and queues*, Probab. Theory Related Fields **119** (2001), no. 2, 256–274. MR1818248 (2002a:60165)

[BB68] R. M. Baer and P. Brock, *Natural sorting over permutation spaces*, Math. Comp. **22** (1968), 385–410. MR0228216 (37 #3800)

[BB92] B. Bollobás and G. Brightwell, *The height of a random partial order: concentration of measure*, Ann. Appl. Probab. **2** (1992), no. 4, 1009–1018. MR1189428 (94b:06005)

[BBD08] J. Baik, R. Buckingham, and J. DiFranco, *Asymptotics of Tracy-Widom distributions and the total integral of a Painlevé II function*, Comm. Math. Phys. **280** (2008), no. 2, 463–497. MR2395479 (2009e:33068)

[BBDS06] J. Baik, A. Borodin, P. Deift, and T. Suidan, *A model for the bus system in Cuernavaca (Mexico)*, J. Phys. A **39** (2006), no. 28, 8965–8975. MR2240467 (2007d:82034)

[BBS$^+$06] E. Bachmat, D. Berend, L. Sapir, S. Skiena, and N. Stolyarov, *Analysis of aeroplane boarding via spacetime geometry and random matrix theory*, J. Phys. A **39** (2006), no. 29, L453–L459. MR2247486 (2007h:60008)

[BBSS07] E. Bachmat, D. Berend, L. Sapir, and S. Skiena, *Optimal boarding policies for thin passengers*, Adv. in Appl. Probab. **39** (2007), no. 4, 1098–1114. MR2381590 (2008m:49161)

[BC14] A. Borodin and I. Corwin, *Macdonald processes*, Probab. Theory Related Fields **158** (2014), no. 1-2, 225–400. MR3152785

[BD02] A. Borodin and P. Deift, *Fredholm determinants, Jimbo-Miwa-Ueno $\tau$-functions, and representation theory*, Comm. Pure Appl. Math. **55** (2002), no. 9, 1160–1230. MR1908746 (2005f:35234)

[BDJ99] J. Baik, P. Deift, and K. Johansson, *On the distribution of the length of the longest increasing subsequence of random permutations*, J. Amer. Math. Soc. **12** (1999), no. 4, 1119–1178. MR1682248 (2000e:05006)

[BDJ00] ———, *On the distribution of the length of the second row of a Young diagram under Plancherel measure*, Geom. Funct. Anal. **10** (2000), no. 4, 702–731. MR1791137 (2001m:05258a)

[BDT88] R. Beals, P. Deift, and C. Tomei, *Direct and inverse scattering on the line*, Mathematical Surveys and Monographs, vol. 28, American Mathematical Society, Providence, RI, 1988. MR954382 (90a:58064)

[Bee97] C. Beenakker, *Random-matrix theory of quantum transport*, Rev. Mod. Phys. **69** (1997), no. 3, 731–808.

[BF08] A. Borodin and P. L. Ferrari, *Large time asymptotics of growth models on space-like paths. I. PushASEP*, Electron. J. Probab. **13** (2008), no. 50, 1380–1418. MR2438811 (2009d:82104)

[BFS08] A. Borodin, P. L. Ferrari, and T. Sasamoto, *Large time asymptotics of growth models on space-like paths. II. PNG and parallel TASEP*, Comm. Math. Phys. **283** (2008), no. 2, 417–449. MR2430639 (2009k:82090)

[BG] A. Borodin and V. Gorin, *Lectures on integrable probability*, arXiv:1212.3351.

[BHKY] R. Bauerschmidt, J. Huang, A. Knowles, and H.-T. Yau, *Bulk eigenvalue statistics for random regular graphs*, arXiv:1505.06700.

[BI99] P. Bleher and A. Its, *Semiclassical asymptotics of orthogonal polynomials, Riemann-Hilbert problem, and universality in the matrix model*, Ann. of Math. (2) **150** (1999), no. 1, 185–266. MR1715324 (2000k:42033)

[Bia09] P. Biane, *Matrix valued Brownian motion and a paper by Pólya*, Séminaire de probabilités XLII, Lecture Notes in Math., vol. 1979, Springer, Berlin, 2009, pp. 171–185. MR2599210 (2011b:11123)

[Bil99] P. Billingsley, *Convergence of probability measures*, second ed., Wiley Series in Probability and Statistics: Probability and Statistics, John Wiley & Sons, Inc., New York, 1999, A Wiley-Interscience Publication. MR1700749 (2000e:60008)

[BJ02] P. Bougerol and T. Jeulin, *Paths in Weyl chambers and random matrices*, Probab. Theory Related Fields **124** (2002), no. 4, 517–543. MR1942321 (2004d:15033)

[BK97] A. Böttcher and Y. I. Karlovich, *Carleson curves, Muckenhoupt weights, and Toeplitz operators*, Progress in Mathematics, vol. 154, Birkhäuser Verlag, Basel, 1997. MR1484165 (98m:47031)

[BKMM07] J. Baik, T. Kriecherbauer, K. T.-R. McLaughlin, and P. D. Miller, *Discrete orthogonal polynomials*, Annals of Mathematics Studies, vol. 164, Princeton University Press, Princeton, NJ, 2007, Asymptotics and applications. MR2283089 (2011b:42081)

[Blo09] G. Blower, *Random matrices: high dimensional phenomena*, London Mathematical Society Lecture Note Series, vol. 367, Cambridge University Press, Cambridge, 2009. MR2566878 (2012a:60007)

[BM05] T. Bodineau and J. Martin, *A universality property for last-passage percolation paths close to the axis*, Electron. Comm. Probab. **10** (2005), 105–112 (electronic). MR2150699 (2006a:60189)

[BO00] A. Borodin and A. Okounkov, *A Fredholm determinant formula for Toeplitz determinants*, Integral Equations Operator Theory **37** (2000), no. 4, 386–396. MR1780118 (2001g:47042a)

[BOO00] A. Borodin, A. Okounkov, and G. Olshanski, *Asymptotics of Plancherel measures for symmetric groups*, J. Amer. Math. Soc. **13** (2000), no. 3, 481–515 (electronic). MR1758751 (2001g:05103)

[Bor03] A. Borodin, *Discrete gap probabilities and discrete Painlevé equations*, Duke Math. J. **117** (2003), no. 3, 489–542. MR1979052 (2004g:39030)

[Böt01] A. Böttcher, *One more proof of the Borodin-Okounkov formula for Toeplitz determinants*, Integral Equations Operator Theory **41** (2001), no. 1, 123–125.

[Böt02] ______, *On the determinant formulas by Borodin, Okounkov, Baik, Deift and Rains*, Toeplitz matrices and singular integral equations (Pobershau, 2001), Oper. Theory Adv. Appl., vol. 135, Birkhäuser, Basel, 2002, pp. 91–99.

[BP08] A. Borodin and S. Péché, *Airy kernel with two sets of parameters in directed percolation and random matrix theory*, J. Stat. Phys. **132** (2008), no. 2, 275–290. MR2415103 (2009b:82045)

[BR00] J. Baik and E. M. Rains, *Limiting distributions for a polynuclear growth model with external sources*, J. Statist. Phys. **100** (2000), no. 3-4, 523–541. MR1788477 (2001h:82067)

[BR01a] ———, *The asymptotics of monotone subsequences of involutions*, Duke Math. J. **109** (2001), no. 2, 205–281. MR1845180 (2003e:60016)

[BR01b] ———, *Symmetrized random permutations*, Random matrix models and their applications, Math. Sci. Res. Inst. Publ., vol. 40, Cambridge Univ. Press, Cambridge, 2001, pp. 1–19. MR1842780 (2002i:82038)

[BS95] A.-L. Barabási and H. E. Stanley, *Fractal concepts in surface growth*, Cambridge University Press, Cambridge, 1995. MR1600794 (99b:82072)

[BS99] A. Böttcher and B. Silbermann, *Introduction to large truncated Toeplitz matrices*, Universitext, Springer-Verlag, New York, 1999. MR1724795 (2001b:47043)

[BS05] J. Baik and T. M. Suidan, *A GUE central limit theorem and universality of directed first and last passage site percolation*, Int. Math. Res. Not. (2005), no. 6, 325–337. MR2131383 (2006c:60025)

[BS10] Z. Bai and J. W. Silverstein, *Spectral analysis of large dimensional random matrices*, second ed., Springer Series in Statistics, Springer, New York, 2010. MR2567175 (2011d:60014)

[BW00] E. L. Basor and H. Widom, *On a Toeplitz determinant identity of Borodin and Okounkov*, Integral Equations Operator Theory **37** (2000), no. 4, 397–401. MR1780119 (2001g:47042b)

[CEP96] H. Cohn, N. Elkies, and J. Propp, *Local statistics for random domino tilings of the Aztec diamond*, Duke Math. J. **85** (1996), no. 1, 117–166. MR1412441 (97k:52026)

[CG81] K. F. Clancey and I. Gohberg, *Factorization of matrix functions and singular integral operators*, Operator Theory: Advances and Applications, vol. 3, Birkhäuser Verlag, Basel-Boston, Mass., 1981. MR657762 (84a:47016)

[CH14] I. Corwin and A. Hammond, *Brownian Gibbs property for Airy line ensembles*, Invent. Math. **195** (2014), no. 2, 441–508. MR3152753

[Cha13] S. Chatterjee, *The universal relation between scaling exponents in first-passage percolation*, Ann. of Math. (2) **177** (2013), no. 2, 663–697. MR3010809

[CJY] S. Chhita, K. Johansson, and B. Young, *Asymptotic domino statistics in the Aztec diamond*, arXiv:1212.5414.

[Cor12] I. Corwin, *The Kardar-Parisi-Zhang equation and universality class*, Random Matrices Theory Appl. **1** (2012), no. 1, 1130001, 76. MR2930377

[Dei78] P. A. Deift, *Applications of a commutation formula*, Duke Math. J. **45** (1978), no. 2, 267–310. MR495676 (81g:47001)

[Dei99a] ———, *Integrable operators*, Differential operators and spectral theory, Amer. Math. Soc. Transl. Ser. 2, vol. 189, Amer. Math. Soc., Providence, RI, 1999, pp. 69–84. MR1730504 (2001d:47073)

[Dei99b] ———, *Orthogonal polynomials and random matrices: a Riemann-Hilbert approach*, Courant Lecture Notes in Mathematics, vol. 3, New York University, Courant Institute of Mathematical Sciences, New York; American Mathematical Society, Providence, RI, 1999. MR1677884 (2000g:47048)

[Dei07] ———, *Universality for mathematical and physical systems*, International Congress of Mathematicians. Vol. I, Eur. Math. Soc., Zürich, 2007, pp. 125–152. MR2334189 (2008g:60024)

[DG07] P. A. Deift and D. Gioev, *Universality at the edge of the spectrum for unitary, orthogonal, and symplectic ensembles of random matrices*, Comm. Pure Appl. Math. **60** (2007), no. 6, 867–910. MR2306224 (2008e:60089)

[Dic30] K. Dickman, *On the frequency of numbers containing prime factors of certain relative magnitude*, Arkiv fur Mathematik, Astronomi och Fysik **22** (1930), 1–14.

[DIK08] P. Deift, A. Its, and I. Krasovsky, *Asymptotics of the Airy-kernel determinant*, Comm. Math. Phys. **278** (2008), no. 3, 643–678. MR2373439 (2008m:47061)

[DIK11] ———, *Asymptotics of Toeplitz, Hankel, and Toeplitz+Hankel determinants with Fisher-Hartwig singularities*, Ann. of Math. (2) **174** (2011), no. 2, 1243–1299. MR2831118 (2012h:47063)

[DKM$^+$99a] P. Deift, T. Kriecherbauer, K. T.-R. McLaughlin, S. Venakides, and X. Zhou, *Strong asymptotics of orthogonal polynomials with respect to exponential weights*, Comm. Pure Appl. Math. **52** (1999), no. 12, 1491–1552. MR1711036 (2001f:42037)

[DKM$^+$99b] ———, *Uniform asymptotics for polynomials orthogonal with respect to varying exponential weights and applications to universality questions in random matrix theory*, Comm. Pure Appl. Math. **52** (1999), no. 11, 1335–1425. MR1702716 (2001g:42050)

[DÖ06] P. Deift and J. Östensson, *A Riemann-Hilbert approach to some theorems on Toeplitz operators and orthogonal polynomials*, J. Approx. Theory **139** (2006), no. 1-2, 144–171. MR2220037 (2007e:42026)

[Dur70] P. L. Duren, *Theory of $H^p$ spaces*, Pure and Applied Mathematics, Vol. 38, Academic Press, New York-London, 1970. MR0268655 (42 #3552)

[Dur96] R. Durrett, *Probability: theory and examples*, second ed., Duxbury Press, Belmont, CA, 1996. MR1609153 (98m:60001)

[DVJ03] D. J. Daley and D. Vere-Jones, *An introduction to the theory of point processes. Vol. I*, second ed., Probability and its Applications (New York), Springer-Verlag, New York, 2003, Elementary theory and methods. MR1950431 (2004c:60001)

[DVZ97] P. Deift, S. Venakides, and X. Zhou, *New results in small dispersion KdV by an extension of the steepest descent method for Riemann-Hilbert problems*, Internat. Math. Res. Notices (1997), no. 6, 286–299. MR1440305 (98b:35155)

[Dys62] F. J. Dyson, *A Brownian-motion model for the eigenvalues of a random matrix*, J. Mathematical Phys. **3** (1962), 1191–1198. MR0148397 (26 #5904)

[DZ] P. A. Deift and X. Zhou, *A Treatise on Riemann-Hilbert Problems, Vol. I*, Cambridge University Presss, Cambridge, United Kingdom, to appear.

[DZ93] ———, *A steepest descent method for oscillatory Riemann-Hilbert problems. Asymptotics for the MKdV equation*, Ann. of Math. (2) **137** (1993), no. 2, 295–368. MR1207209 (94d:35143)

[DZ95] ———, *Asymptotics for the Painlevé II equation*, Comm. Pure Appl. Math. **48** (1995), no. 3, 277–337. MR1322812 (96d:34004)

[DZ99] J.-D. Deuschel and O. Zeitouni, *On increasing subsequences of I.I.D. samples*, Combin. Probab. Comput. **8** (1999), no. 3, 247–263. MR1702546 (2000i:60022)

[DZ03] P. A. Deift and X. Zhou, *Long-time asymptotics for solutions of the NLS equation with initial data in a weighted Sobolev space*, Comm. Pure Appl. Math. **56** (2003), no. 8, 1029–1077, Dedicated to the memory of Jürgen K. Moser. MR1989226 (2004k:35349)

[EKLP92a] N. Elkies, G. Kuperberg, M. Larsen, and J. Propp, *Alternating-sign matrices and domino tilings. I*, J. Algebraic Combin. **1** (1992), no. 2, 111–132. MR1226347 (94f:52035)

[EKLP92b] ———, *Alternating-sign matrices and domino tilings. II*, J. Algebraic Combin. **1** (1992), no. 3, 219–234. MR1194076 (94f:52036)

[EKYY12] L. Erdős, A. Knowles, H.-T. Yau, and J. Yin, *Spectral statistics of Erdős-Rényi Graphs II: Eigenvalue spacing and the extreme eigenvalues*, Comm. Math. Phys. **314** (2012), no. 3, 587–640. MR2964770

[EPR⁺10] L. Erdős, S. Péché, J. A. Ramírez, B. Schlein, and H.-T. Yau, *Bulk universality for Wigner matrices*, Comm. Pure Appl. Math. **63** (2010), no. 7, 895–925. MR2662426 (2011c:60022)

[Erd56] A. Erdélyi, *Asymptotic expansions*, Dover Publications, Inc., New York, 1956. MR0078494 (17,1202c)

[ERS⁺10] L. Erdős, J. Ramírez, B. Schlein, T. Tao, V. Vu, and H.-T. Yau, *Bulk universality for Wigner Hermitian matrices with subexponential decay*, Math. Res. Lett. **17** (2010), no. 4, 667–674. MR2661171 (2011j:60018)

[ES35] P. Erdős and G. Szekeres, *A combinatorial problem in geometry*, Compositio Math. **2** (1935), 463–470. MR1556929

[FIK91] A. S. Fokas, A. R. Its, and A. V. Kitaev, *Discrete Painlevé equations and their appearance in quantum gravity*, Comm. Math. Phys. **142** (1991), no. 2, 313–344. MR1137067 (93a:58080)

[FIKN06] A. S. Fokas, A. R. Its, A. A. Kapaev, and V. Y. Novokshenov, *Painlevé transcendents*, Mathematical Surveys and Monographs, vol. 128, American Mathematical Society, Providence, RI, 2006, The Riemann-Hilbert approach. MR2264522 (2010e:33030)

[Fis84] M. E. Fisher, *Walks, walls, wetting, and melting*, J. Statist. Phys. **34** (1984), no. 5-6, 667–729. MR751710 (85j:82022)

[Fla74] H. Flaschka, *The Toda Lattice. I. Existence of integrals*, Phys. Rev. B (3) **9** (1974), 1924–1925. MR0408647 (53 #12411)

[FMZ92] A. S. Fokas, Ugurhan Muğan, and Xin Zhou, *On the solvability of Painlevé* I, III *and* V, Inverse Problems **8** (1992), no. 5, 757–785. MR1185598 (93h:35154)

[For01] P. J. Forrester, *Random walks and random permutations*, J. Phys. A **34** (2001), no. 31, L417–L423. MR1862639 (2002h:82043)

[For10] ———, *Log-gases and random matrices*, London Mathematical Society Monographs Series, vol. 34, Princeton University Press, Princeton, NJ, 2010. MR2641363 (2011d:82001)

[Fri91] A. Frieze, *On the length of the longest monotone subsequence in a random permutation*, Ann. Appl. Probab. **1** (1991), no. 2, 301–305. MR1102322 (92e:60020)

[Fro00] F. Frobenius, *Über die Charaktere der symmetrischen Gruppe*, Preuss. Akad. Wiss. Sitz. (1900), 516–534.

[FRT54] J. S. Frame, G. de B. Robinson, and R. M. Thrall, *The hook graphs of the symmetric groups*, Canadian J. Math. **6** (1954), 316–324. MR0062127 (15,931g)

[FS86] M. Fukushima and D. Stroock, *Reversibility of solutions to martingale problems*, Probability, statistical mechanics, and number theory, Adv. Math. Suppl. Stud., vol. 9, Academic Press, Orlando, FL, 1986, pp. 107–123. MR875449 (88h:60140)

[FS11] P. L. Ferrari and H. Spohn, *Random growth models*, The Oxford handbook of random matrix theory, Oxford Univ. Press, Oxford, 2011, pp. 782–801. MR2932658

[Ful97] W. Fulton, *Young tableaux*, London Mathematical Society Student Texts, vol. 35, Cambridge University Press, Cambridge, 1997, With applications to representation theory and geometry. MR1464693 (99f:05119)

[Gar07] J. B. Garnett, *Bounded analytic functions*, first ed., Graduate Texts in Mathematics, vol. 236, Springer, New York, 2007. MR2261424 (2007e:30049)

[GC79] J. S. Geronimo and K. M. Case, *Scattering theory and polynomials orthogonal on the unit circle*, J. Math. Phys. **20** (1979), no. 2, 299–310. MR519213 (80f:81100)

[Ges90] I. M. Gessel, *Symmetric functions and P-recursiveness*, J. Combin. Theory Ser. A **53** (1990), no. 2, 257–285. MR1041448 (91c:05190)

[GI71] B. L. Golinskiĭ and I. A. Ibragimov, *A limit theorm of G. Szegő*, Izv. Akad. Nauk SSSR Ser. Mat. **35** (1971), 408–427. MR0291713 (45 #804)

[GNW79] C. Greene, Albert N., and H. S. Wilf, *A probabilistic proof of a formula for the number of Young tableaux of a given shape*, Adv. in Math. **31** (1979), no. 1, 104–109. MR521470 (80b:05016)

[Gol69] G. M. Goluzin, *Geometric theory of functions of a complex variable*, Translations of Mathematical Monographs, Vol. 26, American Mathematical Society, Providence, R.I., 1969. MR0247039 (40 #308)

[Gra99] D. J. Grabiner, *Brownian motion in a Weyl chamber, non-colliding particles, and random matrices*, Ann. Inst. H. Poincaré Probab. Statist. **35** (1999), no. 2, 177–204. MR1678525 (2000i:60091)

[Gre74] C. Greene, *An extension of Schensted's theorem*, Advances in Math. **14** (1974), 254–265. MR0354395 (50 #6874)

[GTW01] J. Gravner, C. A. Tracy, and H. Widom, *Limit theorems for height fluctuations in a class of discrete space and time growth models*, J. Statist. Phys. **102** (2001), no. 5-6, 1085–1132. MR1830441 (2002d:82065)

[GV85] I. Gessel and G. Viennot, *Binomial determinants, paths, and hook length formulae*, Adv. in Math. **58** (1985), no. 3, 300–321. MR815360 (87e:05008)

[GW91] P. W. Glynn and W. Whitt, *Departures from many queues in series*, Ann. Appl. Probab. **1** (1991), no. 4, 546–572. MR1129774 (92i:60162)

[GWW98] I. Gessel, J. Weinstein, and H. S. Wilf, *Lattice walks in* $\mathbf{Z}^d$ *and permutations with no long ascending subsequences*, Electron. J. Combin. **5** (1998), Research Paper 2, 11 pp. (electronic). MR1486395 (98j:05007)

[Ham72] J. M. Hammersley, *A few seedlings of research*, Proceedings of the Sixth Berkeley Symposium on Mathematical Statistics and Probability (Univ. California, Berkeley, Calif., 1970/1971), Vol. I: Theory of statistics, Univ. California Press, Berkeley, Calif., 1972, pp. 345–394. MR0405665 (53 #9457)

[His96] M. Hisakado, *Unitary Matrix Models and Painlevé III*, Modern Phys. Lett. A **11** (1996), no. 38, 3001–3010.

[HM80] S. P. Hastings and J. B. McLeod, *A boundary value problem associated with the second Painlevé transcendent and the Korteweg-de Vries equation*, Arch. Rational Mech. Anal. **73** (1980), no. 1, 31–51. MR555581 (81i:34024)

[IIKS90] A. R. Its, A. G. Izergin, V. E. Korepin, and N. A. Slavnov, *Differential equations for quantum correlation functions*, Proceedings of the Conference on Yang-Baxter Equations, Conformal Invariance and Integrability in Statistical Mechanics and Field Theory, vol. 4, 1990, pp. 1003–1037. MR1064758 (91k:82009)

[Inc44] E. L. Ince, *Ordinary Differential Equations*, Dover Publications, New York, 1944. MR0010757 (6,65f)

[JMMS80] M. Jimbo, T. Miwa, Y. Môri, and M. Sato, *Density matrix of an impenetrable Bose gas and the fifth Painlevé transcendent*, Phys. D **1** (1980), no. 1, 80–158. MR573370 (84k:82037)

[JN06] K. Johansson and E. Nordenstam, *Eigenvalues of GUE minors*, Electron. J. Probab. **11** (2006), no. 50, 1342–1371. MR2268547 (2008d:60066a)

[Joh98] K. Johansson, *The longest increasing subsequence in a random permutation and a unitary random matrix model*, Math. Res. Lett. **5** (1998), no. 1-2, 63–82. MR1618351 (99e:60033)

[Joh00a] ———, *Shape fluctuations and random matrices*, Comm. Math. Phys. **209** (2000), no. 2, 437–476. MR1737991 (2001h:60177)

[Joh00b] ———, *Transversal fluctuations for increasing subsequences on the plane*, Probab. Theory Related Fields **116** (2000), no. 4, 445–456. MR1757595 (2001e:60210)

[Joh01a] ———, *Discrete orthogonal polynomial ensembles and the Plancherel measure*, Ann. of Math. (2) **153** (2001), no. 1, 259–296. MR1826414 (2002g:05188)

[Joh01b] ———, *Universality of the local spacing distribution in certain ensembles of Hermitian Wigner matrices*, Comm. Math. Phys. **215** (2001), no. 3, 683–705. MR1810949 (2002j:15024)

[Joh02] ———, *Non-intersecting paths, random tilings and random matrices*, Probab. Theory Related Fields **123** (2002), no. 2, 225–280. MR1900323 (2003h:15035)

[Joh03] ———, *Discrete polynuclear growth and determinantal processes*, Comm. Math. Phys. **242** (2003), no. 1-2, 277–329. MR2018275 (2004m:82096)

[Joh05a] ———, *The arctic circle boundary and the Airy process*, Ann. Probab. **33** (2005), no. 1, 1–30. MR2118857 (2005k:60304)

[Joh05b] ———, *Non-intersecting, simple, symmetric random walks and the extended Hahn kernel*, Ann. Inst. Fourier (Grenoble) **55** (2005), no. 6, 2129–2145. MR2187949 (2006k:60081)

[Joh06] ———, *Random matrices and determinantal processes*, Mathematical statistical physics, Elsevier B. V., Amsterdam, 2006, pp. 1–55. MR2581882 (2011c:60317)

[Joh08] ———, *On some special directed last-passage percolation models*, Integrable systems and random matrices, Contemp. Math., vol. 458, Amer. Math. Soc., Providence, RI, 2008, pp. 333–346. MR2411916 (2009k:60210)

[JPS] W. Jockusch, J. Propp, and P. Shor, *Random domino tilings and the arctic circle theorem*, arXiv:math.CO/9801068.

[Kam93] S. Kamvissis, *On the long time behavior of the doubly infinite Toda lattice under initial data decaying at infinity*, Comm. Math. Phys. **153** (1993), no. 3, 479–519. MR1218930 (94c:58086)

[Kar88] S. Karlin, *Coincident probabilities and applications in combinatorics*, J. Appl. Probab. (1988), no. Special Vol. 25A, 185–200, A celebration of applied probability. MR974581 (90d:60060)

[Kat04] Y. Katznelson, *An introduction to harmonic analysis*, third ed., Cambridge Mathematical Library, Cambridge University Press, Cambridge, 2004. MR2039503 (2005d:43001)

[Ker93] S. V. Kerov, *Transition probabilities of continual Young diagrams and the Markov moment problem*, Funktsional. Anal. i Prilozhen. **27** (1993), no. 2, 32–49, 96. MR1251166 (95g:82045)

[Kim96] J. H. Kim, *On increasing subsequences of random permutations*, J. Combin. Theory Ser. A **76** (1996), no. 1, 148–155. MR1405997 (97f:60024)

[Kin73] J. F. C. Kingman, *Subadditive ergodic theory*, Ann. Probability **1** (1973), 883–909, With discussion by D. L. Burkholder, Daryl Daley, H. Kesten, P. Ney, Frank Spitzer and J. M. Hammersley, and a reply by the author. MR0356192 (50 #8663)

[KK10] T. Kriecherbauer and J. Krug, *A pedestrian's view on interacting particle systems, KPZ universality and random matrices*, J. Phys. A **43** (2010), no. 40, 403001, 41. MR2725545 (2011m:60298)

[KM59] S. Karlin and J. McGregor, *Coincidence probabilities*, Pacific J. Math. **9** (1959), 1141–1164.

[KM00] A. B. J. Kuijlaars and K. T-R McLaughlin, *Generic behavior of the density of states in random matrix theory and equilibrium problems in the presence of real analytic external fields*, Comm. Pure Appl. Math. **53** (2000), no. 6, 736–785. MR1744002 (2001f:31003)

[KN07] R. Killip and I. Nenciu, *CMV: the unitary analogue of Jacobi matrices*, Comm. Pure Appl. Math. **60** (2007), no. 8, 1148–1188. MR2330626 (2008m:47042)

[KOR02] W. König, N. O'Connell, and S. Roch, *Non-colliding random walks, tandem queues, and discrete orthogonal polynomial ensembles*, Electron. J. Probab. **7** (2002), no. 5, 24 pp. (electronic). MR1887625 (2003e:60174)

[Kry95] N. V. Krylov, *Introduction to the theory of diffusion processes*, Translations of Mathematical Monographs, vol. 142, American Mathematical Society, Providence, RI, 1995, Translated from the Russian manuscript by Valim Khidekel and Gennady Pasechnik. MR1311478 (96k:60196)

[KS] R. Koekeok and R. Swarttouw, *The Askey-scheme of hypergeometric orthogonal polynomials and its q-analogue*, `http://aw.twi.tudelft.nl/~koekoek/askey`.

[KS91] I. Karatzas and S. E. Shreve, *Brownian motion and stochastic calculus*, second ed., Graduate Texts in Mathematics, vol. 113, Springer-Verlag, New York, 1991. MR1121940 (92h:60127)

[KS92] J. Krug and H. Spohn, *Kinetic roughening of growing surfaces*, Solids Far from Equilibrium, Collection Aléa-Saclay: Monographs and Texts in Statistical Physics, 1, Cambridge Univ. Press, Cambridge, 1992, pp. 479–582.

[KS99] N. M. Katz and P. Sarnak, *Zeroes of zeta functions and symmetry*, Bull. Amer. Math. Soc. (N.S.) **36** (1999), no. 1, 1–26. MR1640151 (2000f:11114)

[KŠ00] M. Krbálek and P. Šeba, *Statistical properties of the city transport in Cuernevaca (Mexico) and random matrix theory*, J. Phys. A **33** (2000), no. 26, L229–L234.

[KV86] C. Kipnis and S. R. S. Varadhan, *Central limit theorem for additive functionals of reversible Markov processes and applications to simple exclusions*, Comm. Math. Phys. **104** (1986), no. 1, 1–19. MR834478 (87i:60038)

[Lax02] P. D. Lax, *Functional analysis*, Pure and Applied Mathematics (New York), Wiley-Interscience [John Wiley & Sons], New York, 2002. MR1892228 (2003a:47001)

[Len73] A. Lenard, *Correlation functions and the uniqueness of the state in classical statistical mechanics*, Comm. Math. Phys. **30** (1973), 35–44. MR0323270 (48 #1628)

[Len75a] ———, *States of classical statistical mechanical systems of infinitely many particles. I*, Arch. Rational Mech. Anal. **59** (1975), no. 3, 219–239. MR0391830 (52 #12649)

[Len75b] ———, *States of classical statistical mechanical systems of infinitely many particles. II. Characterization of correlation measures*, Arch. Rational Mech. Anal. **59** (1975), no. 3, 241–256. MR0391831 (52 #12650)

[Lig99] T. M. Liggett, *Stochastic interacting systems: contact, voter and exclusion processes*, Grundlehren der Mathematischen Wissenschaften [Fundamental Principles of Mathematical Sciences], vol. 324, Springer-Verlag, Berlin, 1999. MR1717346 (2001g:60247)

[Lin73] B. Lindström, *On the vector representations of induced matroids*, Bull. London Math. Soc. **5** (1973), 85–90. MR0335313 (49 #95)

[LM01] M. Löwe and F. Merkl, *Moderate deviations for longest increasing subsequences: the upper tail*, Comm. Pure Appl. Math. **54** (2001), no. 12, 1488–1520. MR1852980 (2002f:60045)

[LMR02] M. Löwe, F. Merkl, and S. Rolles, *Moderate deviations for longest increasing subsequences: the lower tail*, J. Theoret. Probab. **15** (2002), no. 4, 1031–1047. MR1937784 (2003j:60009)

[LNP96] C. Licea, C. M. Newman, and M. S. T. Piza, *Superdiffusivity in first-passage percolation*, Probab. Theory Related Fields **106** (1996), no. 4, 559–591. MR1421992 (98a:60151)

[LS77] B. F. Logan and L. A. Shepp, *A variational problem for random Young tableaux*, Advances in Math. **26** (1977), no. 2, 206–222. MR1417317 (98e:05108)

[LS87] G. S. Litvinchuk and I. M. Spitkovskii, *Factorization of measurable matrix functions*, Operator Theory: Advances and Applications, vol. 25, Birkhäuser Verlag, Basel, 1987, Translated from the Russian by Bernd Luderer, With a foreword by Bernd Silbermann. MR1015716 (90g:47030)

[LY14] J. O. Lee and J. Yin, *A necessary and sufficient condition for edge universality of Wigner matrices*, Duke Math. J. **163** (2014), no. 1, 117–173. MR3161313

[Man74] S. V. Manakov, *Complete integrability and stochastization of discrete dynamical systems*, Ž. Èksper. Teoret. Fiz. **67** (1974), no. 2, 543–555. MR0389107 (52 #9938)

[Mar04] J. B. Martin, *Limiting shape for directed percolation models*, Ann. Probab. **32** (2004), no. 4, 2908–2937. MR2094434 (2005i:60198)

[McK69] H. P. McKean, Jr., *Stochastic integrals*, Probability and Mathematical Statistics, No. 5, Academic Press, New York-London, 1969. MR0247684 (40 #947)

[Mea98] P. Meakin, *Fractals, scaling and growth far from equilibrium*, Cambridge Nonlinear Science Series, vol. 5, Cambridge University Press, Cambridge, 1998. MR1489739 (98m:82004)

[Meh04] M. L. Mehta, *Random matrices*, third ed., Pure and Applied Mathematics (Amsterdam), vol. 142, Elsevier/Academic Press, Amsterdam, 2004. MR2129906 (2006b:82001)

[MFQR13] G. Moreno Flores, J. Quastel, and D. Remenik, *Endpoint distribution of directed polymers in* $1+1$ *dimensions*, Comm. Math. Phys. **317** (2013), no. 2, 363–380. MR3010188

[Mil02] P. D. Miller, *Asymptotics of semiclassical soliton ensembles: rigorous justification of the WKB approximation*, Int. Math. Res. Not. (2002), no. 8, 383–454. MR1884077 (2003j:30060)

[NP95] C. M. Newman and M. S. T. Piza, *Divergence of shape fluctuations in two dimensions*, Ann. Probab. **23** (1995), no. 3, 977–1005. MR1349159 (96g:82052)

[NSU91] A. F. Nikiforov, S. K. Suslov, and V. B. Uvarov, *Classical orthogonal polynomials of a discrete variable*, Springer Series in Computational Physics, Springer-Verlag, Berlin, 1991, Translated from the Russian. MR1149380 (92m:33019)

[Odl] A. Odlyzko, *The 1020-th zero of the Riemann zeta function and 70 million of its neighbors*, ATT preprint, 1989.

[Oko00] A. Okounkov, *Random matrices and random permutations*, Internat. Math. Res. Notices (2000), no. 20, 1043–1095. MR1802530 (2002c:15045)

[Oko01] ———, *Infinite wedge and random partitions*, Selecta Math. (N.S.) **7** (2001), no. 1, 57–81. MR1856553 (2002f:60019)

[OR00] A. M. Odlyzko and E. M. Rains, *On longest increasing subsequences in random permutations*, Analysis, geometry, number theory: the mathematics of Leon Ehrenpreis (Philadelphia, PA, 1998), Contemp. Math., vol. 251, Amer. Math. Soc., Providence, RI, 2000, pp. 439–451. MR1771285 (2001d:05003)

[OR03] A. Okounkov and N. Reshetikhin, *Correlation function of Schur process with application to local geometry of a random 3-dimensional Young diagram*, J. Amer. Math. Soc. **16** (2003), no. 3, 581–603 (electronic). MR1969205 (2004b:60033)

[OR07] ———, *Random skew plane partitions and the Pearcey process*, Comm. Math. Phys. **269** (2007), no. 3, 571–609. MR2276355 (2008c:60007)

[Pel80] V. V. Peller, *Hankel operators of class* $\mathfrak{S}_p$ *and their applications (rational approximation, Gaussian processes, the problem of majorization of operators)*, Mat. Sb. (N.S.) **113(155)** (1980), no. 4(12), 538–581, 637. MR602274 (82g:47022)

[PS90] V. Periwal and D. Shevitz, *Unitary-matrix models as exactly solvable string theories*, Phys. Rev. Lett. **64** (1990), no. 12, 1326–1329.

[PS97] L. Pastur and M. Shcherbina, *Universality of the local eigenvalue statistics for a class of unitary invariant random matrix ensembles*, J. Statist. Phys. **86** (1997), no. 1-2, 109–147. MR1435193 (98b:82037)

[PS00] M. Prähofer and H. Spohn, *Statistical self-similarity of one-dimensional growth processes*, Phys. A **279** (2000), no. 1-4, 342–352, Statistical mechanics: from rigorous results to applications. MR1797145 (2001j:82098)

[PS02] ———, *Scale invariance of the PNG droplet and the Airy process*, J. Statist. Phys. **108** (2002), no. 5-6, 1071–1106, Dedicated to David Ruelle and Yasha Sinai on the occasion of their 65th birthdays. MR1933446 (2003i:82050)

[Rai98] E. M. Rains, *Increasing subsequences and the classical groups*, Electron. J. Combin. **5** (1998), Research Paper 12, 9 pp. (electronic). MR1600095 (98k:05146)

[Rob38] G. de B. Robinson, *On the Representations of the Symmetric Group*, Amer. J. Math. **60** (1938), no. 3, 745–760. MR1507943

[Rom15] D. Romik, *The surprising mathematics of longest increasing subsequences*, Institute of Mathematical Statistics Textbooks, Cambridge University Press, 2015.

[Roy88] H. L. Royden, *Real analysis*, third ed., Macmillan Publishing Company, New York, 1988. MR1013117 (90g:00004)

[RS78] M. Reed and B. Simon, *Methods of modern mathematical physics. IV. Analysis of operators*, Academic Press [Harcourt Brace Jovanovich, Publishers], New York-London, 1978. MR0493421 (58 #12429c)

[RS79] ———, *Methods of modern mathematical physics. III*, Academic Press [Harcourt Brace Jovanovich, Publishers], New York-London, 1979, Scattering theory. MR529429 (80m:81085)

[RS80] ———, *Methods of modern mathematical physics. I*, second ed., Academic Press, Inc. [Harcourt Brace Jovanovich, Publishers], New York, 1980, Functional analysis. MR751959 (85e:46002)

[RY99] D. Revuz and M. Yor, *Continuous martingales and Brownian motion*, third ed., Grundlehren der Mathematischen Wissenschaften [Fundamental Principles of Mathematical Sciences], vol. 293, Springer-Verlag, Berlin, 1999. MR1725357 (2000h:60050)

[Sag01] B. E. Sagan, *The symmetric group*, second ed., Graduate Texts in Mathematics, vol. 203, Springer-Verlag, New York, 2001, Representations, combinatorial algorithms, and symmetric functions. MR1824028 (2001m:05261)

[Sch61] C. Schensted, *Longest increasing and decreasing subsequences*, Canad. J. Math. **13** (1961), 179–191. MR0121305 (22 #12047)

[Sch63] M. P. Schützenberger, *Quelques remarques sur une construction de Schensted*, Math. Scand. **12** (1963), 117–128. MR0190017 (32 #7433)

[Sep96] T. Seppäläinen, *A microscopic model for the Burgers equation and longest increasing subsequences*, Electron. J. Probab. **1** (1996), no. 5, approx. 51 pp. (electronic). MR1386297 (97d:60162)

[Sep97] ———, *A scaling limit for queues in series*, Ann. Appl. Probab. **7** (1997), no. 4, 855–872. MR1484787 (99j:60160)

[Sep98] ———, *Large deviations for increasing sequences on the plane*, Probab. Theory Related Fields **112** (1998), no. 2, 221–244. MR1653841 (2000b:60071)

[Ser77] J.-P. Serre, *Linear representations of finite groups*, Springer-Verlag, New York-Heidelberg, 1977, Translated from the second French edition by

Leonard L. Scott, Graduate Texts in Mathematics, Vol. 42. MR0450380 (56 #8675)

[SI04] T. Sasamoto and T. Imamura, *Fluctuations of the one-dimensional polynuclear growth model in half-space*, J. Statist. Phys. **115** (2004), no. 3-4, 749–803. MR2054161 (2005d:82112)

[Sim05a] B. Simon, *Orthogonal polynomials on the unit circle. Part 1*, American Mathematical Society Colloquium Publications, vol. 54, American Mathematical Society, Providence, RI, 2005, Classical theory. MR2105088 (2006a:42002a)

[Sim05b] ———, *Orthogonal polynomials on the unit circle. Part 2*, American Mathematical Society Colloquium Publications, vol. 54, American Mathematical Society, Providence, RI, 2005, Spectral theory. MR2105089 (2006a:42002b)

[Sim05c] ———, *Trace ideals and their applications*, second ed., Mathematical Surveys and Monographs, vol. 120, American Mathematical Society, Providence, RI, 2005. MR2154153 (2006f:47086)

[Sin94] Y. G. Sinaĭ, *Topics in ergodic theory*, Princeton Mathematical Series, vol. 44, Princeton University Press, Princeton, NJ, 1994. MR1258087 (95j:28017)

[Sos99] A. Soshnikov, *Universality at the edge of the spectrum in Wigner random matrices*, Comm. Math. Phys. **207** (1999), no. 3, 697–733. MR1727234 (2001i:82037)

[Sos00] ———, *Determinantal random point fields*, Uspekhi Mat. Nauk **55** (2000), no. 5(335), 107–160. MR1799012 (2002f:60097)

[Spo06] H. Spohn, *Exact solutions for KPZ-type growth processes, random matrices, and equilibrium shapes of crystals*, Phys. A **369** (2006), no. 1, 71–99. MR2246567 (2007f:82077)

[SS98a] Y. G. Sinaĭ and A. Soshnikov, *Central limit theorem for traces of large random symmetric matrices with independent matrix elements*, Bol. Soc. Brasil. Mat. (N.S.) **29** (1998), no. 1, 1–24. MR1620151 (99f:60053)

[SS98b] ———, *A refinement of Wigner's semicircle law in a neighborhood of the spectrum edge for random symmetric matrices*, Funktsional. Anal. i Prilozhen. **32** (1998), no. 2, 56–79, 96. MR1647832 (2000c:82041)

[Sta99] R. P. Stanley, *Enumerative combinatorics. Vol. 2*, Cambridge Studies in Advanced Mathematics, vol. 62, Cambridge University Press, Cambridge, 1999, With a foreword by Gian-Carlo Rota and appendix 1 by Sergey Fomin. MR1676282 (2000k:05026)

[Sui06] T. Suidan, *A remark on a theorem of Chatterjee and last passage percolation*, J. Phys. A **39** (2006), no. 28, 8977–8981. MR2240468 (2007d:82040)

[SV79] D. W. Stroock and S. R. S. Varadhan, *Multidimensional diffusion processes*, Grundlehren der Mathematischen Wissenschaften [Fundamental Principles of Mathematical Sciences], vol. 233, Springer-Verlag, Berlin-New York, 1979. MR532498 (81f:60108)

[Sze75] G. Szegő, *Orthogonal polynomials*, fourth ed., American Mathematical Society, Providence, R.I., 1975, American Mathematical Society, Colloquium Publications, Vol. XXIII. MR0372517 (51 #8724)

[Tao12] T. Tao, *Topics in random matrix theory*, Graduate Studies in Mathematics, vol. 132, American Mathematical Society, Providence, RI, 2012. MR2906465 (2012k:60023)

[TV10] T. Tao and V. Vu, *Random matrices: universality of local eigenvalue statistics up to the edge*, Comm. Math. Phys. **298** (2010), no. 2, 549–572. MR2669449 (2011f:60012)

[TV11] ———, *Random matrices: universality of local eigenvalue statistics*, Acta Math. **206** (2011), no. 1, 127–204. MR2784665 (2012d:60016)

[TW94] C. A. Tracy and Harold Widom, *Level-spacing distributions and the Airy kernel*, Comm. Math. Phys. **159** (1994), no. 1, 151–174. MR1257246 (95e:82003)

[TW01] C. A. Tracy and H. Widom, *On the distributions of the lengths of the longest monotone subsequences in random words*, Probab. Theory Related Fields **119** (2001), no. 3, 350–380. MR1821139 (2002a:60013)

[Ula61] S. M. Ulam, *Monte Carlo calculations in problems of mathematical physics*, Modern mathematics for the engineer: Second series, McGraw-Hill, New York, 1961, pp. 261–281. MR0129165 (23 #B2202)

[Vie77] G. Viennot, *Une forme géométrique de la correspondance de Robinson-Schensted*, Combinatoire et représentation du groupe symétrique (Actes Table Ronde CNRS, Univ. Louis-Pasteur Strasbourg, Strasbourg, 1976), Springer, Berlin, 1977, pp. 29–58. Lecture Notes in Math., Vol. 579. MR0470059 (57 #9827)

[VK77] A. M. Veršik and S. V. Kerov, *Asymptotic behavior of the Plancherel measure of the symmetric group and the limit form of Young tableaux*, Dokl. Akad. Nauk SSSR **233** (1977), no. 6, 1024–1027. MR0480398 (58 #562)

[VK85] ———, *Asymptotic behavior of the maximum and generic dimensions of irreducible representations of the symmetric group*, Funktsional. Anal. i Prilozhen. **19** (1985), no. 1, 25–36, 96. MR783703 (86k:11051)

[VŠ77] A. M. Veršik and A. A. Šmidt, *Limit measures that arise in the asymptotic theory of symmetric groups. I*, Teor. Verojatnost. i Primenen. **22** (1977), no. 1, 72–88. MR0448476 (56 #6782)

[Wal82] P. Walters, *An introduction to ergodic theory*, Graduate Texts in Mathematics, vol. 79, Springer-Verlag, New York-Berlin, 1982. MR648108 (84e:28017)

[Wid76] H. Widom, *Asymptotic behavior of block Toeplitz matrices and determinants. II*, Advances in Math. **21** (1976), no. 1, 1–29. MR0409512 (53 #13266b)

[Wid02] ———, *On convergence of moments for random Young tableaux and a random growth model*, Int. Math. Res. Not. (2002), no. 9, 455–464. MR1884467 (2002m:60018)

[You02] A. Young, *Qualitative substitutional analysis II*, Proc. London Math. Soc. (1) **34** (1902), 361–397.

[Zei83] D. Zeilberger, *André's reflection proof generalized to the many-candidate ballot problem*, Discrete Math. **44** (1983), no. 3, 325–326.

# Index

## Selected Published Titles in This Series

172 **Jinho Baik, Percy Deift, and Toufic Suidan,** Combinatorics and Random Matrix Theory, 2016
170 **Donald Yau,** Colored Operads, 2016
169 **András Vasy,** Partial Differential Equations, 2015
168 **Michael Aizenman and Simone Warzel,** Random Operators, 2015
167 **John C. Neu,** Singular Perturbation in the Physical Sciences, 2015
166 **Alberto Torchinsky,** Problems in Real and Functional Analysis, 2015
165 **Joseph J. Rotman,** Advanced Modern Algebra: Third Edition, Part 1, 2015
164 **Terence Tao,** Expansion in Finite Simple Groups of Lie Type, 2015
163 **Gérald Tenenbaum,** Introduction to Analytic and Probabilistic Number Theory, Third Edition, 2015
162 **Firas Rassoul-Agha and Timo Seppäläinen,** A Course on Large Deviations with an Introduction to Gibbs Measures, 2015
161 **Diane Maclagan and Bernd Sturmfels,** Introduction to Tropical Geometry, 2015
160 **Marius Overholt,** A Course in Analytic Number Theory, 2014
159 **John R. Faulkner,** The Role of Nonassociative Algebra in Projective Geometry, 2014
158 **Fritz Colonius and Wolfgang Kliemann,** Dynamical Systems and Linear Algebra, 2014
157 **Gerald Teschl,** Mathematical Methods in Quantum Mechanics: With Applications to Schrödinger Operators, Second Edition, 2014
156 **Markus Haase,** Functional Analysis, 2014
155 **Emmanuel Kowalski,** An Introduction to the Representation Theory of Groups, 2014
154 **Wilhelm Schlag,** A Course in Complex Analysis and Riemann Surfaces, 2014
153 **Terence Tao,** Hilbert's Fifth Problem and Related Topics, 2014
152 **Gábor Székelyhidi,** An Introduction to Extremal Kähler Metrics, 2014
151 **Jennifer Schultens,** Introduction to 3-Manifolds, 2014
150 **Joe Diestel and Angela Spalsbury,** The Joys of Haar Measure, 2013
149 **Daniel W. Stroock,** Mathematics of Probability, 2013
148 **Luis Barreira and Yakov Pesin,** Introduction to Smooth Ergodic Theory, 2013
147 **Xingzhi Zhan,** Matrix Theory, 2013
146 **Aaron N. Siegel,** Combinatorial Game Theory, 2013
145 **Charles A. Weibel,** The $K$-book, 2013
144 **Shun-Jen Cheng and Weiqiang Wang,** Dualities and Representations of Lie Superalgebras, 2012
143 **Alberto Bressan,** Lecture Notes on Functional Analysis, 2013
142 **Terence Tao,** Higher Order Fourier Analysis, 2012
141 **John B. Conway,** A Course in Abstract Analysis, 2012
140 **Gerald Teschl,** Ordinary Differential Equations and Dynamical Systems, 2012
139 **John B. Walsh,** Knowing the Odds, 2012
138 **Maciej Zworski,** Semiclassical Analysis, 2012
137 **Luis Barreira and Claudia Valls,** Ordinary Differential Equations, 2012
136 **Arshak Petrosyan, Henrik Shahgholian, and Nina Uraltseva,** Regularity of Free Boundaries in Obstacle-Type Problems, 2012
135 **Pascal Cherrier and Albert Milani,** Linear and Quasi-linear Evolution Equations in Hilbert Spaces, 2012
134 **Jean-Marie De Koninck and Florian Luca,** Analytic Number Theory, 2012

For a complete list of titles in this series, visit the
AMS Bookstore at **www.ams.org/bookstore/gsmseries/**.